# annual reports
# in organic
# synthesis -1993

# ANNUAL REPORTS IN ORGANIC SYNTHESIS

**ANNUAL REPORTS IN ORGANIC SYNTHESIS-1970**
John McMurry and R. Bryan Miller, Eds.

**ANNUAL REPORTS IN ORGANIC SYNTHESIS-1972**
John McMurry and R. Bryan Miller, Eds.

**ANNUAL REPORTS IN ORGANIC SYNTHESIS-1973**
R. Bryan Miller and Louis S. Hegedus, Eds.
John McMurry, Series Editor

**ANNUAL REPORTS IN ORGANIC SYNTHESIS-1974**
Louis S. Hegedus and Stephen R. Wilson, Eds.
R. Bryan Miller, Series Editor

**ANNUAL REPORTS IN ORGANIC SYNTHESIS-1975**
R. Bryan Miller and L. G. Wade, Jr., Eds.

**ANNUAL REPORTS IN ORGANIC SYNTHESIS-1976**
R. Bryan Miller and L. G. Wade, Jr., Eds.

**ANNUAL REPORTS IN ORGANIC SYNTHESIS-1978**
L. G. Wade, Jr., and Martin J. O'Donnell, Eds.

**ANNUAL REPORTS IN ORGANIC SYNTHESIS-1980**
L. G. Wade, Jr., and Martin J. O'Donnell, Eds.

**ANNUAL REPORTS IN ORGANIC SYNTHESIS-1981**
L. G. Wade, Jr., and Martin J. O'Donnell, Eds.

**ANNUAL REPORTS IN ORGANIC SYNTHESIS-1982**
L. G. Wade, Jr., and Martin J. O'Donnell, Eds.

**ANNUAL REPORTS IN ORGANIC SYNTHESIS-1983**
Martin J. O'Donnell and Louis Weiss, Eds.

**ANNUAL REPORTS IN ORGANIC SYNTHESIS-1984**
Martin J. O'Donnell and Louis Weiss, Eds.

**ANNUAL REPORTS IN ORGANIC SYNTHESIS-1985**
Martin J. O'Donnell and Eric F. V. Scriven, Eds.

**ANNUAL REPORTS IN ORGANIC SYNTHESIS-1986**
Eric F. V. Scriven and Kenneth Turnbull, Eds.

**ANNUAL REPORTS IN ORGANIC SYNTHESIS-1987**
Eric F. V. Scriven and Kenneth Turnbull, Eds.

**ANNUAL REPORTS IN ORGANIC SYNTHESIS-1989**
Kenneth Turnbull and Daniel M. Ketcha, Eds.

**ANNUAL REPORTS IN ORGANIC SYNTHESIS-1990**
Kenneth Turnbull, Philip M. Weintraub, Daniel M. Ketcha,
and James Keay, Eds.

**ANNUAL REPORTS IN ORGANIC SYNTHESIS-1991**
Philip M. Weintraub and Kenneth Turnbull, Eds.

**ANNUAL REPORTS IN ORGANIC SYNTHESIS-1992**
Philip M. Weintraub, Kenneth Turnbull,
Daniel M. Ketcha, and Raymond Gross, Eds.

# annual reports
# in organic
# synthesis – 1993

edited by

**Philip M. Weintraub**
*Merrell Dow Research Institute*
*Cincinnati, Ohio*

**Kenneth Turnbull**
*Wright State University*
*Dayton, Ohio*

**Daniel M. Ketcha**
*Wright State University*
*Dayton, Ohio*

**Raymond Gross**
*Merrell Dow Institute*
*Cincinnati, Ohio*

**Tony Yantao Zhang**
*Reilly Industries, Inc.*
*Indianapolis, Indiana*

**ACADEMIC PRESS, INC.**
*A Division of Harcourt Brace & Company*
San Diego   New York   Boston   London   Sydney   Tokyo   Toronto

Academic Press Rapid Manuscript Reproduction

Academic Press, Inc.
1250 Sixth Avenue, San Diego, California 92101-4311

*United Kingdom Edition published by*
ACADEMIC PRESS LIMITED
24-28 Oval Road, London NW1 7DX

International Standard Serial Number: 0066-409X

International Standard Book Number: 0-12-040823-6 (alk. paper)

PRINTED IN THE UNITED STATES OF AMERICA

Transferred to Digital Printing 2008

# Contents

# VII. REVIEWS

# *PREFACE*

One of the most difficult problems facing chemists today is that of "keeping up with the literature." For several reasons, the problem is particularly severe for the synthetic organic chemist. Bits of information of potential use are scattered throughout common chemistry journals and can be found in any paper, not just those dealing strictly with synthesis. Thus, synthetic chemists must read a large number of journals and must organize and index what they read to make the information available for future reference. All synthetic chemists do this, but the task is becoming more difficult each year as the flow of information increases.

The problem, however, is shared to some extent by all. Most organic chemists are at some time faced with the problem of synthesizing a desired material, and for many the problems are formidable. Non specialists faced with the synthetic problem are not likely to have kept pace with the developments in synthetic chemistry that may well solve their problems, and they will not have the necessary information in their files.

Thus, we felt that an organized annual review of synthetically useful information would prove beneficial to nearly all organic chemists, both specialists and non specialists in synthesis. It should help relieve some of the information storage burden of the specialist and should enable the non specialist who is seeking help with a specific problem to rapidly become aware of recent synthetic advances. Ideally also, it should appear as promptly as possible after the close of the abstracting period. As in the past years, we have placed particular emphasis on keeping the abstracts as concise as possible, while indicating the generality of the reactions involved. We have tried to combine similar publications into inclusive abstracts, particularly in Chapters I and IV. This practice has allowed us to include a larger number of references without a substantial increase in the book's length. It should be noted that where multiple references are included in the abstract, the first mentioned refers to the equation presented. The remaining references are similar but not identical. To further aid the readers, we have tried to separate less similar references from those represented by the graphic by the phrase "see also:". We have allowed for two such separations per graphic. One more change was instituted this year: we have omitted the year from each reference as they are presumably all 1991. The references from 1990 (from journals received our cutoff date) are noted appropriately.

In producing *Annual Reports in Organic Chemistry–1992* we have abstracted 47 primary chemistry journals, selecting useful synthetic advances. We have tried to present the information in an organized manner, emphasizing rapid visual retrieval. Only the common journals received by our libraries have been abstracted. Any journal received after February 1, 1991 will be covered in the next volume. We have also exercised selectivity in choosing which papers to abstract. Our general guidelines have been to include all reactions and methods that are new, synthetically useful, and reasonably general. The purpose of this emphasis is to aid the reader in scanning the book. The mind is capable of absorbing a whole picture in an instant, but is considerably slowed by having to read sentences. If the pictures presented catch the reader's interest, he or she should then seek details from the original paper.

We have included an author index based on the name of the senior author or sometimes the first author. No subject index is included because we feel the Table of Contents serves that function. Chapters I–III are organized by reaction type and, hopefully, the organization is self-explanatory; thus, there should be no difficulty in locating a new method of oxidation or a new cyclopropanation procedure. Chapter IV deals with methods of synthesizing heterocyclic systems. Where fused ring systems bearing multiple heterocyclic rings are synthesized, we have chosen to categorize the heterocyclic system by the ring formed in the reaction. Chapter V covers the use of new protecting groups. Chapter VI covers those synthetically useful transformations that do not fit easily into the first three chapters. Chapter VII has been divided into sections in order to help the reader to quickly find a review on a specific topic. Heterocyclic reviews may be found in Chapter IV.

Any undertaking of this type involves a series of compromises. We have chosen to emphasize reasonable cost and rapid visual retrieval of information  at the admitted expense of detail and beauty.

The task of typing and preparing the graphics was done by Marcia Ketcha and the editors. We hope the readers will forgive the inevitable typos and other minor "glitches" some are of our doing whereas others can be attributed to the computer or to our inability to master the nasty little black box.

If the reader has any comments (negative or preferably positive) or suggestions they will be well received by the senior editor.

*Senior and Contributing Editor*
Philip M. Weintraub

*Contributing Editors*
Kenneth Turnbull
Daniel M. Ketcha
Raymond S. Gross
Tony Y. Zhang

# JOURNALS ABSTRACTED

Accounts of Chemical Research
Acta Chemica Scandinavia
Aldrichimica Acta
Angewandte Chemie
  International Edition in
  English
Australian Journal of Chemistry
Bulletin of the Chemical Society
  of Japan
Bulletin de Societies Chimiques
Belges
Bulletin de la Societie Chimique
  de France
Canadian Journal of Chemistry
Chemical and Pharmaceutical
  Bulletin
Chemical Reviews
Chemical Society Reviews
Chemische Berichte
Chemistry and Industry
Chemistry Letters
Collection of Czechoslovakian
Chemical Communications
Gazzetta Chimica Italiana
Helvetica Chimica Acta
Heterocycles
Indian Journal of Chemistry
Journal of the American
  Chemical Society
Journal of Chemical Research
  (S)
Journal of the Chemical Society
Chemical Communications

Journal of the Chemical Society
  (Perkin I)
Journal of the Chemical Society
  (Perkin II)
Journal of Heterocyclic
Chemistry
Journal of Medicinal Chemistry
Journal of Organic Chemistry
Journal of Organic Chemistry
  (USSR)
Journal of Organometallic
  Chemistry
Journal fur Practische Chemie
Liebigs Annalen der Chemie
Monatschefte fur Chemie
Organic Preparations and
  Procedures International
Organic Synthesis
Organometallics
Pure and Applied Chemistry
Recueil des Traveaix Chimiques
  des Pays-bas
Russian Chemical Reviews
Steroids
Synlett
Synthesis
Synthetic Communications
Tetrahedron
Tetrahedron Letters
Topics in Current Chemistry

# GLOSSARY OF ABBREVIATIONS

9-BBN 9-borabicyclo[3.3.1]-nonane
18-Cr-6 8-C-6 18-crown-6
AA amino acid
Ac acetyl
acac acetonylacetone
ad adamantanyl
AIBN azobisisobutyronitrile
All allyl
Alloc = ALOC alloyloxycarbonyl
An p-anisyl
aq aqueous
Ar aryl
BDPP (2R, 4R) or (2S, 4S) 2,4-bis(diphenylphosphino)-pentane
BINAP = DINAP 2,2'-bis-(diphenyl-phosphino)-1,1'-binaphthyl
Bn benzyl
Boc t-butyloxycarbonyl
BOM benzyloxymethyl
BPO benzoyl peroxide
BPPM t-butoxycarbonyl-
bpy bipyridyl
BQ benzoquinone
BSA bovine serum albumin
BSA N,O-bis-silylacetamide
Bt 1- or 2-benzotriazolyl
BTEAC benzyl triethyl-ammonium chloride
BTFP 2-bromotrifluoroisoprene
BTMA benzyltrimethyl ammonium
Bu butyl
Bz benzoyl
CAN ceric ammonium nitrate
cat. catalyst
Cbz benzyloxycarbonyl
cod 1,5-cyclooctadiene
cot cyclooctatriene
Cp cyclopentadienyl

Cr-PILC chromium-pillared clay catalyst
CRA complex reducing agent
CSA camphor sulfonic acid
CTAB cetyl trimethyl-ammonium bromide
CTMS = TMCS chlorotrimethyl-silyl
Cy cyclohexyl
Δ heat
d day
DABCO 1,4-diazabicyclo[2.2.2]-octane
DAST diethylaminosulfur trifluoride
dba dibenzylidene acetone
DBU 1,5-diazabicyclo[5.4.0]-undec-5-ene
DCA 9,10-dicyanoanthracene
DCB dichlorobenzene
DCC dicyclohexylcarbodiimide
DCE 1,2-dichloroethane
DCME α,α-dichloromethyl methyl sulfide
DDQ 2,3-dichloro-5,6-dicyano-benzoquinone
de d.e. diastereomeric excess
DEAD diethyl azodicarboxylate
DEPC diethyl cyanophos phoridate
DET diethyl tartrate
DIBAH DIBAL diisobutyl-aluminum hydride
DIOP 2,3-O-isopropylidene-2,3-dihydroxy-1,4-bis-(diphenyl-phosphino)-butane
DMA N,N-dimethylaceamide
DMAD dimethyl acetylene dicarboxylate
DMAP dimethylamino-pyridine
DMD dimethyl dioxirane

DME  dimethoxyethane
DMF  dimethylformamide
DMPS  dimethylphenylsilyl
DMPU  *N,N'*-dimethyl-
propyleneurea
DMSO  dimethylsulfoxide
DPDC  di-*iso*-propyl peroxydi-
carbonate
DPDM  diphenyl diazomethane
DPEDA  1,2-diphenylethane-
1.2-diamine
dppb  bis(1,4-diphenyl-
phosphino)butane
DPPE  dppe  diphenyl-
phosphinoethane
dppf  dichloro[1,1'-bis-
(diphenylphosphino-
ferrocene
DPPP  1,3-(diphenyl-
phosphino)propane
dr  diastereomeric ratio
ds  diastereoselectivity
E  general electrophile
EDTA  edetic acid
ee  e.e. enantiomeric excess
en  ethylene diamine
Et  ethyl
Et$_2$0  diethyl ether
Et$_3$N  triethylamine
EWG  electron withdrawing
group
F$_c$  ferrocenyl
FDP  fructose-1,6-biphosphste
fl  flavin
Fmoc  9-fluorenylmethoxy-
carbonyl
fod  6,6,7,7,8,8,8-heptafluoro-
2,2-dimethyl-3,5-octane-
dione
FTT  1-fluoro-2,4,6-trimethyl-
pyridinium triflate
FVP  flash vapor pyrolysis
Gr  graphite
h  hours

Hap  hydroxyapatite
HGK  4-hydroxy-2-ketoglu-
tarate
HMDS  1,1,1,3,3,3-
hexamethyldisilazane
HMPA  HMPT  hexamethyl-
phosphoramide
hv  irradiation with light
Ipc  isopinocamphenyl
L-selectride®  lithium tri-
$^s$butylborohydride
L.R.  Lawesson's reagent
LAH  lithium aluminum
hydride
LDA  lithium diisopropylamide
LDBB  lithium 4,4'-*t*-butylbi-
phenylide
liq.  liquid
LTMP  lithium 2,2,6,6-tetra-
methylpiperidide
MABR  methyl aluminum
bis(4-bromo-2,6-di-$^t$ butyl-
phenoxide)
MAD  methylaluminum bis-(2,6-
di-*tert*-butyl-4-methyl-
phenoxide)
MCPBA  *m*-chloroperbenzoic
acid
Me  methyl
Mek  methyl ethyl ketone
MEM  β-methoxyethoxymethyl
Mes  mesityl
MMPP  magnesium mono-
peroxyphthalate
MOM  methoxymethyl
MPM  methoxy(phenylthio)-
methyl
MS  molecular sieves
Ms  methanesulfonyl
MSA  methanesulfonic acid
MTPA  methoxy-α-trifluoro-
methylphenylacetyl
MV$^{2+}$  methyl viologen
Naph  Np  naphthyl

NBS  N-bromosuccinimide
NCS  N-chlorosuccinimide
NFOBS  N-fluoro-O-benzenedi-
  sulfonimide
NIS  N-iodosuccinimide
NMO  N-methylmorpholine-N-
  oxide
NpS  2-nitrophenyl sulfenyl
NR  no reaction
Nuc.  general nucleophile
PCC  pyridinium chloro-
  chromate
PDC  pyridium dichromate
PEG  polyethylene glycol
Ph  phenyl
Ph-H  benzene
Ph-Me  toluene
PLAP  porcine liver acetone
  powder
PMB  p-methoxybenzyl
PMP  p-methoxyphenyl
PPA  polyphosphoric acid
PPTS  pyridinium p-toluene-
  sulfonate
Pr  propyl
psi  pounds per square inch
PTC  phase transfer catalysis
PTSA  p-toluenesulfonic acid
pyr  pyridine
rac  racemic
RaNi  Raney nickel
Rf  perfluorinated alkyl
rt  room temperature
Salen  N,N'-ethylenebis-
  (salicylideneiminato)
SEM  TEOC  β-trimethylsilyl-
  ethoxymethyl
Sia  Siamyl
SMEAH  sodium bis(2-
  methoxyethoxy)aluminum
  hydride
TBAB  tetrabutyl ammonium
  bromide

TBAF  tetrabutylammonium
  fluoride
TBDMS  = TBS  t-butyldimethyl-
  silyl
TBDPS  $^t$butyldiphenylsilyl
TBDPS  tert-butyl diphenylsilyl
Tbfmoc  Tetrabenzo[a,c,g,i]
  fluorenyl-17-methyloxy-
  carbonyl
TBHP  tert-butyl hydroper-
  oxide
TBME  t-butyl methyl ether
TBSOP  N-tert-butylcarbonyl-
  2-(tert-butyldimethyl-
  siloxy)pyrrole
TBTH  tributyltin hydride
TCAA  trichloroacetyl
  anhydride
TCF  trichloromethyl chloro-
  formate
TCNE  tetracyanoethylene
TCNEO  tetracyanoethylene
  oxide
TDS  dimethyl thexylsilyl
TEA  triethylamine
TEOC  SEM  β-trimethylsilyl-
  ethoxymethyl
Tf  trifluoromethanesulfonyl
TFA  trifluoroacetic acid
TFAA  trifluoroacetic
  anhydride
TFPZ  trifluoroisopropenyl zinc
THF  tetrahydrofuran
THP  tetrahydropyranyl
TIPS  tri-i-propylsilyl
TMAO  = TMANO
  trimethylamine N-oxide
TMEDA  tetramethylethylene-
  diamine
TMG  tetramethylguanidine
Tmob  2,4,6-trimethoxybenzyl
TMP  2,2,6,6-tetramethyl-
  piperine
TMS  trimethylsilyl

TMSDEA  N,N-diethyltrimethyl-
   silylamine
TMU  tetramethylurea
Tol  tolyl
Tos  Ts  p-toluenesulfonyl
TPP  Tetraphenylporphyrin
TPPTS  m-sulfonated tri-
   phenylphosphine
Tr  trityl
TT Co(II) Pc  tetrabutyl-
   ammonium cobalt(II)
   phthalocyanine-5,12,19,26-
   tetrasulfate

TTS  p-tolylsulphonate
wk  week
Z  benzyloxycarbonyl
[O]  general oxidation
Ⓟ  polymeric support
《《ᶜ·  US ultrasound

# I

# CARBON–CARBON BOND FORMING REACTIONS

## I.A.  Carbon - Carbon Single Bonds

(see also:  I.E., I.F., I.G., I.H.)

## I.A.1.  Alkylations of Aldehydes, Ketones and Their Derivatives

**I.A.1-1**  E. Diez-Barra et al., *J. Chem. Soc., Perkin Trans. 1*, 2427; M.N. Mattson and P. Helquist, *Organometallics*, **11**, 4; J.W. Patterson, *Synth. Commun.*, **22**, 1959; D. Caine and B. Stanhope, *Tetrahedron*, **48**, 33.

$$X = Br, I$$

80-96%

---

**I.A.1-2**  G. Bartoli et al., *Synlett*, 64.

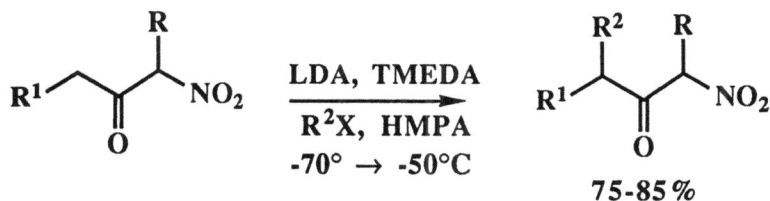

75-85%

**I.A.1-3**  D. Mukherjee et al., *Tetrahedron Lett.*, **33**, 1229.

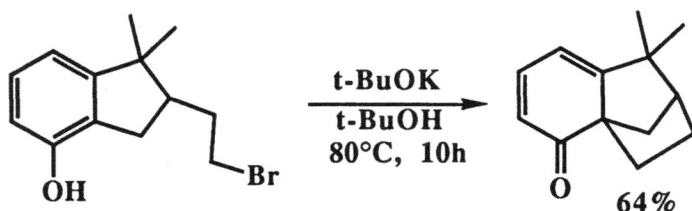

**I.A.1-4**  S.-H. Lee and M. Hulce, *Synlett*, 485.

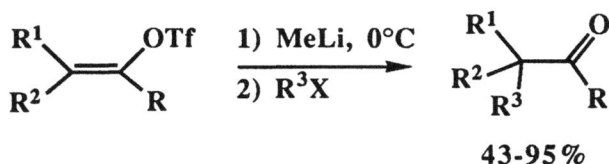

**I.A.1-5**  I. Fleming and J.J. Lewis, *J. Chem. Soc., Perkin Trans. 1*, 3257; **see also:** idem, *ibid.*, 3267; I. Fleming et al., *ibid.*, 3277 and 3295; **see also:** I. Fleming, *ibid.*, 3363.

**The Diastereoselectivity of Electrophilic Attack on Trigonal Carbon Adjacent to a Stereogenic Centre: Diastereoselective Alkylation and Protonation of Open-chain Enolates having a Stereogenic Centre at the ß-Position**

**I.A.1-6**  G. Bartoli et al., *J. Chem. Soc., Perkin Trans. 2*, 649 and *J. Chem. Soc., Perkin Trans. 1*, 2095.

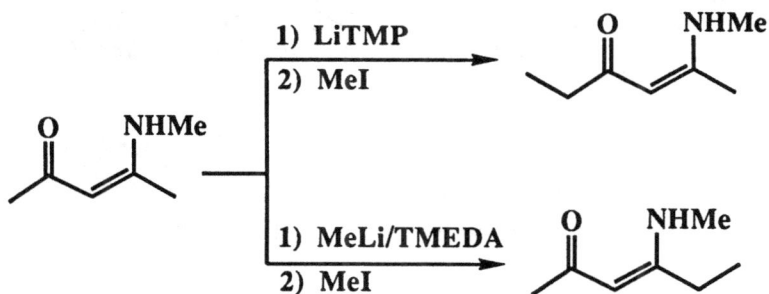

---

**I.A.1-7**  J.F.G.A. Jansen and B.L. Feringa, *Synth. Commun.*, 22, 1367

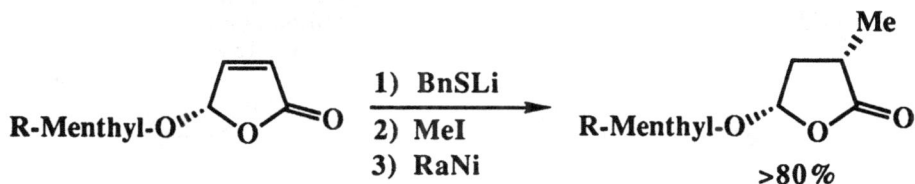

---

**I.A.1-8**  B.R. Langlois et al., *Tetrahedron Lett.*, 33, 1291.

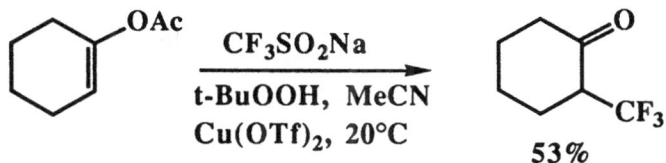

**I.A.1-9** H. Yamamoto et al., *J. Am. Chem. Soc.*, **114**, 4422; C.W. Jefford et al., *Tetrahedron Lett.*, **33**, 1855; A.J. Laurent and S. Lesniak, *Tetrahedron Lett.*, **33**, 8091; T. Benneche et al., *Acta Chem. Scand.*, **46**, 172; R.J. Linderman and T.V. Anklekar, *J. Org. Chem.*, **57**, 5078; C. Gajola and J.R. Green, *Synlett*, 973; Y. Kita et al., *J. Chem. Soc., Perkin Trans. 1*, 1795.

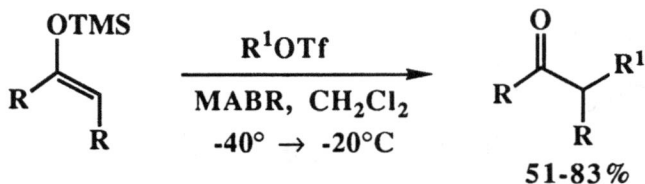

**other leaving groups and Lewis acids used similarly**

---

**I.A.1-10** E. Tagliavini, A. Umani-Ronchi et al., *Tetrahedron*, **48**, 1299; K. Conde-Frieboes and D. Hoppe, *Tetrahedron*, **48**, 6011; H. Sugimura and K. Yoshida, *Bull. Chem. Soc. Jpn.*, **65**, 3209; I. Kuwajima et al., *Tetrahedron Lett.*, **33**, 6979; T. Kubota et al., *Tetrahedron Lett.*, **33**, 1351.

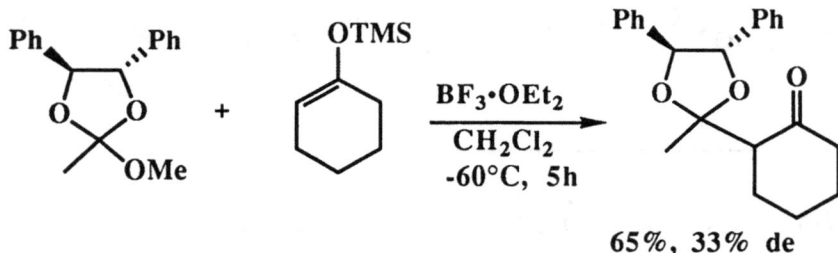

**similar reactions with acetals and analogs and other Lewis acids**

**I.A.1-11**  K. Takeda et al., *Tetrahedron Lett.*, **33**, 951.

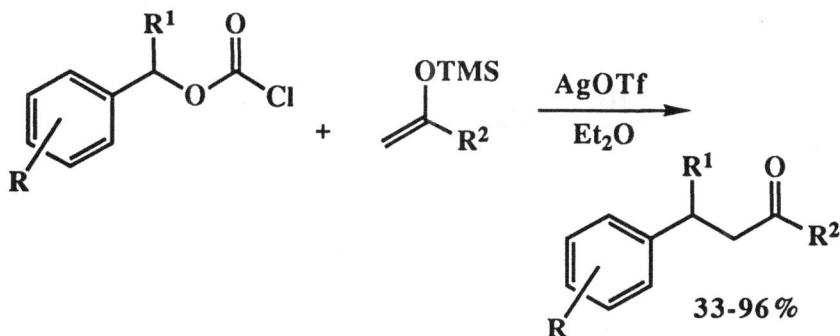

**I.A.1-12**  K. Narasaka et al., *Chem. Lett.*, **32**, 1229.

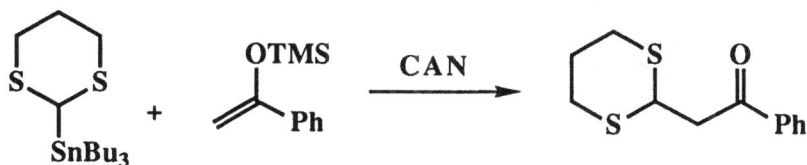

**I.A.1-13**  D. Enders et al., *Angew. Chem., Int. Ed. Engl.*, **31**, 618; I.R. Trehan et al., *Ind. J. Chem.*, **30B**, 793 (1991).

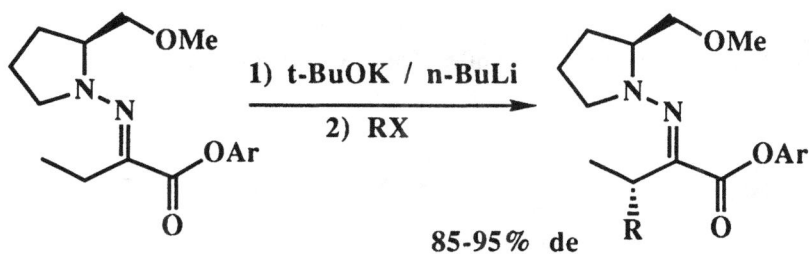

**I.A.1-14**  T. Fujii, T. Hirao and Y. Ohshiro, *Tetrahedron Lett.*, **33**, 5823.

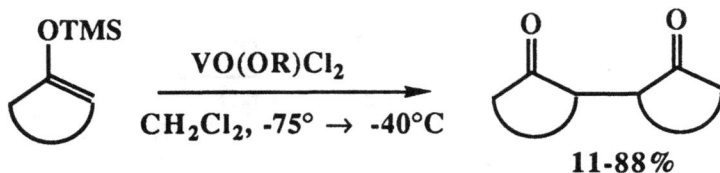

11-88%

---

**I.A.1-15**  K.I. Booker-Milburn, *Synlett*, 809.

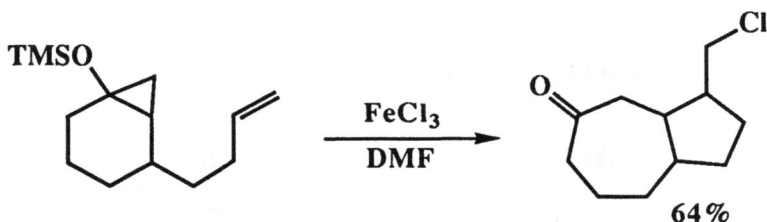

64%

## I.A.2.  Alkylations of Nitriles, Acids and Acid Derivatives

**I.A.2-1**  A. DeNicola, J. Einhorn and J.-L. Luche, *Tetrahedron Lett.*, **33**, 6461; M. Majewski and G.W. Bantle, *Synth. Commun.*, **22**, 23.

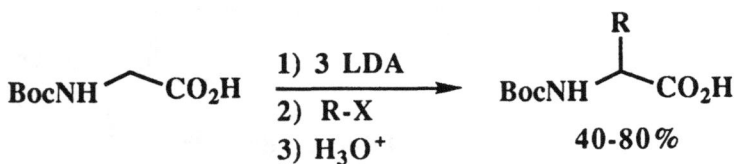

40-80%

**I.A.2-2** J.A. Clase and T. Money, *Can. J. Chem.*, **70**, 1536; K. Sakai et al., *Tetrahedron Lett.*, **33**, 247; W.A. Donaldson et al., *Tetrahedron Lett.*, **33**, 3967; K. Mori and K. Fukamatsu, *Liebigs Ann. Chem.*, 489.

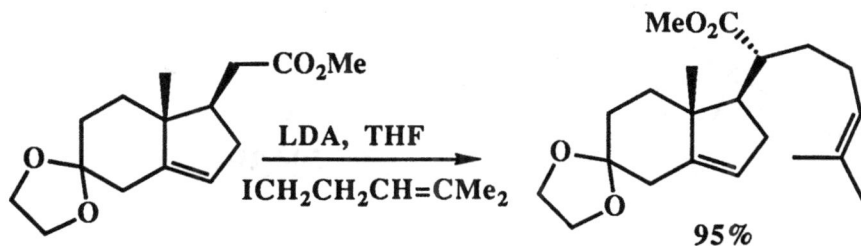

**95%**

---

**I.A.2-3** K. Sakai et al., *Tetrahedron Lett.*, **33**, 3481.

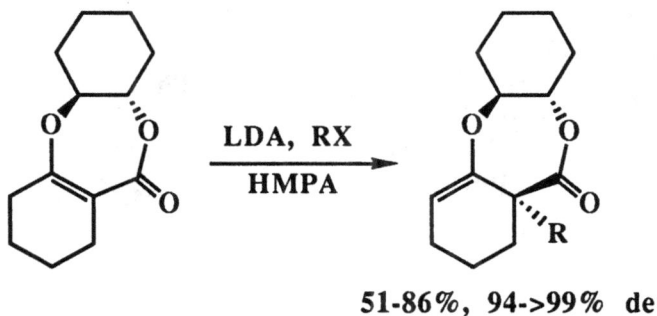

**51-86%, 94->99% de**

---

**I.A.2-4** J. Seyden-Penne et al., *Tetrahedron*, **48**, 6253.

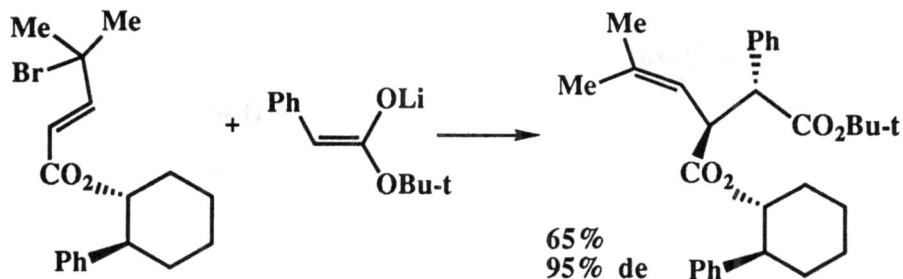

**65%**
**95% de**

**I.A.2-5**　A.G. Schultz and R.E. Taylor, *J. Am. Chem. Soc.*, **114**, 3937.

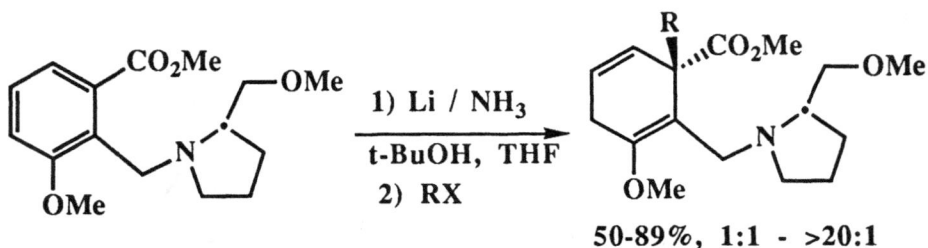

1) Li / NH$_3$
t-BuOH, THF
2) RX

50-89%, 1:1 - >20:1

---

**I.A.2-6**　R.M. Kellogg et al., *Rec. Trav. Chim.*, **111**, 129.

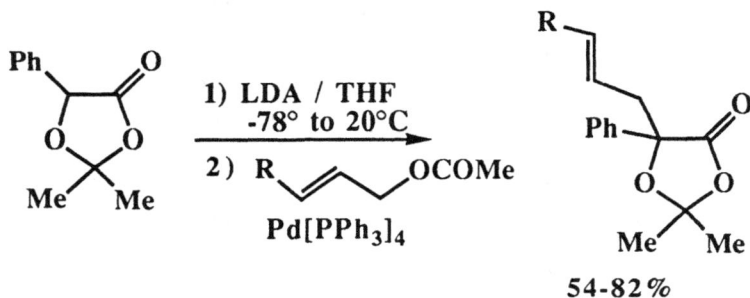

1) LDA / THF
-78° to 20°C
2) R⌒OCOMe
Pd[PPh$_3$]$_4$

54-82%

---

**I.A.2-7**　I. Ojima et al., *Chem. Lett.*, 1591

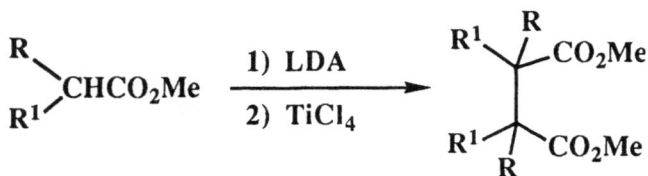

1) LDA
2) TiCl$_4$

**I.A.2-8**  W.H. Pearson and J.M. Schkeryantz, *J. Org. Chem.*, **57**, 2986.

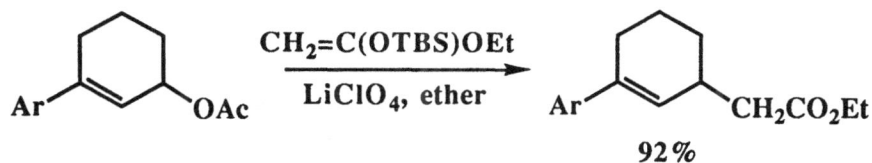

$$CH_2=C(OTBS)OEt$$

$$LiClO_4, \text{ ether}$$

92%

---

**I.A.2-9**  J.P. Konopelski et al., *J. Am. Chem. Soc.*, **114**, 1800; S.G. Davies and A.A. Mortlock, *Tetrahedron Lett.*, **33**, 1117; A. Fadel, *Synlett*, 48; S. Hanessian et al., *Synlett*, 33.

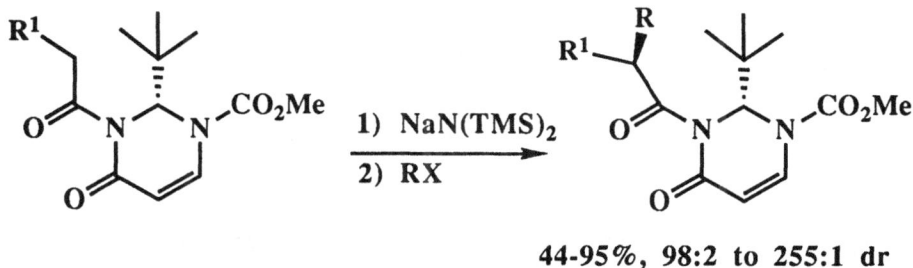

1) NaN(TMS)$_2$
2) RX

44-95%, 98:2 to 255:1 dr

**other chiral auxiliaries used similarly**

---

**I.A.2-10**  Y. Nagao et al., *J. Org. Chem.*, **57**, 4232, 4238, 4243.

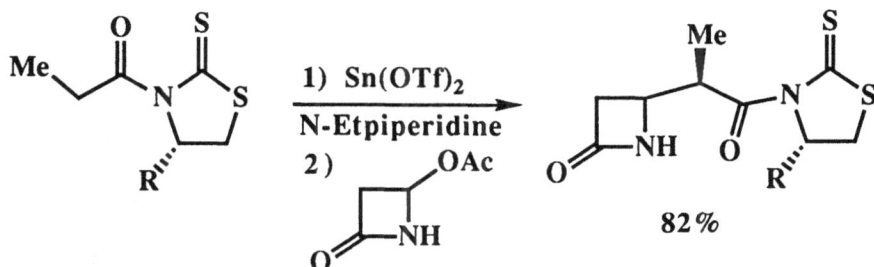

1) Sn(OTf)$_2$

N-Etpiperidine

2)

82%

**I.A.2-11** A. Mourino et al., *Tetrahedron Lett.*, **33**, 105; S.D. Rychnovsky and G. Griesgraber, *J. Org. Chem.*, **57**, 1559; J. Royer et al. *J. Org. Chem.*, **57**, 4211; T. Takahashi et al., *Tetrahedron Lett.*, **33**, 369.

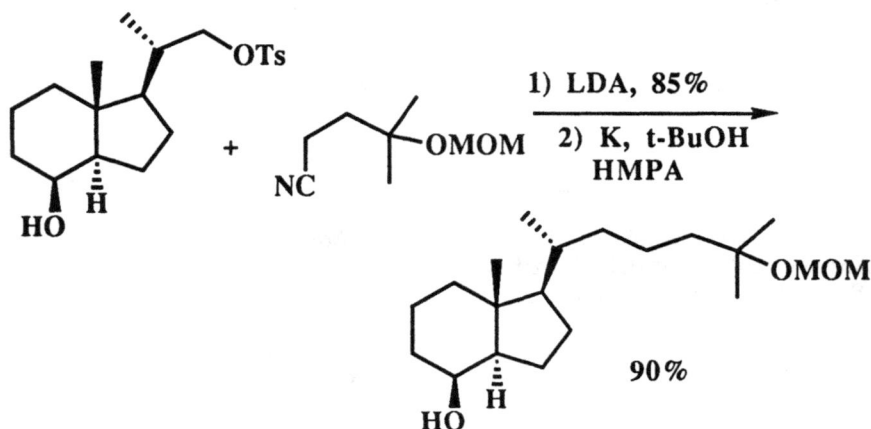

## I.A.3.  Alkylations of β-Dicarbonyl, β-Cyanocarbonyl Systems and Other Active Methylene Compounds

**I.A.3-1** M.A. Sanner et al., *Tetrahedron Lett.*, **33**, 5287; F. Bonadies, A. Scettri et al., *Gazz. Chim. Ital.*, **122**, 237; B.C. Ranu, S. Bhar et al., *J. Chem. Soc., Perkin Trans. 1*, 365; N. Yoshihara et al., *Bull. Chem. Soc. Jpn.*, **65**, 1185.

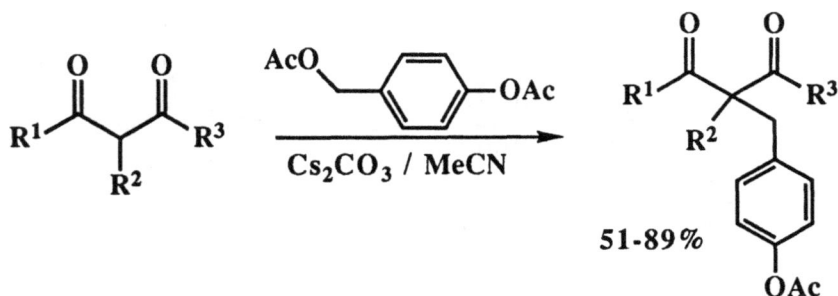

other bases and leaving groups used similarly

**I.A.3-2**  A.J. Pearson and K. Srinivasan, *Tetrahedron Lett.*, **33**, 7295.

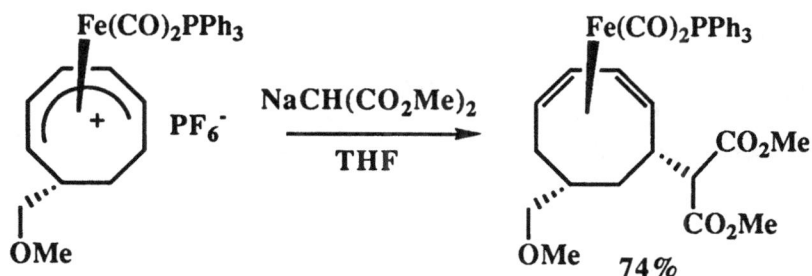

**I.A.3-3**  R.V. Hoffman and H.-O. Kim, *Tetrahedron Lett.*, **33**, 3579.

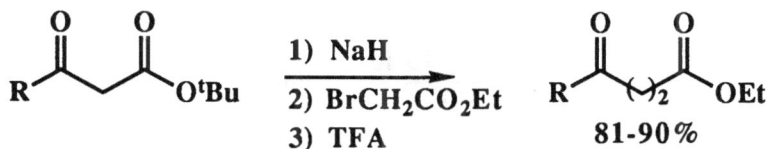

**I.A.3-4**  S. Kang et al., *Tetrahedron Asym.*, **3**, 1139.

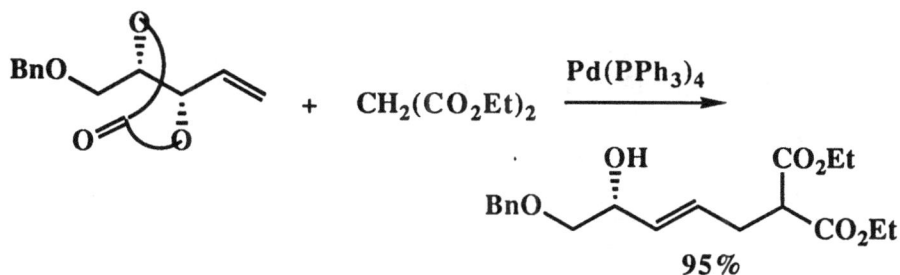

**I.A.3-5** H. Nishino et al., *J. Org. Chem.*, **57**, 3551; B.B. Snider and Q. Zhang, *Tetrahedron Lett.*, **33**, 5921.

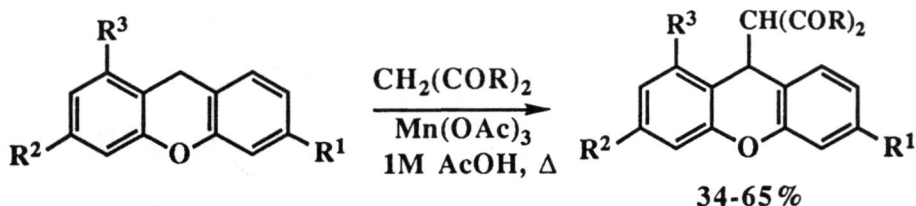

**I.A.3-6** T. Gallagher et al., *J. Chem. Soc., Perkin Trans. 1*, 2169; A.G. Fallis et al., *Can. J. Chem.*, **70**, 1531.

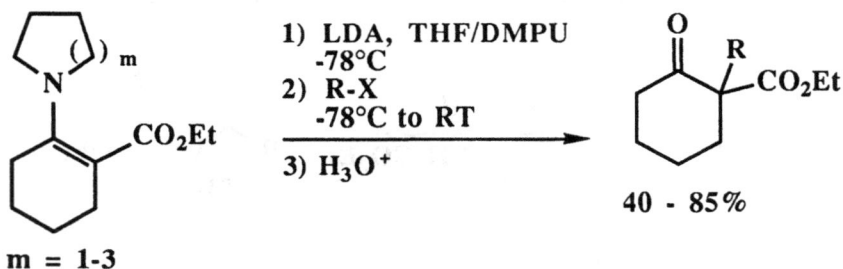

**I.A.3-7** I. Shimizu et al., *Synlett*, 301; V.K. Datcheva and B.A. Marples, *J. Chem. Res. (S)*, 238; G.W. Klumpp et al., *Tetrahedron Lett.*, **33**, 475.

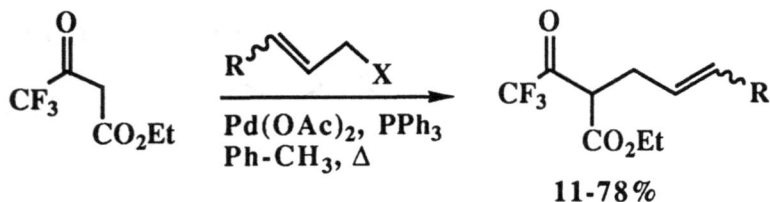

**I.A.3-8**  A. de Meijere et al., *J. Am. Chem. Soc.*, **114**, 4051 and *Synlett*, 558; C. Moberg, K. Nordstrom and P. Helquist, *Synthesis*, 685; L. Geng and X. Lu, *J. Chem. Soc., Perkin Trans. 1*, 17.

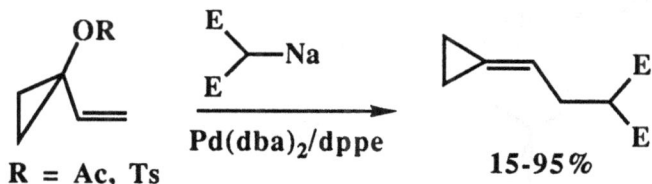

**R = Ac, Ts**

**Pd(dba)$_2$/dppe**

**15-95%**

**various substrates and leaving groups employed similarly**

---

**I.A.3-9**  D. Gravel et al., *Tetrahedron Lett.*, **33**, 1403 and 1407.

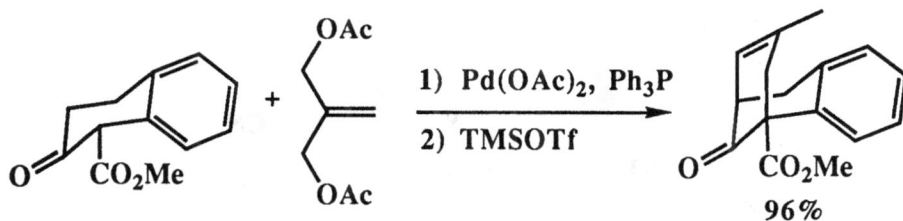

1) Pd(OAc)$_2$, Ph$_3$P

2) TMSOTf

**96%**

---

**I.A.3-10**  Y. Ito et al., *J. Am. Chem. Soc.*, **114**, 2586.

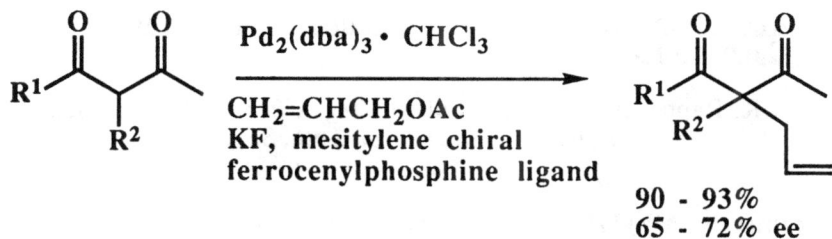

Pd$_2$(dba)$_3$ · CHCl$_3$

CH$_2$=CHCH$_2$OAc
KF, mesitylene chiral
ferrocenylphosphine ligand

**90 - 93%**
**65 - 72% ee**

**I.A.3-11**  C. Cativiela et al., *Bull. Chem. Soc. Jpn.*, **65**, 1657 and *Tetrahedron Asymm.*, **3**, 1141.

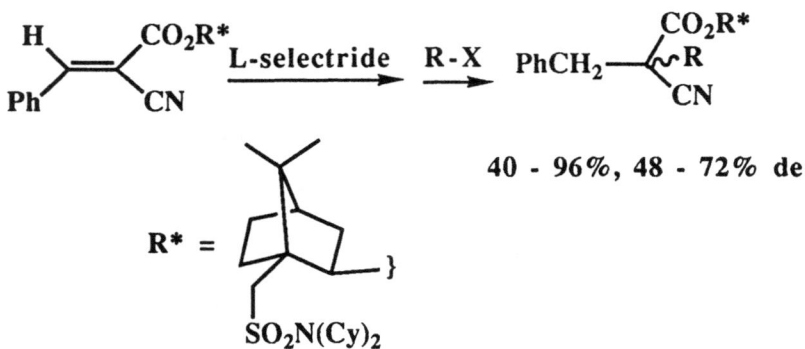

40 - 96%, 48 - 72% de

---

**I.A.3-12**  D.P. Curran and G. Thoma, *J. Am. Chem. Soc.*, **114**, 4436.

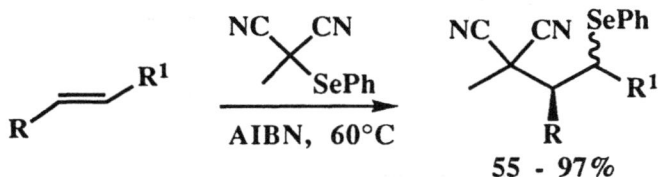

55 - 97%

## I.A.4. Alkylations of N-, P-, S-, Se and Similar Stabilized Carbanions

**I.A.4-1**  A.K. Rappe, A.I. Meyers et al., *J. Org. Chem.*, **57**, 3819; **see also:** P.I. Dalko and Y. Langlois, *Tetrahedron Lett.*, **33**, 5213.

Asymmetric Alkylations on Chiral Formamidines.  Molecular Mechanics Studies Relating to the Facial Selectivity of the Lithiated Intermediates

**I.A.4-2**  Y. Jiang et al., *Synth. Commun.*, **22**, 265.

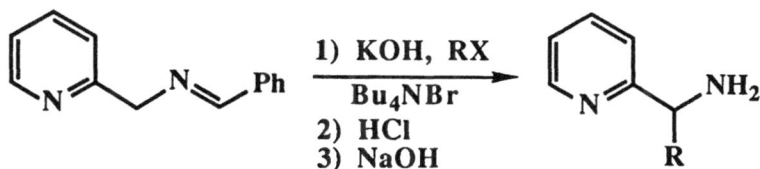

1) KOH, RX
Bu$_4$NBr
2) HCl
3) NaOH

---

**I.A.4-3**  J. Barluenga et al., *Tetrahedron Lett.*, **33**, 7573.

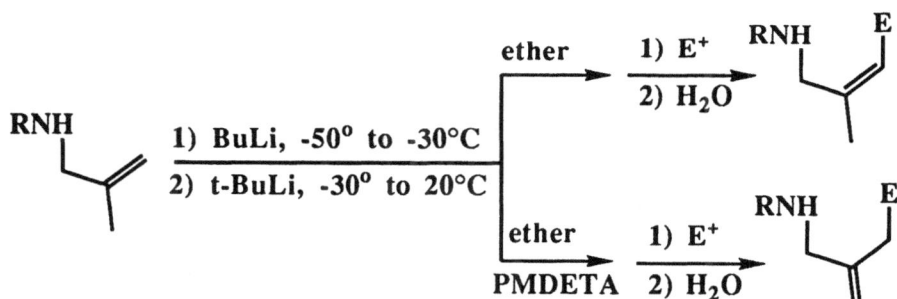

1) BuLi, -50° to -30°C
2) t-BuLi, -30° to 20°C

ether    1) E$^+$
         2) H$_2$O

ether    1) E$^+$
PMDETA   2) H$_2$O

**PMDETA  =  N,N,N',N",N"'-pentamethyldiethylene-    57-94%**
            **triamine**

---

**I.A.4-4**  G. Jommi et al., *Synth. Commun.*, **22**, 107; C. Maury et al.,
*Tetrahedron Lett.*, **33**, 6127.

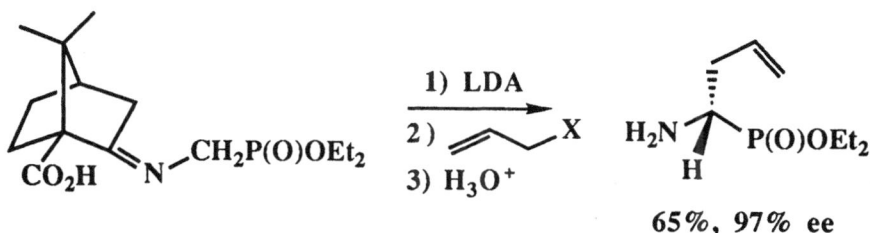

1) LDA
2) $\overset{}{=}\!\!\!\diagdown$ X
3) H$_3$O$^+$

65%, 97% ee

**I.A.4-5**  S. Cabiddu et al., *J. Organomet. Chem.*, **441**, 197.

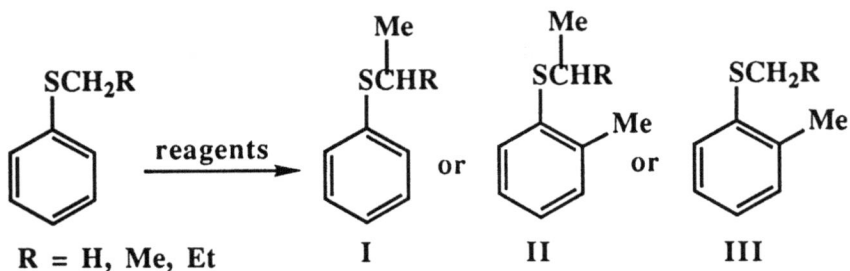

reagents = 1.15eq. BuLi / ᵗBuOK; MeI        69-80% I
2.3 eq. BuLi / ᵗBuOK; MeI        58-78% II
1.15eq. BuLi ; MeI                70-75% III

---

**I.A.4-6**  H. Miyaoka and M. Kajiwara, *Chem. Pharm. Bull.*, **40**, 1659; P. Jankowski and J. Wicha, *J. Chem. Soc., Chem. Commun.*, 802; A.R. Battersby et al., *Tetrahedron*, **48**, 7519; S.S. Magar and P.L. Fuchs, *Tetrahedron Lett.*, **33**, 745.

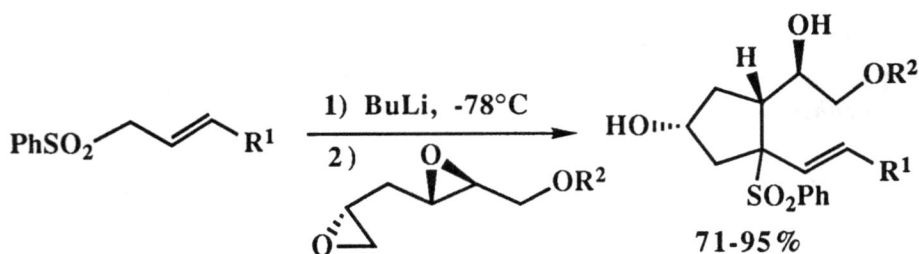

other sulfone alkylations also reported

**I.A.4-7** B.M. Trost et al., *Tetrahedron Lett.*, **33**, 717.

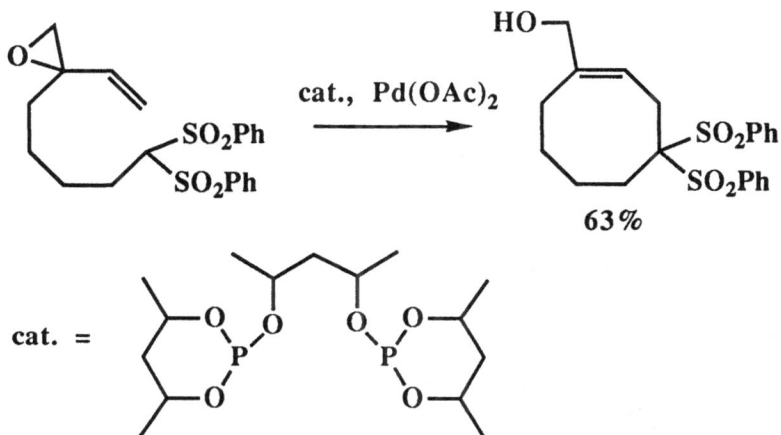

**I.A.5. Alkylations of Organometallic Reagents**

(see also: I.B.3., I.B.5., I.F., I.G.)

**I.A.5-1** S.D. Rychnovsky and D.J. Skalitzky, *J. Org. Chem.*, **57**, 4336.

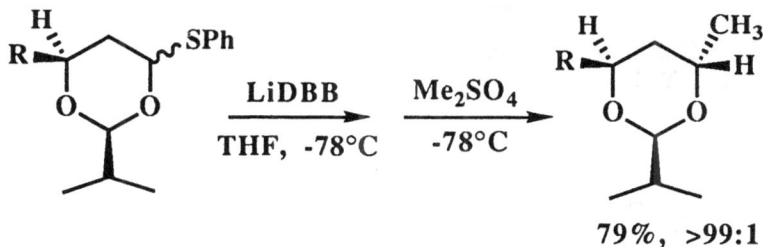

**I.A.5-2** W.F. Bailey et al., *J. Am. Chem. Soc.*, **114**, 8053; A. Krief et al., *Synlett*, 907.

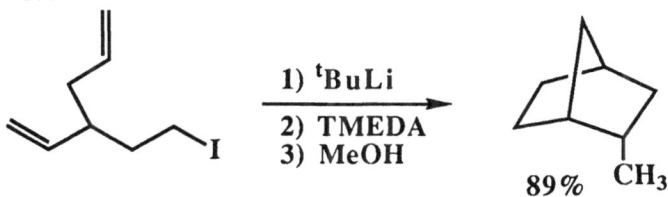

**I.A.5-3** B.A. Keay et al., *J. Chem. Soc., Perkin Trans. 1*, 2729; A.R. Katritzky et al., *Synthesis*, 911.

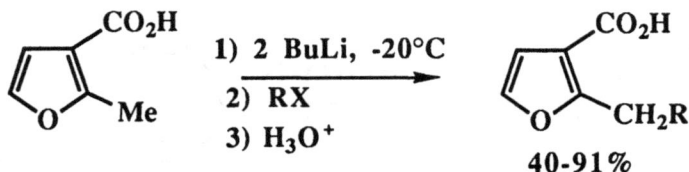

1) 2 BuLi, -20°C
2) RX
3) $H_3O^+$

40-91%

**similarly with a different heterocycle**

---

**I.A.5-4** W.A. Donaldson and M.A. Hossain, *Tetrahedron Lett.*, 33, 4107; S.-S.P. Chou, C.-H. Hsu and M.-C.P. Yeh, *Tetrahedron Lett.*, 33, 643; A.J. Pearson and M.P. Burello, *Organometallics*, 11, 448.

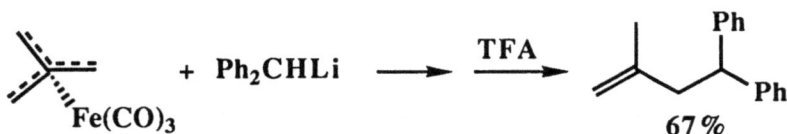

$+ Ph_2CHLi \longrightarrow \xrightarrow{TFA}$

67%

**other nucleophiles used for similar transformations**

---

**I.A.5-5** P. Rollin et al., *Tetrahedron Lett.*, 33, 4575; **see also:** T.-M. Yuan and T.-Y. Luh, *J. Org. Chem.*, 57, 4550.

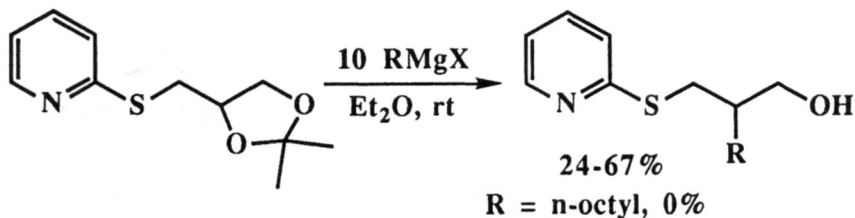

$\xrightarrow[\text{Et}_2\text{O, rt}]{\text{10 RMgX}}$

24-67%

R = n-octyl, 0%

**I.A.5-6** K. Higashiyama et al., *Heterocycles*, **33**, 17; J.-L. Parrain et al. *J. Organomet. Chem.*, **437**, C19; C. Mioskowski, J.R. Falck et al., *Tetrahedron Lett.*, **33**, 5201; **see also:** P.J. Kocienski et al., *J. Chem. Soc., Perkin Trans. 1*, 3419 and 3431.

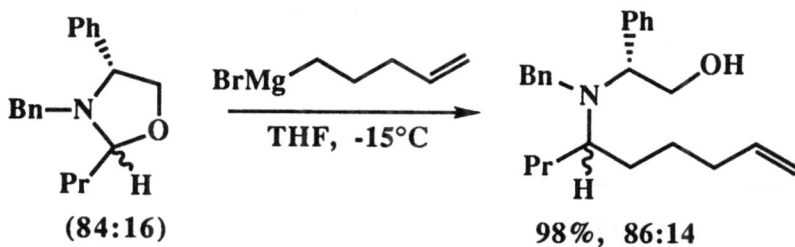

**(84:16)**

**98%, 86:14**

**similarly with organocuprates and acetal derivatives or Grignards and vinyl ethers with (Ph$_3$P)$_2$NiCl$_2$**

---

**I.A.5-7** J.R. Falck, C. Mioskowski et al., *Tetrahedron Lett.*, **33**, 4885; M. Santelli et al., *ibid.*, **33**, 7515; T. Waglund and A. Claesson, *Acta Chem. Scand.*, **46**, 73.

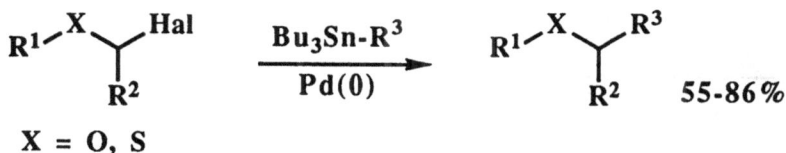

X = O, S

55-86%

**similarly with different leaving groups and catalysts**

---

**I.A.5-8** A.R. Katritzky et al., *Organometallics*, **11**, 1381.

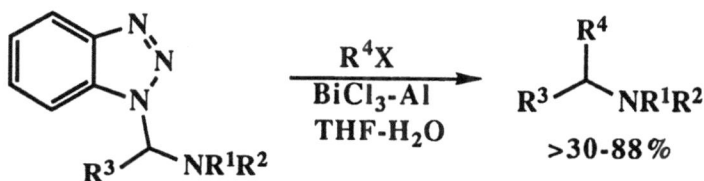

>30-88%

**I.A.5-9**  R.M. Waymouth et al., *Tetrahedron Lett.*, **33**, 7735; A.H. Hoffmann et al., *J. Am. Chem. Soc.*, **114**, 6692.

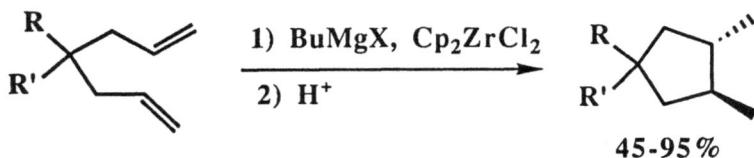

$$
\begin{array}{c}
\text{R}\diagdown\diagup\diagup \\
\text{R}'
\end{array}
\quad
\xrightarrow[\text{2) H}^+]{\text{1) BuMgX, Cp}_2\text{ZrCl}_2}
\quad
\begin{array}{c}
\text{R} \\
\text{R}'
\end{array}
$$

**45-95%**

---

**I.A.5-10**  H.M.R. Hoffmann and U. Karama, *Chem. Ber.*, **125**, 2803 and 2809; **see also:** D.J. Burton et al., *Synlett*, 141.

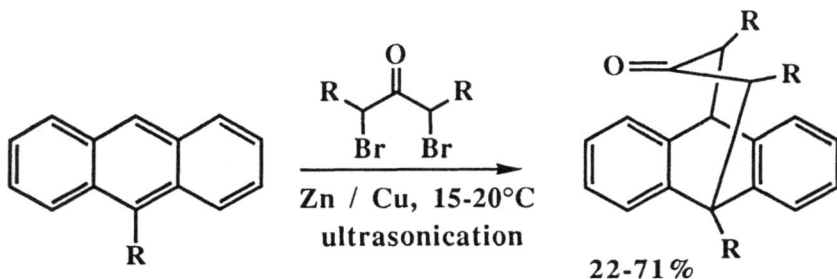

$$
\xrightarrow[\substack{\text{Zn / Cu, 15-20°C} \\ \text{ultrasonication}}]{\substack{\text{O} \\ \text{R}\diagup\diagdown\text{R} \\ \text{Br}\;\;\text{Br}}}
$$

**22-71%**

---

**I.A.5-11**  T. Takeda et al., *Tetrahedron Lett.*, **33**, 5381.

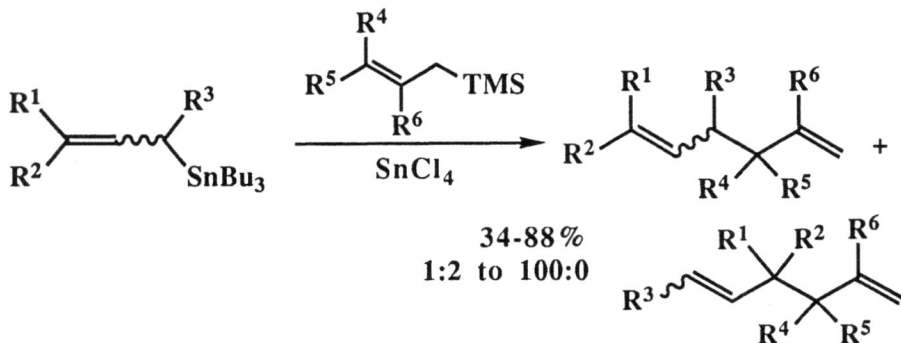

$$
\begin{array}{c}
\text{R}^1 \\
\text{R}^2
\end{array}\!\!\diagup\!\!\!\sim\!\!\!
\begin{array}{c}
\text{R}^3 \\
\text{SnBu}_3
\end{array}
\quad
\xrightarrow[\text{SnCl}_4]{\substack{\text{R}^4 \\ \text{R}^5\diagdown\diagup\text{TMS} \\ \text{R}^6}}
$$

**34-88%**
**1:2 to 100:0**

**I.A.5-12**  R. Fischer et al., *J. Organomet. Chem.*, **427**, 395.

**I.A.5-13**  H. Molines et al., *J. Org. Chem.*, **57**, 5530.

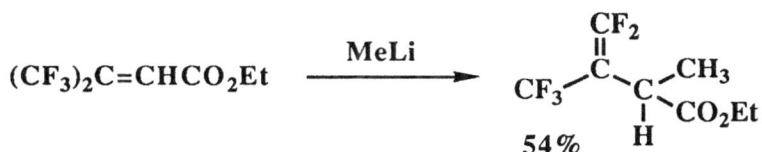

**I.A.5-14**  S. Woo and B.A. Keay, *Tetrahedron Lett.*, **33**, 2661; M. Lautens and R.K. Belter, *ibid.*, **33**, 2617; P.N. Rao and I.B. Taraporewala, *Steroids*, **57**, 154.

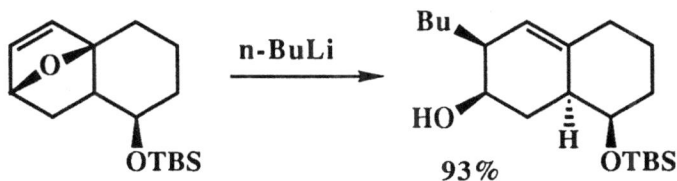

similarly with organocuprates and allyl epoxides

**I.A.5-15**  L.N. Pridgen et al., *J. Org. Chem.*, **57**, 1237.

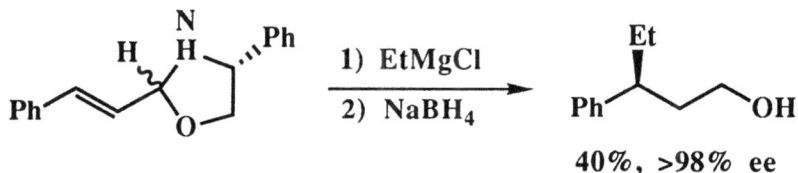

**I.A.5-16** T. Ibuka, N. Fujii, Y. Yamamoto et al., *Tetrahedron Lett.*, **33**, 3783; Y. Yamamoto et al., *J. Org. Chem.*, **57**, 1025; J. Mathew, *ibid.*, **57**, 2753; M.J. Dunn and R.F.W. Jackson, *J. Chem. Soc., Chem. Commun.*, 319.

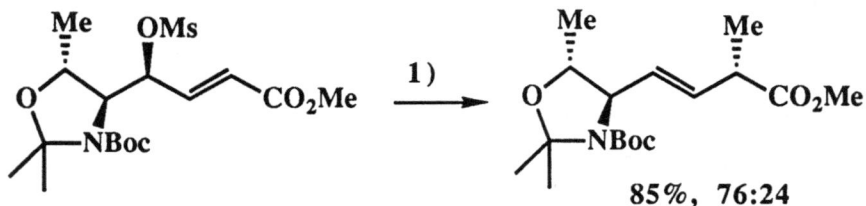

85%, 76:24

1) $Me_2Cu(CN)(ZnCl)_2 \cdot 2MgBrCl \cdot 3LiCl$

**halide leaving groups and copper-lithio species also employed**

---

**I.A.5-17** V. Calo et al., *Tetrahedron*, **48**, 6051; S. Valverde et al., *J. Org. Chem.*, **57**, 4546.

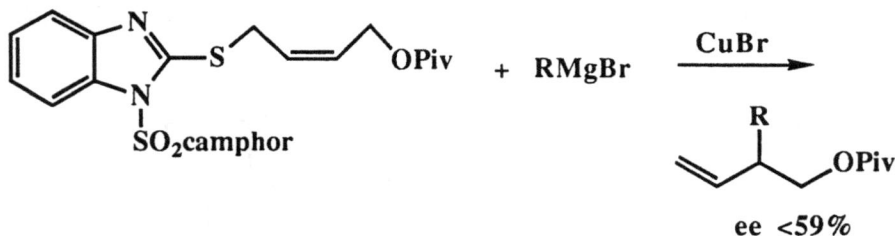

+ RMgBr    $\xrightarrow{CuBr}$

ee <59%

---

**I.A.5-18** E. Nakamura et al., *Tetrahedron*, **48**, 5709.

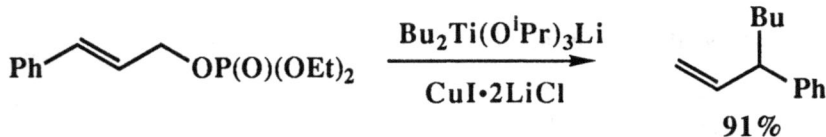

$\xrightarrow[CuI \cdot 2LiCl]{Bu_2Ti(O^iPr)_3Li}$

91%

**I.A.5-19**  J.C. Jaszberenyi et al., *Tetrahedron*, **48**, 8881.

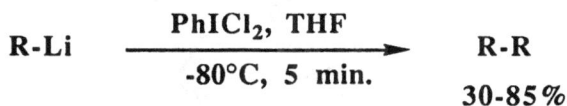

$$\text{R-Li} \quad \xrightarrow[\text{-80°C, 5 min.}]{\text{PhICl}_2, \text{ THF}} \quad \begin{array}{c} \text{R-R} \\ \textbf{30-85\%} \end{array}$$

## I.A.6.  Other Alkylation Procedures

**I.A.6-1**  J.A. Ciaccio et al., *Tetrahedron Lett.*, **33**, 1431; M. Bessodes et al., *J. Org. Chem.*, **57**, 4441.

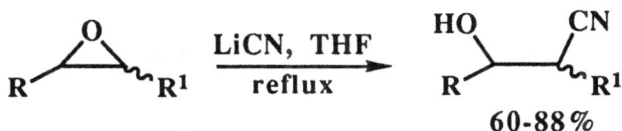

**I.A.6-2**  S. Torii et al., *Tetrahedron Lett.*, **33**, 3499 and 3503.

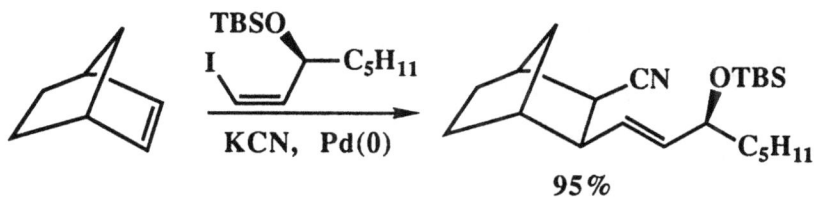

**I.A.6-3**  D.H.R. Barton et al., *Tetrahedron*, **48**, 2613.

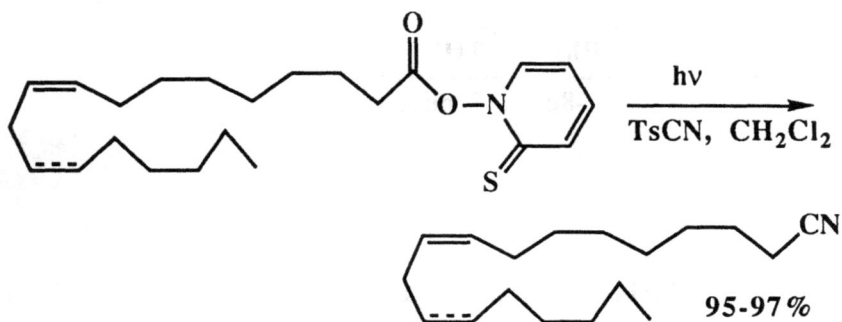

95-97%

---

**I.A.6-4**  D. Hoppe et al., *Tetrahedron Lett.*, **33**, 5323 and 5327.

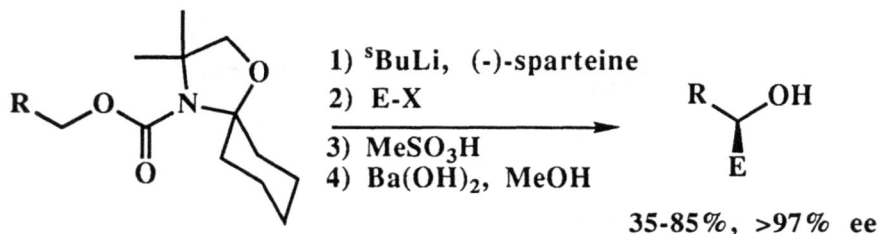

1) $^s$BuLi, (-)-sparteine
2) E-X
3) MeSO$_3$H
4) Ba(OH)$_2$, MeOH

35-85%, >97% ee

---

**I.A.6-5**  M. Yus et al., *Tetrahedron Lett.*, **33**, 5597.

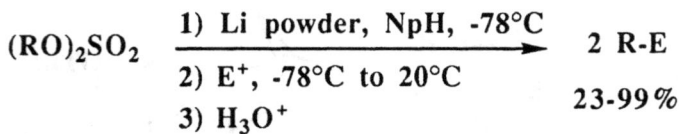

$$(RO)_2SO_2 \xrightarrow{\begin{array}{l} 1)\ \text{Li powder, NpH, -78°C} \\ 2)\ E^+,\ \text{-78°C to 20°C} \\ 3)\ H_3O^+ \end{array}} 2\ R\text{-}E$$

23-99%

---

**I.A.6-6**  S. Takano et al., *Synlett*, 70.

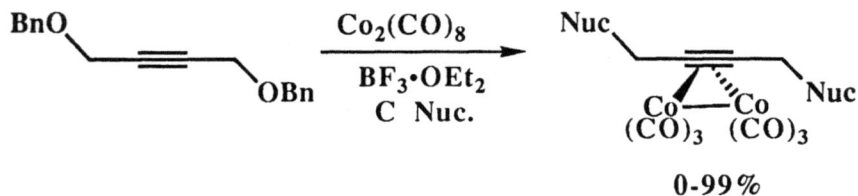

Co$_2$(CO)$_8$
BF$_3$·OEt$_2$
C Nuc.

0-99%

# I.A.7. Nucleophilic Addition to Electrophilic Carbon

## I.A.7.a.1a. Intermolecular Aldol-Type 1,2-Additions

**I.A.7.a.1a-1** M. Majewski and D.M. Gleave, *J. Org. Chem.*, **57**, 3599.

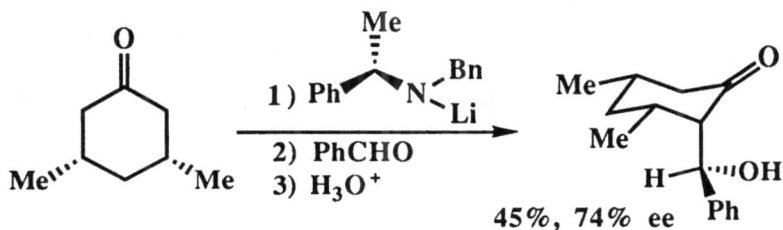

45%, 74% ee

**similarly with other chiral lithium amide bases**

---

**I.A.7.a.1a-2** E.J. Corey and G.A. Reichard, *J. Am. Chem. Soc.*, **114**, 10677.

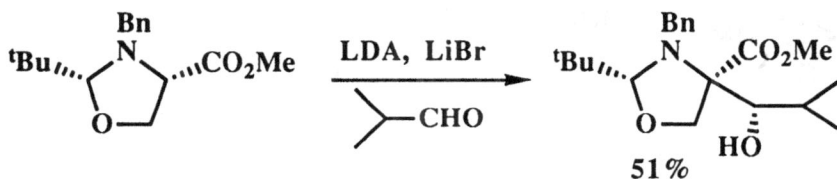

51%

---

**I.A.7.a.1a-3** S. Brandange and H. Leijonmarck, *Tetrahedron Lett.*, **33**, 3025; **see also:** T.N. Nanninga et al., *Tetrahedron Lett.*, **33**, 2279.

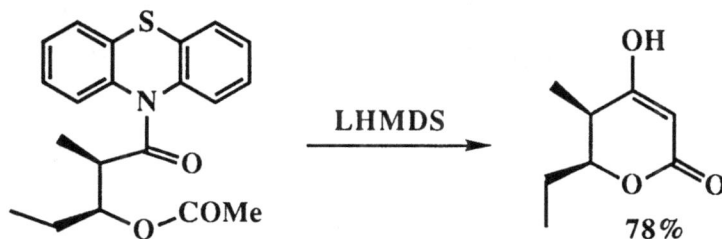

78%

**I.A.7.a.1a-4**   K. Chibale and S. Warren, *Tetrahedron Lett.*, **33**, 4369; **see also:** I. Paterson, *Pure Appl. Chem.*, **64**, 1821; **see also:** J.M. Goodman, *Tetrahedron Lett.*, **33**, 7223; C. Gennari, A. Bernardi et al., *J. Org. Chem.*, **57**, 5173.

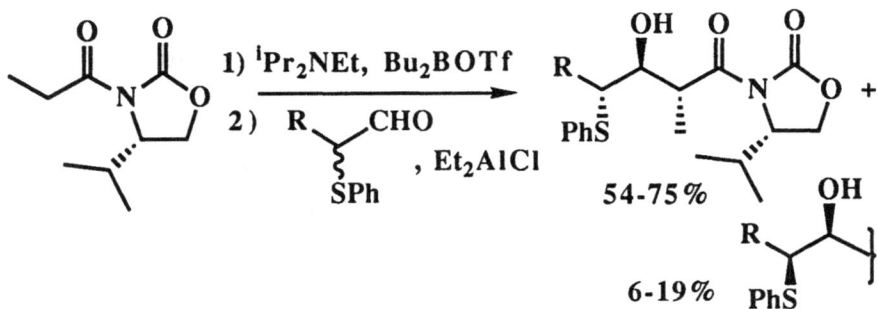

**I.A.7.a.1a-5**   M. Uemura et al., *J. Org. Chem.*, **57**, 5590.

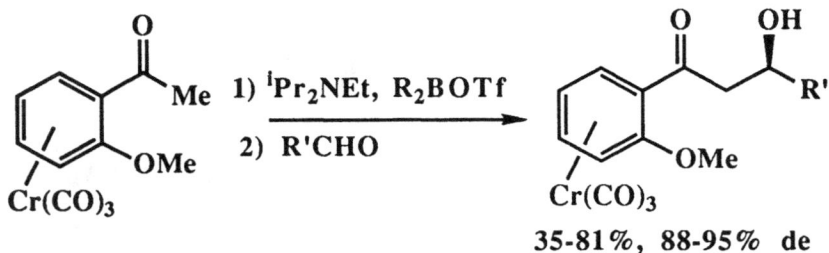

**I.A.7.a.1a-6**   I. Paterson and M.V. Perkins, *Tetrahedron Lett.*, **33**, 801.

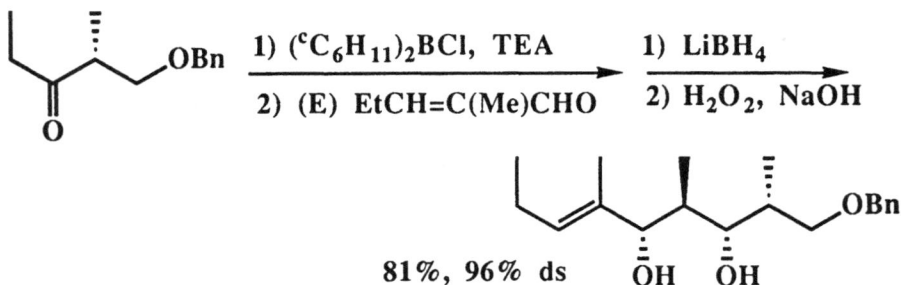

**I.A.7.a.1a-7** K.H. Ahn et al., *J. Org. Chem.*, **57**, 5065 and *Tetrahedron Lett.*, **33**, 6661; **see also:** A. Choudhury and E.R. Thornton, *Tetrahedron*, **48**, 5701.

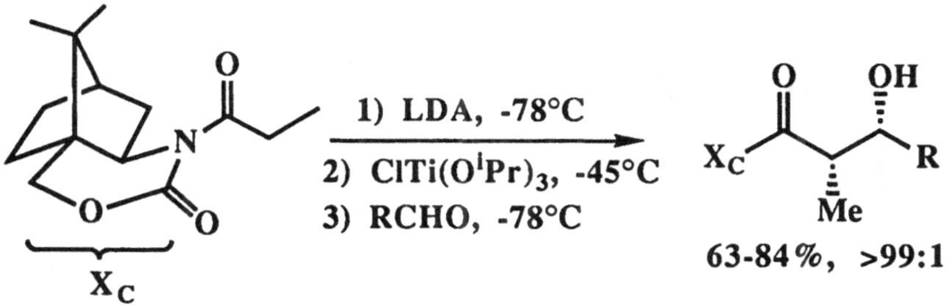


**another chiral auxiliary used similarly**

---

**I.A.7.a.1a-8** M. Cinquini, F. Cozzi et al., *J. Org. Chem.*, **57**, 6339; M. Hanaoka et al., *Tetrahedron Asym.*, 3, 1007.

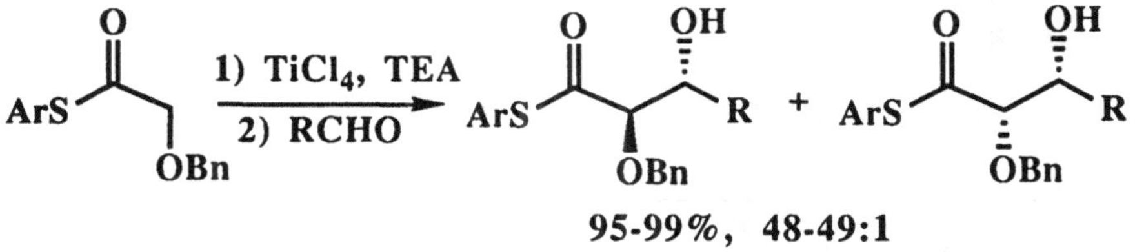


**similarly with a formylarene•Cr(CO)3 species**

---

**I.A.7.a.1a-9** I. Paterson and R.D. Tillyer, *Tetrahedron Lett.*, **33**, 4233.

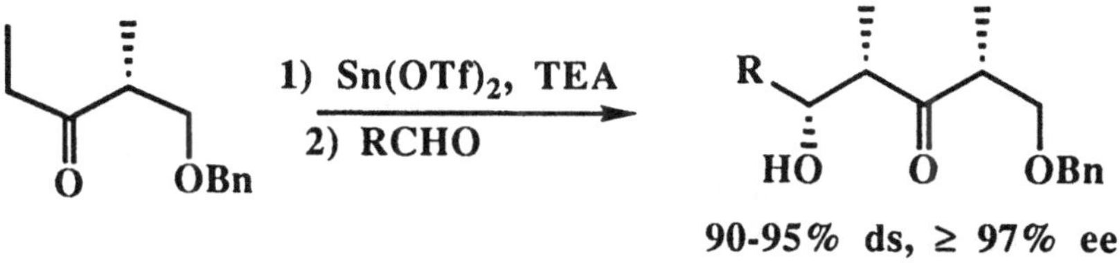

**I.A.7.a.1a-10**  E.J. Enholm and S. Jiang, *Tetrahedron Lett.*, **33**, 313, 6069 and *Heterocycles*, **34**, 2247; see also: E. Dunach et al., *Synlett*, 293.

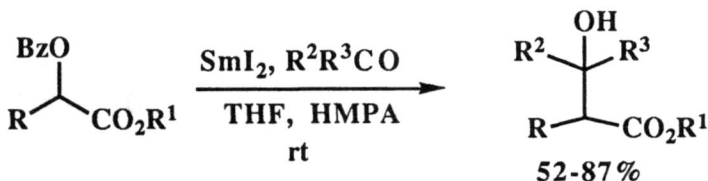

$$\begin{array}{c} BzO \\ R \overset{|}{\underset{}{\diagdown}} CO_2R^1 \end{array} \xrightarrow[\text{rt}]{\substack{SmI_2,\ R^2R^3CO \\ \text{THF, HMPA}}} \begin{array}{c} OH \\ R^2 \overset{|}{\underset{}{\diagdown}} R^3 \\ R \diagdown CO_2R^1 \end{array}$$

52-87%

---

**I.A.7.a.1a-11**  N.J. Turner et al., *J. Chem. Soc., Perkin Trans. 1*, 1085; E.J. Toone et al., *J. Org. Chem.*, **57**, 426; see also: C.-H. Wong et al., *J. Am. Chem. Soc.*, **114**, 741..

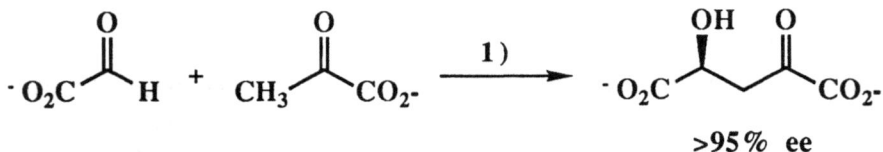

>95% ee

1) HGK Aldolase

various aldolases used similarly

---

**I.A.7.a.1a-12**  B.P. Maliakel and W. Schmid, *Tetrahedron Lett.*, **33**, 3297; G.M. Whitesides et al., *Liebigs Ann. Chem.*, 95; see also: F. Effenberger et al., *ibid.*, 1297.

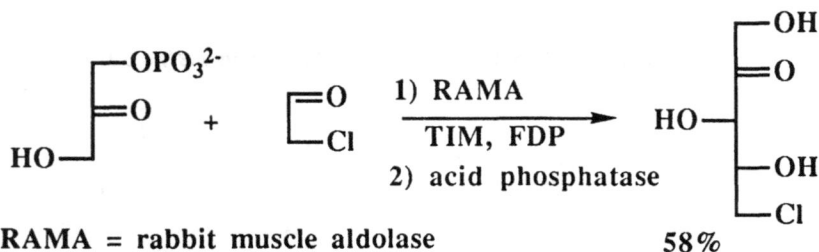

1) RAMA

TIM, FDP

2) acid phosphatase

58%

RAMA = rabbit muscle aldolase
TIM = triosephosphate isomerase

**I.A.7.a.1a-13**  M.J. Munchhof and C.H. Heathcock, *Tetrahedron Lett.*, **33**, 8005; F. Tanaka and K. Fuji *Tetrahedron Lett.*, **33**, 7885.

### Control of silyl enol ether stereochemistry

---

**I.A.7.a.1a-14**  A.G. Myers et al., *J. Am. Chem. Soc.*, **114**, 7922.

### Silicon-Directed Aldol Reactions.  Rate Acceleration by Small Rings

---

**I.A.7.a.1a-15**  W. Oppolzer and C. Starkman, *Tetrahedron Lett.*, **33**, 2439.

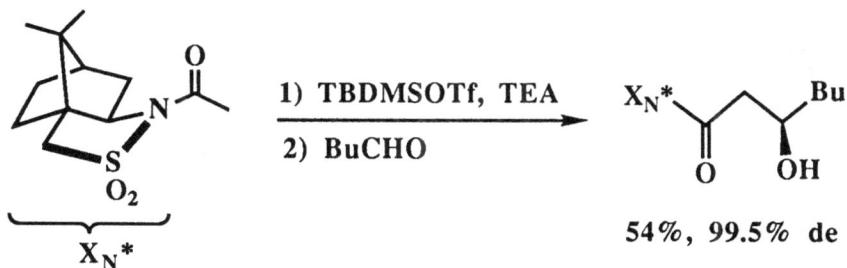

1) TBDMSOTf, TEA

2) BuCHO

54%, 99.5% de

---

**I.A.7.a.1a-16**  M. Hanaoka et al., *J. Org. Chem.*, **57**, 2034.

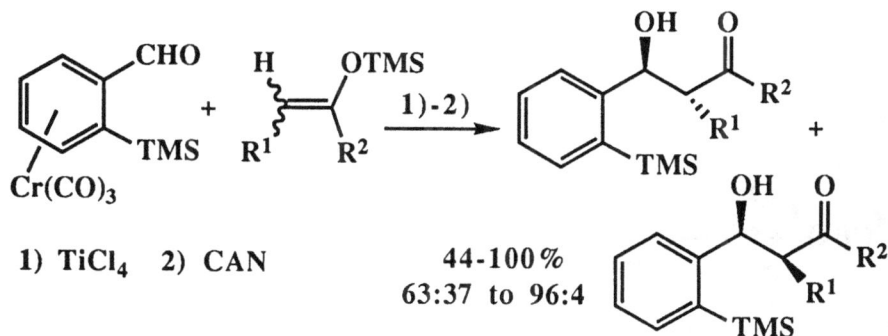

1) TiCl$_4$   2) CAN

44-100%
63:37 to 96:4

**I.A.7.a.1a-17**   A. Kamimura et al., *J. Org. Chem.*, **57**, 5403.

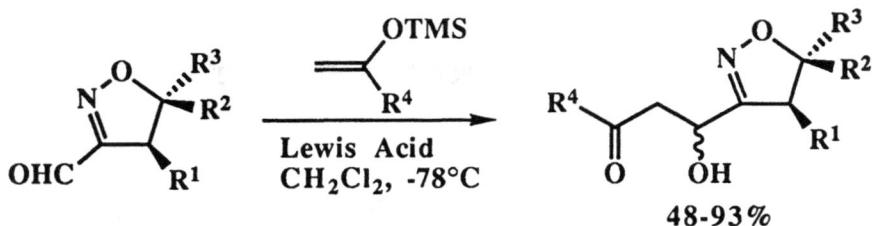

Lewis Acid = $TiCl_4$  -  primarily syn
              = $BF_3 \cdot OEt_2$  -  primarily anti

---

**I.A.7.a.1a-18**   J. Dubac et al., *Tetrahedron Lett.*, **33**, 1093; B. Bosnich et al., *ibid.*, **33**, 5729 and 6423; M.T. Reetz et al., *J. Chem. Soc., Chem. Commun.*, 1634.

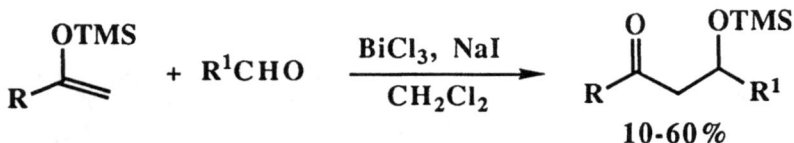

similarly with $[Ru(salen)(NO)H_2O]SbF_6$ or $TiCp_2(OTf)_2$ + $ZrCp_2(OTf)_2THF$ or mono and binuclear iron complexes

---

**I.A.7.a.1a-19**   S.-i. Kiyooka et al., *Tetrahedron Lett.*, **33**, 4927; E.J. Corey et al., *ibid.*, **33**, 6907.

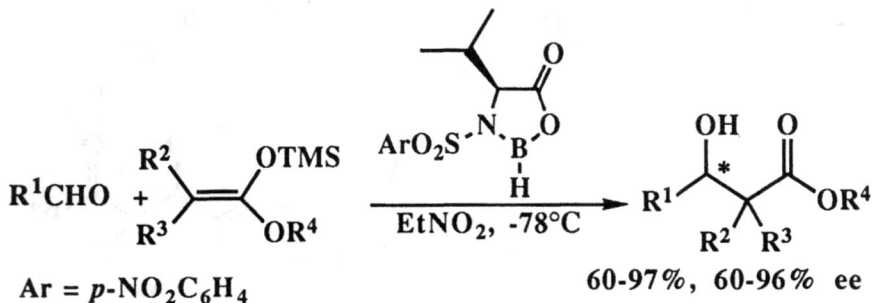

Ar = $p\text{-}NO_2C_6H_4$

60-97%, 60-96% ee

**I.A.7.a.1a-20** S. Kobayashi and I. Hachiya, *Tetrahedron Lett.*, **33**, 1625 and *Chem. Lett.*, 2187 (1991); **see also:** T. Nakai et al., *ibid.*, 29.

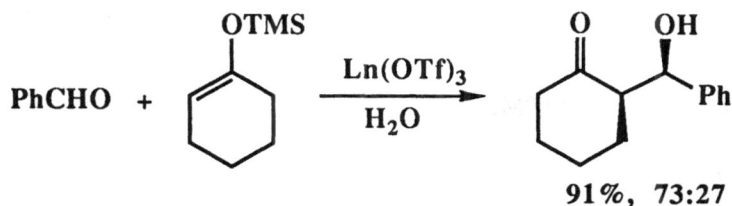

**91%, 73:27**

**similarly with Eu(fod)₃**

---

**I.A.7.a.1a-21** S. Kobayashi et al., *J. Org. Chem.*, **57**, 1324 and *Chem. Lett.*, 373; **see also:** S. Masamune et al., *Tetrahedron Lett.*, **33**, 1729.

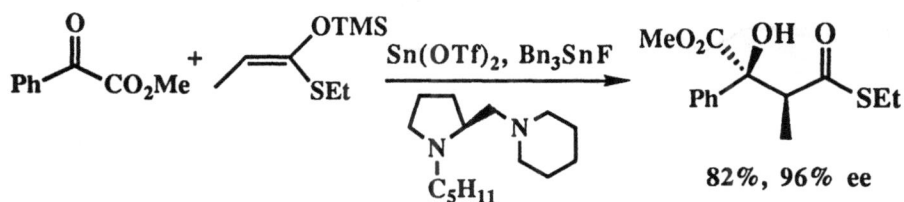

**82%, 96% ee**

---

**I.A.7.a.1a-22** H. Hagiwara et al., *J. Chem. Soc., Perkin Trans. 1*, 693; **see also:** I. Paterson and J.D. Smith, *J. Org. Chem.*, **57**, 3261.

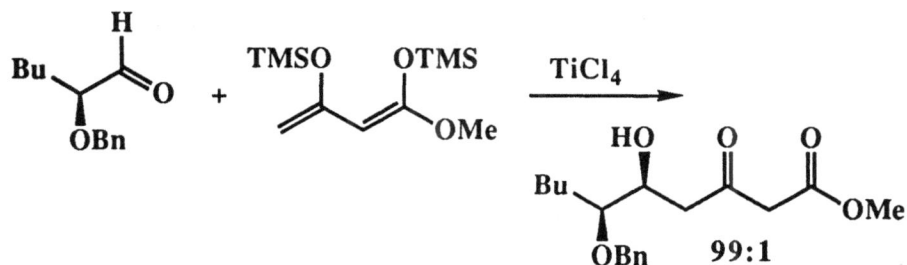

**I.A.7.a.1a-23**  K. Saigo et al., *Chem. Lett.*, 1445.

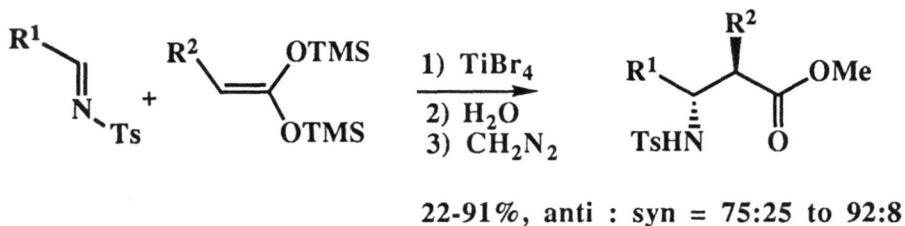

22-91%, anti : syn = 75:25 to 92:8

---

**I.A.7.a.1a-24**  D.P. Curran et al., *J. Org. Chem.*, **57**, 4341.

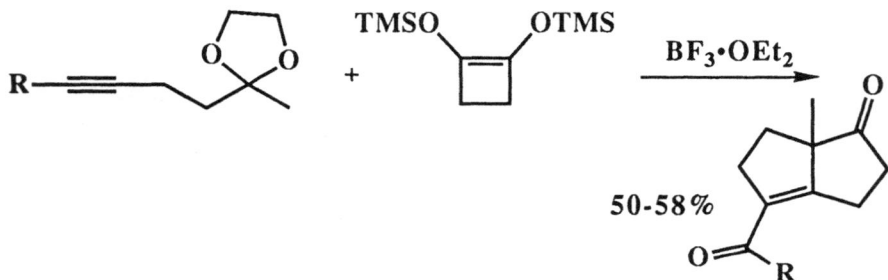

50-58%

## I.A.7.a.1b.  Intramolecular Aldol-Type 1,2-Additions

**I.A.7.a.1b-1**  K.-i. Tadano et al., *Tetrahedron*, **48**, 4283; J.-C. Blazejewski et al., *Tetrahedron Lett.*, **33**, 499.

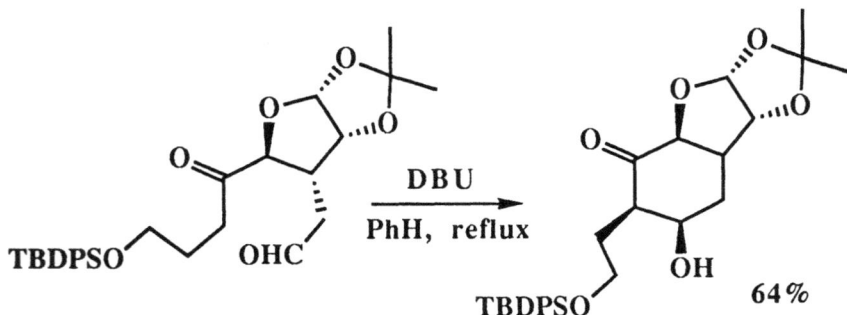

64%

## I.A.7.a.2. Addition of N-, P-, S-, Se and Similar Stabilized Carbanions

**I.A.7.a.2-1**  P. Zhang and R.E. Gawley, *Tetrahedron Lett.*, **33**, 2945.

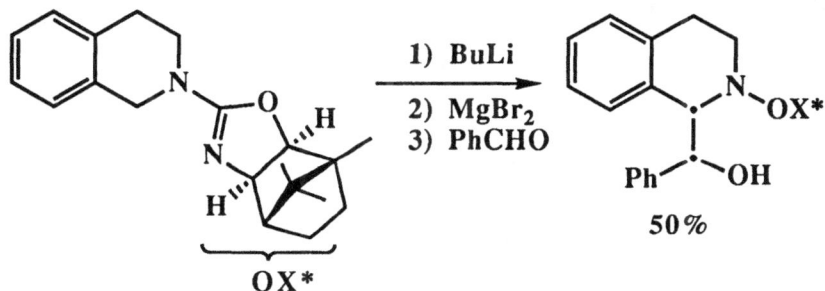

**I.A.7.a.2-2**  K. Kojima and S. Saito, *Synthesis*, 949; S. Halazy and V. Gross-Berges,  *J. Chem. Soc., Chem. Commun.*, 743.

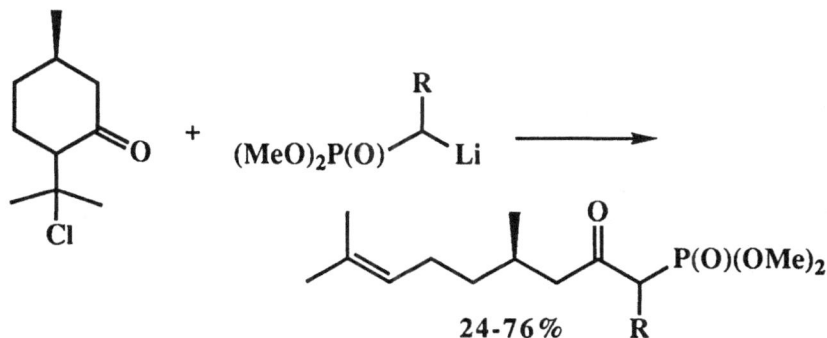

**I.A.7.a.2-3**  C.R. Holmquist and E.J. Roskamp, *Tetrahedron Lett.*, **33**, 1131.

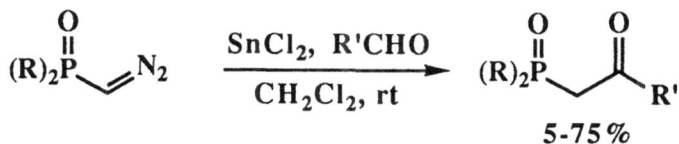

**I.A.7.a.2-4**  R.W. Hoffmann and M. Bewersdorf, *Liebigs Ann. Chem.*, 643.

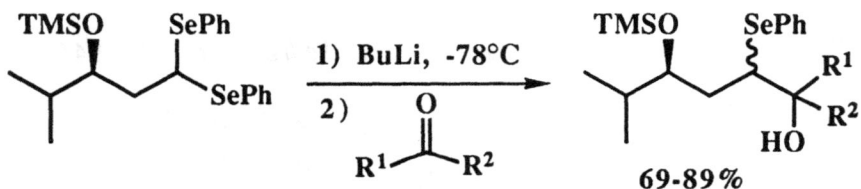

**I.A.7.a.2-5**  J. Voss et al., *Chem. Ber.*, **125**, 1611; P.R. Jenkins and M.M.R. Selim, *J. Chem. Res. (S)*, 85; H. Chikashita et al., *Chem. Lett.*, 1457.

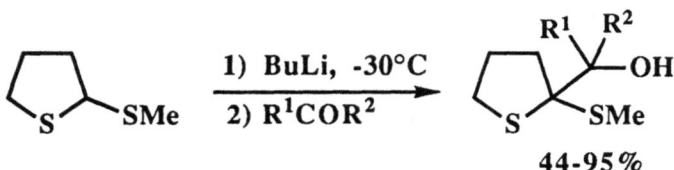

**I.A.7.a.2-6**  P. Venturello et al., *Tetrahedron*, **48**, 2501; X. Bai and E.L. Eliel, *J. Org. Chem.*, **57**, 5162; R.N. Young et al., *Tetrahedron Lett.*, **33**, 725.

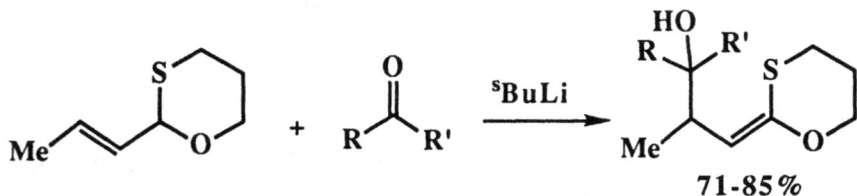

**I.A.7.a.2-7** M. Casey et al., *Tetrahedron Lett.*, **33**, 127; G. Solladie et al., *ibid.*, **33**, 4561; M. Wills et al., *ibid.*, **33**, 5427; M. Shimazaki and A. Ohta, *Synthesis*, 957; C. Maignan et al., *ibid.*, 547; see also: S.G. Pyne and A.R. Hajipour, *Tetrahedron*, **48**, 9385.

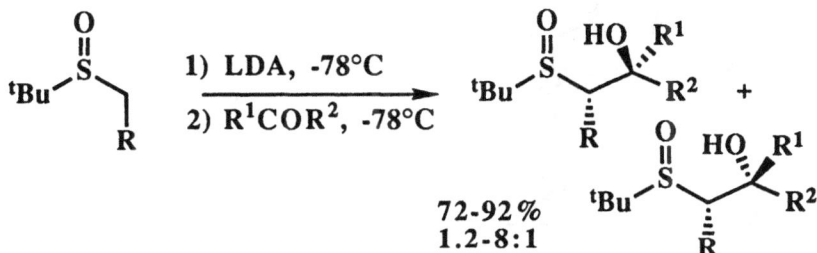

72-92%
1.2-8:1

**I.A.7.a.2-8** P. Bonete and C. Najera, *Tetrahedron Lett.*, **33**, 4065; E.V. Sadanandan and P.C. Srinivasan, *Synthesis*, 648.

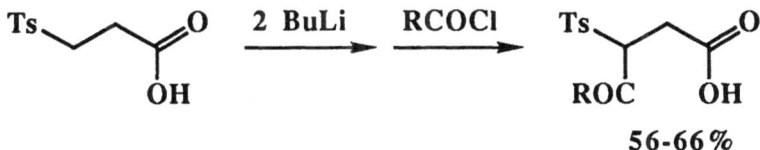

56-66%

## I.A.7.a.3. Addition of Organometallic and Related Species

**I.A.7.a.3-1** H. Chikashita et al., *Chem. Lett.*, 439; see also: D.J. Ramon and M. Yus, *Tetrahedron Lett.*, **33**, 2217; see also: J.P. Cherkauskas and T. Cohen, *J. Org. Chem.*, **57**, 6.

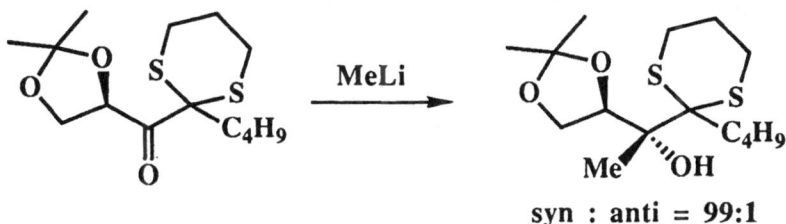

syn : anti = 99:1

**I.A.7.a.3-2**  D.C. Harrowven, *Tetrahedron Lett.*, **33**, 2879.

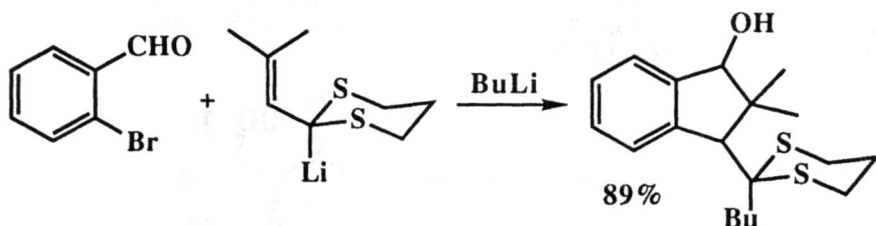

**I.A.7.a.3-3**  K. Higashiyama et al., *Tetrahedron Lett.*, **33**, 235; **see also:** W.R. Baker and S.L. Condon, *ibid.*, **33**, 1581; S. Itsuno et al., *Synth. Commun.*, **22**, 3229.

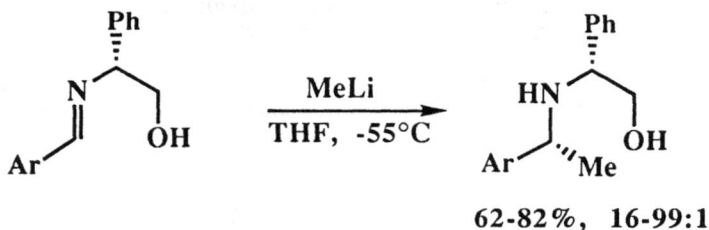

**I.A.7.a.3-4**  I.S. Aidhen and J.R. Ahuja, *Tetrahedron Lett.*, **33**, 5431; M.P. Sibi et al., *ibid.*, **33**, 1941.

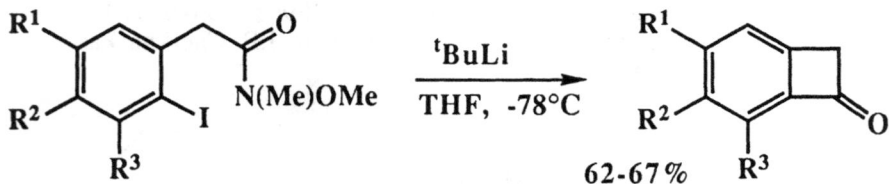

**I.A.7.a.3-5**  B. Weber and D. Seebach, *Angew. Chem., Int. Ed. Engl.*, **31**, 84; **see also:** M.T. Reetz and N. Harmat, *ibid.*, **31**, 342.

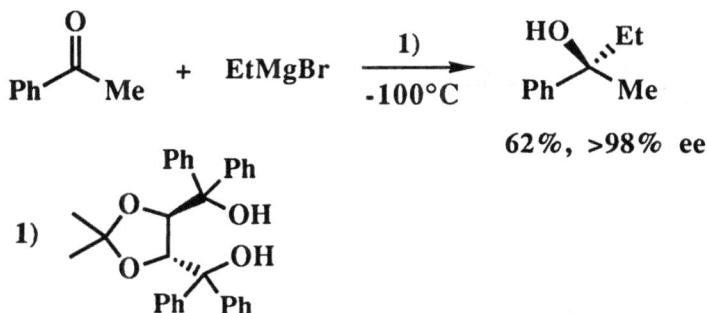

---

**I.A.7.a.3-6**  Y. Tamai et al., *J. Chem. Soc., Chem. Commun.*, 687; **see also:** G. Procter et al., *Tetrahedron Lett.*, **33**, 3351 and 3355.

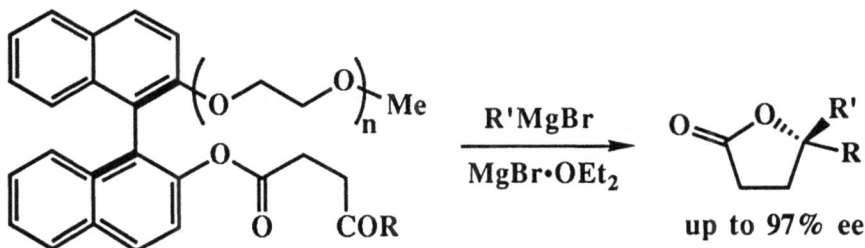

---

**I.A.7.a.3-7**  T. Gracza and V. Jager, *Synlett*, 191; R. Bloch and C. Brillet, *Tetrahedron Asym.*, **3**, 333.

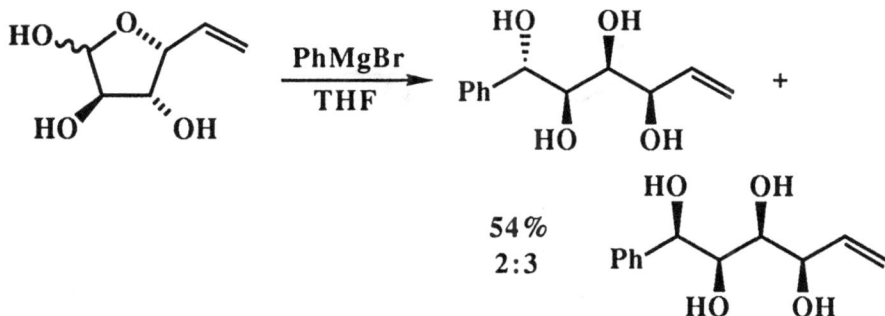

**I.A.7.a.3-8**  H. Xiong and R.D. Rieke, *J. Am. Chem. Soc.*, **114**, 4415; T. Shono et al., *J. Org. Chem.*, **57**, 5561.

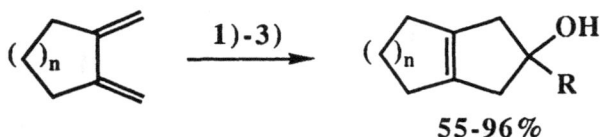

55-96%

**1) Mg, THF   2) RCO$_2$R'   3) H$_3$O$^+$**

---

**I.A.7.a.3-9**  M.T. Reetz et al., *Angew. Chem., Int. Ed. Engl.*, **31**, 1626; G. Cainelli et al., *Tetrahedron Lett.*, **33**, 7783.

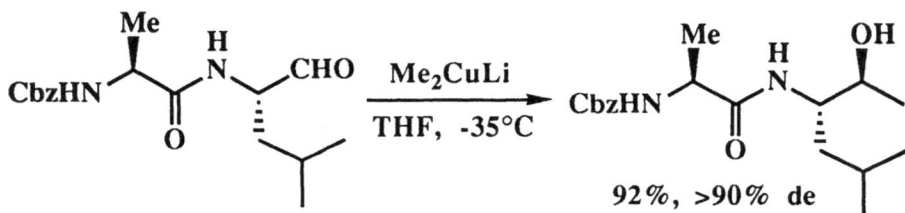

92%, >90% de

**similarly with an N-silylimine**

---

**I.A.7.a.3-10**  D.E. Stack and R.D. Rieke, *Tetrahedron Lett.*, **33**, 6575.

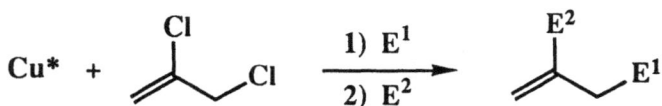

**I.A.7.a.3-11**  R.D. Rieke et al., *J. Am. Chem. Soc.*, **114**, 5112.

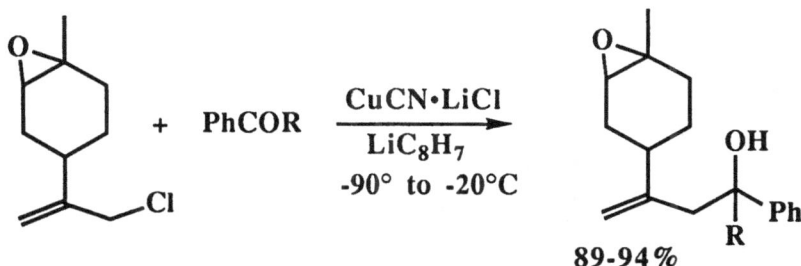

PhCOR  $\xrightarrow[\substack{LiC_8H_7 \\ -90° \text{ to } -20°C}]{CuCN \cdot LiCl}$

89-94%

---

**I.A.7.a.3-12**  J. Yaozhong et al., *Tetrahedron Asymm.*, **3**, 1467; F. Sato et al., *ibid.*, **3**, 5; T. Kunieda et al., *Tetrahedron Lett.*, **33**, 3147; S.B. Heaton and G.B. Jones, *ibid.*, **33**, 1693; C. Bolm et al., *Chem. Ber.*, **125**, 1191 and 1205; S. Inoue et al., *Synlett*, 427; T. Katsuki et al., *ibid.*, 573; W. Oppolzer and R.N. Radinov, *Helv. Chim. Acta*, **75**, 170; K. Soai et al., *Bull. Chem. Soc. Jpn.*, **65**, 1734 and *J. Chem. Soc., Chem. Commun.*, 927.

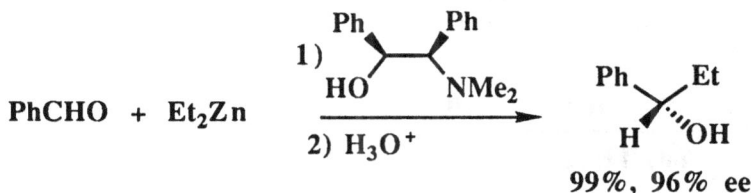

PhCHO  +  Et$_2$Zn  $\xrightarrow[\text{2) } H_3O^+]{\text{1)}}$

99%, 96% ee

**various other chiral ligands employed similarly**

---

**I.A.7.a.3-13**  D. Seebach et al., *Tetrahedron* , **48**, 5719; S. Kobayashi et al., *ibid.* , **48**, 5691; **see also:** M. Reetz et al., *ibid.*, **48**, 5731.

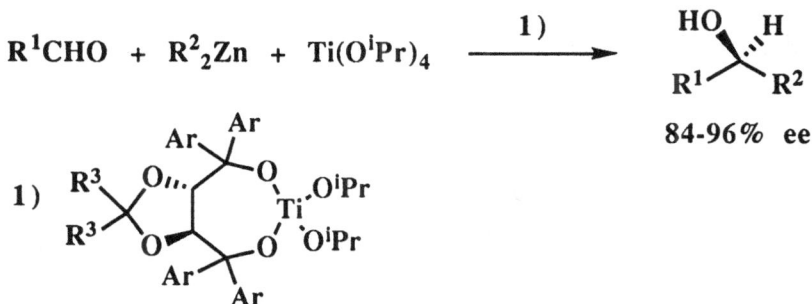

R$^1$CHO  +  R$^2_2$Zn  +  Ti(O$^i$Pr)$_4$  $\xrightarrow{\text{1)}}$

84-96% ee

1)

**I.A.7.a.3-14** Y. Oda et al., *Tetrahedron Lett.*, **33**, 97; C. Najera and J.M. Sansano, *Tetrahedron.*, **48**, 5179; K. Oshima et al., *Chem. Lett.*, 2135.

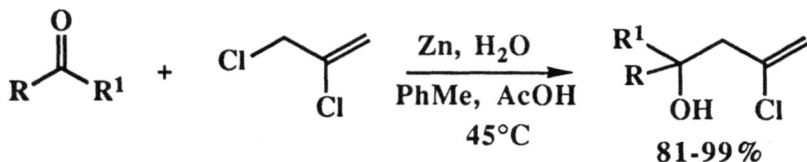

**I.A.7.a.3-15** P. Knochel et al., *J. Org. Chem.*, **57**, 1956; **see also:** E.J. Corey et al., *Tetrahedron Lett.*, **33**, 3435.

$$(AcO(CH_2)_5)_2Zn \ + \ PhCOCl \ \longrightarrow \ PhCO(CH_2)_5OAc$$

$$87\%$$

**I.A.7.a.3-16** Y. Ding and G. Zhao, *J. Chem. Soc., Chem. Commun.*, 941; H. Schick et al., *J. Org. Chem.*, **57**, 4013; K. Soai et al., *Tetrahedron Asym.*, **3**, 677.

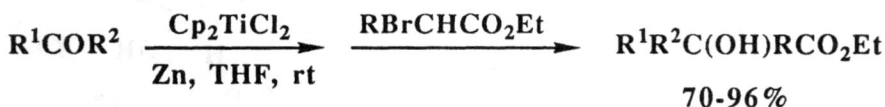

$$R^1COR^2 \xrightarrow[\text{Zn, THF, rt}]{Cp_2TiCl_2} \xrightarrow{RBrCHCO_2Et} R^1R^2C(OH)RCO_2Et$$

$$70\text{-}96\%$$

**electrochemically generated Reformatsky reagents and others with a chiral catalyst also reported**

**I.A.7.a.3-17** M. Taddei et al., *Tetrahedron Lett.*, **33**, 5621; J.S. Panek et al., *J. Org. Chem.*, **57**, 5790; K. Yamakawa et al., *Heterocycles*, **33**, 97; A.P. Davis and M. Jaspars, *J. Chem. Soc., Perkin Trans. 1*, 2111; **see also:** I.E. Marko and A. Mekhalfia, *Tetrahedron Lett.*, **33**, 1799.

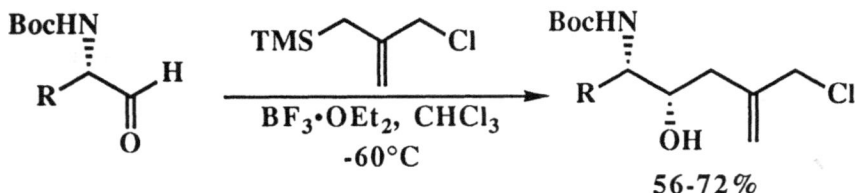

**I.A.7.a.3-18** L.F. Tietze et al., *Angew. Chem., Int. Ed. Engl.*, **31**, 1372; A.P. Davis and M. Jaspars, *ibid.*, **31**, 470.

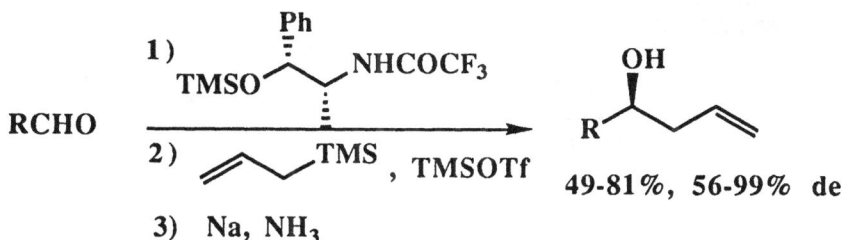

RCHO

1) [structure: Ph, TMSO, NHCOCF₃]

2) [allyl-TMS structure], TMSOTf

3) Na, NH₃

[product: OH, R, allyl]

49-81%, 56-99% de

---

**I.A.7.a.3-19** Y. Yamamoto et al., *J. Org. Chem.*, **57**, 7003; **see also:** B.W. Gung et al., *Tetrahedron*, **48**, 5455; P.A. Grieco et al., *Tetrahedron Lett.*, **33**, 1817; **see also:** Y. Kita et al., *J. Chem. Soc., Perkin Trans. 1*, 2813.

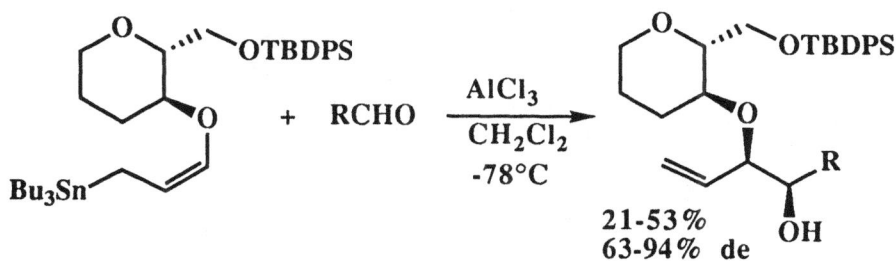

[structure with OTBDPS, Bu₃Sn]  +  RCHO  $\xrightarrow[\text{-78°C}]{\begin{array}{c}\text{AlCl}_3\\\text{CH}_2\text{Cl}_2\end{array}}$  [product structure with OTBDPS, R, OH]

21-53%
63-94% de

---

**I.A.7.a.3-20** J.A. Marshall and Y. Tang, *Synlett*, 653.

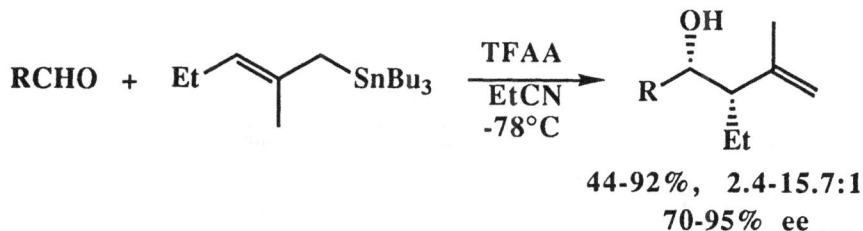

RCHO  +  [Et structure with SnBu₃]  $\xrightarrow[\text{-78°C}]{\begin{array}{c}\text{TFAA}\\\text{EtCN}\end{array}}$  [product: OH, R, Et]

44-92%, 2.4-15.7:1
70-95% ee

**I.A.7.a.3-21**  N. Greeves and L. Lyford, *Tetrahedron Lett.*, **33**, 4759;  H.-J. Liu and B.-Y. Zhu, *Can. J. Chem.*, **69**, 2008 (1991); **see also:** S.E. Denmark and J. Kim, *Synthesis*, 229; **see also:** S. Fukuzawa and S. Sakai, *Bull. Chem. Soc. Jpn.*, **65**, 3308.

$$CeCl_3 \cdot THF \quad \xrightarrow[\substack{\text{2) } R^1COR^2 \\ \text{3) } NH_4Cl}]{\text{1) } R^3Li, \ -78°C}}$$

HO   R³
$\diagdown$ /
$R^1$ $R^2$

77-97%

---

**I.A.7.a.3-22**  K. Oshima, K. Utimoto et al., *Tetrahedron Lett.*, **33**, 4597.

40-72%
96:4 to 99:1

---

**I.A.7.a.3-23**  J.D. Armstrong, III et al., *Tetrahedron Lett.*, **33**, 6599; **see also:** Y. Ding and G. Zhao, *ibid.*, **33**, 8117.

non-chelation
product favored

**I.A.7.a.3-24**  P. Knochel et al., *J. Org. Chem.*, **57**, 6384; D.M. Hodgson and C. Wells, *Tetrahedron Lett.*, **33**, 4761; **see also:** T. Kauffmann et al., *Chem. Ber.*, **125**, 157.

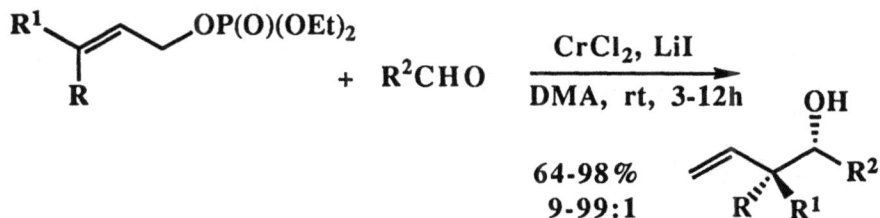

**similarly with allyl halides**

---

**I.A.7.a.3-25**  G. Cahiez et al., *Tetrahedron Lett.*, **33**, 4439 and 5245; **see also:** K. Oshima, K. Utimoto et al., *ibid.*, **33**, 4353.

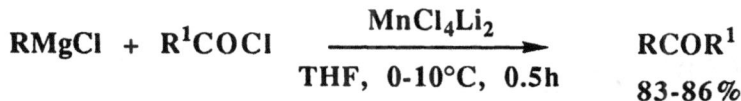

$$RMgCl \ + \ R^1COCl \ \xrightarrow[\text{THF, 0-10°C, 0.5h}]{MnCl_4Li_2} \ RCOR^1$$

83-86%

---

**I.A.7.a.3-26**  T. Kauffmann et al., *Chem. Ber.*, **125**, 899.

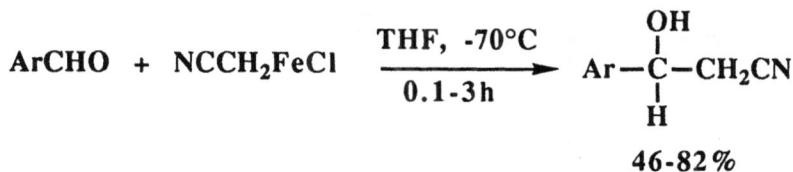

**I.A.7.a.3-27** T. Taguchiet al., *Tetrahedron Lett.*, **33**, 7873, 1295, 3769 and 4469; Y. Hamada et al., *ibid.*, **33**, 2031; K. Suzuki et al., *ibid.*, **33**, 5965 and 5969; T. Takahashi et al., *Chem. Lett.*, 331; see also: P. Wipf and W. Xu, *Synlett*, 718.

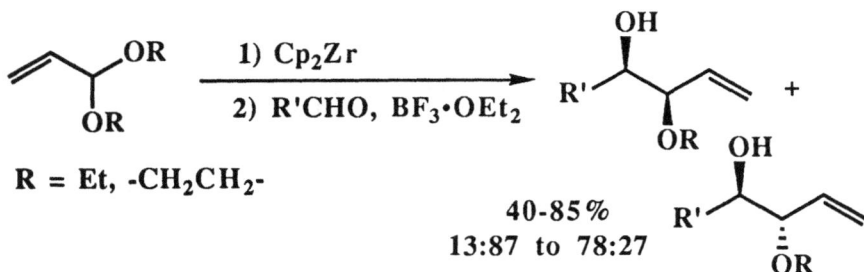

$$R = Et, -CH_2CH_2-$$

40-85%
13:87 to 78:27

---

**I.A.7.a.3-28** Y. Masuyama et al., *J. Am. Chem. Soc..*, **114**, 2577 and *J. Chem. Soc., Chem. Commun.*, 1102; see also: A.M. Echavarren et al., *J. Org. Chem.*, **57**, 5047.

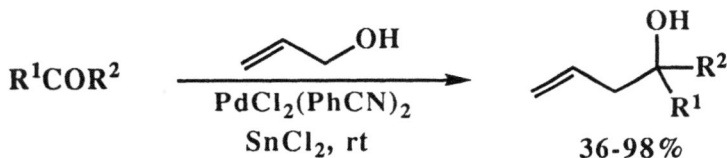

$R^1COR^2$

$PdCl_2(PhCN)_2$
$SnCl_2$, rt

36-98%

---

**I.A.7.a.3-29** Y. Butsugan et al., *Bull. Chem. Soc. Jpn.*, **65**, 1736 and *J. Chem. Soc., Perkin Trans. 1*, 711; T.-H. Chan and C.-J. Li, *J. Chem. Soc., Chem. Commun.*, 747; see also: P. Mosset et al., *Tetrahedron Lett.*, **33**, 5959.

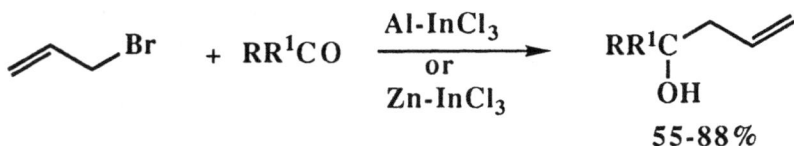

$+ RR^1CO$

$Al-InCl_3$
or
$Zn-InCl_3$

55-88%

**I.A.7.a.3-30** H. Yamamoto et al., *Synlett.*, 593.

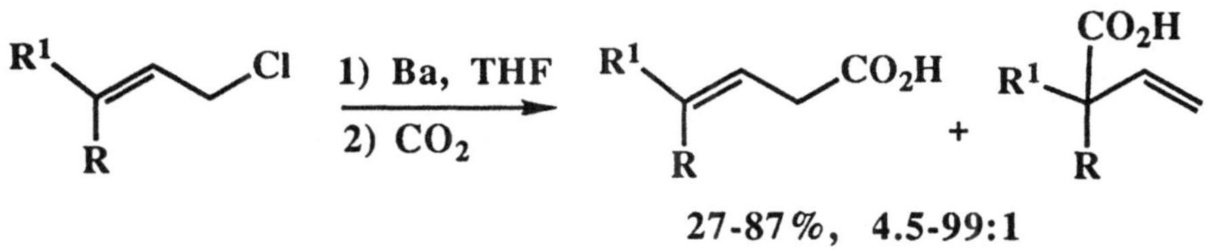
R1
Cl
R
1) Ba, THF
2) CO2
R1
CO2H
R
+
CO2H
R1
R
27-87%, 4.5-99:1

---

**I.A.7.a.3-31** K. Takai, K. Utimoto et al., *J. Org. Chem.*, **57**, 1973 and *Chem. Lett.*, 99.

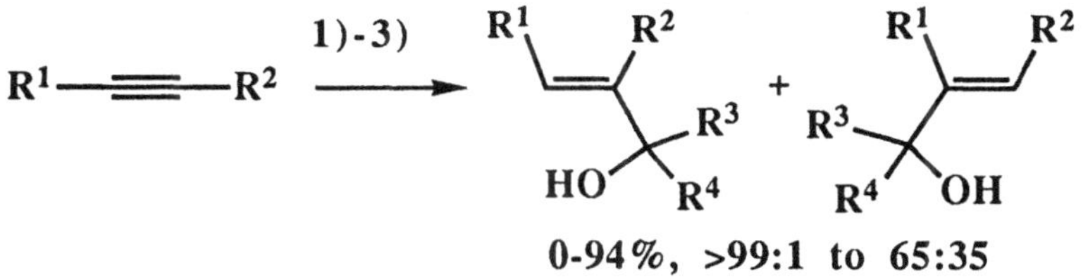
R1
R2
1)-3)
R1
R2
R3
HO
R4
+
R1
R2
R3
R4
OH
0-94%, >99:1 to 65:35

1) $TaCl_5$ / Zn 2) $R^3R^4C{=}O$ 3) NaOH / $H_2O$

---

**I.A.7.a.3-32** Y. Ito et al., *J. Org. Chem.*, **57**, 793; see also: H.B. Kagan et al., *Tetrahedron Lett.*, **33**, 2973.

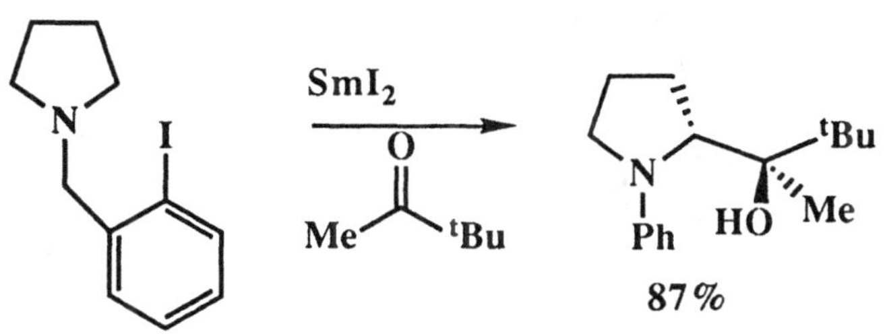
N
I
SmI2
O
Me
tBu
N
Ph
tBu
HO
Me
87%

**I.A.7.a.3-33**  T. Furuta and Y. Yamamoto, *J. Chem. Soc., Chem. Commun.*, 863 and *J. Org. Chem.*, **57**, 2981; Y. Yamamoto et al., *Tetrahedron*, **48**, 5587.

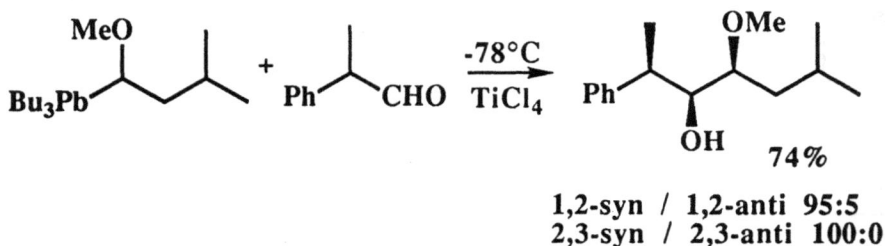

74%

**1,2-syn / 1,2-anti  95:5**
**2,3-syn / 2,3-anti  100:0**

## I.A.7.a.4.  Other 1,2-Additions

**I.A.7.a.4-1**  K. Narasaka et al., *Bull. Chem. Soc. Jpn.*, **65**, 1392.

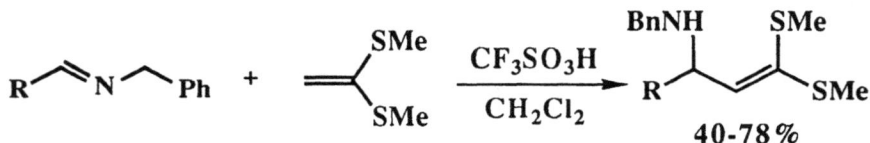

40-78%

---

**I.A.7.a.4-2**  I. Kuwajima et al., *Tetrahedron Lett.*, **33**, 1337 and *Chem. Lett.*, 1425; S. Hanessian and S. Beaudoin, *Tetrahedron Lett.*, **33**, 7659; F.T. van der Meer and B.L. Feringa, *ibid.*, **33**, 6695; G.B. Gill et al., *J. Chem. Soc., Perkin Trans. 1*, 2355; I. Kumadaki et al., *Heterocycles*, **33**, 51 and *Chem. Pharm. Bull.*, **40**, 593.

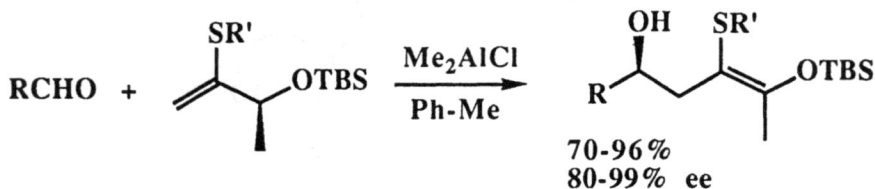

70-96%
80-99% ee

**other similar ene reactions reported**

**I.A.7.a.4-3**  D. Seyferth et al., *J. Org. Chem.*, **57**, 5620.

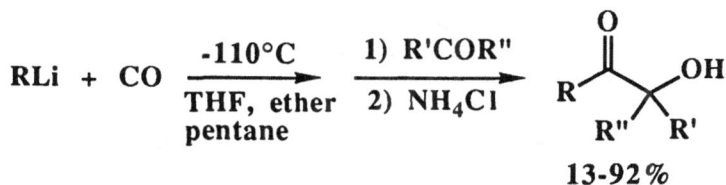

**13-92%**

---

**I.A.7.a.4-4**  N.S. Simpkins et al., *J. Chem. Soc., Perkin Trans. 1*, 2471.

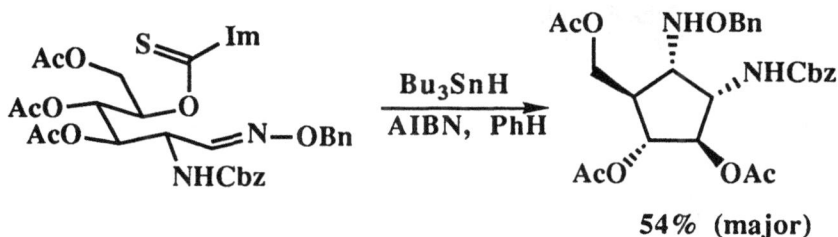

**54% (major)**

---

**I.A.7.a.4-5**  Jack E. Baldwin et al., *Tetrahedron*, **48**, 3385.

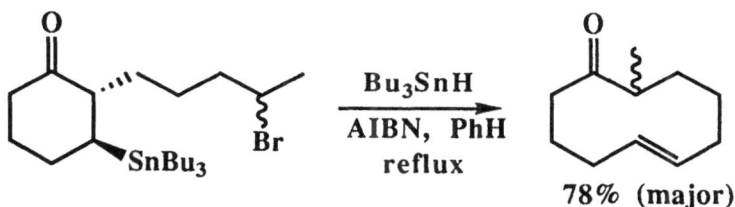

**78% (major)**

---

**I.A.7.a.4-6**  K. Sakai et al., *Tetrahedron Lett.*, **33**, 6331.

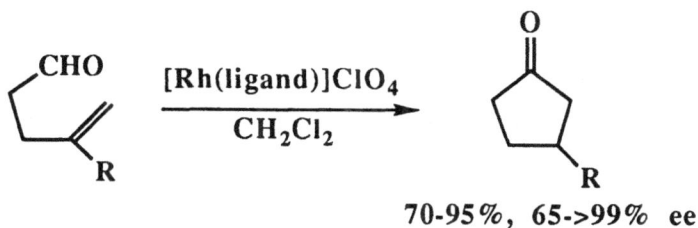

**70-95%, 65->99% ee**

**I.A.7.a.4-7**  J. Oda et al., *Bull. Chem. Soc. Jpn.*, **65**, 111; T. Huuhtanen and L.T. Kanerva, *Tetrahedron Asym.*, **3**, 1223; **see also:** S. Terashima et al., *Bull. Chem. Soc. Jpn.*, **65**, 360.

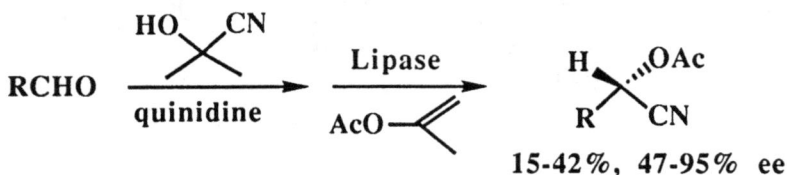

**15-42%, 47-95% ee**

**I.A.7.a.4-8**  T. Nakai et al., *Chem. Lett.*, 1169; N. Oguni et al., *J. Chem. Soc., Perkin Trans. 1*, 3135; **see also:** S. Inoue et al., *J. Am. Chem. Soc.*, **114**, 7969.

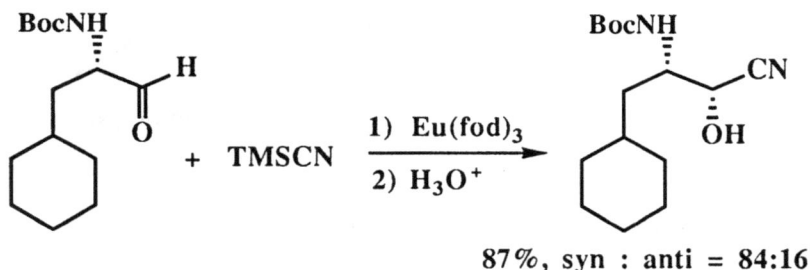

**87%, syn : anti = 84:16**

**I.A.7.a.4-9**  T. Matsumoto et al., *J. Chem. Soc., Chem. Commun.*, 610; H.N.C. Wong et al., *J. Org. Chem.*, **57**, 4033.

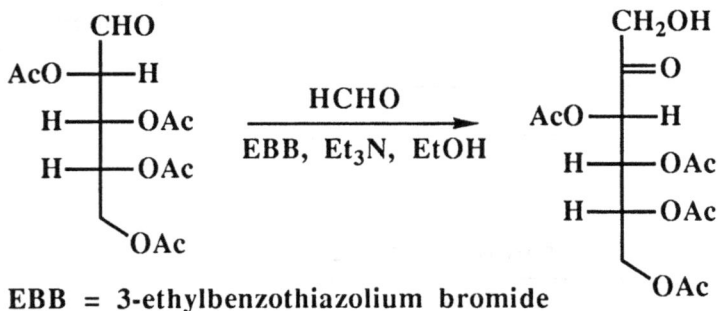

**EBB = 3-ethylbenzothiazolium bromide**

**similarly with KCN**

**I.A.7.a.4-10** A.G. Myers and P.S. Dragovich, *J. Am. Chem. Soc.*, **114**, 5859; K.C. Nicolaou, R.B. Bergman et al., *Angew. Chem., Int. Ed. Engl.*, **31**, 1044.

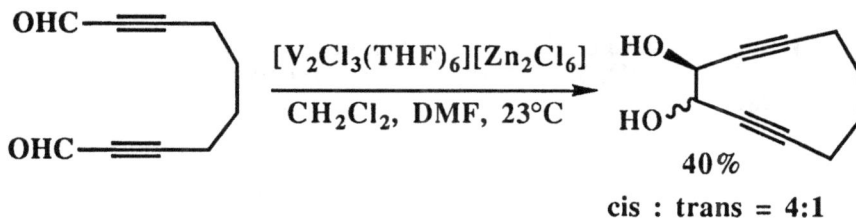

cis : trans = 4:1

similarly with SmI₂ / Ti*

---

**I.A.7.a.4-11** S.M. Ruder et al., *Tetrahedron Lett.*, **33**, 2621; R.J.K. Taylor et al., *J. Chem. Soc., Perkin Trans. 1*, 2657; T. Shono et al., *J. Org. Chem.*, **57**, 7175.

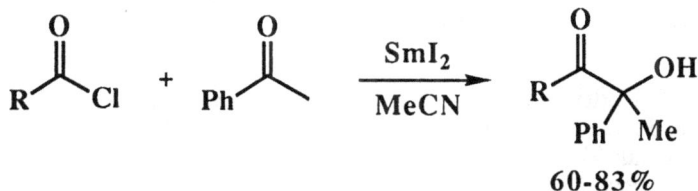

60-83%

---

**I.A.7.a.4-12** A. Furstner and D.N. Jumbam, *Tetrahedron*, **48**, 5991.

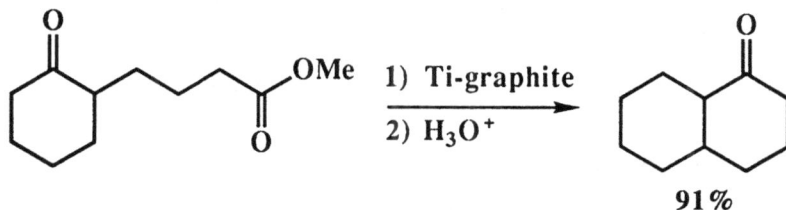

91%

## I.A.7.b.  Conjugate Additions

### I.A.7.b.1.  Enolate-Type Carbanions

**I.A.7.b.1-1**  A. Bernardi et al., *Tetrahedron*, **48**, 5597; T. Kitazume et al., *Chem. Lett.*, 2171 (1991); Y. Yokoyama and K. Tsuchikura, *Tetrahedron Lett.*, **33**, 2823.

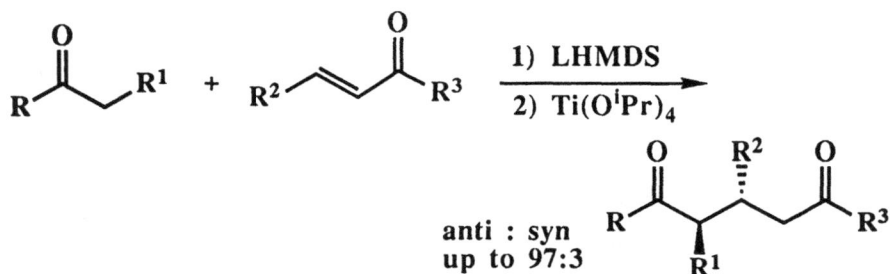

anti : syn
up to 97:3

---

**I.A.7.b.1-2**  J. Rodriguez et al., *Angew. Chem., Int. Ed. Engl.*, **31**, 1651; S. Girard and P. Deslongchamps, *Can. J. Chem.*, **70**, 1265; see also: B.C. Ranu and S. Bhar, *Tetrahedron*, **48**, 1327; see also: K. Fukumoto et al., *J. Chem. Soc., Chem. Commun.*, 976.

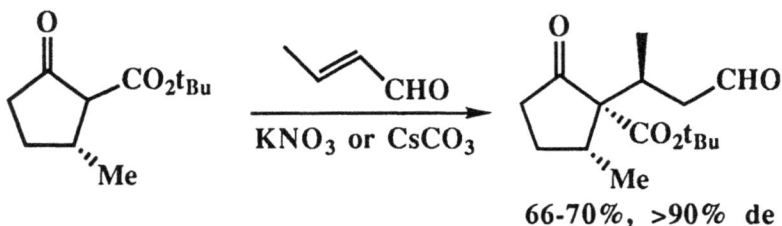

66-70%, >90% de

similarly with DMAP as base or using $Al_2O_3$

**I.A.7.b.1-3**  K. Koga et al., *Heterocycles*, **33**, 493.

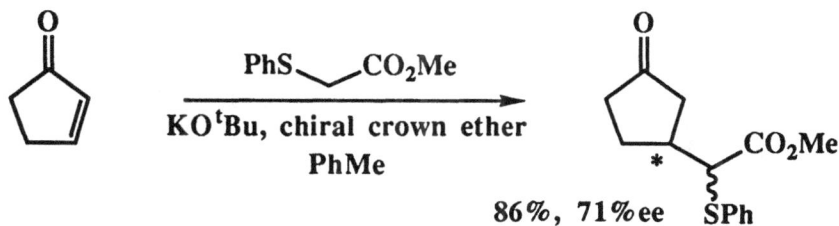

PhS⏜CO$_2$Me

KO$^t$Bu, chiral crown ether
PhMe

86%, 71%ee    SPh

---

**I.A.7.b.1-4**  D. Seebach et al., *Liebigs Ann. Chem.*, 51 and 1145; **see also:**J. Mulzer et al., *Angew. Chem., Int. Ed. Engl.*, **31**, 870

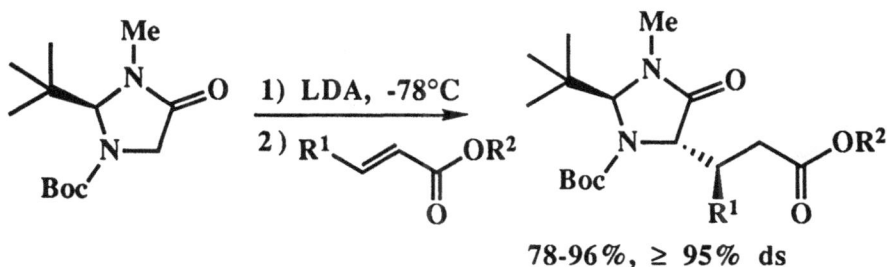

1) LDA, -78°C

2) R$^1$⏜OR$^2$

78-96%, ≥ 95%  ds

---

**I.A.7.b.1-5**  G.H. Posner et al., *Tetrahedron*, **48**, 4677; M.M. Al-Arab et al., *Synthesis*, 1003; **see also:** K. Fukumoto et al., *Tetrahedron*, **48**, 5089.

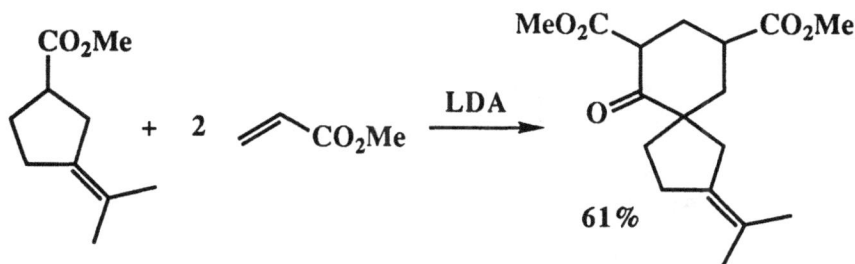

CO$_2$Me

+  2  ⏜CO$_2$Me

LDA

MeO$_2$C⏜CO$_2$Me

61%

**I.A.7.b.1-6** D.B. Reddy et al., *Org. Prep. Proced., Int. Ed. Engl.*, **24**, 21 and *Ind. J. Chem.*, **31B**, 407.

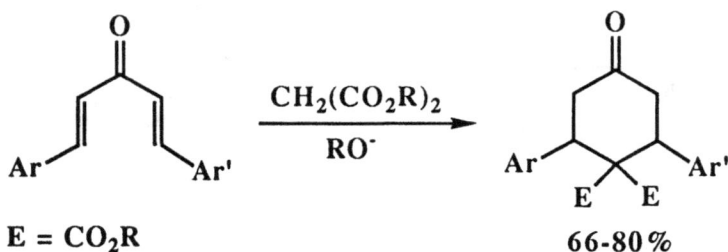

E = CO₂R

$$E = CO_2R$$

66-80%

---

**I.A.7.b.1-7** R. Ballini et al., *Synthesis*, 355.

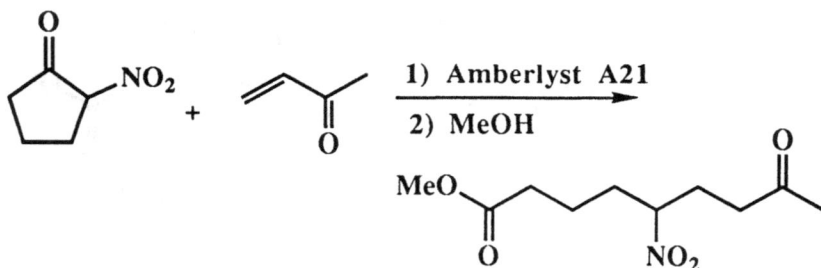

1) Amberlyst A21

2) MeOH

---

**I.A.7.b.1-8** D. Enders and W. Karl, *Synlett*, 895.

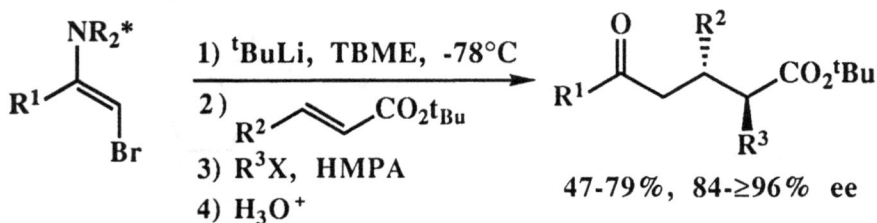

1) ᵗBuLi, TBME, -78°C

2) $R^2$ ⌒ $CO_2{}^tBu$

3) $R^3X$, HMPA

4) $H_3O^+$

47-79%, 84-≥96% ee

R* = a chiral auxiliary

**I.A.7.b.1-9** E. Valentin et al., *Gazz. Chim. Ital.*, **122**, 85 and *J. Chem. Soc., Perkin Trans. 1*, 2331; **see also:** G. Kaupp et al., *Angew. Chem., Int. Ed. Engl.*, **31**, 768.

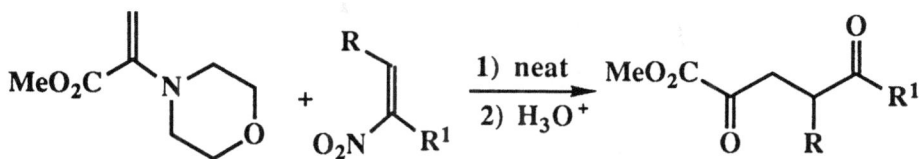

---

**I.A.7.b.1-10** A. Padwa et al., *J. Org. Chem.*, **57**, 298.

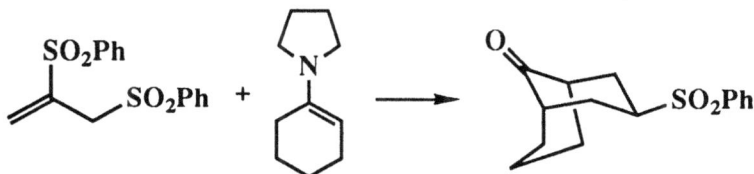

---

**I.A.7.b.1-11** P. Duhamel et al., *J. Chem. Soc., Perkin Trans. 1*, 387.

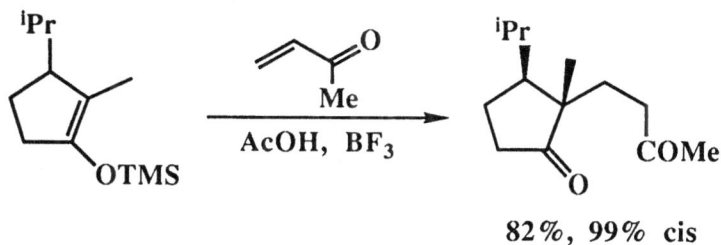

82%, 99% cis

**I.A.7.b.1-12**  G. Rousseau et al., *J. Org. Chem.*, **57**, 6890.

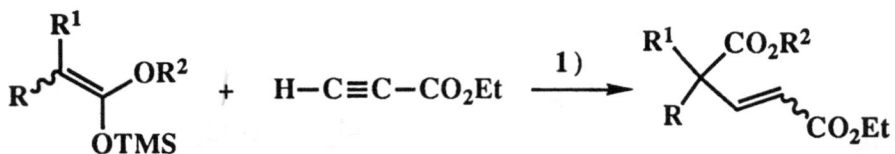

1) TiCl$_4$, CH$_2$Cl$_2$, -78°C

47-88%, E:Z = 2.3-99:1

---

**I.A.7.b.1-13**  J. Wolf et al., *Tetrahedron Lett.*, **33**, 1741; N. Langlois et al., *ibid.*, **33**, 1743.

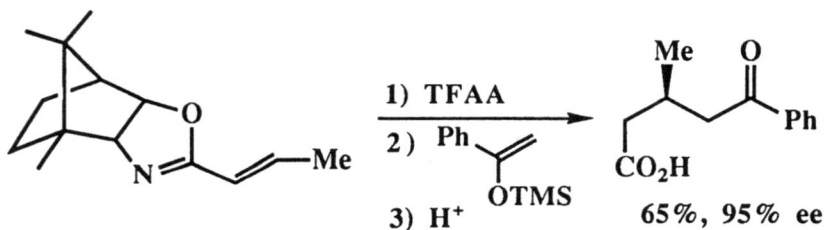

1) TFAA
2) Ph (OTMS)
3) H$^+$

65%, 95% ee

---

**I.A.7.b.1-14**  P.G. Klimko and D.A. Singleton, *J. Org. Chem.*, **57**, 1733; **see also:** H. Hagiwara et al., *J. Chem. Soc., Chem. Commun.*, 866.

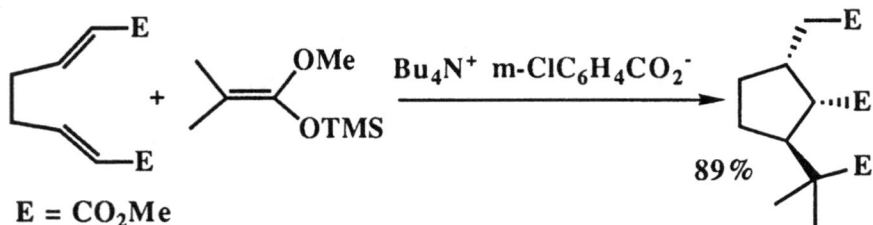

Bu$_4$N$^+$ m-ClC$_6$H$_4$CO$_2$$^-$

89%

E = CO$_2$Me

**I.A.7.b.1-15**  S. Kobayashi et al., *Tetrahedron Lett.*, **33**, 6815.

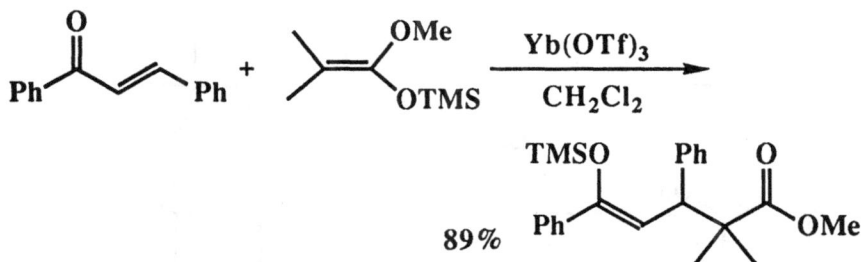

**I.A.7.b.2.  Organometallic and Related Reagents**

**I.A.7.b.2-1**  M.P. Cooke, Jr., *J. Org. Chem.*, **57**, 1495.

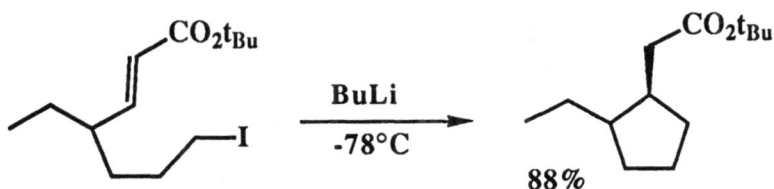

**I.A.7.b.2-2**  J. Otera et al., *Tetrahedron Lett.*, **33**, 3655; **see also:** T. Cohen et al., *J. Org. Chem.*, **57**, 1968.

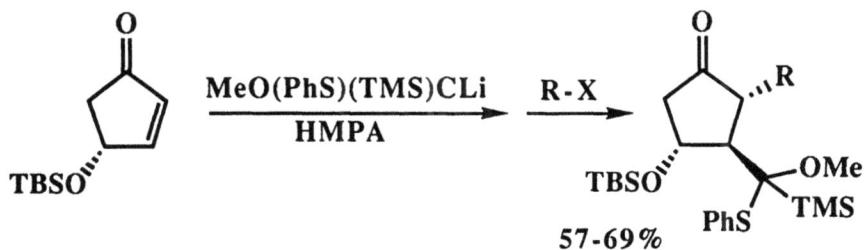

**I.A.7.b.2-3**  T.G. Gant and A.I. Meyers, *J. Am. Chem. Soc.*, **114**, 1010;
J.A. Seijas et al., *J. Org. Chem.*, **57**, 5283.

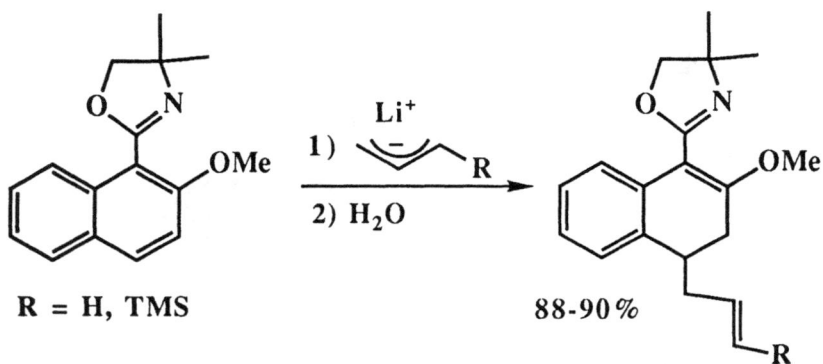

R = H, TMS                                          88-90%

---

**I.A.7.b.2-4**  K. Tomioka et al., *Tetrahedron Lett.*, **33**, 7193; **see also:** K.
Tanaka et al., *J. Chem. Soc., Perkin Trans. 1*, 1193.

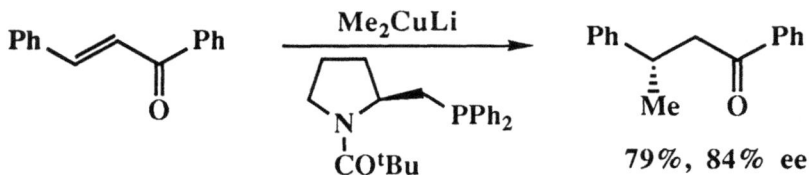

79%, 84% ee

**I.A.7.b.2-5**  D.C. Liotta, *Synthesis*, 127; H. Takei et al., *Tetrahedron Lett.*, **33**, 3145; **see also:** T.E. Goodwin et al., *J. Org. Chem.*, **57**, 2469.

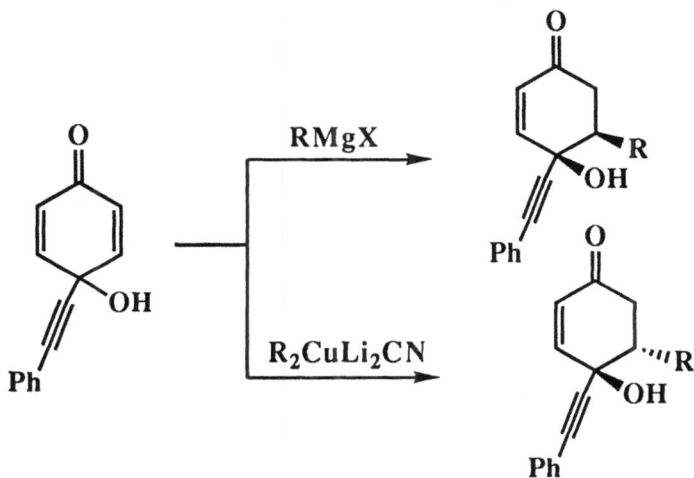

**I.A.7.b.2-6**  Y. Yamamoto et al., *J. Am. Chem. Soc.*, **114**, 7652; J.-F. Normant et al., *J. Organomet. Chem.*, **423**, 281.

**Conjugate addition of organocopper reagents to alkoxy substituted α,β–unsaturated carbonyl derivatives**

**I.A.7.b.2-7**  P. Cresson et al., *Tetrahedron*, **48**, 841; V.J. Hruby et al., *Tetrahedron Lett.*, **33**, 7491; K. Ruck and H. Kunz, *Synlett*, 343; A.I. Meyers and L. Snyder, *J. Org. Chem.*, **57**, 3814; **see also:** K. Sakai et al., *ibid.*, **57**, 4300; **see also:** M. Uemura, M. Shiro et al., *Organometallics*, **11**, 3705; **see also:** C. Tamm et al., *Angew. Chem., Int. Ed. Engl.*, **31**, 193.

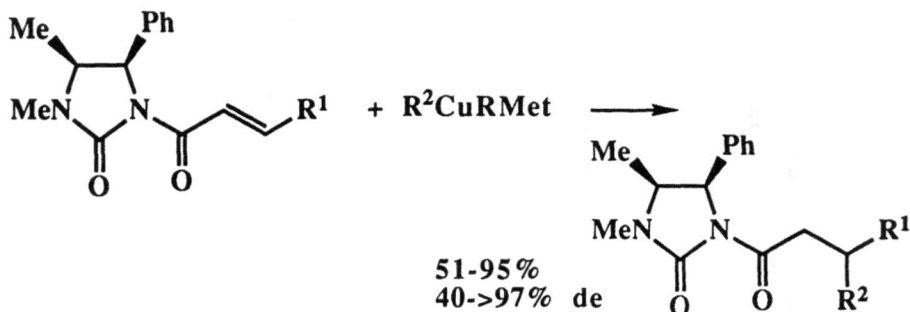

51-95%
40->97% de

**other chiral auxiliaries used similarly**

---

**I.A.7.b.2-8**  R. Tamura et al., *J. Org. Chem.*, **57**, 4895.

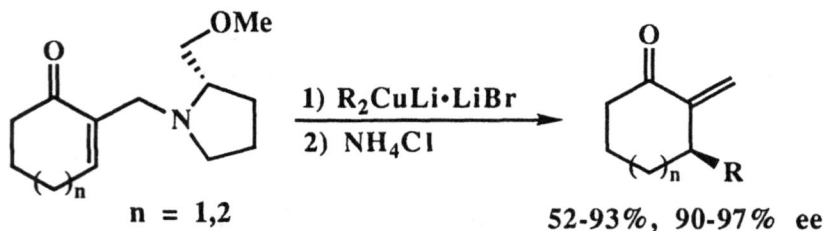

n = 1,2

52-93%, 90-97% ee

---

**I.A.7.b.2-9**  J.P. Marino et al., *Tetrahedron Lett.*, **33**, 49; **see also:** B.H. Lipshutz and R. Keil, *J. Am. Chem. Soc.*, **114**, 7919.

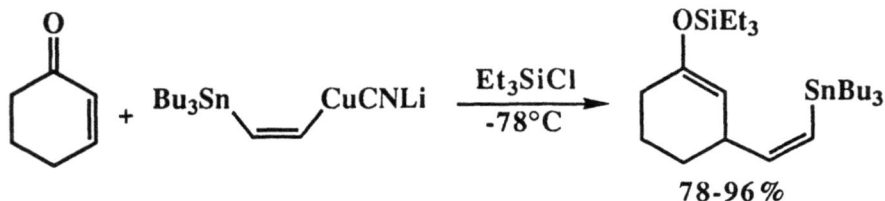

78-96%

**I.A.7.b.2-10**  P.B. Mackenzie et al., *J. Am. Chem. Soc.*, **114**, 5160.

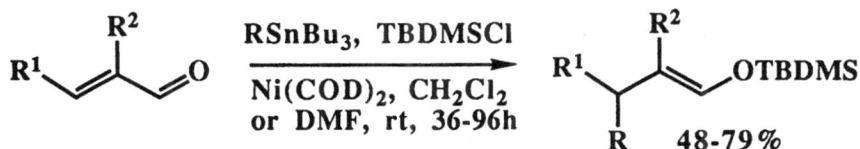

$$R^1 \diagdown \diagup^{R^2} O \xrightarrow[\substack{Ni(COD)_2, \ CH_2Cl_2 \\ or \ DMF, \ rt, \ 36\text{-}96h}]{RSnBu_3, \ TBDMSCl}}$$

$R^1$  $R^2$  OTBDMS

R    48-79%

---

**I.a.7.b.2-11**  P. Knochel et al., *J. Org. Chem.*, **57**, 5431 and 5425 and *Tetrahedron Lett.*, **33**, 3717; **see also:** S. Sibille et al., *J. Chem. Soc., Chem. Commun.*, 283; **see also:** H.I. Tashtoush and R. Sustmann, *Chem. Ber.*, **125**, 287.

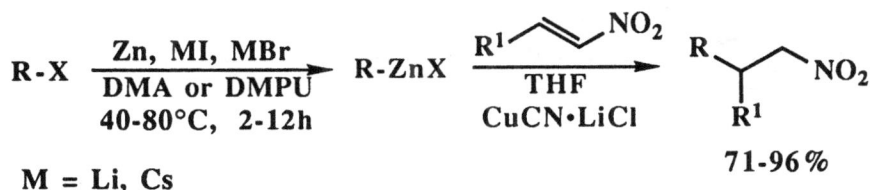

$$R\text{-}X \xrightarrow[\substack{DMA \ or \ DMPU \\ 40\text{-}80°C, \ 2\text{-}12h}]{Zn, \ MI, \ MBr}} R\text{-}ZnX \xrightarrow[\substack{THF \\ CuCN \cdot LiCl}]{R^1 \diagup\diagdown NO_2}}$$

R    $NO_2$

$R^1$

71-96%

M = Li, Cs

**similarly under electrolytic conditions or with Cr(en)$_2^{2+}$**

---

**I.A.7.b.2-12**  C. Bolm et al., *Chem. Ber.*, **125**, 1205 and *Synlett*, 439; J.F.G.A. Jansen and B.L. Feringa, *Tetrahedron Asym.*, **3**, 581; M. Uemura et al., *ibid.*, **3**, 713.

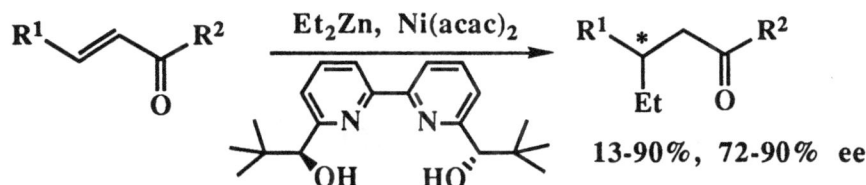

$$R^1 \diagup\diagdown\diagup_O^{R^2} \xrightarrow{Et_2Zn, \ Ni(acac)_2}$$

$R^1$  *  $R^2$

Et   O

13-90%, 72-90% ee

**various other chiral ligands used similarly**

**I.A.7.b.2-13**  K.-T. Kang et al., *Tetrahedron Lett.*, **33**, 3495; D. Schinzer et al., *Liebigs Ann. Chem.*, 139 and *Synlett*, 766; L.F. Tietze and M. Rischer, *Angew. Chem., Int. Ed. Engl.*, **31**, 1221.

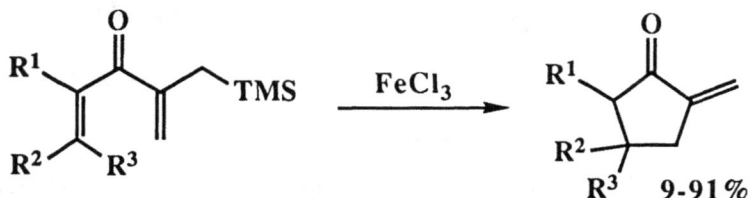

**other Lewis acids used for similar transformations**

---

**I.A.7.b.2-14**  J. Ipaktschi et al., *Angew. Chem., Int. Ed. Engl.*, **31**, 313.

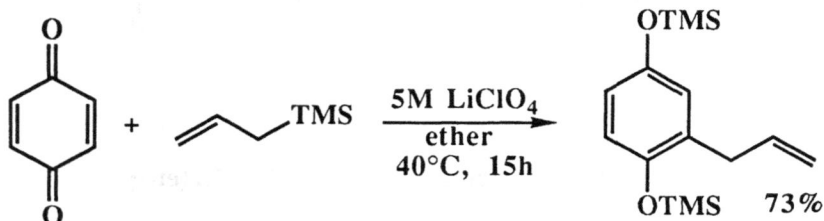

---

**I.A.7.b.2-15**  L.A. Paquette et al., *J. Am. Chem. Soc.*, **114**, 7375; B.M. Trost and T.A. Grese, *J. Org. Chem.*, **57**, 686.

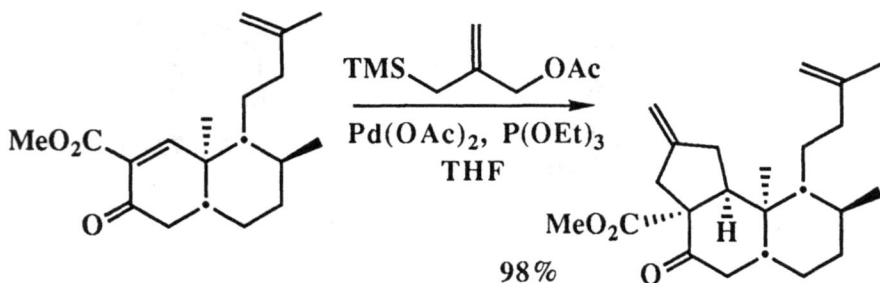

**I.A.7.b.2-16**  I.E. Marko and F. Rebiere, *Tetrahedron Lett.*, **33**, 1763.

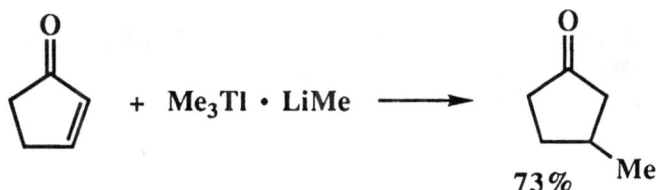

$+$  Me$_3$Tl • LiMe  $\longrightarrow$

73%   Me

---

**I.A.7.b.2-17**  T. Toru et al., *Tetrahedron Lett.*, **33**, 4037.

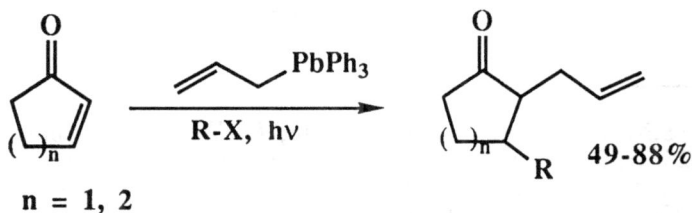

PbPh$_3$

R-X, hν

49-88%

n = 1, 2

---

**I.A.7.b.2-18**  P. Wipf et al., *J. Org. Chem.*, **57**, 1740; Y. Ding et al., *Tetrahedron Lett.*, **33**, 8119.

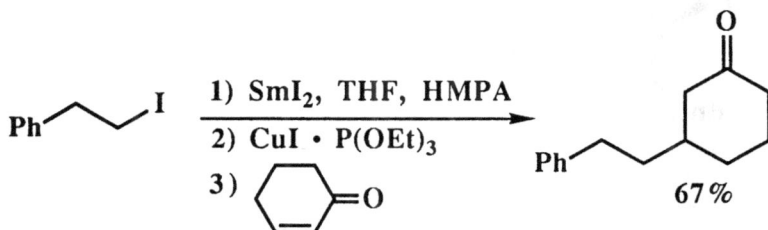

Ph        I     1) SmI$_2$, THF, HMPA

2) CuI • P(OEt)$_3$

3)

Ph

O

67%

**similarly with  Zn / LnCl$_3$**

## I.A.7.b.3.   Other Conjugate Additions

**I.A.7.b.3-1**  G. Pitacco et al., *J. Chem. Res. (S)*, 86; J.W. Huffman et al., *Tetrahedron*, **48**, 8213; **see also:** J.-M. Lassaletta and R. Fernandez, *Tetrahedron Lett.*, **33**, 3691.

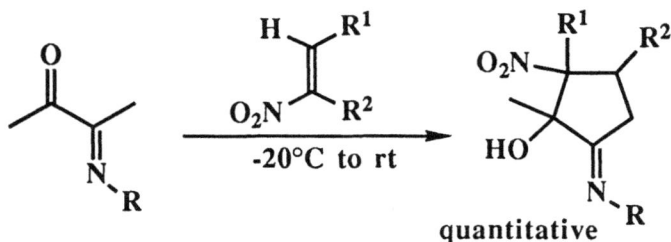

quantitative

**I.A.7.b.3-2**  S.M. Roberts and K.A. Shoberu, *J. Chem. Soc., Perkin Trans. 1*, 2625; G. Pattenden et al., *ibid.*, 1313 and 1323; M.E. Jung et al., *Tetrahedron Lett.*, **33**, 6719; T.B. Lowinger and L. Weiler, *J. Org. Chem.*, **57**, 6099; B. Quiclet-Sire et al., *Tetrahedron*, **48**, 1627; H. Kessler et al., *Angew. Chem., Int. Ed. Engl.*, **31**, 902.

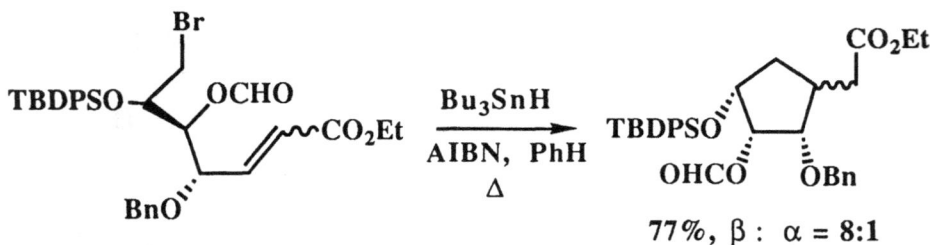

77%, β : α = 8:1

**I.A.7.b.3-3**  G. Stork and P.J. Franklin, *Aust. J. Chem.*, **45**, 275.

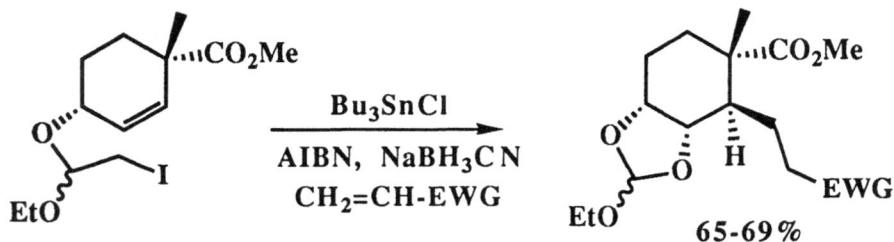

65-69%

**I.A.7.b.3-4**  W.P. Neumann et al., *J. Chem. Soc., Perkin Trans. 1*, 3165.

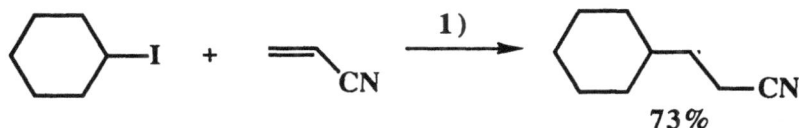

1) PolystyrylBu$_2$SnX, NaBH$_4$, AIBN, EtOH, 78°C

---

**I.A.7.b.3-5**  D.P. Curran, J. Rebek, Jr. et al., *J. Am. Chem. Soc.*, **114**, 7007.

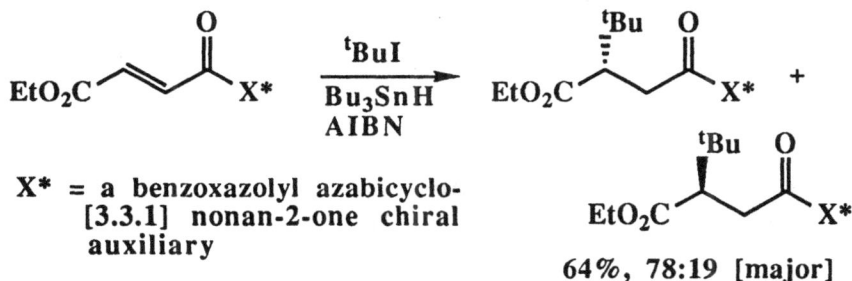

X* = a benzoxazolyl azabicyclo-
[3.3.1] nonan-2-one chiral
auxiliary

64%, 78:19 [major]

---

**I.A.7.b.3-6**  S.P. Green and D.A. Whiting, *J. Chem. Soc., Chem. Commun.*, 1753; **see also:** C.L. Bumgardner and J.F. Burgess, *Tetrahedron Lett.*, **33**, 1683.

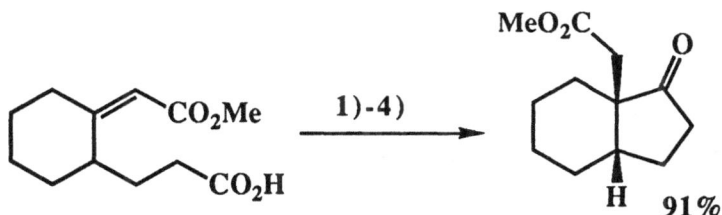

1) (COCl)$_2$;  2) PhSeNa;  3) PhH, Δ;  4) Bu$_3$SnH, AIBN

**I.A.7.b.3-7**  H. Togo et al., *Chem. Lett.*, 2169.

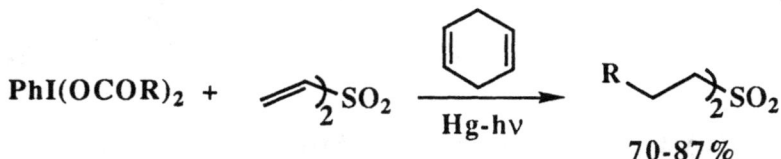

PhI(OCOR)$_2$  +  $\diagup\!\!\!=\!\!\!\diagdown_2$SO$_2$  $\xrightarrow[\text{Hg-h}\nu]{}$  R$\diagdown\!\!\diagup\!\!\diagdown_2$SO$_2$

70-87%

## I.A.8.  Other Carbon-Carbon Single Bond Forming Reactions

**I.A.8-1**  S.Z. Zard et al., *J. Am. Chem. Soc.*, **114**, 7909.

**Novel Radical Chain Reactions Based on O-Alkyl Tin Dithiocarbonates**

---

**I.A.8-2**  N.A. Porter et al., *J. Am. Chem. Soc.*, **114**, 7664.

**Origins of Stereoselectivity in Radical Additions: Reactions of Alkenes and Radicals Bearing Oxazolidine and Thiazolidine Amide Groups**

---

**I.A.8-3**  D.J. Hart and R. Krishnamurthy, *J. Org. Chem.*, **57**, 4457.

**Investigation of a Model for 1,2-Asymmetric Induction in Reactions of α-Carbalkoxy Radicals: A Stereochemical Comparison of Reactions of α-Carbalkoxy Radicals and Ester Enolates**

**I.A.8-4** T. Taguchi et al., *Tetrahedron*, **48**, 8915; P. Bravo et al., *Tetrahedron Asymm.*, **3**, 9; R. Cloux and E. Kovats, *Synthesis*, 409; A.L.J. Beckwith and S. Gerba, *Aust. J. Chem.*, **45**, 289; M. Yamamoto et al., *Bull. Chem. Soc. Jpn.*, **65**, 1550; M. Yus et al., *Tetrahedron*, **48**, 9531; **see also:** M. Shibasaki et al., *Chem. Lett.*, 395.

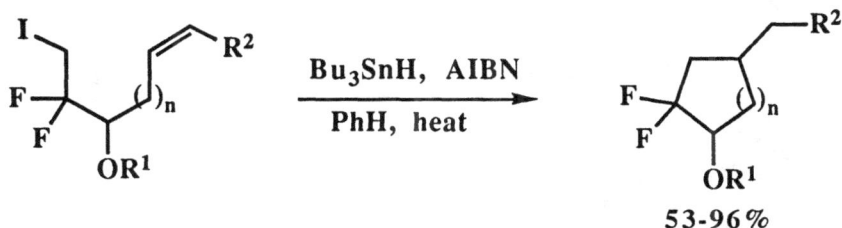

**various other radical additions reported**

---

**I.A.8-5** P. Dowd and W. Zhang, *J. Am. Chem. Soc.*, **114**, 10084.

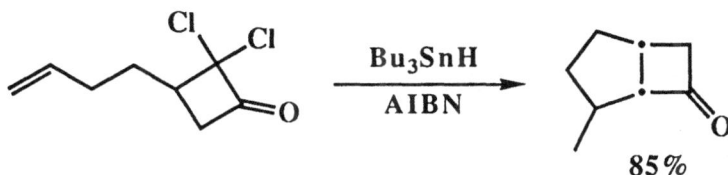

---

**I.A.8-6** K. Oshima, K. Utimoto et al., *Tetrahedron Lett.*, **33**, 7031; J. Chattopadhyaya et al., *Tetrahedron*, **48**, 349.

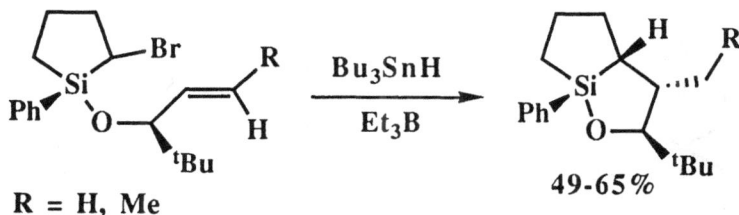

R = H, Me

**I.A.8-7** W.B. Motherwell et al., *Tetrahedron*, **48**, 8031; **see also:** T. Honda et al., *Heterocycles*, **34**, 1515; **see also:** C. Destabel and J.D. Kilburn, *J. Chem. Soc., Chem. Commun.*, 597.

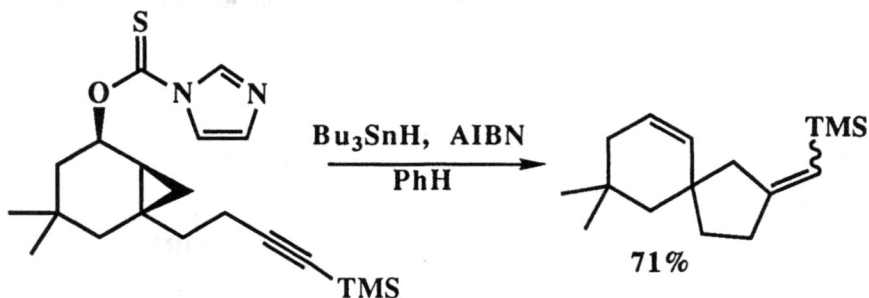

71%

---

**I.A.8-8** V.H. Rawal and V. Krishnamurthy, *Tetrahedron Lett.*, **33**, 3439; S. Kim and J.S. Koh, *ibid.*, **33**, 7391; **see also:** S. Kim and J.S. Koh, *J. Chem. Soc., Chem. Commun.*, 1377.

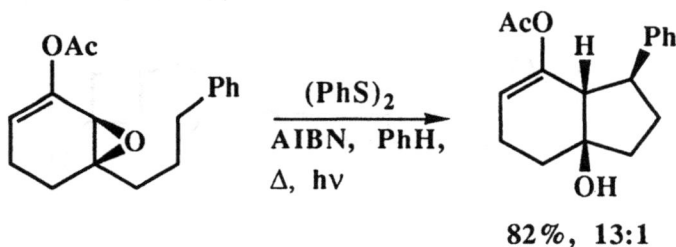

82%, 13:1

---

**I.A.8-9** Y. Chen and W. Lin, *Tetrahedron Lett.*, **33**, 1749.

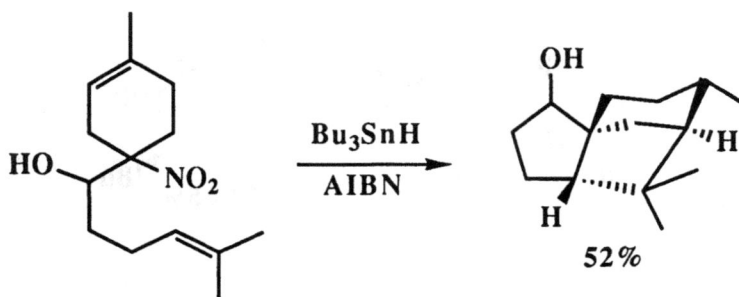

52%

**I.A.8-10**  W. Zhang and P. Dowd, *Tetrahedron Lett.*, **33**, 3285; P. Dowd and S.-W. Choi, *Tetrahedron*, **48**, 4773.

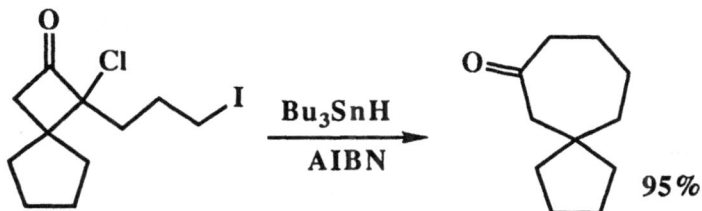

**I.A.8-11**  D.L. Boger and R.J. Mathrink, *J. Org. Chem.*, **57**, 1429; D. Batty and D. Crich, *J. Chem. Soc., Perkin Trans. 1*, 3193 and 3205.

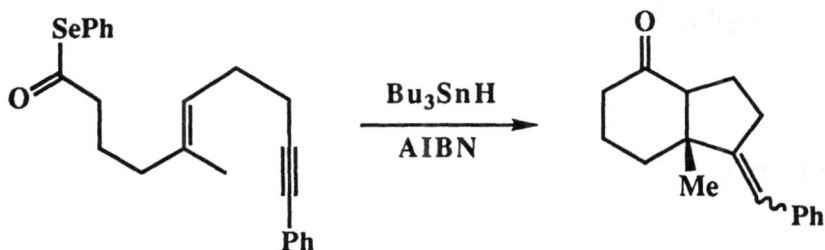

**I.A.8-12**  J.K. Crandall and T.A. Ayers, *Organometallics*, **11**, 473.

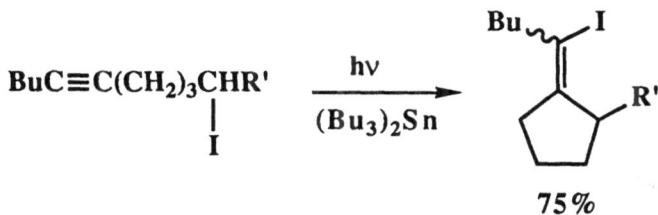

**I.A.8-13**  D.P. Curran and B. Yoo, *Tetrahedron Lett.*, **33**, 6931.

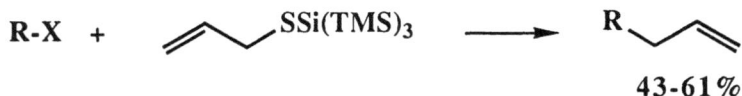

R-X  +  [allyl-SSi(TMS)₃]  $\longrightarrow$  R  

43-61%

---

**I.A.8-14**  B. Giese et al., *Tetrahedron Lett.*, **33**, 4545.

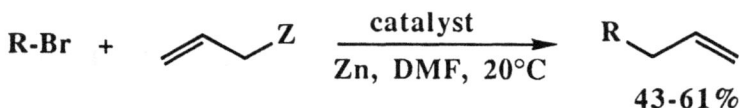

R-Br  +  [allyl-Z]  $\xrightarrow[\text{Zn, DMF, 20°C}]{\text{catalyst}}$  R  

43-61%

Z = SO₂Ph, SPh, SePh, OPO(OPh)₂

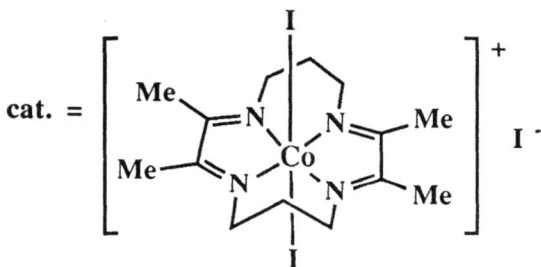

cat. =  

---

**I.A.8-15**  G.A. Molander and J.A. McKie, *J. Org. Chem.*, **57**, 3132; S.C. Suri and K.I. Hardcastle, *ibid.*, **57**, 6357.

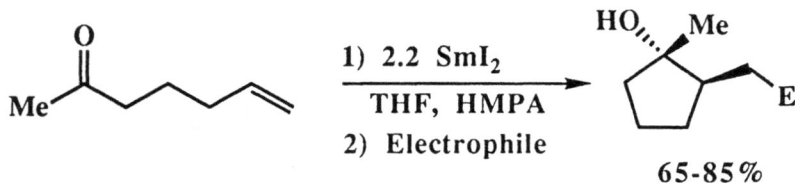

$\xrightarrow[\substack{\text{THF, HMPA} \\ \text{2) Electrophile}}]{\text{1) 2.2 SmI}_2}$

65-85%

**I.A.8-16** I. Fleming and S.K. Ghosh *J. Chem. Soc., Chem. Commun.*, 1775.

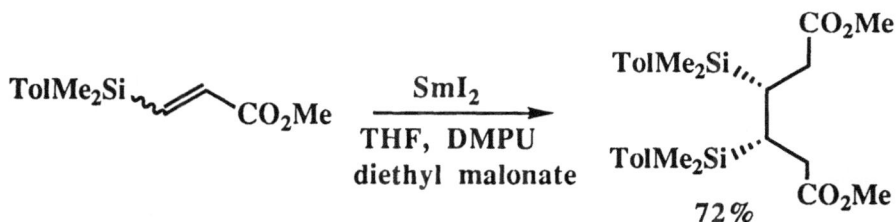

**I.A.8-17** J.M. Aurrecoechea and A. Fernandez-Acebes, *Tetrahedron Lett.*, **33**, 4763.

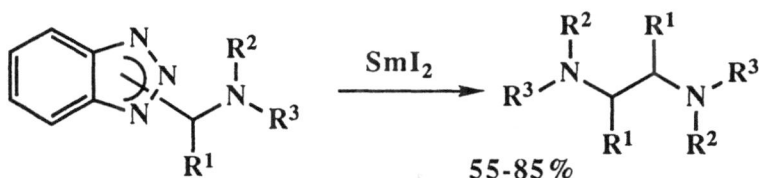

**I.A.8-18** P.A. Zoretic et al., *Tetrahedron Lett.*, **33**, 2637; M.A. Dombroski and B.B. Snider, *Tetrahedron*, **48**, 1417.

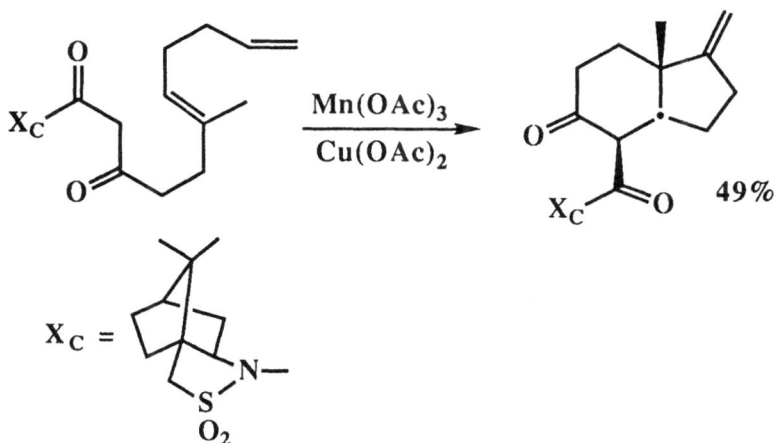

**I.A.8-19**  K. Kato and T. Mukaiyama, *Chem. Lett.*, 1137.

$$R\diagup\!\!\!=\!\!\!\diagdown R \xrightarrow[\substack{\text{BuONO} \\ 10\% \ \text{Fe(acac)}_3}]{\text{PhSiH}_3} \left(\underset{R}{R}\overset{NO}{\diagup}\underset{R}{\diagdown}\right)_2$$

---

**I.A.8-20**  K. Oshima et al., *Chem. Lett.*, 2477; B. Giese et al., *Helv. Chim. Acta*, **75**, 935.

$$\text{MeO}_2\text{C} \diagup\diagdown \xrightarrow[\text{AIBN}]{\text{HSi(TMS)}_3} \text{MeO}_2\text{C}\diagdown\diagup \overset{\text{TMS}}{\underset{\text{TMS}}{\text{Si}}}$$

71%

---

**I.A.8-21**  C. Chuang et al., *Synth. Commun.*, **22**, 3151 and 467.

$$\overset{\text{MeO}_2\text{C} \quad \text{CO}_2\text{Me}}{\diagup\diagdown} \xrightarrow[\text{h}\nu, \ \Delta]{\text{TsSePh}} \overset{\text{MeO}_2\text{C} \quad \text{CO}_2\text{Me}}{\diagdown} \quad \text{Ts}\diagdown \quad \diagup\text{SePh}$$

98%

---

**I.A.8-22**  K.D. Moeller and L.V. Tinao, *J. Am. Chem. Soc.*, **114**, 1033.

$$\underset{R}{R}\diagdown \diagup \diagdown^{\text{OMe}} \xrightarrow[\substack{\text{LiClO}_4 \\ \text{MeOH, MeCN} \\ 2,6\text{-lutidine}}]{\text{Pt anode}} \quad$$

n = 1-3

OMe
OMe
OMe
OMe

50-70%

**I.A.8-23**  J.M. Takacs et al., *J. Am. Chem. Soc.*, **114**, 773.

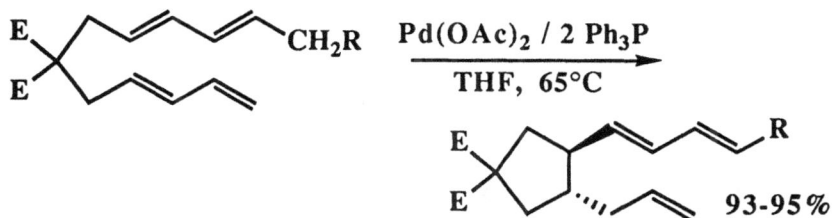

---

**I.A.8-24**  Y. Tsuji and T. Kakehi, *J. Chem. Soc., Chem. Commun.*, 1000.

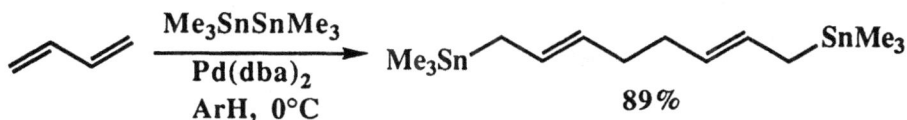

---

**I.A.8-25**  F. Kido et al., *J. Chem. Soc., Perkin Trans. 1*, 229.

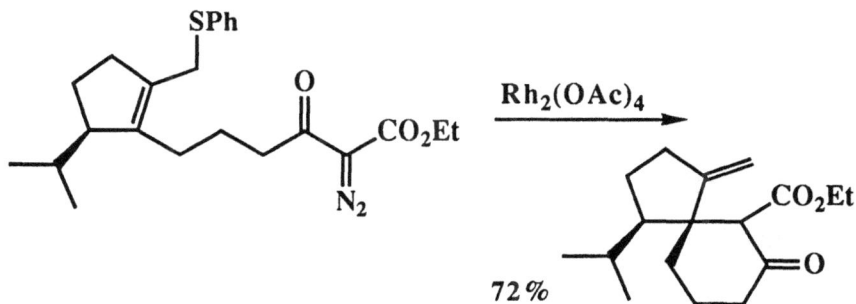

---

**I.A.8-26**  R. Curci et al., *Tetrahedron Lett.*, **33**, 7929.

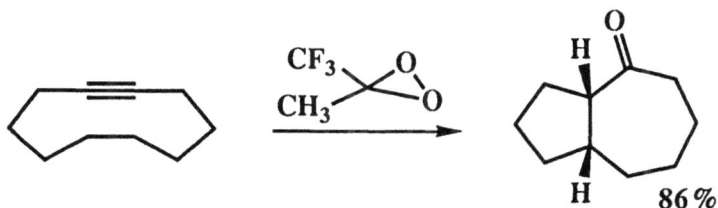

## I.B.  Carbon-Carbon  Double  Bonds

(see also: I.E.1)

## I.B.1  Wittig-Type  Olefination  Reactions

**I.B.1-1**  Y. Le Floc'h et al., *Bull. Soc. Chim. Fr.*, **129**, 62; B.G. Hazra et al., *Tetrahedron Lett.*, **33**, 3375.

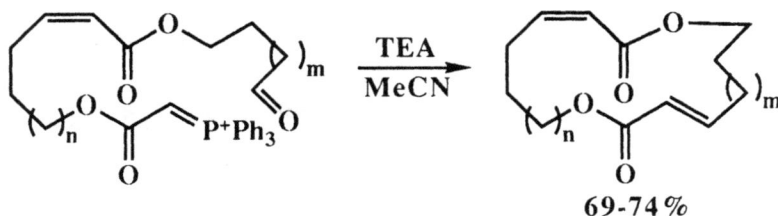

69-74%

**I.B.1-2**  Y. Tsuda et al., *Chem. Pharm. Bull.*, **40**, 1130; H. Yamataka et al, *J. Org. Chem.*, **57**, 2865.

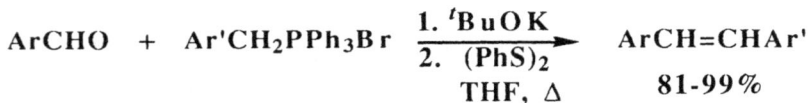

$$ArCHO + Ar'CH_2PPh_3Br \xrightarrow[\substack{2.\ (PhS)_2 \\ THF,\ \Delta}]{1.\ {}^tBuOK} ArCH{=}CHAr'$$

81-99%

**I.B.1-3**  J.D. White and M.S. Jensen, *Tetrahedron Lett.*, **33**,577.

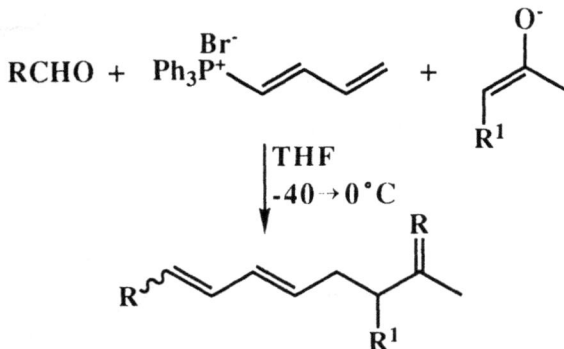

**I.B.1-4** Y.L. Bigot et al., *Synth. Commun.*, **22**, 1421; F. Effenberger and C.-P. Niesert, *Synthesis*, 1137; Y. Hamada, T. Shioiri et al., *Tetrahedron Lett.*, **33**, 4187.

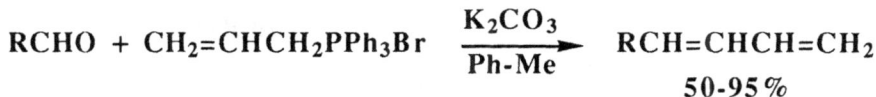

$$RCHO + CH_2=CHCH_2PPh_3Br \xrightarrow[\text{Ph-Me}]{K_2CO_3} RCH=CHCH=CH_2$$

$$50\text{-}95\%$$

---

**I.B.1-5** A. Giannis et al., *Liebigs Ann. Chem.*, 167; W.V. Dahlhoff, *ibid.*, 109.

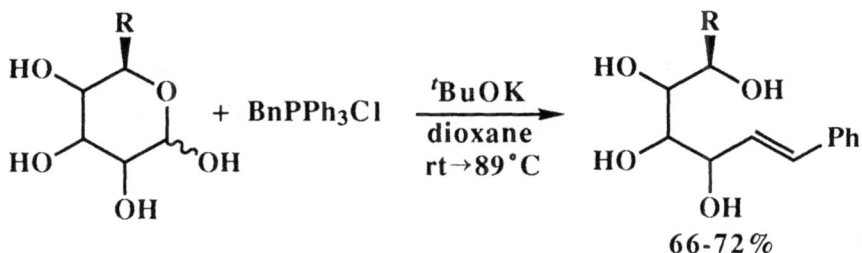

66-72%

---

**I.B.1-6** Y. Shen et al., *Synth. Commun.*, **22**, 1611, 657.

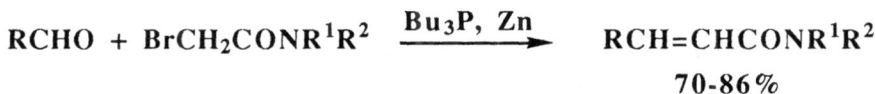

$$RCHO + BrCH_2CONR^1R^2 \xrightarrow{Bu_3P,\ Zn} RCH=CHCONR^1R^2$$

$$70\text{-}86\%$$

---

**I.B.1-7** T. Kubota and M. Yamamoto, *Tetrahedron Lett.*, **33**, 2603.

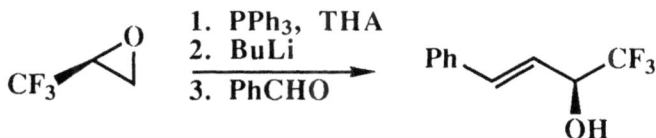

1. PPh$_3$, THA
2. BuLi
3. PhCHO

---

**I.B.1-8** E. Turos et al., *J. Org. Chem.*, **57**, 6667.

Ph$_3$PCH$_2$RBr

or $\qquad$ + R$^1$Li $\xrightarrow{DMF}$ RCH=CHR$^1$

(EtO)$_2$PCH$_2$Ph
$\quad\overset{\|}{O}$

69-95%

**I.B.1-9** P. Vinczer, C. Szantay et al., *Tetrahedron Lett.*, **33**, 683.

$$RCHO \xrightarrow[\text{THF, 24h, rt}]{\text{Ph}_3\text{P, CCl}_4, \text{ Mg}} RCH{=}CCl_2$$

60-85%

---

**I.B.1-10** U. Schmidt et al., *Synthesis*, 487; P. Roth and R. Metternich, *Tetrahedron Lett.*, **33**, 3993; D.F. Netz and J.L. Seidel, *Tetrahedron Lett.*, **33**, 1957; Z. Mouloungui et al., *Tetrahedron*, **48**, 1219; Z. Xu and D.D. DesMarteau, *J. Chem. Soc., Perkin Trans. 1*, 313.

52-99%

---

**I.B.1-11** T. Janecki, *Synth. Commun.*, **22**, 2063; G. Heinisch et al., *Monatsh. Chem.*, **122**, 1055 (1991); S. Warren et al., *J. Chem. Soc., Perkin Trans. 1*, 3407.

31-50%

---

**I.B.1-12** H. Fillion et al., *Tetrahedron Lett.*, **33**, 4909; D. Craig and N.J. Geach, *Synlett*, 299.

33-81%
(0.25-19:1)

**I.B.1-13** S. Hanessian and S. Beaudein, *Tetrahedron Lett.*, **33**, 7655; S.E. Denmark and C. Chen, *J. Am. Chem. Soc.*, **114**, 10674.

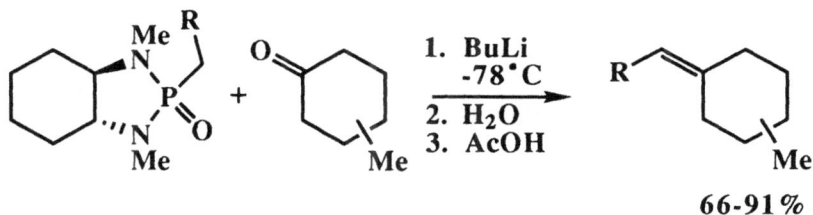

**I.B.1-14** M. Hatanaka et al., *Chem. Lett.*, 2253; H.J. Bestmann et al., *Chem. Ber.*, **125**, 2081.

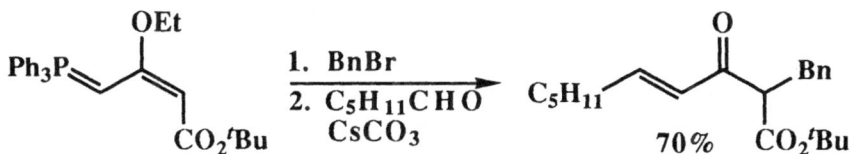

**I.B.1-15** J.R. Hwu et al., *J. Chem. Soc., Perkin Trans. 1*, 3219; M. Muzard and C. Portella, *Synthesis*, 965.

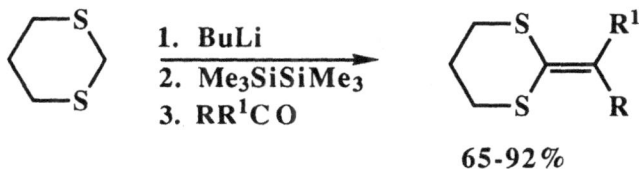

**I.B.1-16** A.R. Katritzky et al., *Gazz. Chim. Ital.*, **121**, 471 (991).

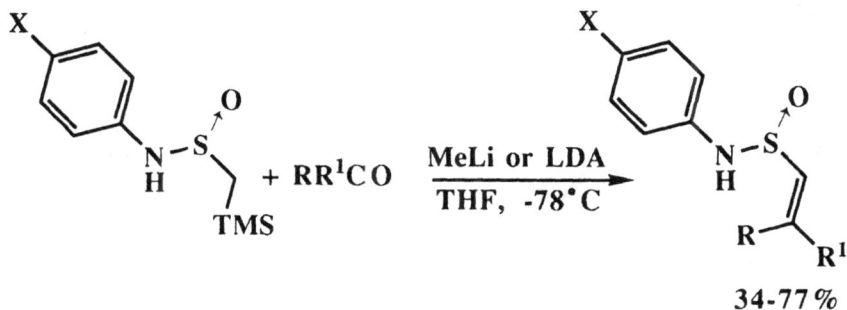

**I.B.1-17** C. Palomo et al., *Tetrahedron Lett.*, **33**, 3903; C. Couret et al., *J. Organomet. Chem.*, **440**, 233; A. Couture et al., *ibid.*, **440**, 7.

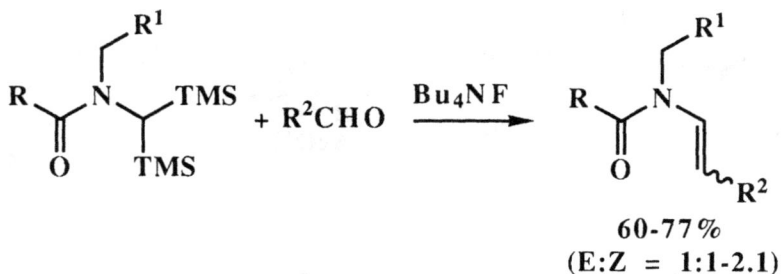

60-77%
(E:Z = 1:1-2.1)

**I.B.1-18** J. Wicha et al., *J. Org. Chem.*, **57**, 6593.

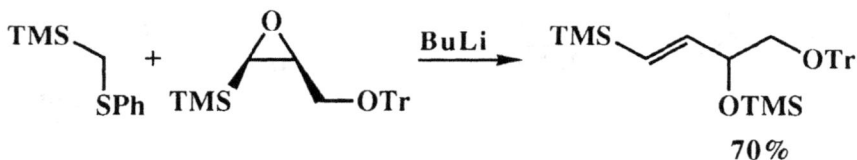

70%

**I.B.1-19** L.S. Hegedus et al., *J. Am. Chem. Soc.*, **114**, 4079.

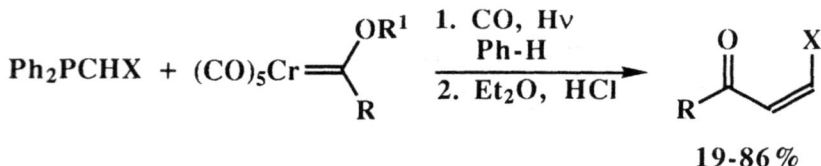

19-86%

**I.B.1-20** N.A. Petasis and I. Akritopoulou, *Synlett.*, 665; N.A. Petasis and E.I. Bzowej, *J. Org. Chem.*, **57**, 1327.

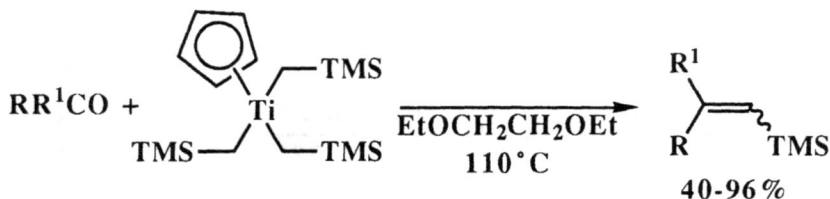

40-96%

**I.B.1-21** A. Zapata et al., *J. Organomet. Chem.*, **424**, C9.

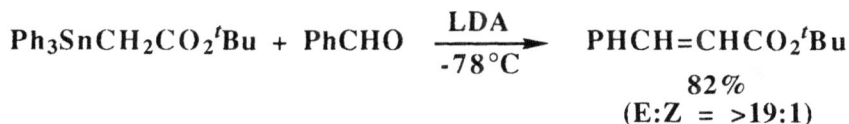

$$Ph_3SnCH_2CO_2{}^tBu + PhCHO \xrightarrow[-78°C]{LDA} PHCH=CHCO_2{}^tBu$$

82%
(E:Z = >19:1)

---

**I.B.1-22** R. Angell et al., *Synlett.*, 599.

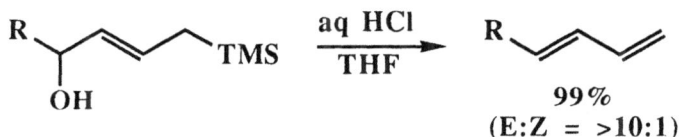

99%
(E:Z = >10:1)

---

**I.B.1-23** E.J. Miller et al., *J. Organomet. Chem.*, **440**, 91.

**The Wittig reaction in the generation of organometallic compounds containing alkenes as side groups**

## I.B.2.    Eliminations

## I.B.2.a   Eliminations of Alcohols and Derivatives

**I.B.2.a-1** D.H.R. Barton and C. Tachdjian, *Tetrahedron*, **48**, 7109; see also: M. Vandewalle et al., *Synlett.*, 51.

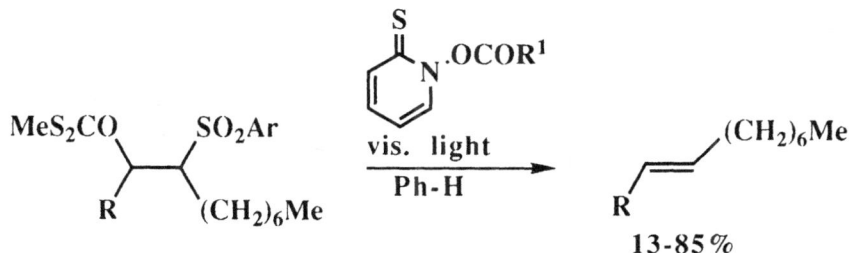

13-85%

**I.B.2.a-2** S.K. Kang et al., *Synth. Commun.*, **22**, 1109 and *J. Chem. Soc., Perkin Trans. 1*, 405; see also: A. Alexakis et al., *Synth. Commun.*, **22**, 1839; see also: H.W. Pinnick et al., *Tetrahedron Lett.*, **33**, 2665.

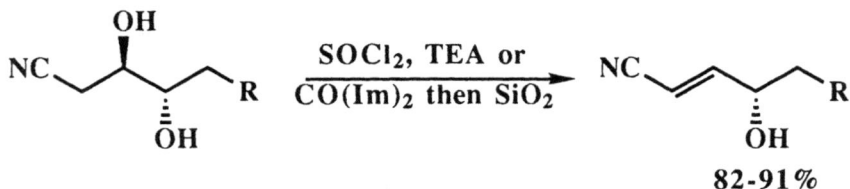

**I.B.2.a-3** S. Takano et al., *Synlett.*, 668; O. Arjona, R.F. de la Pradilla, J. Plumet et al., *J. Org. Chem.*, **57**, 1945; see also: I. Stohrer and H.M.R. Hoffmann, *Tetrahedron*, **48**, 6021.

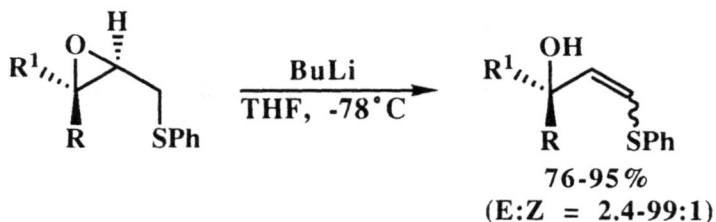

**I.B.2.a-4** B.M. Trost and M.S. Rodriguez, *Tetrahedron Lett.*, **33**, 4675.

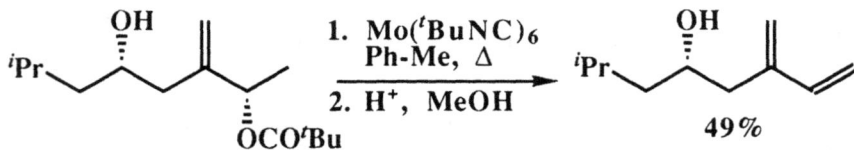

**I.B.2.a-5** T. Mandai et al., *Tetrahedron Lett.*, **33**, 2549.

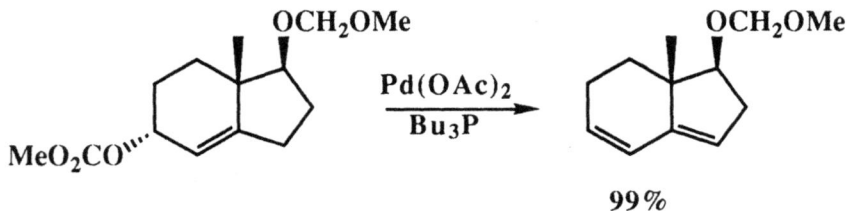

**I.B.1.a-6** F. Berti et al., *Tetrahedron Lett.*, **33**, 8145.

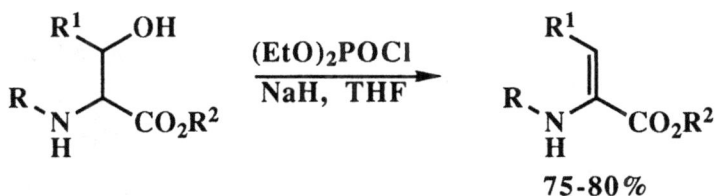

**I.B.2.a-7** R. Ballini et al., *J. Org. Chem.*, **57**, 2160,

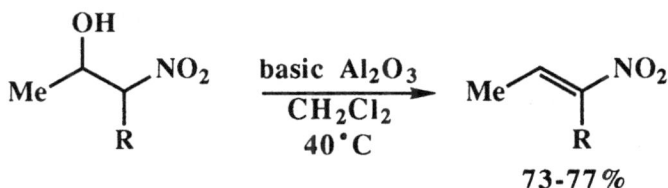

**I.B.2.a-8** V.S. Martin et al., *Tetrahedron Asymm.*, **3**, 573.

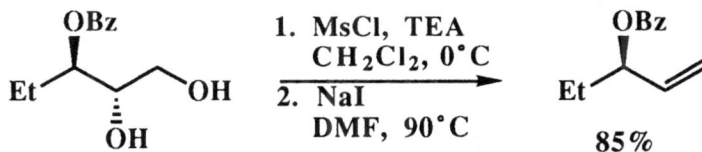

**I.B.2.a-9** G. Solladie and V. Berl, *Tetrahedron Lett.*, **33**, 3477.

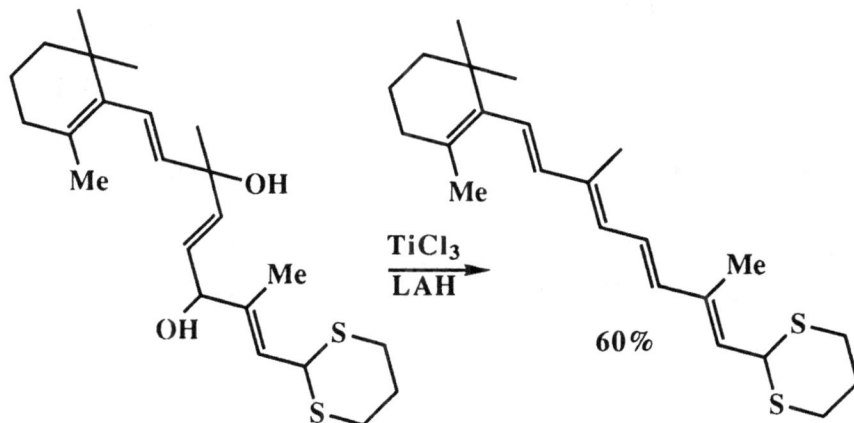

## I.B.2.b.  Eliminations  of  Halides

**I.B.2.b-1** L.A. Swandeses and J.-L. Luche *J. Org. Chem.*, **57**, 2757; see also: H.-J. Borschberg et al., *Tetrahedron Lett.*, **33**, 6449.

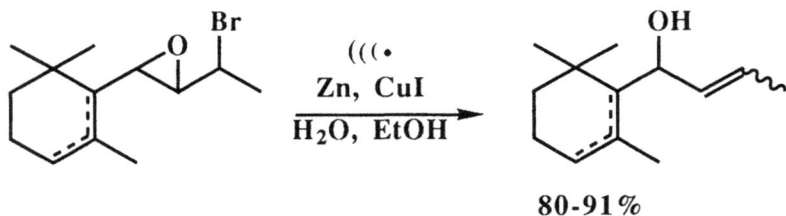

$((( \bullet$
**Zn, CuI**
$\overline{\text{H}_2\text{O, EtOH}}$

**80-91%**

---

**I.B.2.b-2** T. Honda et al., *Chem. Commun.*, 1218.

$\overline{\text{THF/HMPA}}$ **SmI$_2$**

**77%**

---

**I.B.2.b-3** H. Choi and A.R. Pinhas, *Organometallics,* **11**, 442

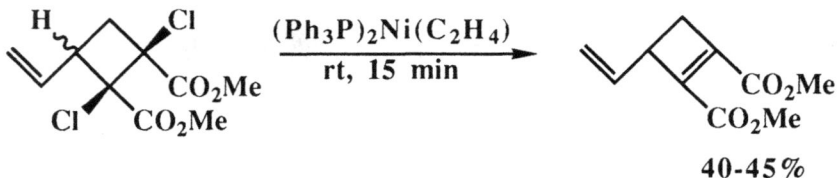

$(\text{Ph}_3\text{P})_2\text{Ni}(\text{C}_2\text{H}_4)$
**rt, 15 min**

**40-45%**

---

**I.B.2.b-4** N. De Kimpe et al., *Tetrahedron Lett.*, **33**, 393.

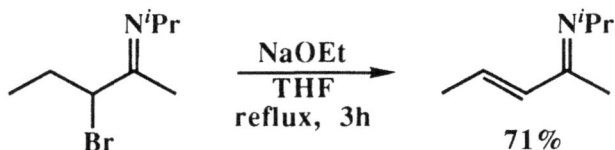

**NaOEt**
**THF**
**reflux, 3h**

**71%**

## I.B.2.c  Other  Eliminations

**I.B.2.c-1**  C. Fehr et al., *Tetrahedron Lett.*, **33**, 2465.

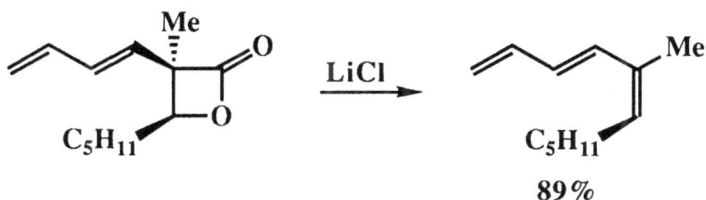

**89%**

---

**I.B.2.c-2**  T. Chou and S.-Y. Chang, *J. Chem. Soc., Perkin Trans. 1*, 1459.

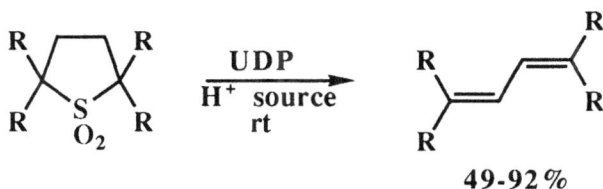

**49-92%**

**UDP = ultrasonically dispersed potassium**

---

**I.B.2.c-3**  A.L. Schwan and D.A. Wilson, *Tetrahedron Lett.*, **33**, 5897.

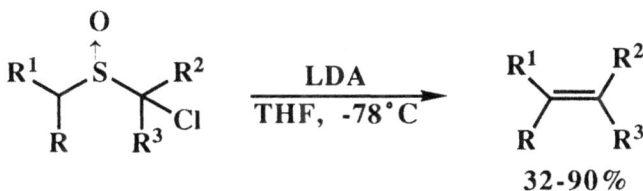

**32-90%**

---

**I.B.2.c-4**  S.A. Julia et al., *Bull. Soc. Chim. Fr.*, **129**, 440.

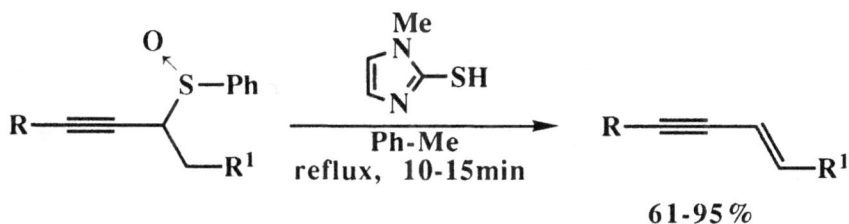

**61-95%**

**I.B.2.c-5** I. Ryu, N. Sonoda et al., *J. Am. Chem. Soc.*, **114**, 1521.

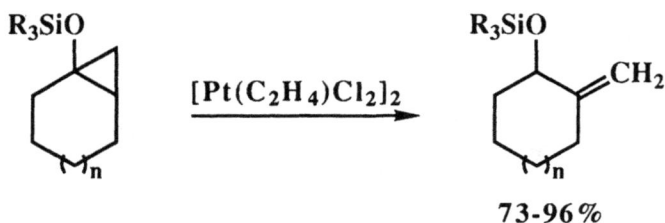

73-96%

---

**I.B.2.c-6** M. Franck-Neumann et al., *Tetrahedron*, **48**, 1911.

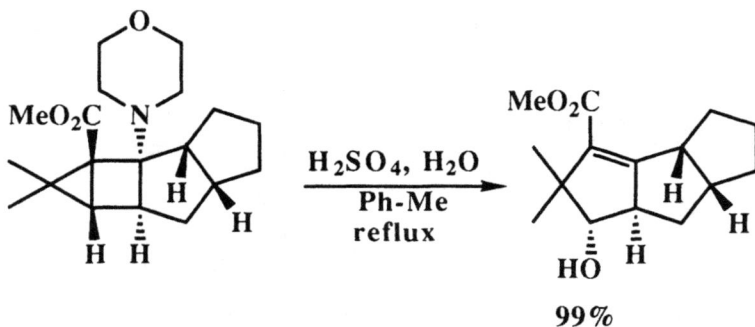

99%

---

**I.B.2.c-7** T.K. Serker and B.K. Ghorai, *Chem. Commun.*, 1184.

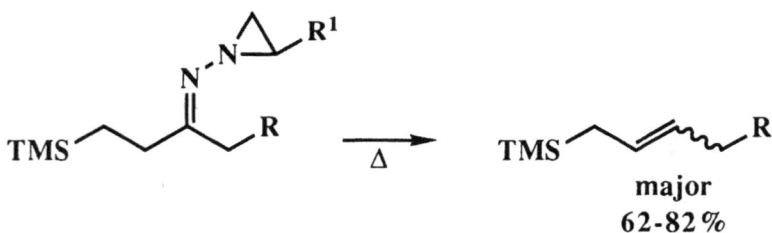

major
62-82%

---

**I.B.2.c-8** C. Chapuis, *Tetrahedron Lett.*, **33**, 2461.

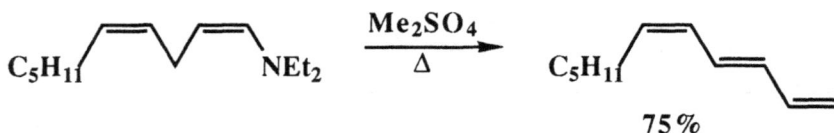

75%

**I.B.2.c-9** M.C. Wang and T.-Y. Luh, *J. Org. Chem.*, **57**, 2178.

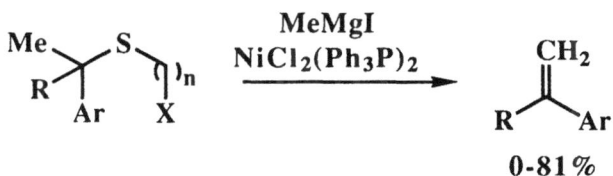

## I.B.3  Other Carbon-Carbon Double Bond Forming Reactions

**I.B.3-1** D. Brillon and G. Sauve, *J. Org. Chem.*, **57**, 1838; T. Saito et al., *Tetrahedron Lett.*, **33**, 7539; D. Villemin et al., *Synth. Commun.*, **22**, 3169.

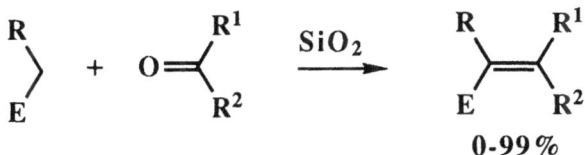

**I.B.3-2** W.B. Motherwell et al., *Tetrahedron Lett.*, **33**, 3899; J. Besancon et al., *J. Organomet. Chem.*, **426**, 325.

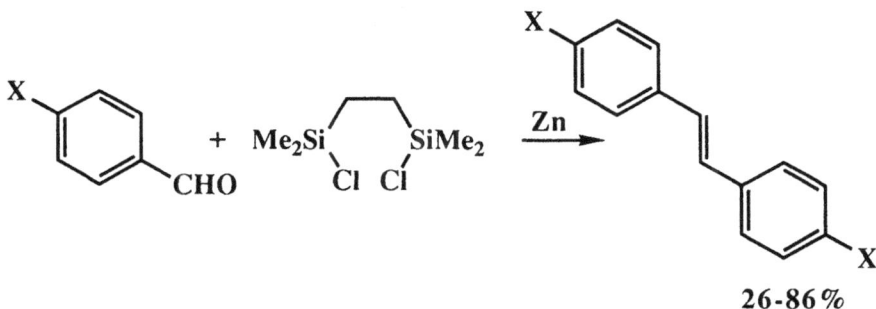

**.B.3-3** G.K. Musorin and S.V. Amosova, *J. Org. Chem. (USSR)*, **28**, 519.

$$(CH_2=CHCH_2)_2Se \xrightarrow[\substack{DMSO \\ 80°C}]{KOH} CH_2=CHCH=CHCH=CH_2$$
$$48\%$$

**I.B.3-4** M. Julia et al., *Synlett.*, 133.

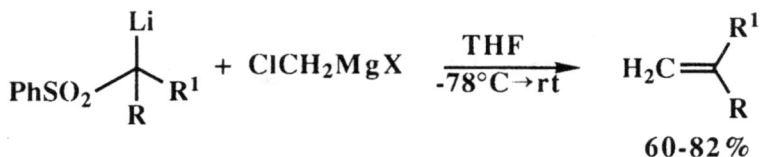

$$PhSO_2 \overset{Li}{\underset{R}{\diagup}} R^1 \quad + \quad ClCH_2MgX \quad \xrightarrow[-78°C \to rt]{THF} \quad H_2C = \overset{R^1}{\underset{R}{\diagdown}}$$

**60-82%**

---

**I.B.3-5** W.B. Motherwell et al., *Tetrahedron Lett.*, **33**, 3367.

**Observations on the selective deoxygenation of epoxides to olefins with chlorotrimethylsilane and zinc.**

---

**I.B.3-6** K.-T. Wong and T.-Y. Luh, *J. Am. Chem. Soc.*, **114**, 7308.

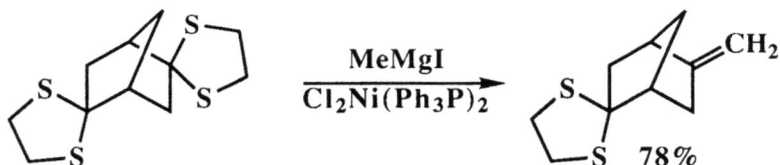

$$\xrightarrow[Cl_2Ni(Ph_3P)_2]{MeMgI}$$

**78%**

---

**I.B.3-7** S. Kagabu et al., *Bull. Soc. Chim. Fr.*, **129**, 435.

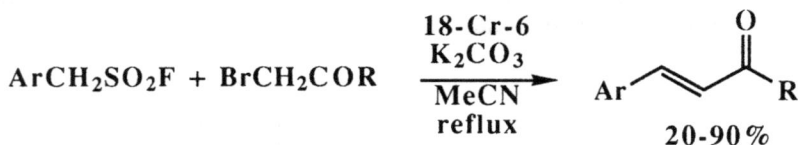

$$ArCH_2SO_2F + BrCH_2COR \quad \xrightarrow[\substack{MeCN \\ reflux}]{\substack{18\text{-}Cr\text{-}6 \\ K_2CO_3}} \quad Ar \diagup\!\!\diagdown\!\!\diagup \overset{O}{\diagdown} R$$

**20-90%**

---

**I.B.3-8** A. Padwa et al., *J. Org. Chem.*, **57**, 4940; M. Curini et al., *Org. Prep. Proced. Int.*, **24**, 497; see also: S. Shatzmiller and S. Bercovici, *Liebigs Ann. Chem.*, 877.

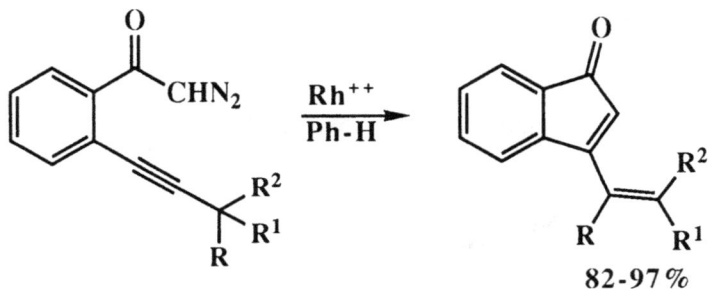

$$\xrightarrow[Ph\text{-}H]{Rh^{++}}$$

**82-97%**

**I.B.3-9** D.J. Pasto and D.E. Alonso, *Tetrahedron Lett.*, **33**, 783.

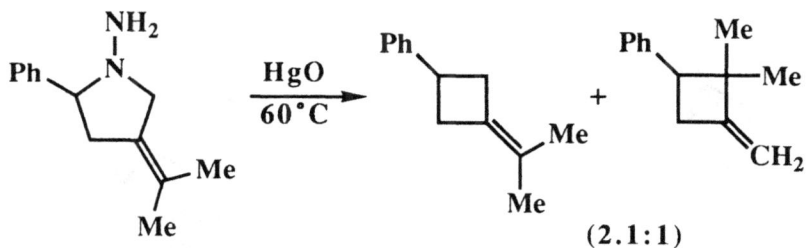

**I.B.3-10** T. Honda et al., *Tetrahedron*, **48**, 79.

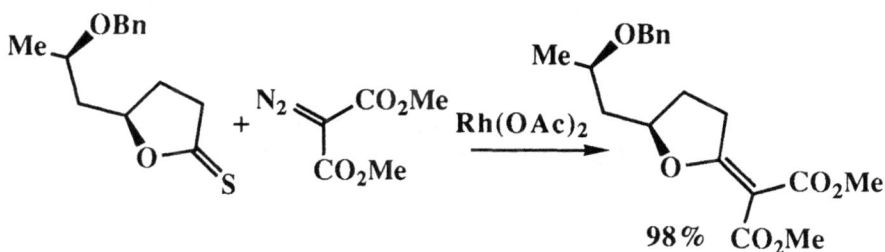

**I.B.3-11** R.C. Cambie et al., *J. Organomet. Chem.*, **429**, 41, 59; J. Vicente et al., *ibid.*, **436**, C9.

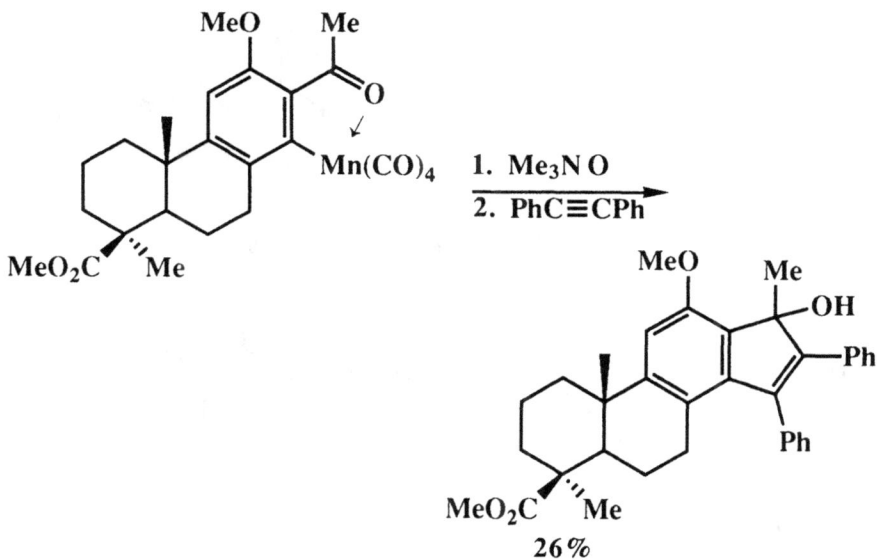

**I.B.3-12** A.E. Harms and J.R. Stille, *Tetrahedron Lett.*, **33**, 6565.

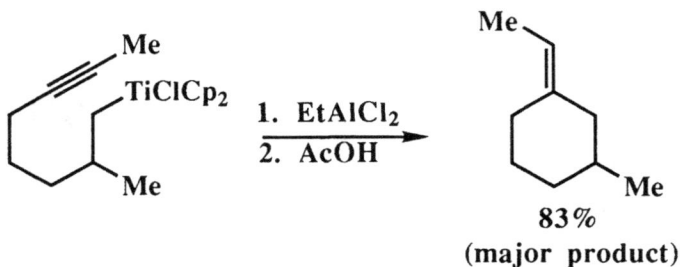

83%
(major product)

---

**I.B.3-13** C. Qian et al., *J. Organomet. Chem.*, **430**, 175; K. Inomata et al., *Bull. Chem. Soc. Jpn.*, **65**, 75; K. Oshima, K. Utimoto et al., *ibid.*, **65**, 349.

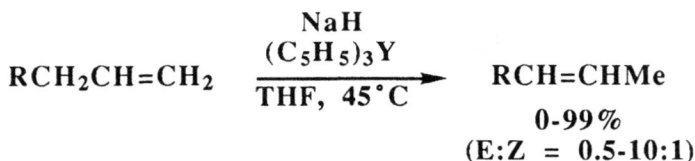

$$RCH_2CH=CH_2 \xrightarrow[\text{THF, 45°C}]{\begin{array}{c}NaH\\(C_5H_5)_3Y\end{array}} RCH=CHMe$$

0-99%
(E:Z = 0.5-10:1)

---

**I.B.3-14** B.M. Trost and U. Kazmaier, *J. Am. Chem. Soc.*, **114**, 7933; C. Gua and X. Lu, *Tetrahedron Lett.*, **33**, 3659; X. Lu et al., *J. Organomet. Chem.*, **428**, 259.

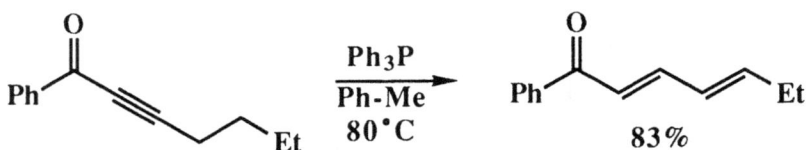

83%

---

**I.B.3-15** K. Narasake et al., *Bull. Chem. Soc. Jpn.*, **65**, 2825.

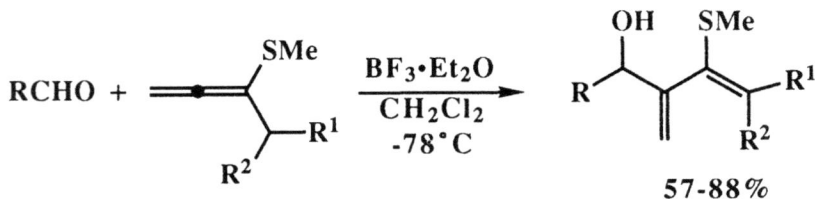

57-88%

**I.B.3-16** Y. Masuyama et al., *Tetrahedron Lett.*, **33**, 6477; Z. Ni and A. Padwa, *Synlett.*, 869; W.S. Johnson et al., *Tetrahedron Lett.*, **33**, 8001.

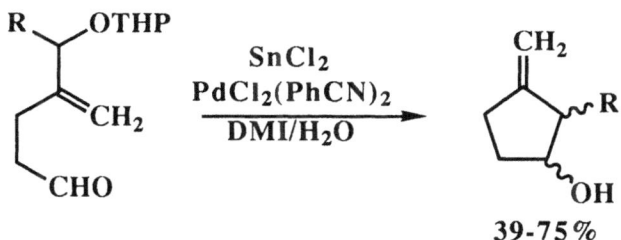

**I.B.3-17** T. Honda et al., *J. Chem. Soc., Perkin Trans. 1*, 1557; F.E. Ziegler et al., *Tetrahedron Lett.*, **33**, 3117; see also: E. Wenkert et al., *J. Am. Chem. Soc.*, **114**, 644.

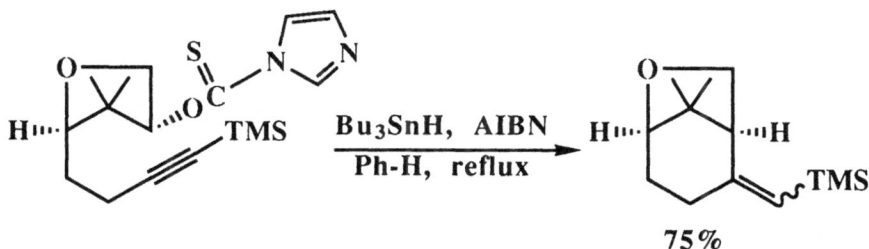

**I.B.3-18** G. Balme et al., *Tetrahedron*, **48**, 10103.

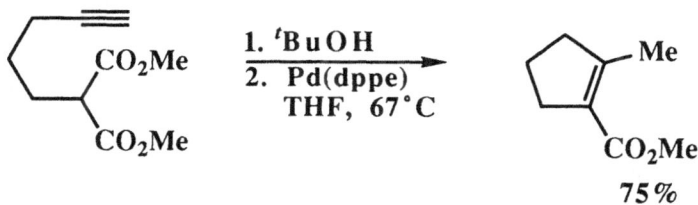

**I.B.3-19** T.W. Wallace et al., *Tetrahedron*, **48**, 515.

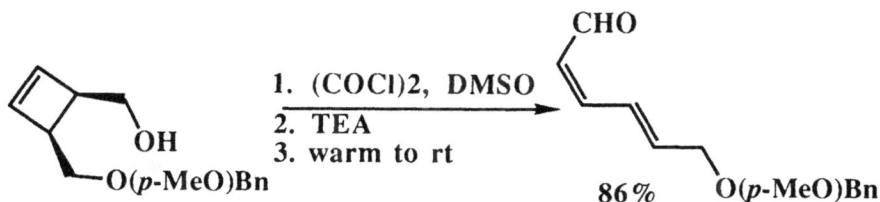

**I.B.3-20** R. Tamura et., *Organometallics*, **11**, 954.

$$(RCOCH_2)_2TeCl_2 + \underset{\text{O}^-\,\,\,\,\,\,\,\,R^1}{\overset{\text{LiO}\,\,\,\,\,\,\,\,R^2}{N=}} \xrightarrow{\text{DMF}} RCOCH=\overset{R^2}{\underset{R^1}{}}$$

22-98%

---

**I.B.3-21** B.M. Trost and Y. Shi, *J. Am. Chem. Soc.*, **114**, 791; B. Trost et al., *ibid.*, **114**, 1923; A. de Meijere et al., *Tetrahedron Lett.*, **33**, 8039; M. Costa et al., *J. Chem. Soc., Perkin Trans. 1*, 1399, 1407.

1. Pd$_2$(dba)$_3$·CHCl$_3$
   Ph$_3$P
2. AcOH
3. 70°C

75%

---

**I.B.3-22** F.-T. Luo et al., *Tetrahedron Lett.*, **33**, 6839.

$$RC\equiv CH \xrightarrow[\substack{2.\ R_1ZnCl \\ Pd(Ph_3P)_4}]{\substack{1.\ \text{TMSCl, NaI} \\ H_2O/MeCN}} \overset{R^1}{\underset{R}{}}=CH_2$$

54-74%

---

**I.B.3-23** P. Parsons et al., *Tetrahedron*, **48**, 9461.

Bu$_3$SnH, AIBN
Ph-H

43%

**I.B.3-24** D. van Leusen and A.M. van Leusen, *Rec. Trav. Chim.*, **110**, 393 (1991) and *ibid.*, **111**, 469.

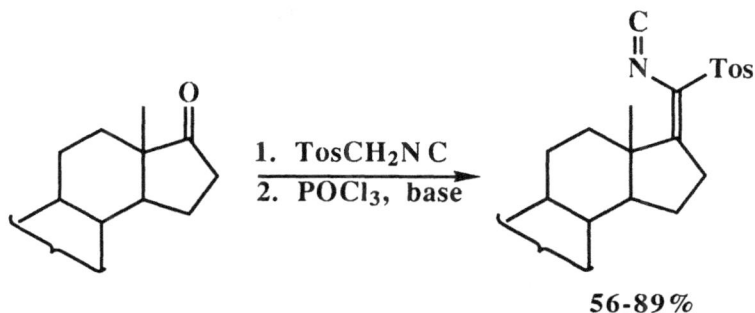

**56-89%**

**I.B.3-25** K. Oshima, K. Utimoto et al., *Bull. Chem. Soc. Jpn.*, **65**, 1513.

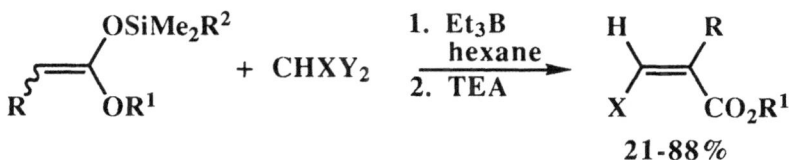

**21-88%**

**I.B.3-26** S. Kim and J.R. Cho, *Synlett.*, 629.

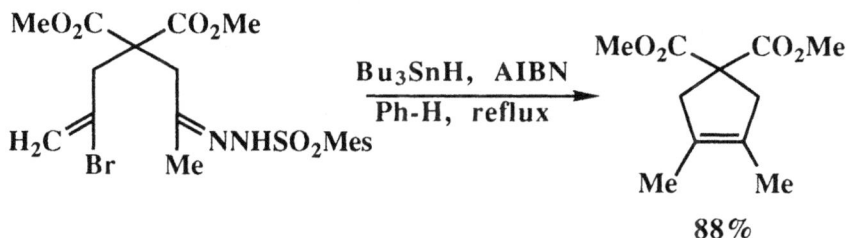

**88%**

## I.B.4 Vinylations

**I.B.4-1** B.D. Wilson, *Synthesis*, 283.

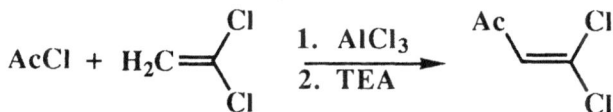

**I.B.4-2**  T. Kauffmann and D. Stach, *Chem. Ber.*, **125**, 913.

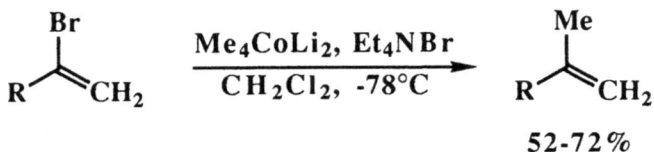

$$\underset{R}{\overset{Br}{\diagdown}}C{=}CH_2 \quad \xrightarrow[\text{CH}_2\text{Cl}_2,\ -78°\text{C}]{\text{Me}_4\text{CoLi}_2,\ \text{Et}_4\text{NBr}} \quad \underset{R}{\overset{Me}{\diagdown}}C{=}CH_2$$

52-72%

---

**I.B.4-3**  T. Toru et al., *J. Org. Chem.*, **57**, 3145.

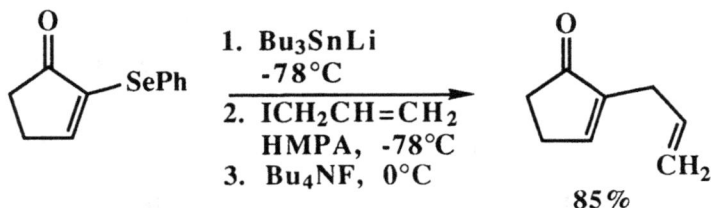

1. Bu$_3$SnLi
   -78°C
2. ICH$_2$CH=CH$_2$
   HMPA, -78°C
3. Bu$_4$NF, 0°C

85%

---

**I.B.4-4**  H.-J. Gais and G. Bulow, *Tetrahedron Lett.*, **33**, 461, 465.

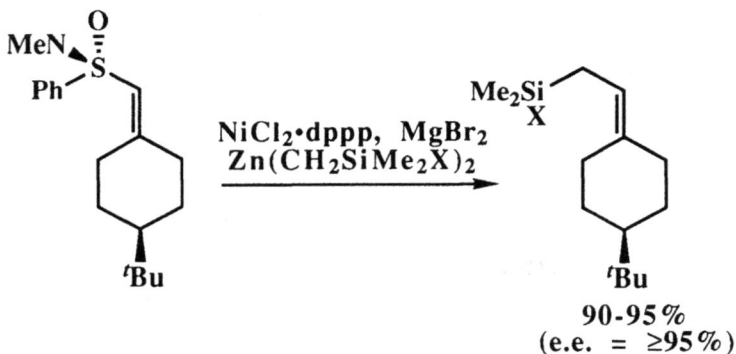

NiCl$_2$·dppp, MgBr$_2$
Zn(CH$_2$SiMe$_2$X)$_2$

90-95%
(e.e. = ≥95%)

---

**I.B.4-5**  J. Ichikawa et al., *Tetrahedron Lett.*, **33**, 3779, 337.

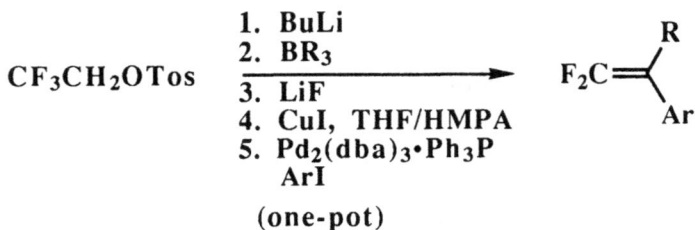

CF$_3$CH$_2$OTos

1. BuLi
2. BR$_3$
3. LiF
4. CuI, THF/HMPA
5. Pd$_2$(dba)$_3$·Ph$_3$P
   ArI

(one-pot)

$$F_2C{=}\underset{Ar}{\overset{R}{\diagup}}$$

**I.B.4-6** D. Villemin and B. Labiad, *Synth. Commun.*, **22**, 3181; see also:
W. Priebe et al., *Tetrahedron Lett.*, **33**, 7681.

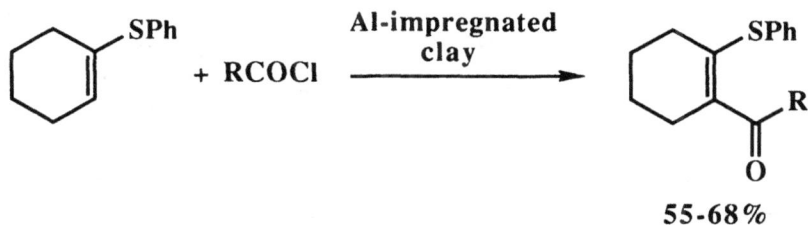

55-68%

**I.B.4-7** A.B. Holmes and G.R. Pooley, *Tetrahedron*, **48**, 7775.

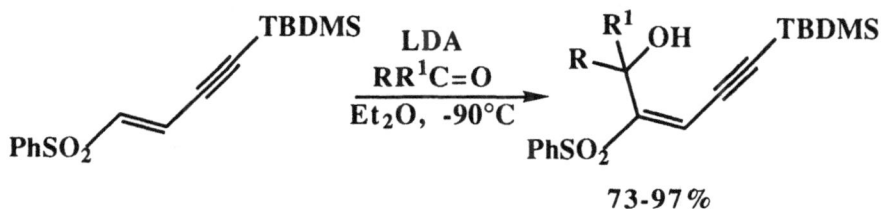

73-97%

**I.B.4-8** T. Mitsudo, Y. Watanabe et al., *J. Organomet. Chem.*, **423**, 405
and *Tetrahedron Lett.*, **33**, 341.

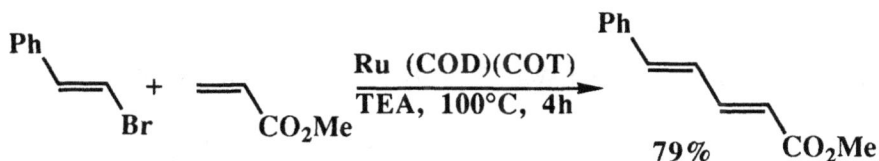

**I.B.4-9** J.P. Genet et al., *Synlett.*, 715; B.M. Choudary et al.,
*Tetrahedron*, **48**, 719; A.-S. Carlstrom and T. Frejd, *Acta Chem. Scand.*,
**46**, 163.

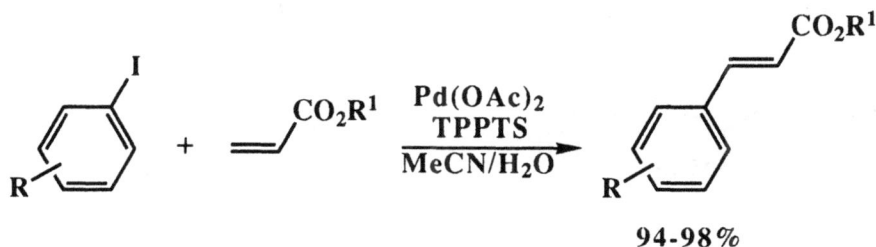

94-98%

**I.B.4-10** J.T. Pinhey et al., *J. Chem. Soc., Perkin Trans. 1*, 1911, 1917; see also: P. Kocienski et al., *Synlett.*, 886.

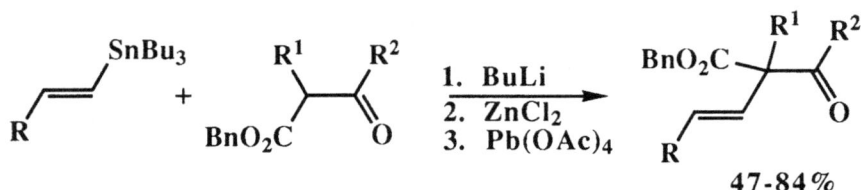

$$47\text{-}84\%$$

**I.B.4-11** E. Piers et al., *Can. J. Chem.*, **70**, 1385; A. Mourino et al., *Tetrahedron Lett.*, **33**, 7589; C.R. Johnson et al., *ibid.*, **33**, 919; M. Lampilas and R. Lett, *ibid.*, **33**, 773; Q. Han and D.F. Wiemer, *J. Am. Chem. Soc.*, **114**, 7692; M. Uemura, T. Hayashi et al., *Tetrahedron Asymm.*, **3**, 213.

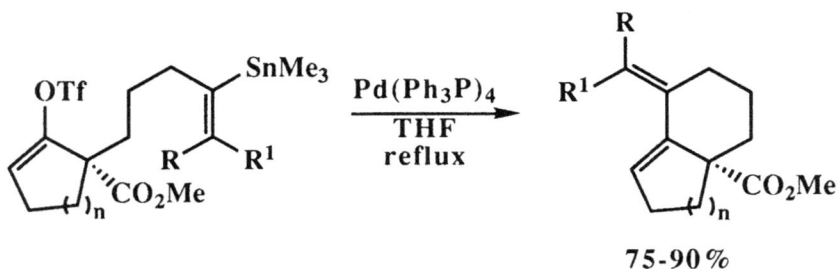

$$75\text{-}90\%$$

**I.B.4-12** R.M. Moriarty and W.R. Epa, *Tetrahedron Lett.*, **33**, 4095.

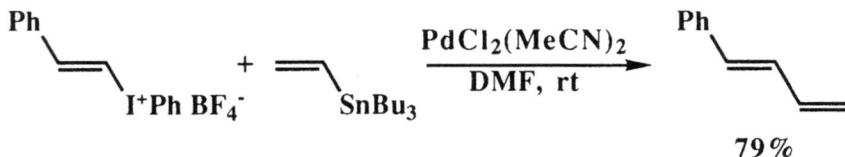

$$79\%$$

**I.B.4-13** G.T. Crisp and P.T. Glink, *Tetrahedron Lett.*, **33**, 4649.

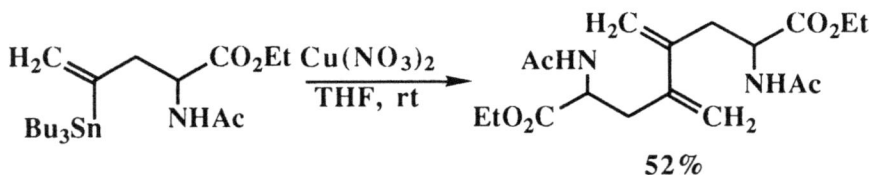

$$52\%$$

**I.B.4-14** S. Liang and L.A. Paquette, *Acta Chem. Scand.*, **46**, 597.

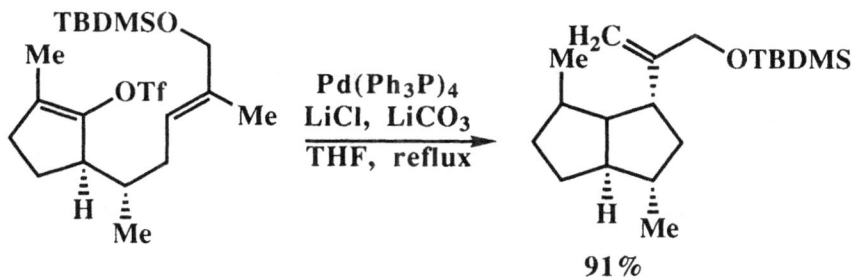

91%

---

**I.B.4-15** B. Jiang and Y. Xu, *Tetrahedron Lett.*, **33**, 511.

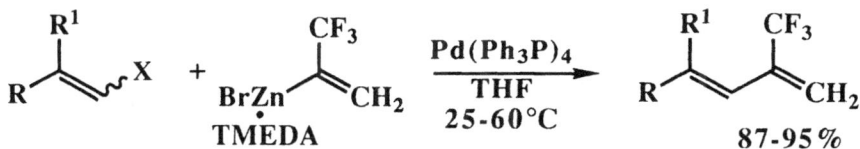

87-95%

---

**I.B.4-16** R. Rossi et al., *Gazz. Chim. Ital.*, **122**, 65.

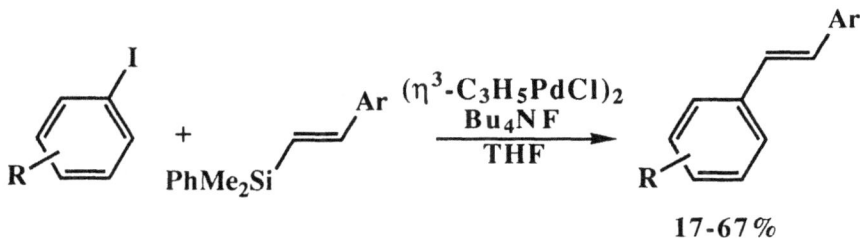

17-67%

---

**I.B.4-17** G. Frater et al., *Tetrahedron Lett.*, **33**, 1045.

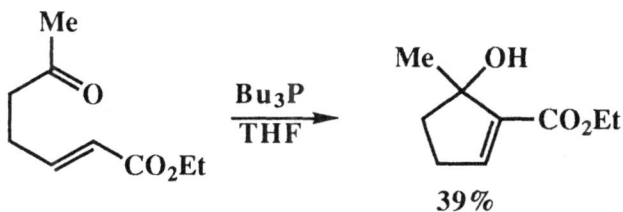

39%

DABCO does not work

## I.B.5    Allene  Forming  Reactions

**I.B.5-1**  A. Pelter et al., *J. Chem. Soc., Perkin Trans. 1*, 747.

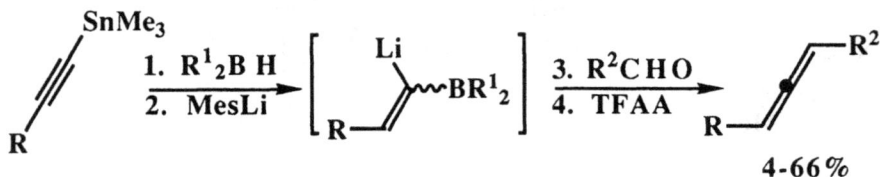

**I.B.5-2**  M. Malaria et al., *Synlett.*, 493.

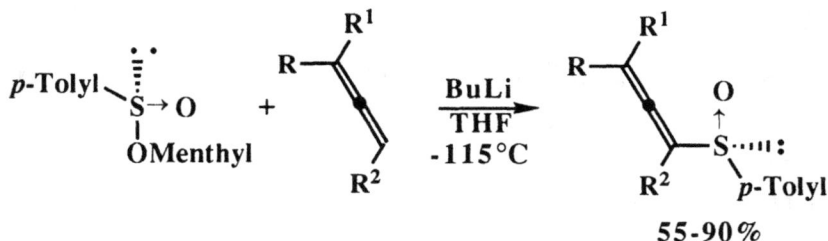

**I.B.5-3**  T. Kurihara et al., *Chem. Pharm. Bull.*, **40**, 21; S.-K. Kang et al., *Tetrahedron Asymm.*, **3**, 1509; R.W. Saalfrank et al., *Angew. Chem., Int. Ed. Engl.*, **31**, 224.

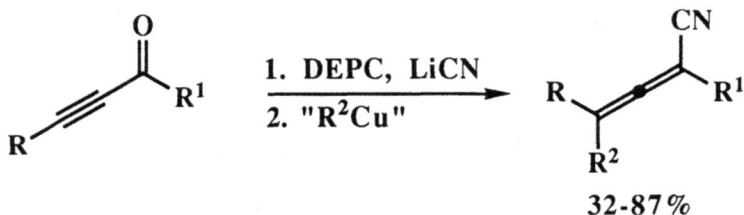

**I.B.5-4**  T. Konoike and Y. Araki, *Tetrahedron Lett.*, **33**, 5093.

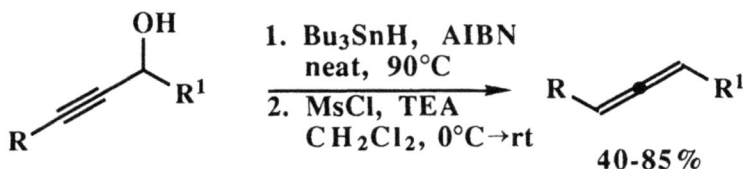

**I.B.5-5** D. Schinzer et al., *Tetrahedron Lett.*, **33**, 8017; M. Laabassi and R. Gree, *Bull. Soc. Chim. Fr.*, **129**, 151; see also: C.R. Bertozzi and M.D. Bednarski, *Tetrahedron Lett.*, **33**, 3109.

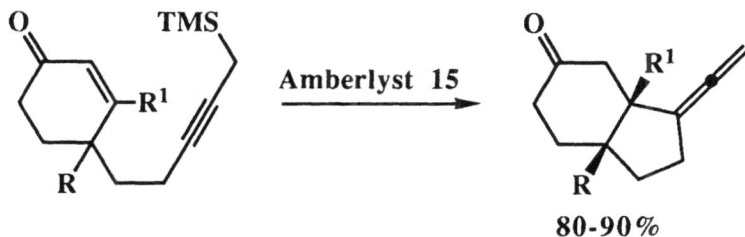

80-90%

**I.B.5-6** R. Herges and C. Hoock, *Angew. Chem., Int. Ed. Engl.*, **31**, 1611.

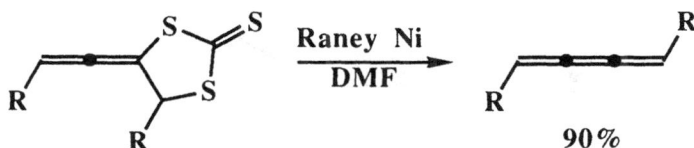

90%

**I.B.5-7** T.G. Back and B.P. Dyck, *Tetrahedron Lett.*, **33**, 4725.

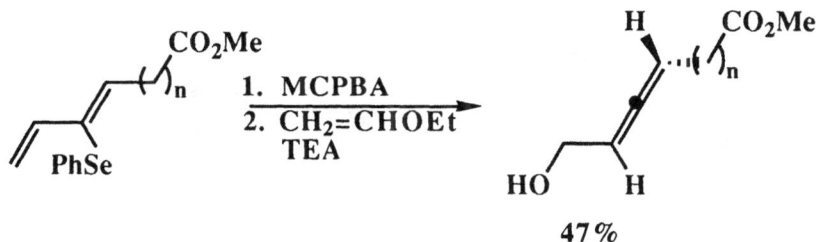

47%

## I.C.  Carbon-Carbon  Triple  Bonds

**I.C.1** P. Vinczer et al., *Org. Prep. Proced. Int.*, **24**, 540; H.L. Holland et al., *Synth. Commun.*, **22**, 1473.

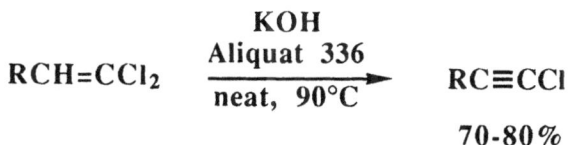

$$RCH=CCl_2 \xrightarrow[\text{neat, 90°C}]{\substack{\text{KOH} \\ \text{Aliquat 336}}} RC\equiv CCl$$

70-80%

**I.C-2** J.S. Yadav et al., *Tetrahedron Lett.*, **33**, 135.

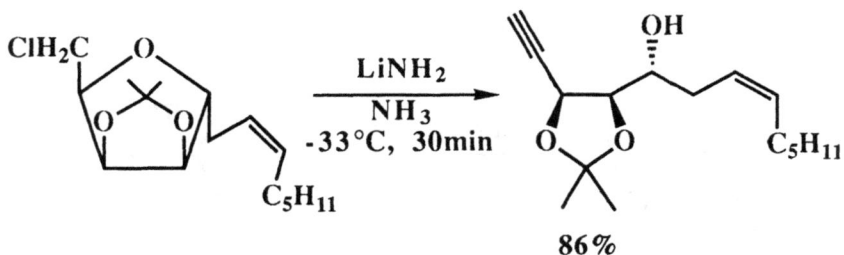

86%

**I.C-3** S. Shiotani and H. Morita, *J. Heterocyclic Chem.*, **29**, 413; A. Lottler and O. Himbert, *Synthesis,* 495.

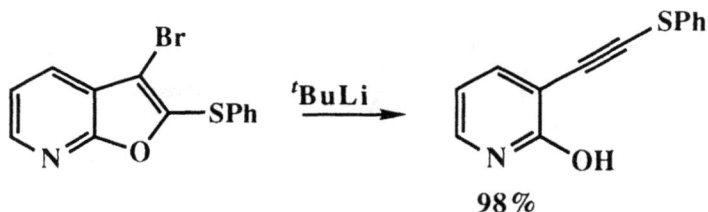

98%

**I.C-4** K.-D. Roth, *Synlett.*, 435.

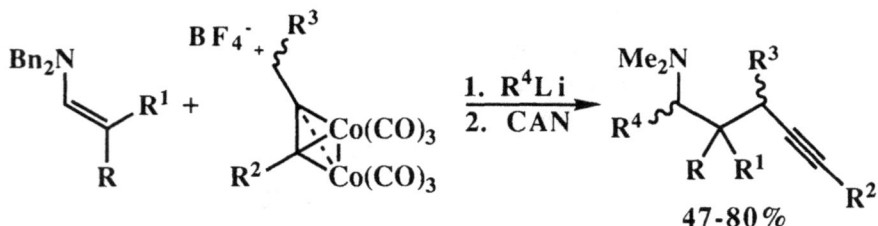

47-80%

**I.C-5** J.R. Hauske et al., *Tetrahedron Lett.*, **33**, 3715; S. Ohira et al., *Chem. Commun.*, 721.

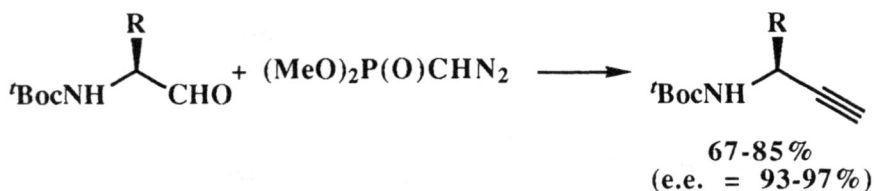

67-85%
(e.e. = 93-97%)

**I.C.6** M. D'Auria and D. Tofani, *Tetrahedron*, **48**, 9315.

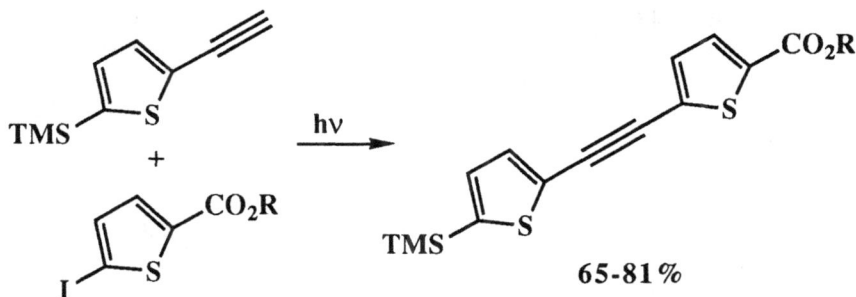

65-81%

**I.C-7** B.J. Wakefield et al., *J. Chem. Res. (S)*, 128.

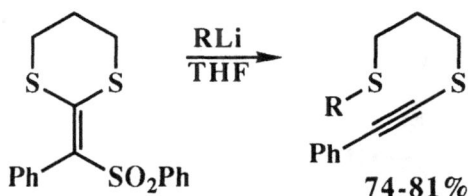

74-81%

**I.C-8** U. Bunz et al., *Angew. Chem., Int. Ed. Engl.*, **31**, 1648; G. Balavoine et al., *J. Organmet. Chem.*, **425**, 113; M.C. Clasby and D. Craig, *Synlett.*, 825.

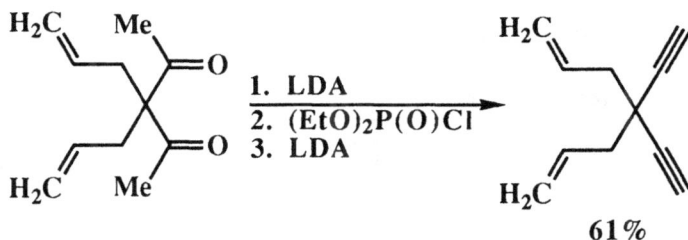

1. LDA
2. (EtO)₂P(O)Cl
3. LDA

61%

**I.C-9** A.R. Katritzky and M.F. Gordeev, *J. Chem. Soc., Perkin Trans. 1*, 1295.

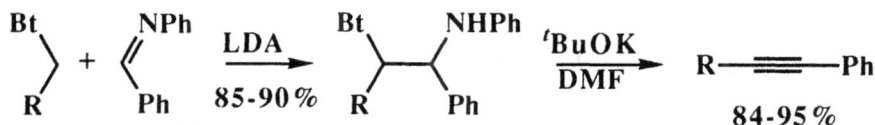

85-90%

'BuOK
DMF

84-95%

**I.C-10** G.T. Crisp and T.A. Robertson, *Tetrahedron*, 48, 3239; B. Das and N.G. Kundu, *J. Chem. Res. (S)*, 364; D. Chemin and G. Linstrumelle, *Tetrahedron*, **48**, 1943; S. Terashima et al., *ibid.*, **48**, 633; Q.-Y. Chen and Z.-T. Li, *J. Chem. Soc., Perkin Trans. 1*, 2931; W.H. Okamura et al., *J. Org. Chem.*, **57**, 4374; R. Bruckner, J. Suffert et al., *Chem. Ber.*, **125**, 271; H. Yamanaka et al., *Synthesis*, 746; R. Bose, K.P.C. Vollhardt et al., *Angew. Chem., Int. Ed. Engl.*, **31**, 1643.

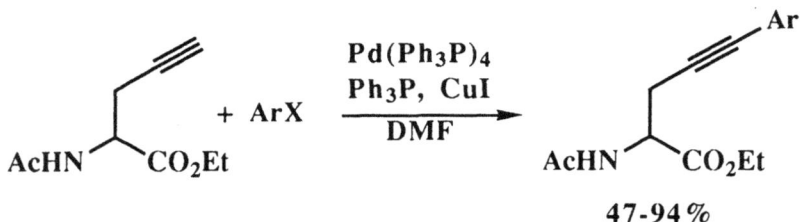

**I.C.11** J.-P. Quintard et al., *Tetrahedron Lett.*, **33**, 3647.

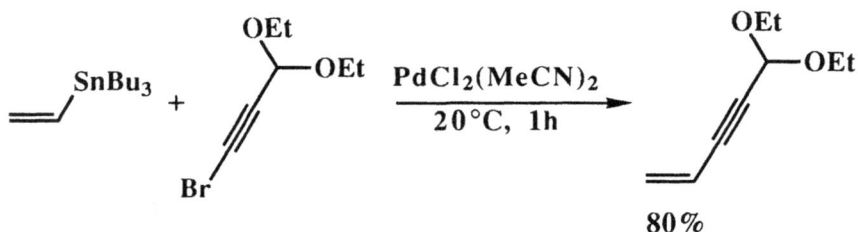

**I.C-12** S. Tani et al., *Chem. Commun.*, 219.

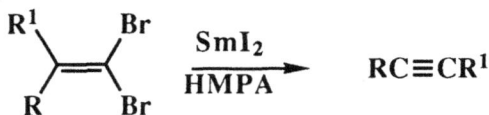

**I.C-13** T. Jeffery et al., *Tetrahedron Lett.*, **33**, 5757; V.V. Grushin and H. Alper, *J. Org. Chem.*, **57**, 2188; M. Miara et al., *Tetrahedron Lett.*, **33**, 5363.

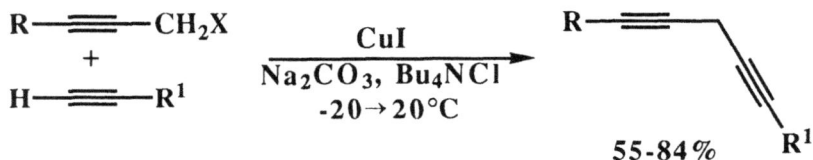

**I.C-14** T. Tsukiyama and M. Isobe, *Tetrahedron Lett.*, **33**, 7911.

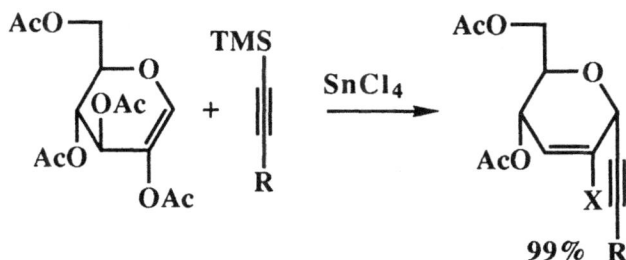

## I.D.   Cyclopropanations

### I.D.1  Carbene or Carbenoid Additions to a Multiple Bond

**I.D.1-1**  G. Descotes, *Synlett.*, 119.

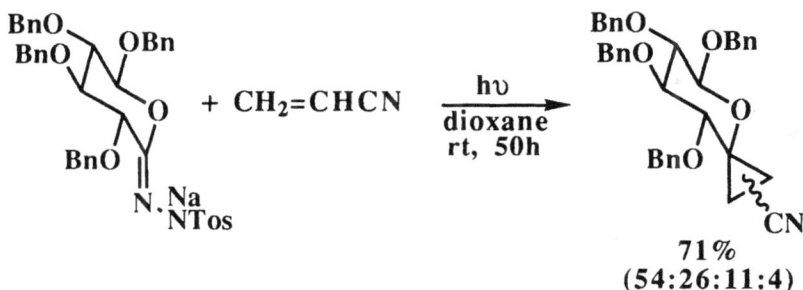

71%
(54:26:11:4)

---

**I.D.1-2** M.P. Doyle, P. Muller et al., *J. Am. Chem. Soc.*, **114**, 2755; O.M. Nefedov et al., *Org. Prep. Proced. Int.*, **24**, 515; K. Pauliniand H.-U. Reissig, *Synlett.*, 505.

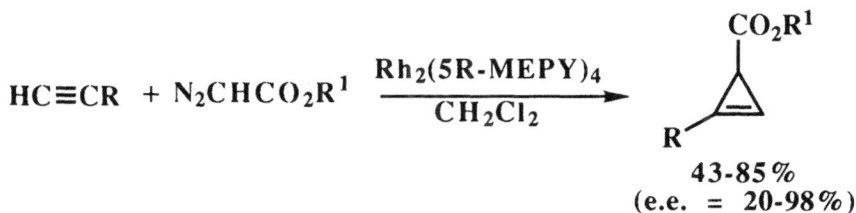

$$HC\equiv CR \ + \ N_2CHCO_2R^1 \xrightarrow[CH_2Cl_2]{Rh_2(5R\text{-}MEPY)_4}$$

43-85%
(e.e. = 20-98%)

**I.D.1-3** D.F. Taber and R.S. Hoerrner, *J. Org. Chem.*, **57**, 441; A. Padwa, M.P. Doyle et al., *J. Am. Chem. Soc.*, **114**, 1874; S.F. Martin et al., *Tetrahedron Lett.*, **33**, 6727; P. Ceccherelli et al., *Tetrahedron*, **48**, 9767;H.R. Sonawane et al., *Ind. J. Chem.*, **31B**, 606; see also: D. Demonceau et al., *Tetrahedron Lett.*, **33**, 2009; I. Basnak et al., *Synth. Commun.*, **22**, 773; see also: D.A. Evans et al., *Angew. Chem., Int. Ed. Engl.*, **31**, 430; M.M. Hossain et al., *Tetrahedron Lett.*, **33**, 7755; D.S. Iyengar et al., *J. Org. Chem.*, **57**, 6684.

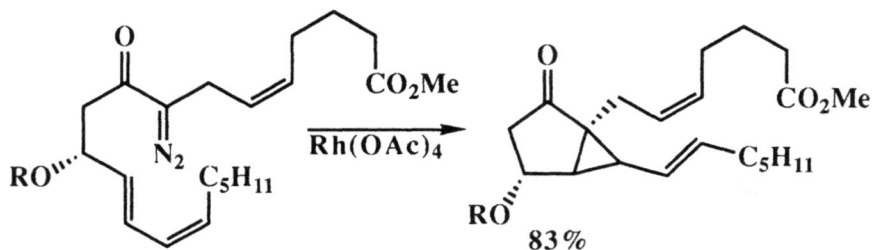

**I.D.1-4** M. Buchert and H.-U. Reissig, *Chem. Ber.*, **125**, 2723; D.L. Harvey et al., *J. Am. Chem. Soc.*, **114**, 8425; see also: D.F. Harvey and M.F. Brown, *J. Org. Chem.*, **57**, 5559; see also: H. Fischer et al., *J. Organmet. Chem.*, **427**, 63.

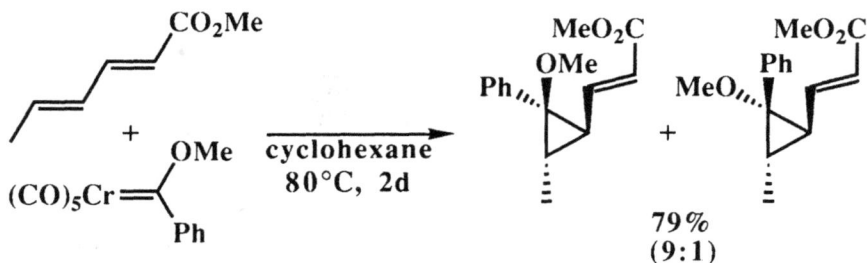

**I.D.1-5** S. Kobayashi et al., *Tetrahedron Lett.*, **33**, 2575; T. Fujisawa et al., *Chem. Lett.*, 61; S. Terashima et al., *Tetrahedron Lett.*, **33**, 3483, 3487; Y. Stenstrom et al., *Synth. Commun.*, **22**, 2801; S.R. Wilson et al., *J. Org. Chem.*, **57**, 2007.

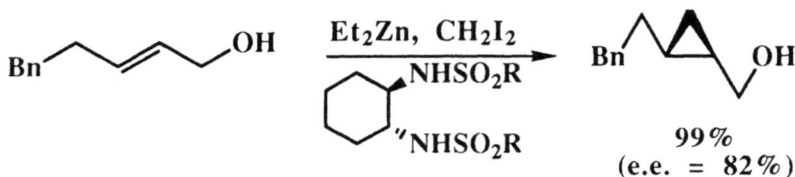

**I.D.1-6** C.P. Casey and L.J.S. Vosejpka, *Organometallics*, **11**, 738; see also: V. Guerchais et al., *ibid.*, **11**, 3926.

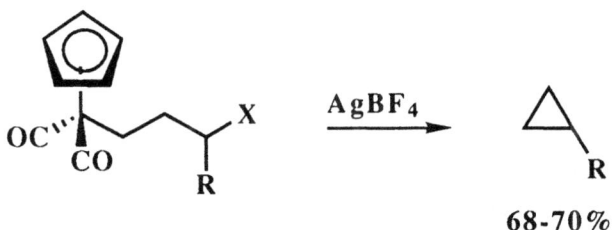

68-70%

---

**I.D.1-7** W.B. Motherwell and L.R. Roberts, *Chem. Commun.*, 1582.

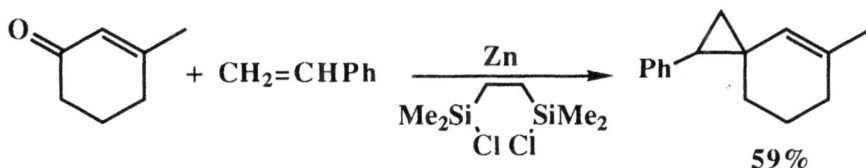

59%

---

**I.D.1-8** T.S. Kuznetsova et al., *J. Org. Chem. (USSR)*, **28**, 256.

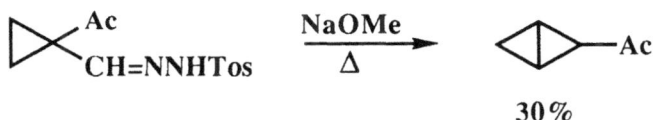

30%

## I.D.2  Other  Cyclopropanations

**I.D.2-1** J.K. Cha et al., *Tetrahedron Lett.*, **33**, 4703; see also: A. de Meijere et al., *Synlett.*, 735; A. Bryson et al., *ibid.*, 723; J. Wicha et al., *Tetrahedron*, **48**, 10201.

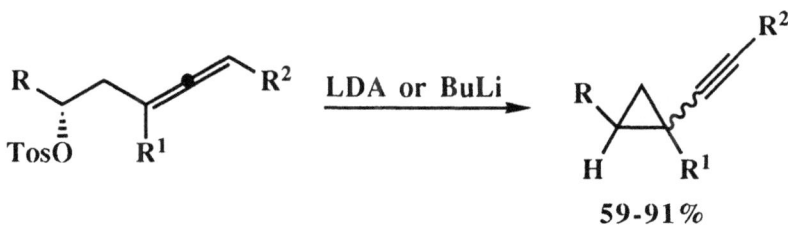

59-91%

**I.D.2-2** P.A. Aristoff and P.D. Johnson, *J. Org. Chem.*, **57**, 6234; D.L. Boger and M.S.S. Palanki, *J. Am. Chem. Soc.*, **114**, 9318.

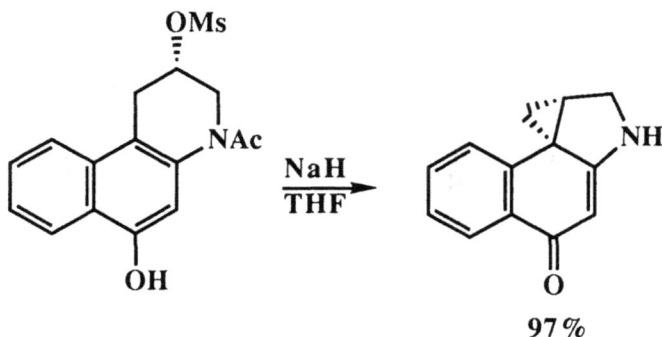

97%

**I.D.2-3** C. Inata et al., *Chem. Commun.*, 269; C. Yam et al., *Synth. Commun.*, **22**, 1651; Q. Zhong et al., *ibid.*, **22**, 489; K. Prasad et al., *J. Org. Chem.*, **57**, 6344; see also: K. Tanaka and H. Suzuki, *J. Chem. Soc., Perkin Trans. 1*, 2071; A.L.J. Beckwith and M.J, Tozer, *Tetrahedron Lett.*, **33**, 4975; see also: I. Yamamoto et al., *J. Chem. Res. (S)*, 84.

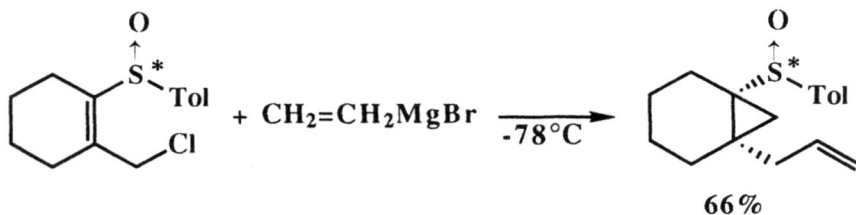

66%

**I.D.2-4** D. Romo and A.I. Meyers, *J. Org. Chem.*, **57**, 6265; D.B. Reddy et al., *Ind. J. Chem.*, **31B**, 503, 620; see also: C. Hamdouchi, *Tetrahedron Lett.*, **33**, 1701.

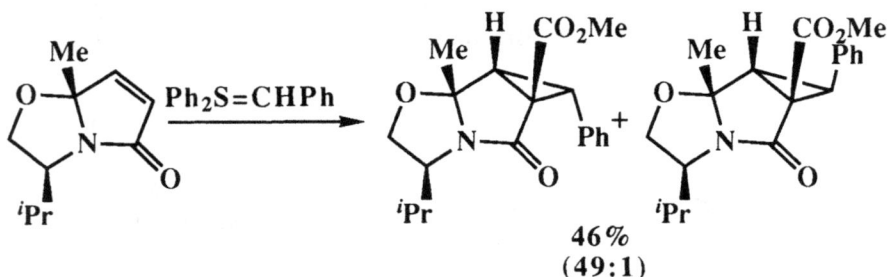

46%
(49:1)

**I.D.2-5** Y. Tamaru et al., *Tetrahedron Lett.*, **33**, 785; J.-L. Luche, *Chem. Commun.*, 798.

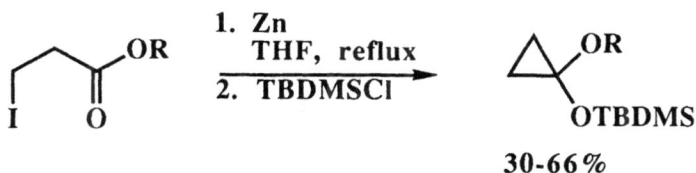

**30-66%**

---

**I.D.2-6** A. Krief and M. Hobe, *Tetrahedron Lett.*, **33**, 6527, 6529 and *Synlett.*, 317.

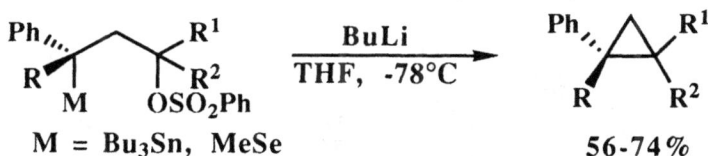

M = Bu₃Sn, MeSe

**56-74%**

---

**I.D.2-7** B.R. Davis et al., *Aust. J. Chem.*, **45**, 865.

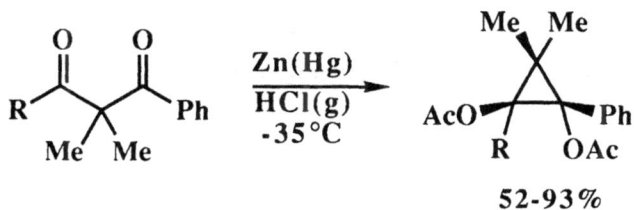

**52-93%**

---

**I.D.2-8** W. Zhang and P. Dowd, *Tetrahedron Lett.*, **33**, 7307.

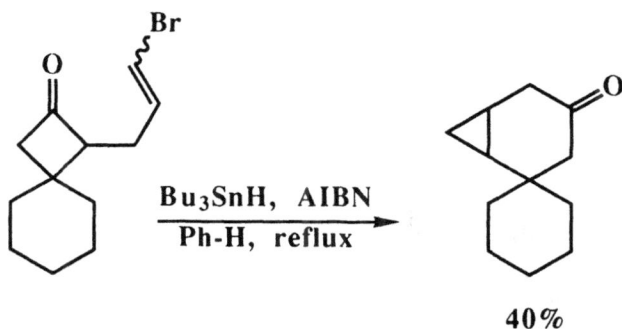

**40%**

**I.D.2-9** T. Yamamoto et al., *Org. Prep. Proced. Int.*, **24**, 548.

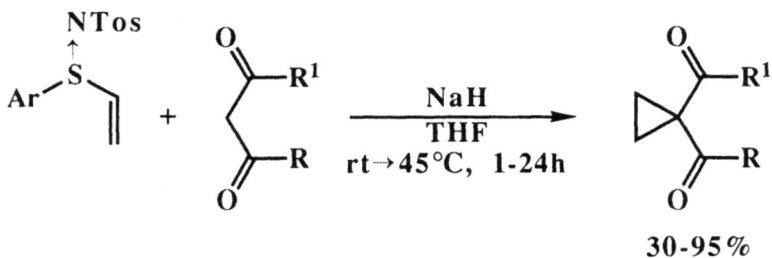

30-95%

---

**I.D.2-10** S. Yamazaki et al., *J. Org. Chem.*, **57**, 4.

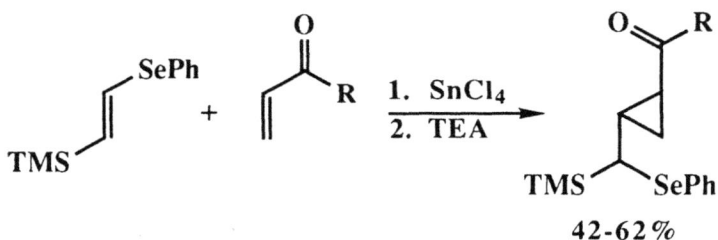

42-62%

---

**I.D.2-11** H.M.R. Hoffman et al., *Angew. Chem., Int. Ed. Engl.*, **31**, 234.

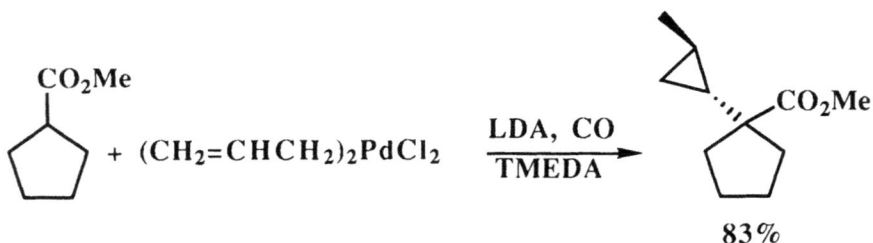

83%

---

**I.D.2-12** E. Nigishi et al., *J. Am. Chem. Soc.*, **114**, 10091.

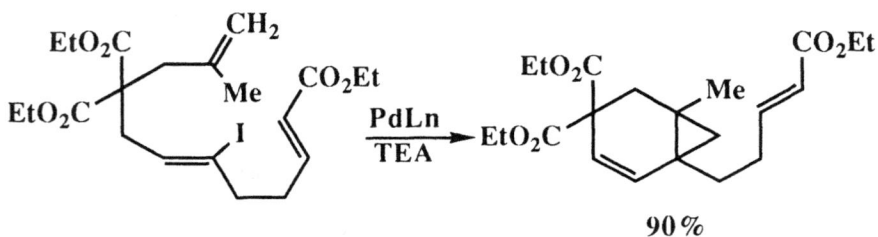

90%

## I.E   Thermal and Photochemical Reactions

### I.E.1.   Cycloadditions

**I.E.1-1**  E. Winterfeldt and V. Wray, *Chem. Ber.*, **125**, 2159; Y. Tsuda et al., *Heterocycles*, **33**, 497; J. d'Angelo et al., *Tetrahedron Lett.*, **33**, 1289.

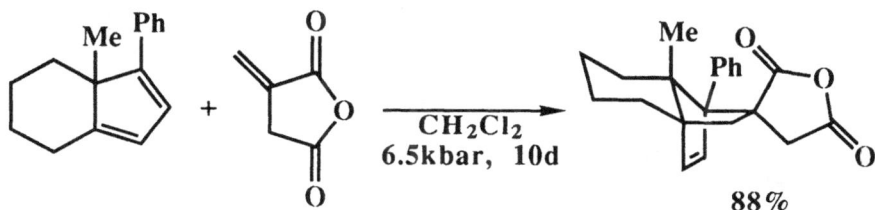

**I.E.1-2**  J.K. Snyder et al., *J. Org. Chem.*, **57**, 5301; W.-B. Wang and E.J. Roskamp, *Tetrahedron Lett.*, **33**, 7631.

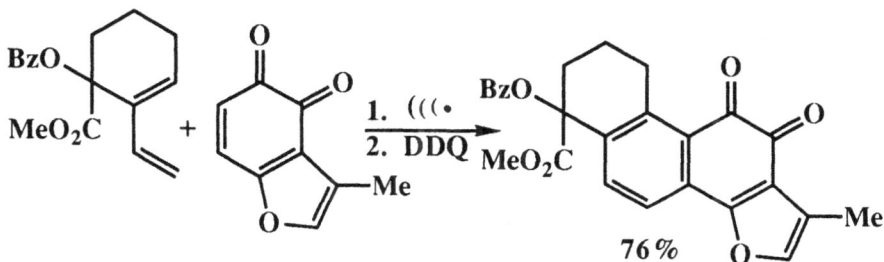

**I.E.1-3**  R.L. Danheiser et al., *Tetrahedron Lett.*, **33**, 1149.

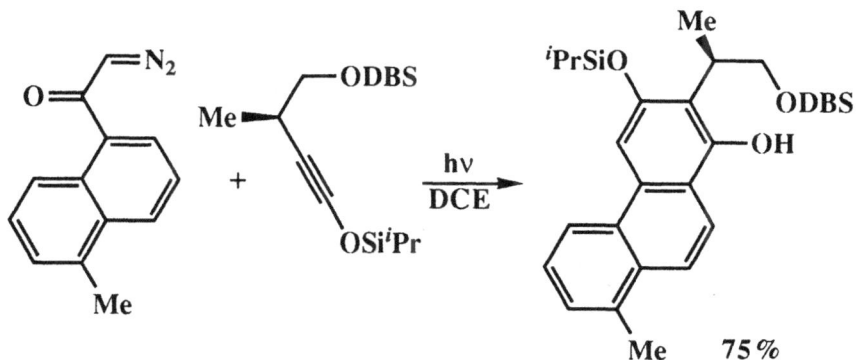

**I.E.1-4** R.C. Storr et al., *Tetrahedron*, **48**, 8101; T. Chou et al., *Chem. Commun.*, 1643.

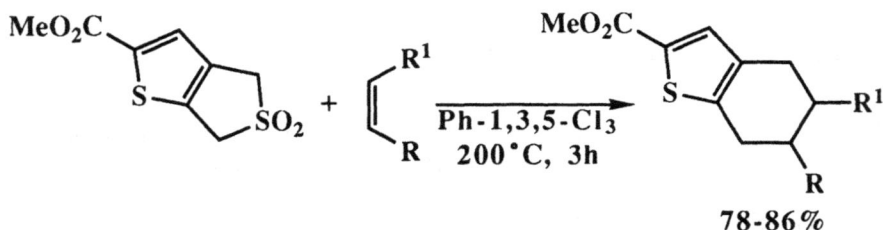

78-86%

---

**I.E.1-5** S.Z. Zard et al., *Tetrahedron Lett.*, **33**, 7853.

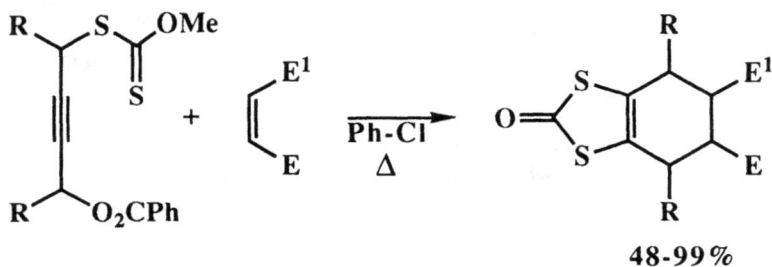

48-99%

---

**I.E.1-6** M. Yamashita et al., *Chem. Lett.*, 1201.

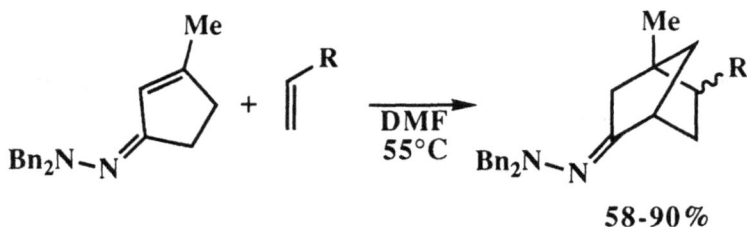

58-90%

---

**I.E.1-7** J. Inanaga et al., *Tetrahedron Lett.*, **33**, 7035.

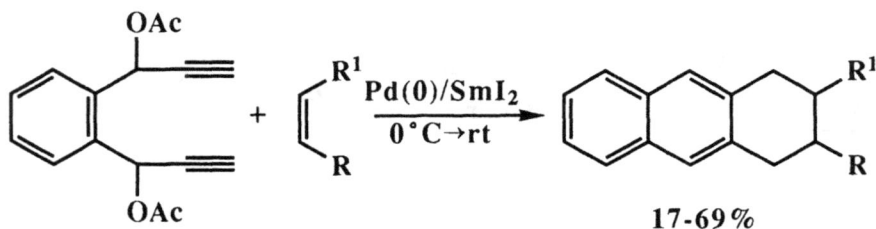

17-69%

**I.E.1-8** G. Bobowski et al., *J. Heterocyclic Chem.*, **29**, 33; E. Roman et al., *J. Chem. Soc., Perkin Trans. 1*, 941.

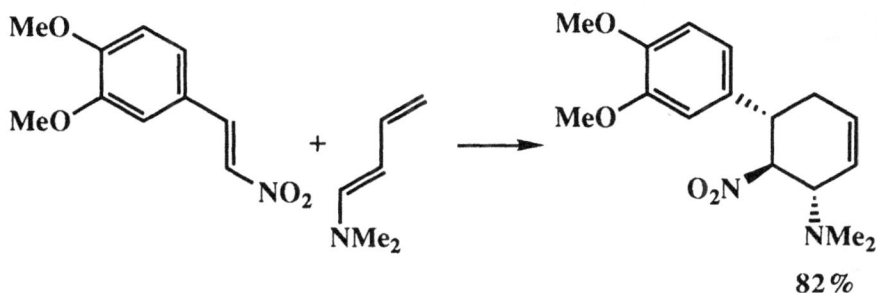

MeO, MeO — CH=CH—NO$_2$  +  diene with NMe$_2$  ⟶  MeO, MeO-substituted cyclohexene with O$_2$N and NMe$_2$

**82%**

---

**I.E.1-9** R.S. Mali and P.G. Jagtop, *Tetrahedron Lett.*, **33**, 1655.

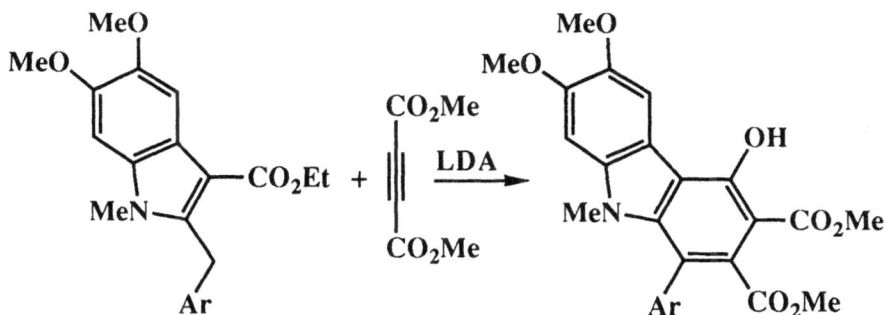

MeO, MeO-indole (MeN, CH$_2$Ar) —CO$_2$Et  +  CO$_2$Me / CO$_2$Me (alkyne)  $\xrightarrow{\text{LDA}}$  carbazole product: MeO, MeO, OH, MeN, Ar, —CO$_2$Me, CO$_2$Me

---

**I.E.1-10** M. Christl et al., *Chem. Ber.*, **125**, 1913; see also: P.M. Jackson and C.J. Moody, *Tetrahedron*, **48**, 7447; see also: K. Afarinkia and G.H. Posner, *Tetrahedron Lett.*, **33**, 7839.

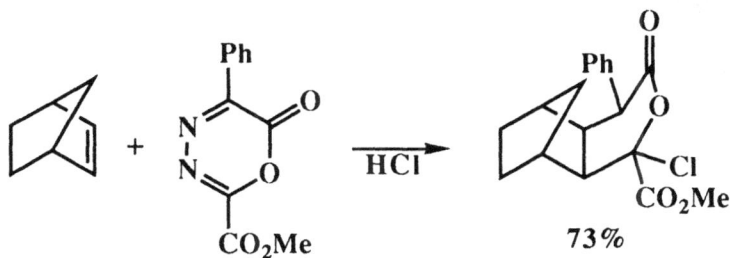

norbornene  +  oxadiazinone (Ph, O, N, N, O, CO$_2$Me)  $\xrightarrow{\text{HCl}}$  bicyclic product (Ph, O, O, Cl, CO$_2$Me)

**73%**

**I.E.1-11** R.A. Olofson et al., *J. Org. Chem.*, **57**, 7122.

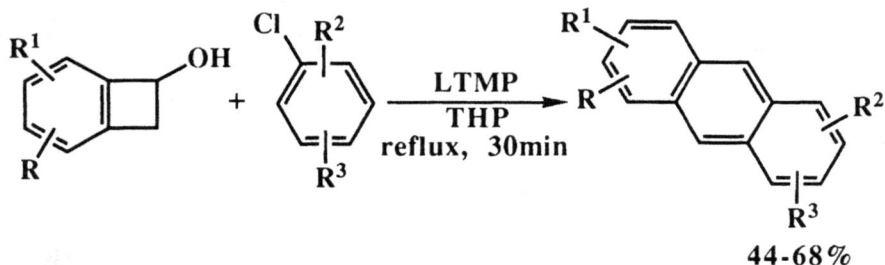

44-68%

---

**I.E.1-12** R.N. Warrener et al., *Aust. J. Chem.*, **45**, 1035; J.L. Bloomer and M.E. Lankin, *Tetrahedron Lett.*, **33**, 2769; see also: J. Leroy, *ibid.*, **33**, 2969; M. Tsukazaki and V. Snieckus, *Heterocycles,* **33**, 533.

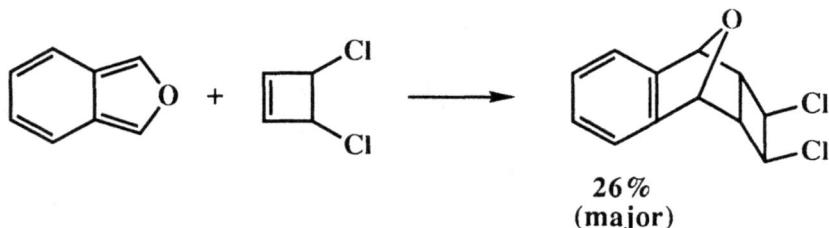

26%
(major)

---

**I.E.1-13** S. Hashmi and G. Szeimies, *Chem. Ber.*, **125**, 1769; M.B. Smith et al., *Chem. Lett.*, 2451; N. Ruiz et al., *Tetrahedron Lett.*, **33**, 2965; D.A. Jaeger and J. Wang, *ibid.*, **33**, 6415; D.A. Singleton and S.-W. Leung, *J. Org. Chem.*, **57**, 4796; S.-J. Lee, T.-S. Chou et al., *Chem. Ber.*, **125**, 499; K.C. Nicolaou et al., *Chem. Commun.*, 1117; S. Takano et al., *ibid.*, 953; J.M. Percy and M.H. Rock, *Tetrahedron Lett.*, **33**, 6177; S. McN. Sieburth and L. Fensterbank, *J. Org. Chem.*, **57**, 5279; A. Riahi et al., *Rec. trav. Chim.*, **111**, 345; M. Hasegawa et al., *Chem. Lett.*, 1353; J.A. Mayoral et al., *Tetrahedron*, **48**, 6467.

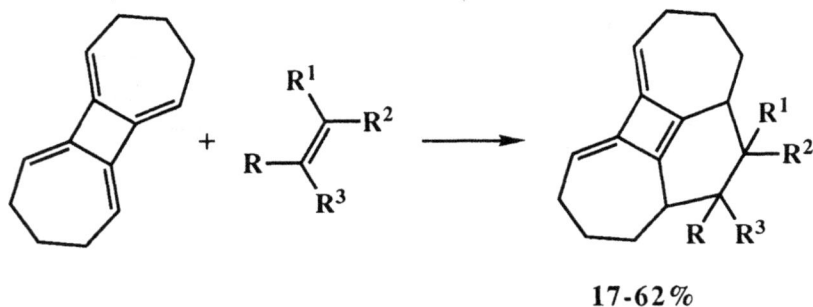

17-62%

**I.E.1-14** J. Blechert and T. Wirth, *Tetrahedron Lett.*, **33**, 6621.

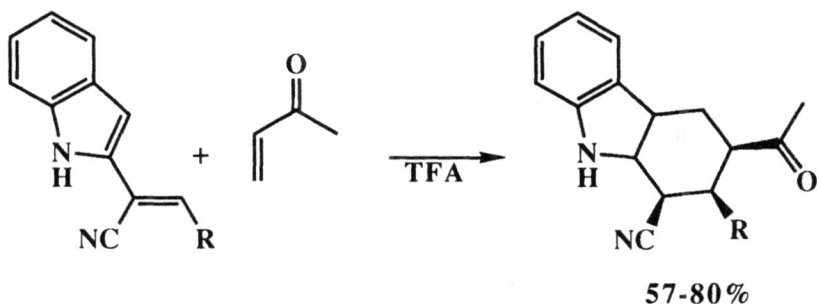

57-80%

**I.E.1-15** H.J. Liu et al., *Can. J. Chem.*, **70**, 1545; M.C. Carreno, J.L.G. Ruano and A. Urbano, *J. Org. Chem.*, **57**, 6870; C. Alexandre et al., *J. Chem. Res. (S)*, 48.

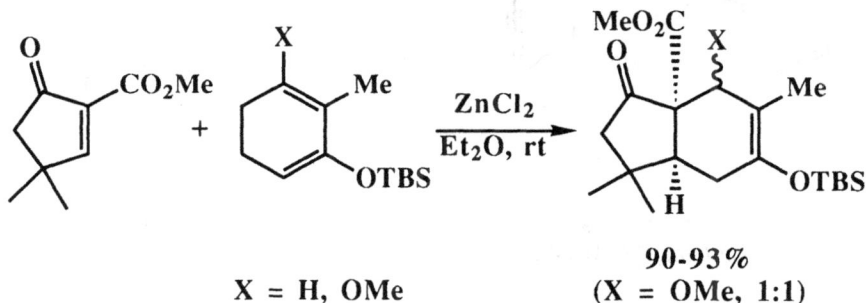

X = H, OMe

90-93%
(X = OMe, 1:1)

**I.E.1-16** V.K. Singh et al., *J. Chem. Soc., Perkin Trans. 1*, 903; V. Singh and B. Thomas, *Chem. Commun.*, 1211; see also: J.-M. Fary et al *J. Chem. Soc., Perkin Trans. 1*, 1209.

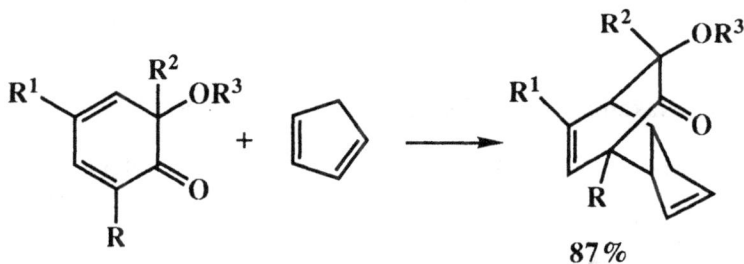

87%

**I.E.1-17** D.A. Singleton et al., *J. Org. Chem.*, **57**, 5768 and *Tetrahedron*, **48**, 5831 and *Tetrahedron Lett.*, **33**, 1017; see also: M. Vaultier et al., *Chem. Commun.*, 1105.

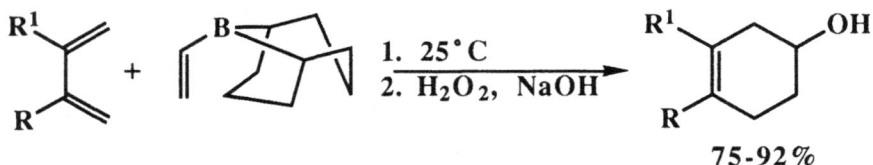

**I.E.1-18** R. Hirschmann, A.B. Smith, III et al., *J. Am. Chem. Soc.*, **114**, 9699; J.R. Gillardand and D.J. Burnell, *Can. J. Chem.*, **70**, 1296.

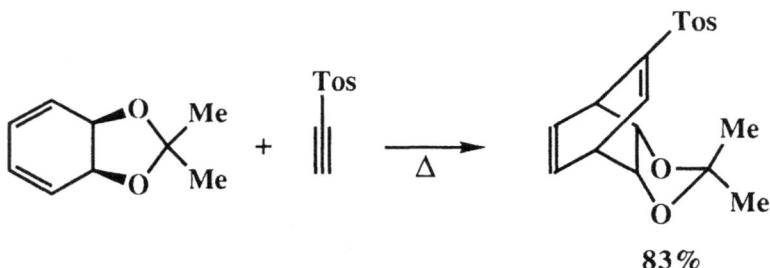

**I.E.1-19** R.P. Kreher et al., *Chem. Ber.*, **125**, 183; C.-K. Sha and J.-F. Yong, *Tetrahedron*, **48**, 10645.

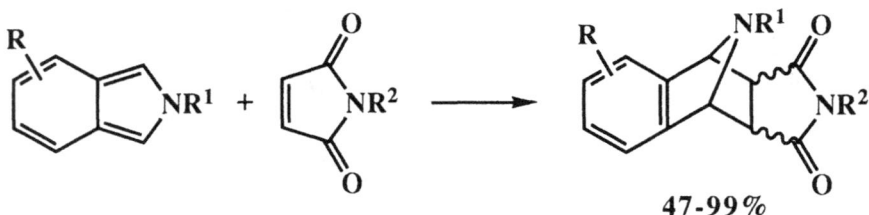

**I.E.1-20** H. Nakamura et al., *Tetrahedron Lett.*, **33**, 8113.

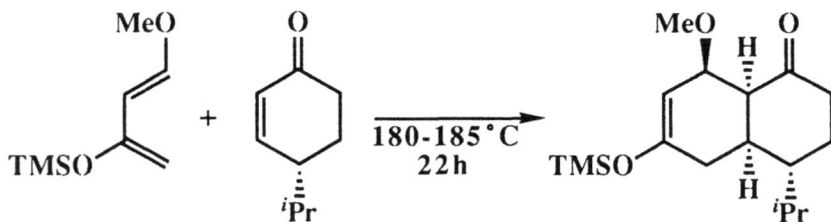

**I.E.1-21** H. Hongo et al., *Heterocycles,* **33,** 195.

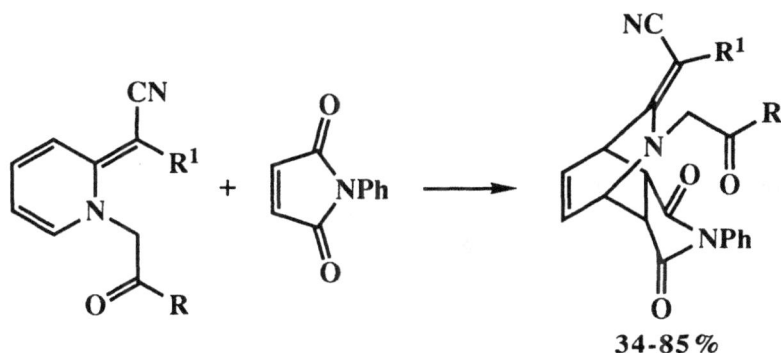

34-85%

---

**I.E.1-22** R.M. Letcher and T.-Y. Yue, *Chem. Commun.,* 1310; J. Backvall et al., *Tetrahedron Lett.,* **33,** 2417.

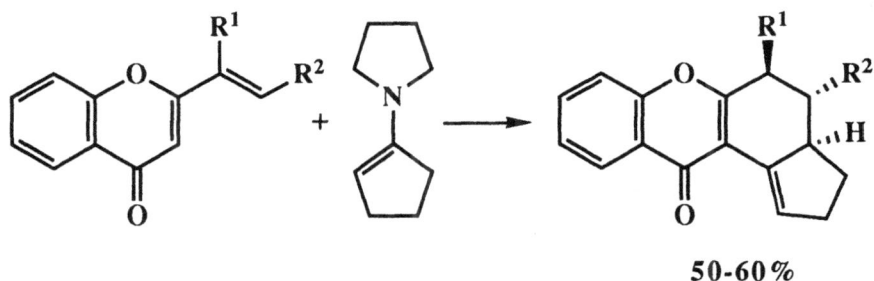

50-60%

---

**I.E.1-23** E.J. Corey and K. Ishihara, *Tetrahedron Lett.,* **33,** 6807; T.A. Engler et al., *ibid.,* **33,** 6731; J.A. Marshall and S. Xie, *J. Org. Chem.,* **57,** 2987; J. Kim and G.B. Shuster, *J. Am. Chem. Soc.,* **114,** 9309; H. Yamamoto et al., *Bull. Chem. Soc. Jpn.,* **65,** 3501.

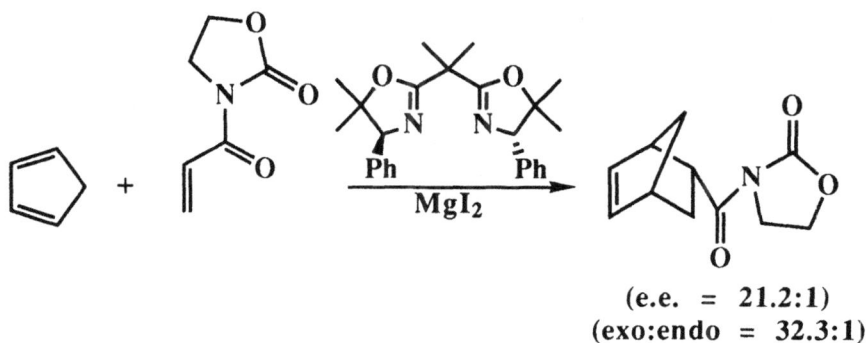

(e.e. = 21.2:1)
(exo:endo = 32.3:1)

**I.E.1-24** K. Kanematsu et al., *Tetrahedron Lett.*, **33**, 5787; R. Bloch and N. Chaptal-Gradoz, *ibid.*, **33**, 6147; D.E. Ward and Y. Gai, *ibid.*, **33**, 1851; H. Yamamoto et al., *J. Am. Chem. Soc.*, **114**, 1089; R. Klein et al., *J. Organmet. Chem.*, **436**, 143; see also: B. Bosnich et al, *Organometallics*, **11**, 2745.

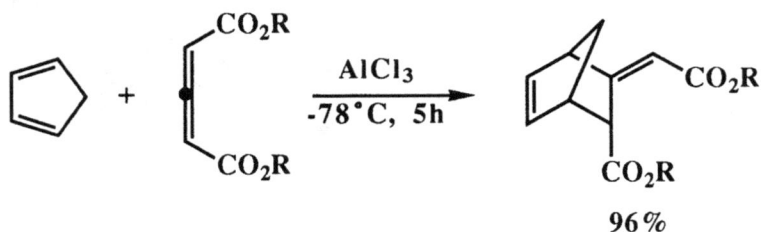

96%

---

**I.E.1-25** (A) J.-L. Gras et al., *Tetrahedron Lett.*, **33**, 3323; (B) M. Natsume et al., *ibid.*, **33**, 4595; (C) K.S. Kim et al., *ibid.*, **33**, 4029; (D) M.T. Reetz et al., *ibid.*, **33**, 3453; (E) O. De Lucchi et al., *Tetrahedron*, **48**, 1485; (F) B. Ronan and H.B. Kagan, *Tetrahedron Asymm.*, **3**, 115; (G) K. Fuji et al., *ibid.*, **3**, 609.

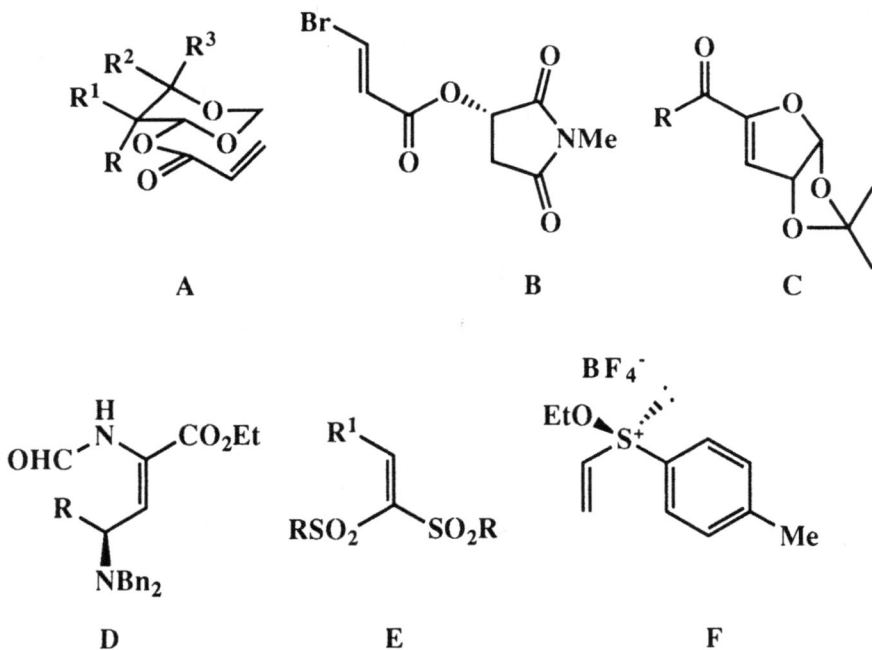

A          B          C

D          E          F

**I.E. 1-26** (A) T. Hierstetter et al., *Tetrahedron Lett.*, **33**, 8019; (B) S. Arseniyadis et al., *Tetrahedron*, **48**, 1255; (C) M. Tada and T. Shimizu, *Bull. Chem. Soc. Jpn.*, **65**, 1252; (D) H. Takayama et al. *Chem. Commun.*, 870; (E) H. Takayama et al., *ibid.*, 1100; (F) T. Chou and C.-Y. Tsai, *Tetrahedron Lett.*, **33**, 4201; (G) R.L. Jarvest and S.A. Readshaw, *Synthesis,* 962.

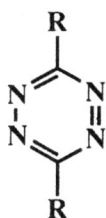

R = CF₃, CO₂Me

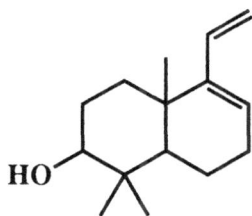

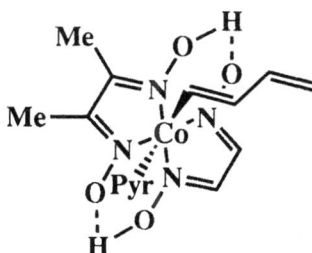

A                         B                         C

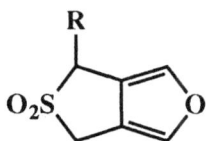

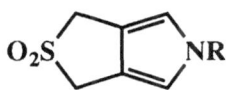

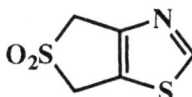

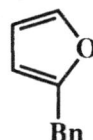

D                    E                    F                    G

---

**I.E.1-27** G.H. Posner et al., *J. Org. Chem.*, **57**, 7012; R. Nouguier et al., *Tetrahedron*, **48**, 6245; T. Koizumi et al., *Tetrahedron Asymm.*, **3**, 535; Z. Chen and R.M. Ortuno, *ibid.*, **3**, 621; R.K. Boeckman, Jr., et al., *J. Am. Chem. Soc.*, **114**, 2258; T. Kunieda et al., *Tetrahedron Lett.*, **33**, 4461; W.R. Roush and B.B. Brown, *J. Org. Chem.*, **57**, 3380.

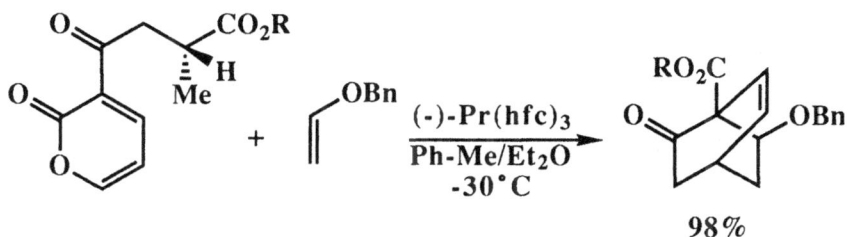

98%

**I.E.1-28** J.-C. Blazejewski et al., *Tetrahedron Lett.*, **33**, 1269; S.V. Kessar et al., *Ind. J. Chem.*, **31B**, 381; see also: C.W. Jefford, K.N. Houk et al., *J. Am. Chem. Soc.*, **114**, 1157.

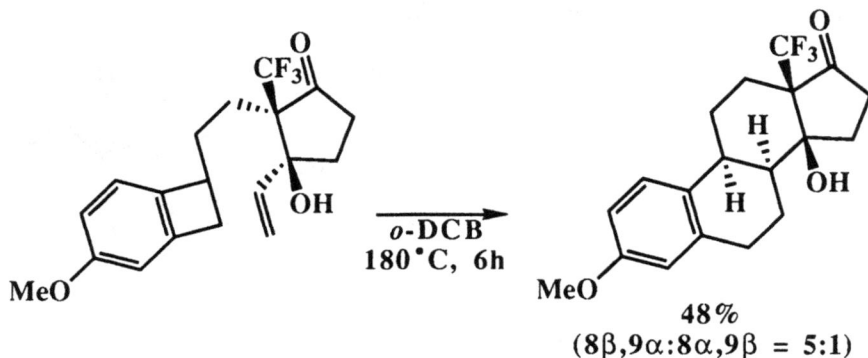

48%
(8β,9α:8α,9β = 5:1)

**I.E.1-29** D. Perez et al., *Tetrahedron Lett.*, **33**, 2407; see also: J.K. Snyder et al., *J. Org. Chem.*, **57**, 5285.

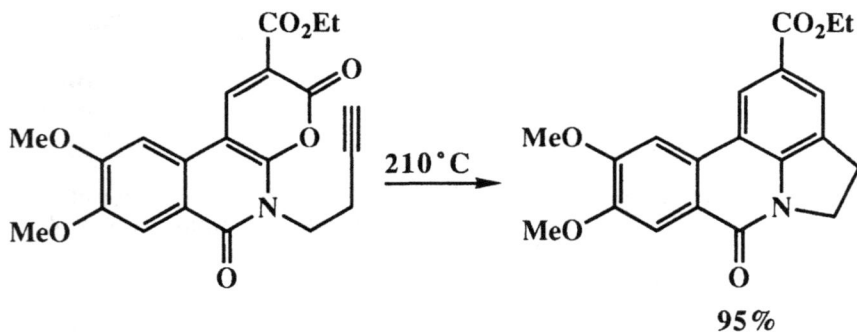

95%

**I.E.1-30** P. Magnus et al., *J. Am. Chem. Soc.*, **114**, 382.

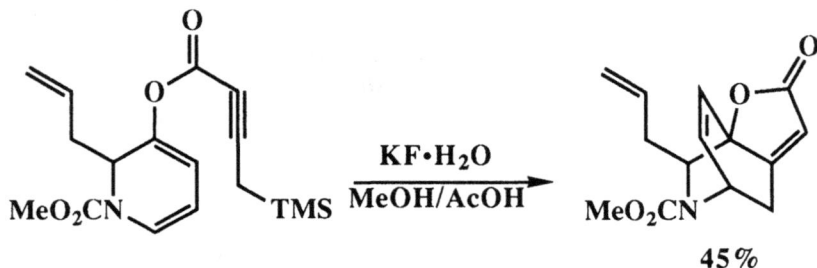

45%

**I.E.1-31** T. Hashimoto et al., *J. Med. Chem.*, **35**, 816.

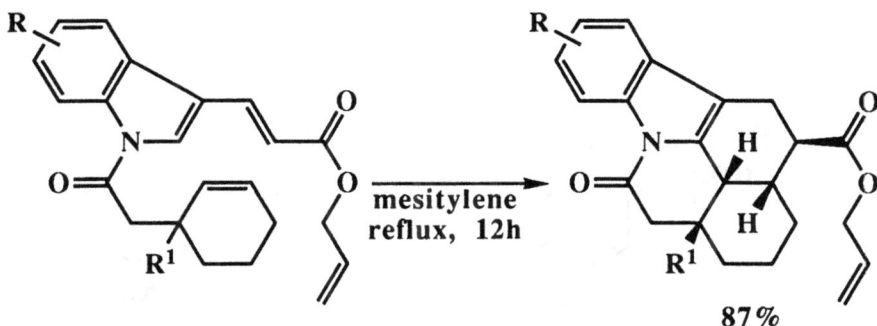

87%

**I.E.1-32** P. Deslongchamps et al., , *Can. J. Chem.*, **70**, 2350 and, *Tetrahedron Lett.*, **33**, 5217 and *Pure Appl. Chem.*, **64**, 1831.

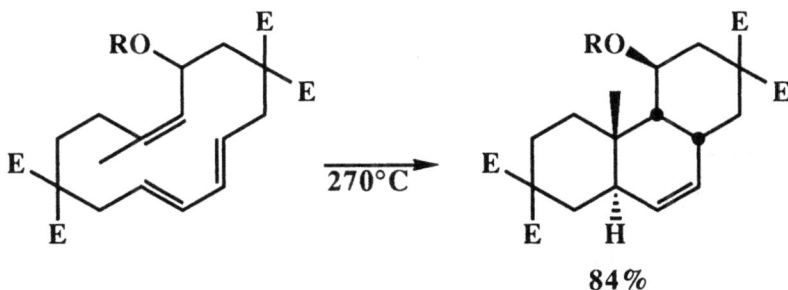

84%

**I.E.1-33** S.R. Wilson and L. Jacob, *J. Org. Chem.*, **57**, 4380; M. Hatanaka et al., *Chem. Commun.*, 1684; K. Fukumoto et al., *Tetrahedron Lett.*, **33**, 4581; K.J. Shea et al., *ibid.*, **33**, 4695; P.C.B. Page and D.C. Jennen, *J. Chem. Soc., Perkin Trans. 1*, 2587.

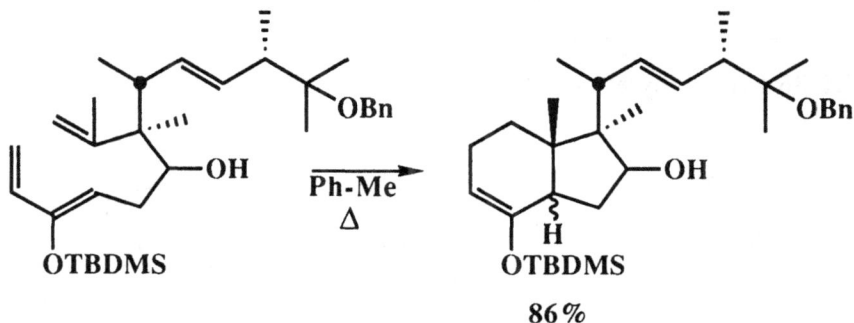

86%

**I.E.1-34** L.F. Tietze and P. Saling, *Synlett.*, 281.

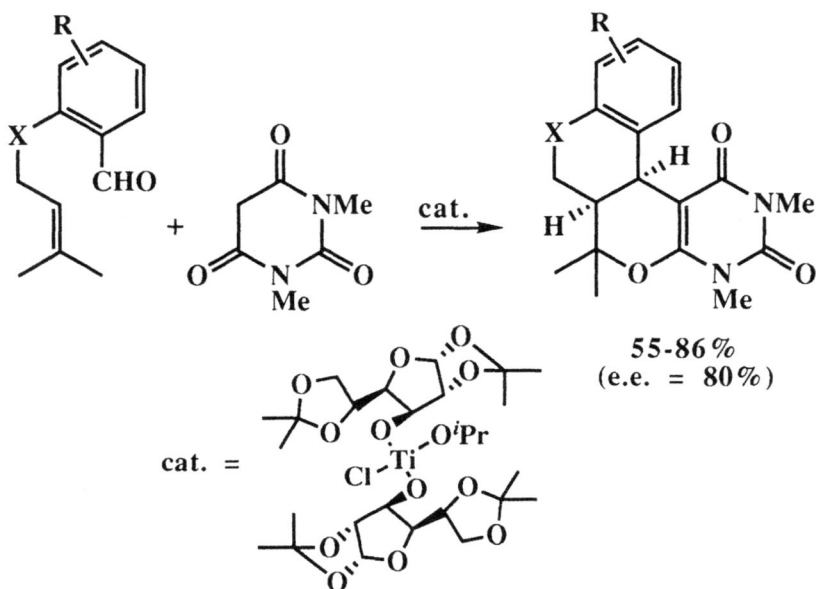

55-86%
(e.e. = 80%)

---

**I.E.1-35** K. Fukumoto et al., *J. Chem. Soc., Perkin Trans. 1*, 2527.

TBDMSOTf
TEA, CH$_2$Cl$_2$
rt, 45min

48%

---

**I.E.1-36** J.H. Rigby et al., *J. Org. Chem.*, **57**, 5290.

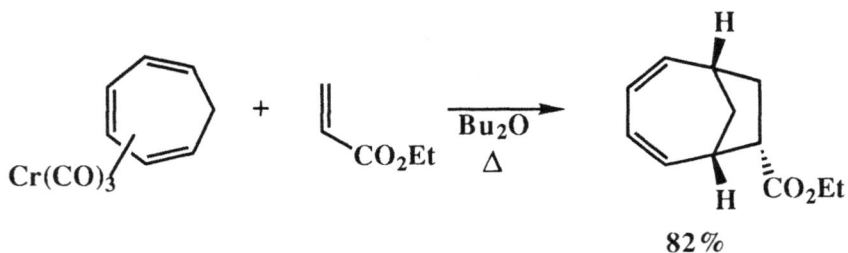

Bu$_2$O
Δ

82%

**I.E.1-37** M.C. Clasby and D. Craig, *Tetrahedron Lett.*, **33**, 3813; D. Craig and J.C. Reader, *ibid.*, **33**, 4073 and *Synlett.*, 757; G.W. Morrow et al., *Synth. Commun.*, **22**, 179.

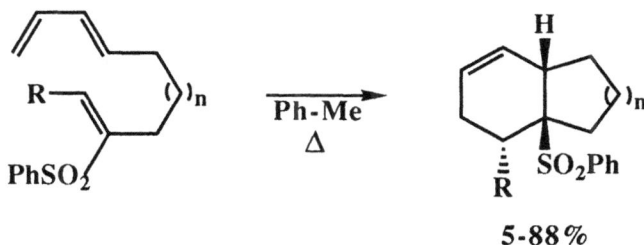

5-88%

**I.E.1-38** A. de Meijere et al., *Synlett.*, 521.

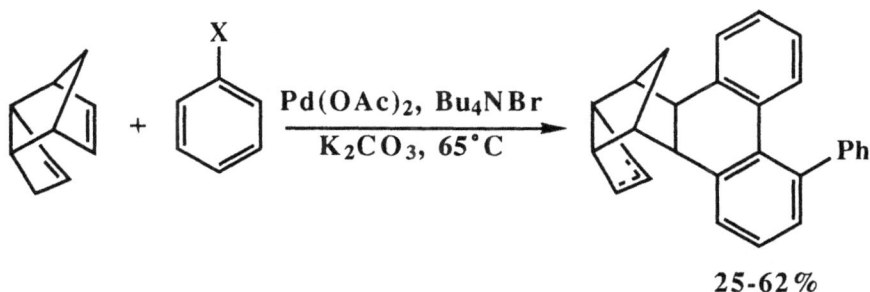

25-62%

**I.E.1-39** S. Rubinsztajn et al., *Tetrahedron Lett.*, **33**, 1821.

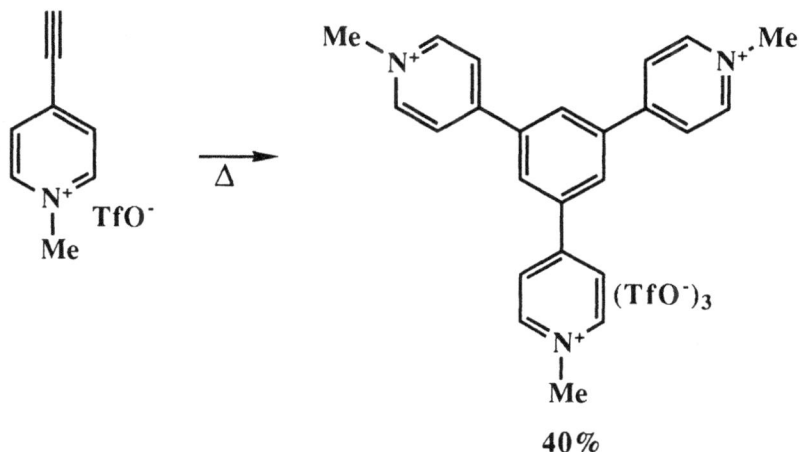

40%

**I.E.1-40** C.N. Lewis et al., *J. Org. Chem.*, **57**, 5596; K. Fukumoto et al., *J. Chem. Soc., Perkin Trans. 1*, 865.

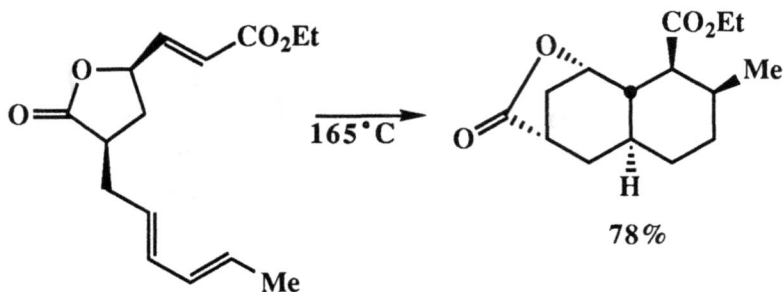

78%

**I.E.1-41** L. Ghosez et al., *Pure Appl. Chem.*, **64**, 1849; C.R. Johnson and R.L. De Jong, *J. Org. Chem.*, **57**, 594; K. Narasaka et al., *J. Am. Chem. Soc.*, **114**, 8869; T.-a. Mitsudo, Y. Watanabe et al., *Tetrahedron Lett.*, **33**, 5533.

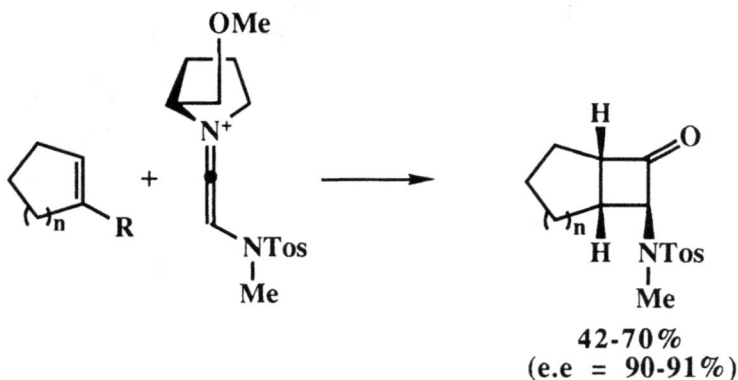

42-70%
(e.e = 90-91%)

**I.E.1-42** P.A. Wender and J.L. Mascarenas, *Tetrahedron Lett.*, **33**, 2115; M. Lautens et al., *J. Org. Chem.*, **57**, 8.

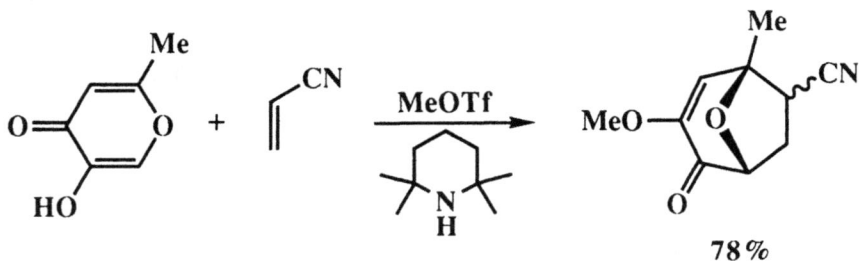

78%

**I.E.1-43** E. Nakamura et al., *J. Am. Chem. Soc.*, **114**, 8707.

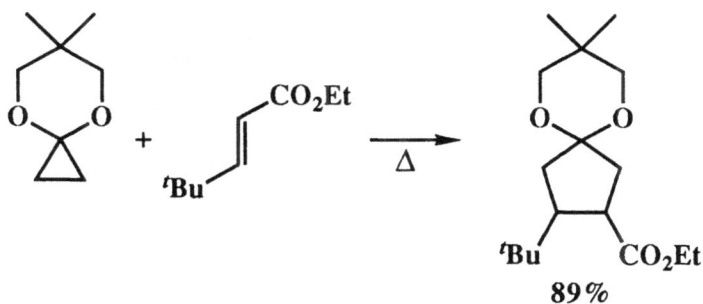

89%

---

**I.E.1-44** S.R. Angle and D.O. Arnaiz, *J. Org. Chem.*, **57**, 5937; C.-Y, Liu and S.-T. Ding, *ibid.*, **57**, 4539.

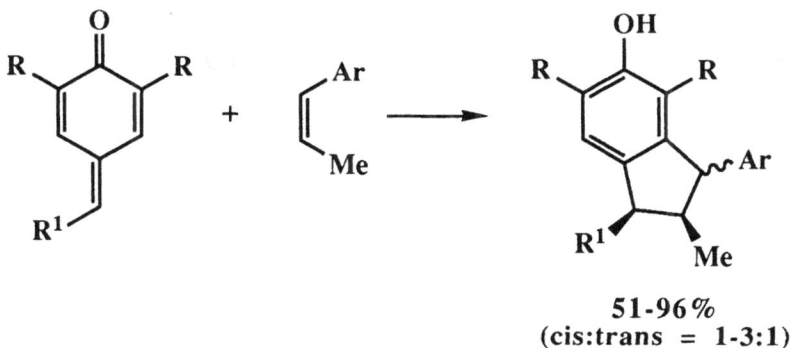

51-96%
(cis:trans = 1-3:1)

---

**I.E.1-45** A.J. Rippert and H.-J. Hansen, *Helv. Chim. Acta,* **75**, 2219, 2211, 2447, 2493.

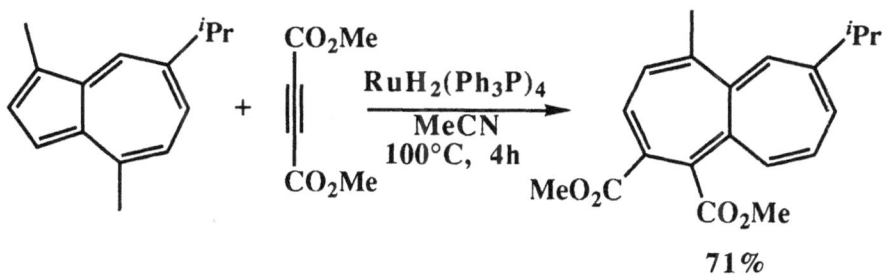

71%

**I.E.1-46** H.T. Dieck et al., *Angew. Chem., Int. Ed. Engl.*, **31**, 305.

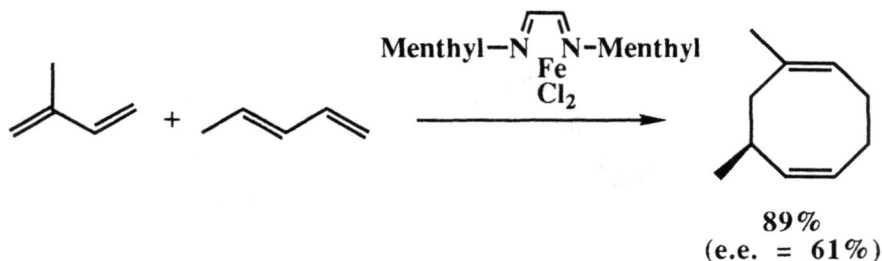

89%
(e.e. = 61%)

**I.E.1-47** E.V. Dehmlow and A.L. Veretenov, *Synthesis*, 939.

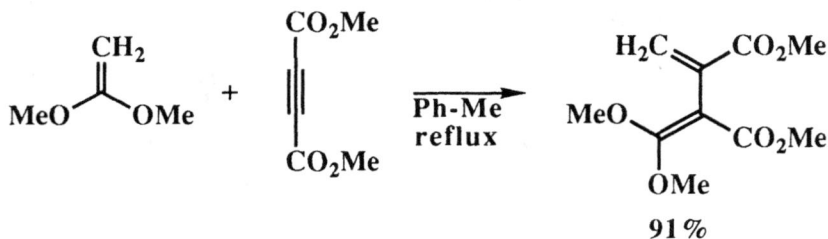

91%

**I.E.1-48** L.A. Paquette et al., *J. Am. Chem. Soc.*, **114**, 7387.

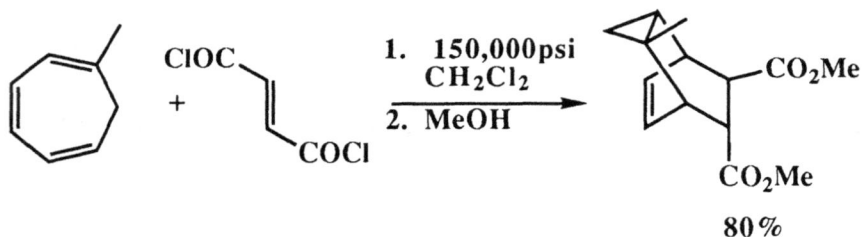

80%

**I.E.1-49** J.H. Rigby et al., *Tetrahedron Lett.*, **33**, 5873.

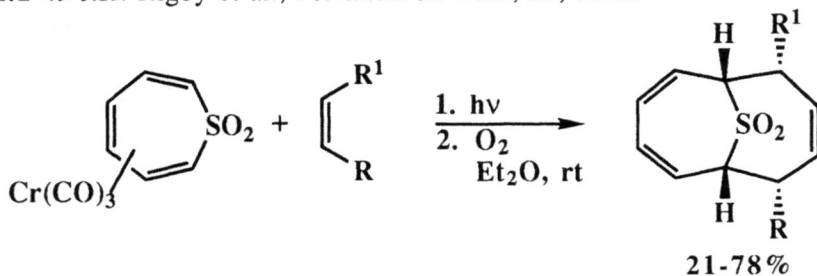

21-78%

## I.E.2.  Other  Thermal  Reactions

**I.E.2-1**  L.T.  Scott et al., *J. Am. Chem. Soc.*, **114**, 1920.

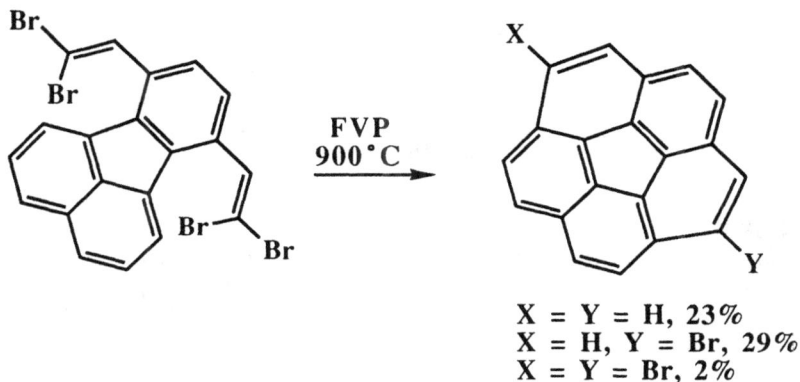

X = Y = H, 23%
X = H, Y = Br, 29%
X = Y = Br, 2%

---

**I.E.2-2**  M.  Yasunami et al., *Bull. Chem. Soc. Jpn.*, **65**, 1527.

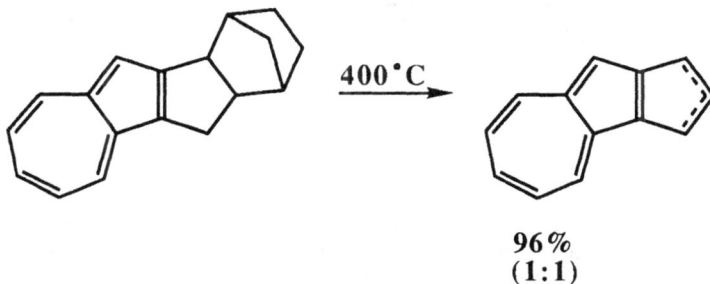

96%
(1:1)

---

**I.E.2-3**  G.  Wulff and P.  Birnbrick, *Chem. Ber.*, **125**, 473.

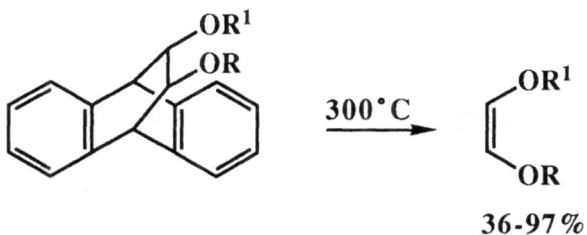

36-97%

**I.E.2-4** K. Banert and S. Groth, *Angew. Chem., Int. Ed. Engl.*, **31**, 866.

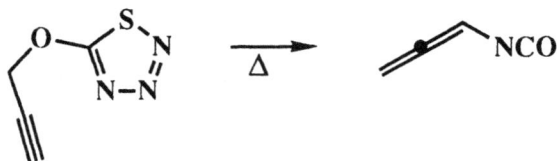

## I.E.3. Photochemical Reactions

**I.E.3-1** A.B. Smith, III et al., *J. Am. Chem. Soc.*, **114**, 2567 and *J. Chem. Soc., Perkin Trans. 1*, 979; F. Tode et al., *ibid.*, 307; M. Yasuda et al., *Tetrahedron Lett.*, **33**, 6465; M. Fetizon et al., *Synth. Commun.*, **22**, 245; H.-G. Henning and G. Mazunaitis, *Monatsh. Chem.*, **123**, 93; M. Buback, J. Bunger and L.F. Tietze, *Chem. Ber.*, **125**, 2577.

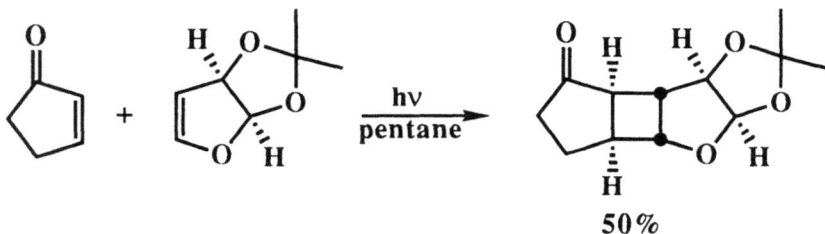

50%

**I.E.3-2** R. Keese et al., *Helv. Chim. Acta,* **75**, 1897 and *Tetrahedron Lett.*, 33, 3987; J. Mattay et al., *Chem. Ber.*, **125**, 2119 and *Liebigs Ann. Chem.*, 257; J.-P. Pete et al., *Tetrahedron Lett.*, **33**, 7347; J.D. White et al., *J. Am. Chem. Soc.*, **114**, 9673; S.A. Fleming and S.C. Ward, *Tetrahedron Lett.*, **33**, 1013; K. Somekawa et al., *J. Org. Chem.*, **57**, 5708.

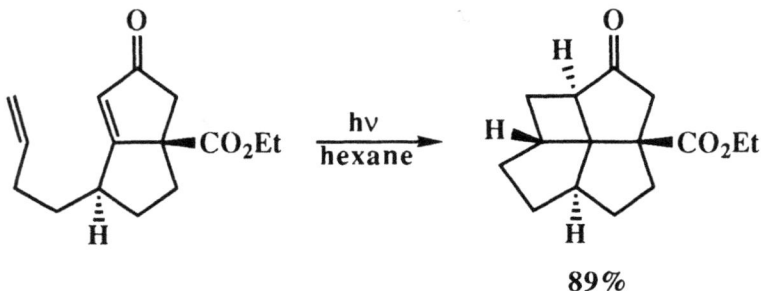

89%

**I.E.3-3** A.G. Schultz and J. Reilly, *J. Am. Chem. Soc.*, **114**, 5068.

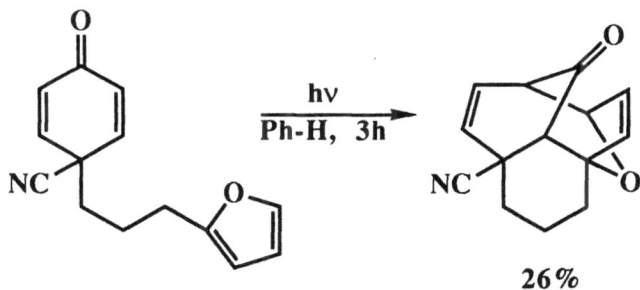

**26%**

---

**I.E.3-4** G. Pattenden and A.J. Smithies, *Synlett.*, 577.

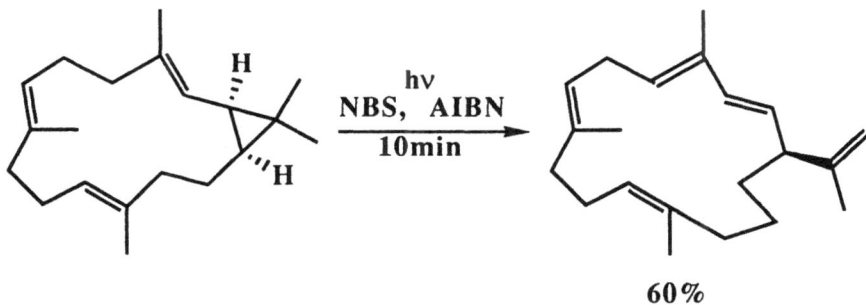

**60%**

---

**I.E.3-5** S. Goyal and M.R. Parthasarathy, *Ind. J. Chem.*, **31B**, 391; L. Jayabalan and P. Shanmugam, *ibid.*, **31B**, 436.

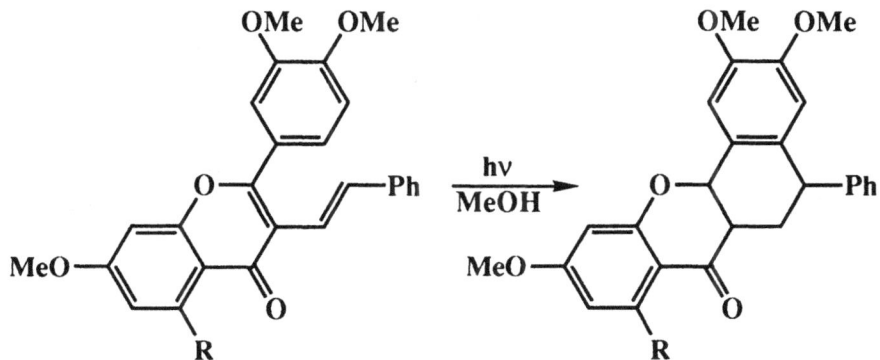

**I.E.3-6** S. Lahiri et al., *J. Chem. Res. (S)*, 372.

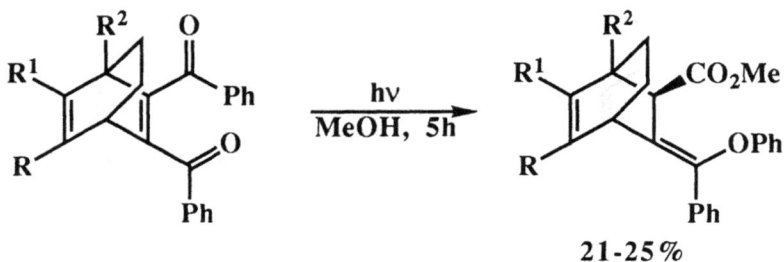

21-25%

---

**I.E.3-7** A. Heidbreder and J. Mattay, *Tetrahedron Lett.*, **33**, 1973.

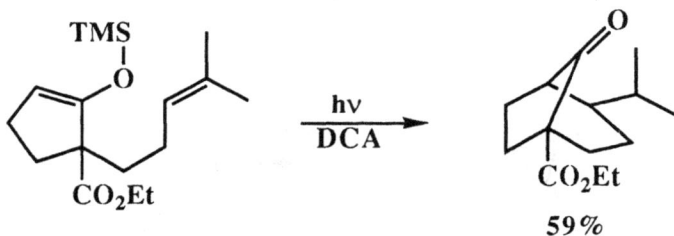

59%

---

**I.E.3-8** M.L. Graziano et al., *J. Chem. Soc., Perkin Trans. 1*, 1269.

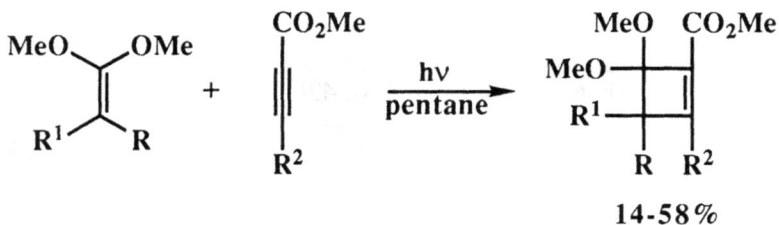

14-58%

---

**I.E.3-9** C. Liao et al., *Tetrahedron Lett.*, **33**, 2155; see also: D. Caine and P.L. Kotian, *J. Org. Chem.*, **57**, 6587.

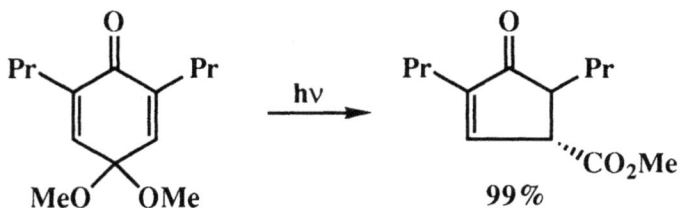

99%

**I.E.3-10** D.C. Blakemore and A. Gilbert, *J. Chem. Soc., Perkin Trans. 1*, 2265.

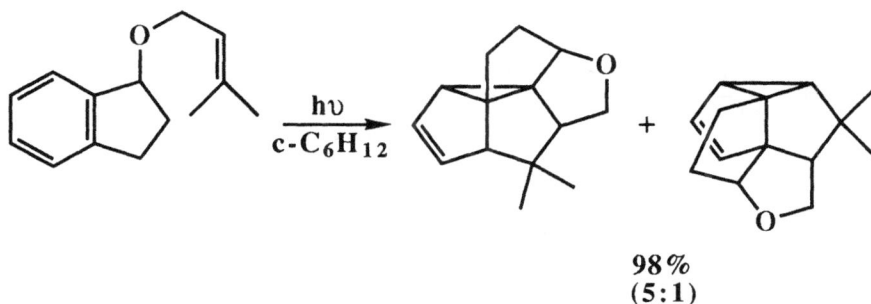

98%
(5:1)

**I.E.3-11** R.H. Schmidt-Rodde and K.P. Vollhardt, *J. Am. Chem. Soc.*, **114**, 9713.

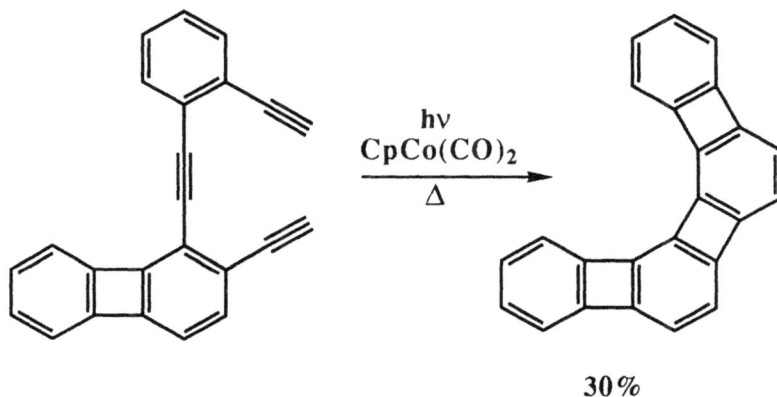

30%

**I.E.3-12** R. Gleiter and B. Treptow, *Angew. Chem., Int. Ed. Engl.*, **31**, 862.

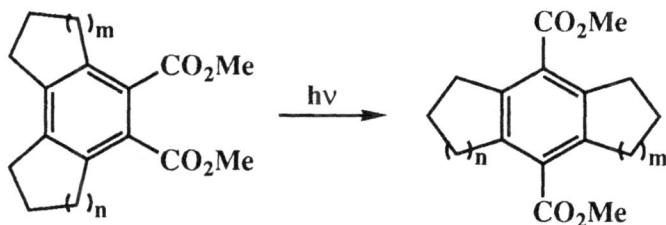

**I.E.3-13** A.G. Schultz and N.J. Green, *J. Am. Chem. Soc.*, **114**, 1824.

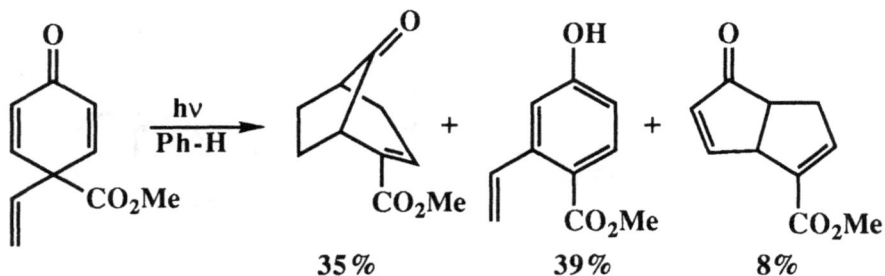

35%            39%            8%

**I.E.3-14** S.-H. Chen et al., *Tetrahedron Lett.*, **33**, 7679.

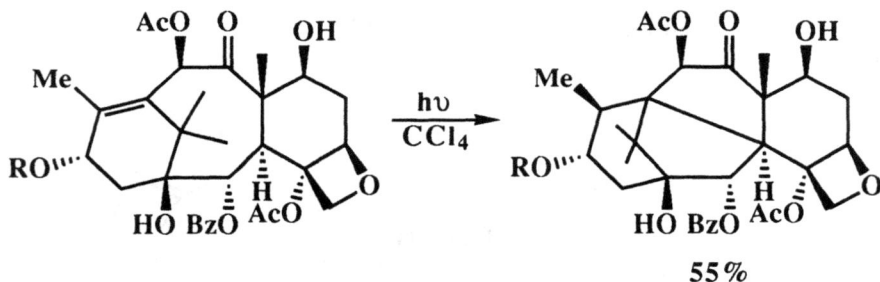

55%

**I.E.3-15** G.A. Kraus and Y. Wu, *J. Am. Chem. Soc.*, **114**, 8705.

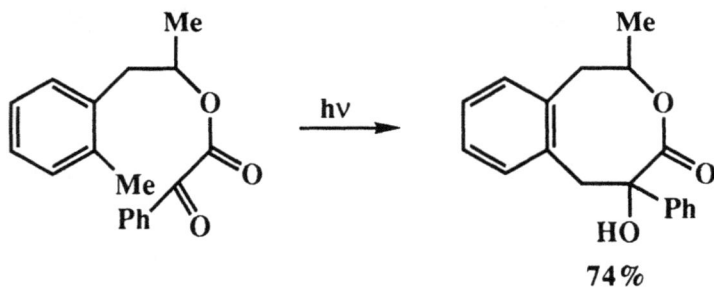

74%

**I.E.3-16** G.A. Kraus and Y. Wu, *J. Org. Chem.*, **57**, 2922.

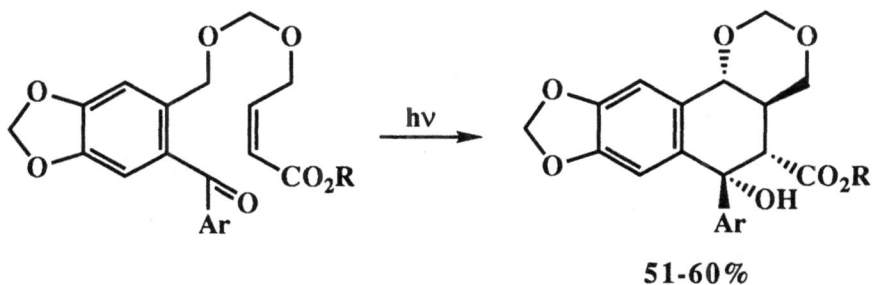

51-60%

**I.E.3-17** G. Quinkert et al., *Tetrahedron Lett.*, **33**, 1977.

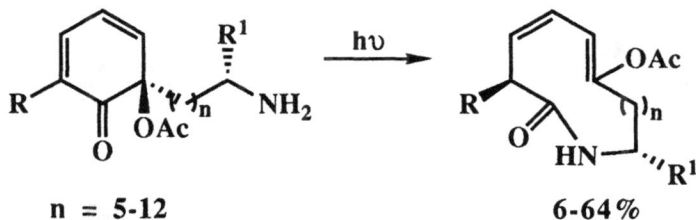

n = 5-12                                          6-64%

**I.E.3-18** A.G. Griesbeck and H. Mander, *Angew. Chem., Int. Ed. Engl.*, **31**, 73.

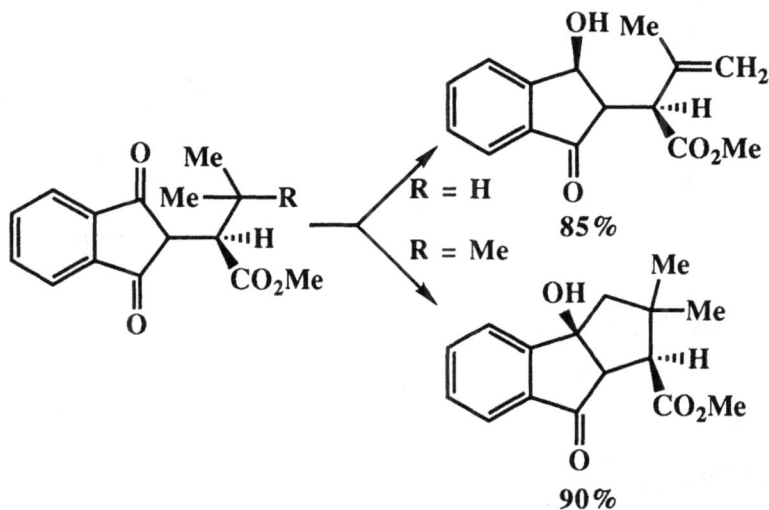

R = H        85%

R = Me       90%

**I.E.3-19** T. Sumathi and K.K. Balasubramanian, *Tetrahedron Lett.*, **33**, 2213; J.M. Saa et al., *J. Org. Chem.*, **57**, 6222.

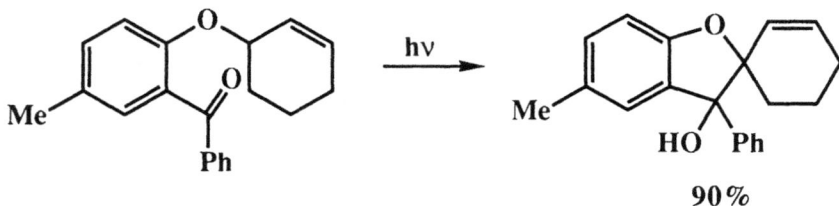

90%

**I.E.3-20** T. Sato et al., *Tetrahedron*, **48**, 9687.

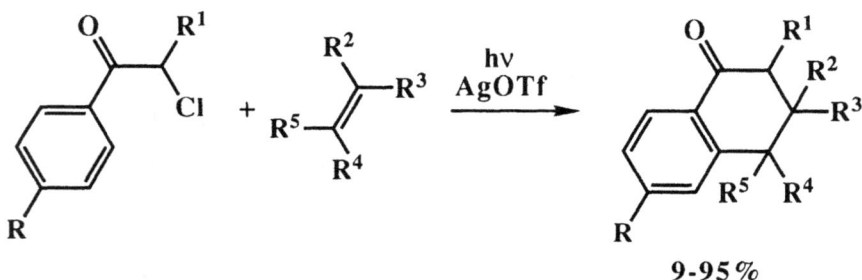

9-95%

**I.E.3-21** S. Hanessian et al., *Tetrahedron Lett.*, **33**, 749; see also: T. Caronna and S. Morrocchi, *J. Heterocyclic Chem.*, **29**, 975.

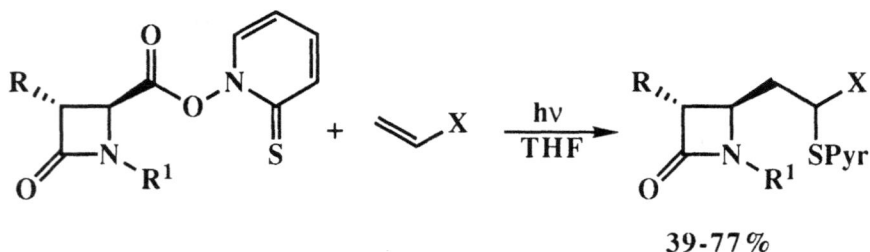

39-77%

**I.E.3-22** J.H. Byers and B.C. Harper,   ; see also: J.C. Jaszberenyi et al., *Tetrahedron*, **48**, 9261.

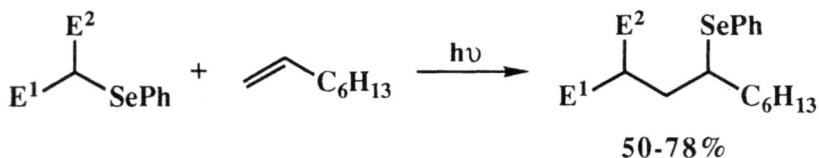

50-78%

**I.E.3-23** H. Seto et al., *Chem. Commun.*, 908.

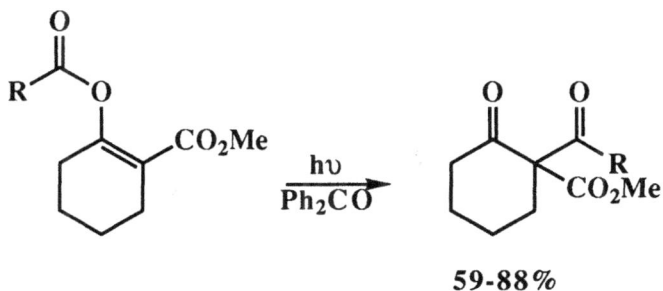

59-88%

**I.E.3-24** M.P. Bertrand et al., *J. Org. Chem.*, **57**, 6118; G. Mills et al., *Tetrahedron Lett.*, **33**, 6779.

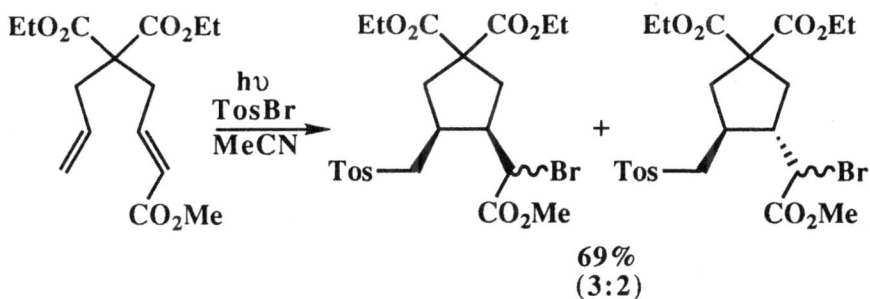

69%
(3:2)

**I.E.3-25** H. Garcia, M.A. Miranda et al., *Tetrahedron*, **48**, 3437.

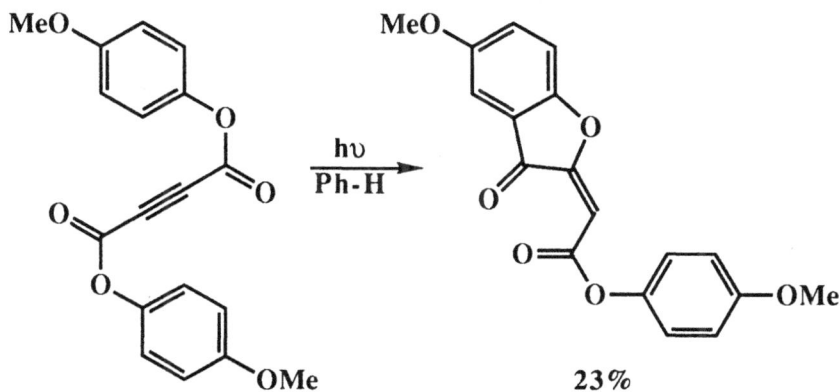

23%

**I.E.3-26** H. Takeshita, *Chem. Lett.*, 1891.

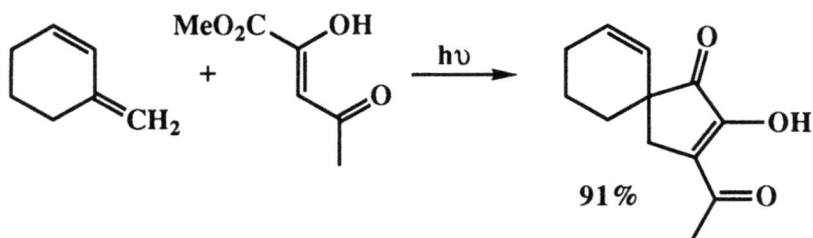

91%

---

**I.E.3-27** T. Momose et al., *Chem. Pharm. Bull.*, **40**, 2524.

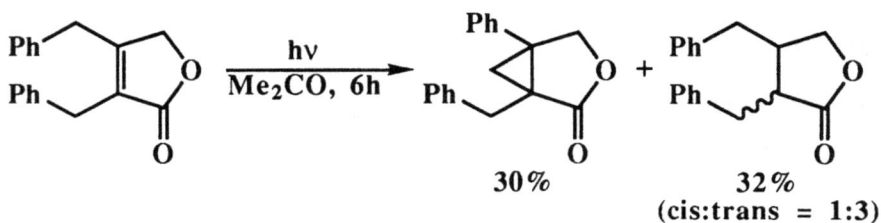

30%                    32%
                (cis:trans = 1:3)

---

**I.E.3-28** P. Garner et al., *J. Am. Chem. Soc.*, **114**, 2767.

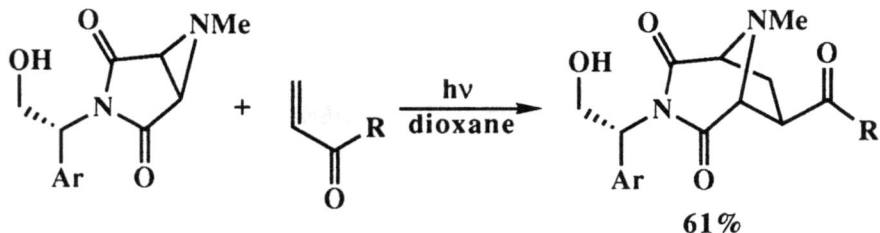

61%

---

**I.E.3-29** V.H. Rawal and S. Iwasa, *Tetrahedron Lett.*, **33**, 4687.

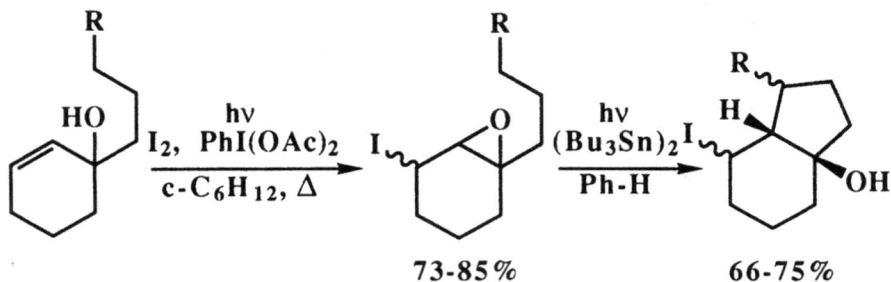

73-85%                    66-75%

**I.E.3-30** K. Oda, *Synlett.*, 603.

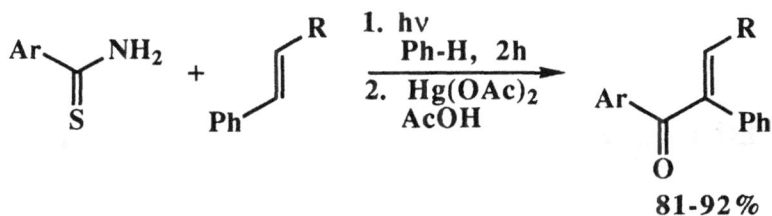

81-92%

---

**I.E.3-31** K. Isobe et al., *Chem. Pharm. Bull.*, **40**, 2188.

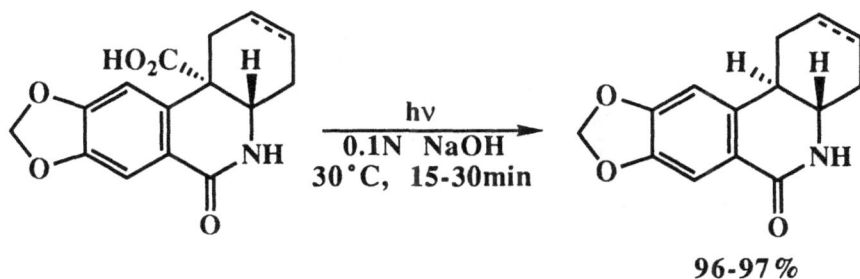

96-97%

---

**I.E.3-32** O. Muraoka and T. Momose et al., *Heterocycles*, **34**, 1093.

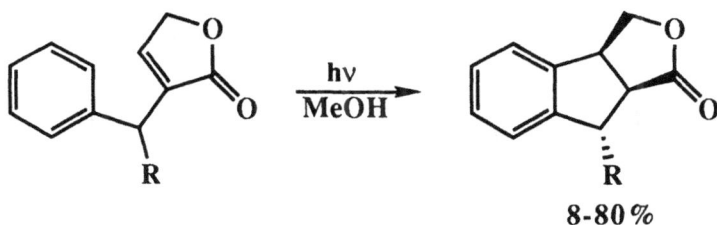

8-80%

---

**I.E.3-33** H. Suginome et al., *Heterocycles*, **33**, 553.

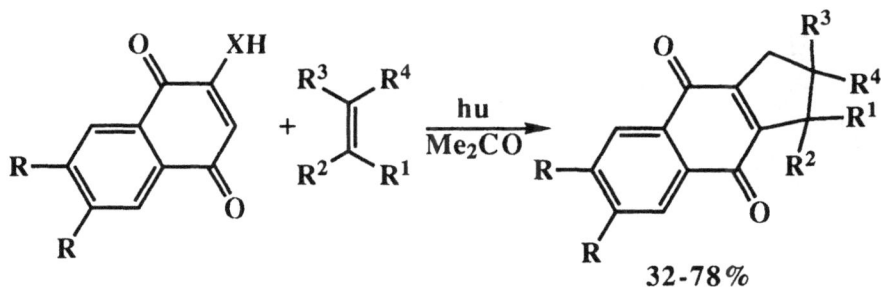

32-78%

## I.F.  Aromatic Substitutions Forming a New Carbon-Carbon Bond

## I.F.1.  Friedel-Crafts Type Aromatic Substitution Reactions

**I.F.1-1** C. Cativiela et al., *Synlett.*, 121; see also: M.A. Rustamov et al., *J. Org. Chem. (USSR)*, **28**, 242; see also: K. Saigo et al., *Tetrahedron Lett.*, **33**, 6351.

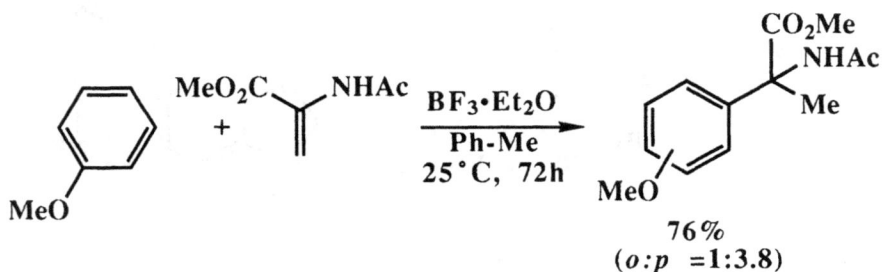

76%
(*o:p* =1:3.8)

---

**I.F.1-2** T. Mukaiyama et al., *Chem. Lett.*, 435.

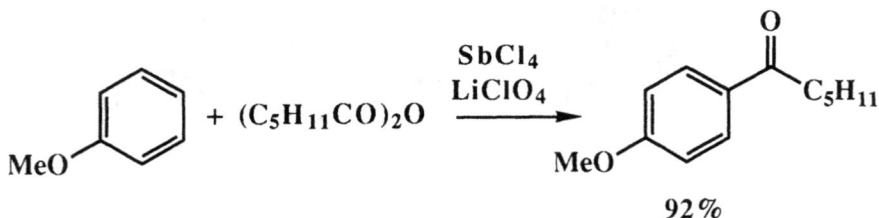

92%

---

**I.F.1-3** O. Itoh et al., *J. Org. Chem.*, **57**, 7334.

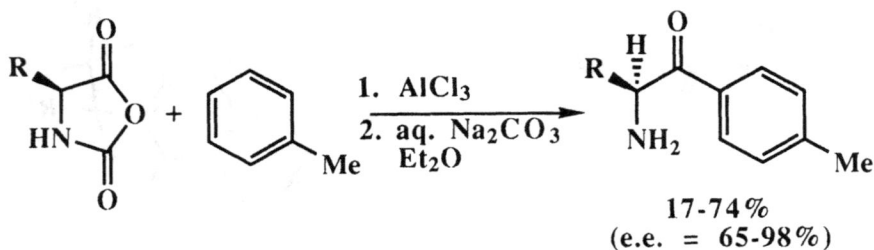

17-74%
(e.e. = 65-98%)

**I.F.1-4** S.P. Tanis et al., *J. Am. Chem. Soc.*, **114**, 8349; see also: C.R. Dalton et al., *Tetrahedron Lett.*, **33**, 5713.

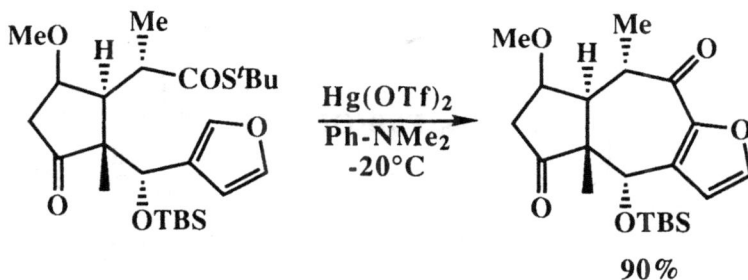

**I.F.1-5** H. Cerfontain et al., *Rec. Trav. Chim*, **111**, 389.

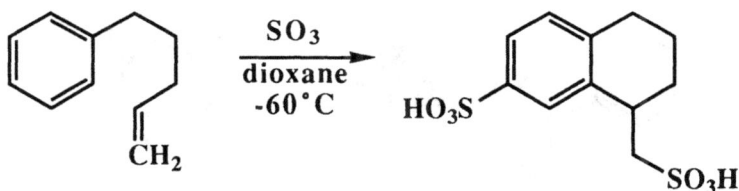

**I.F.1-6** G. Sartori et al., *Tetrahedron Lett.*, **33**, 4771.

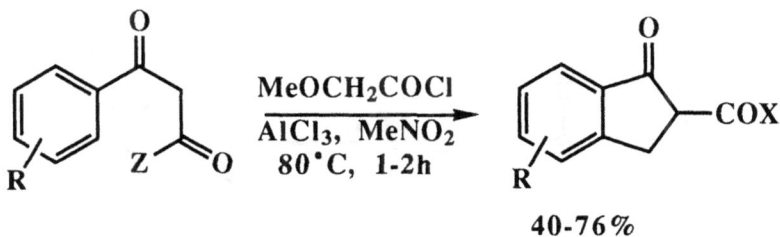

**I.F.1-7** O.A. Attanasi and P. Filippone, *Gazz. Chim. Ital.*, **121**, 487 (1991).

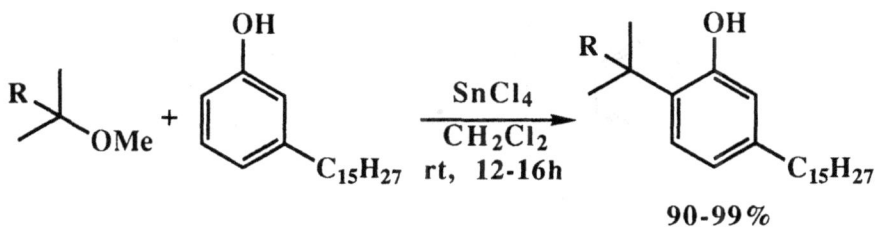

**I.F.1-8** W.S. Murphy et al., *J. Chem. Soc., Perkin Trans.1*, 605.

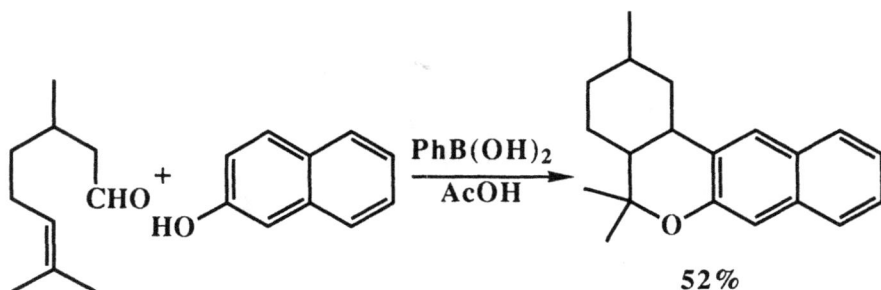

**52%**

---

**I.F.1-9** A.R. Katritzky et al., *J. Chem. Soc., Perkin Trans.1*, 1111 and *Tetrahedron*, **48**, 4971.

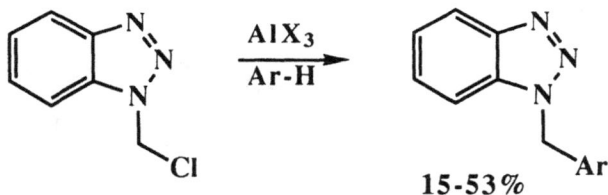

**15-53%**

---

**I.F.1-10** E.K. Ryu and J.N. Kim, *J. Org. Chem.*, **57**, 1088.

**50-90%**

---

**I.F.1-11** H.J. Niclas et al., *Synth. Commun.*, **22**, 2237.

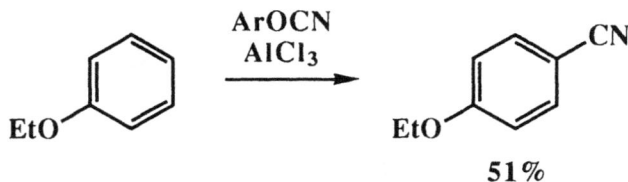

**51%**

**I.F.1-12** K. Toshima et al., *Tetrahedron Lett.*, **33**, 2175.

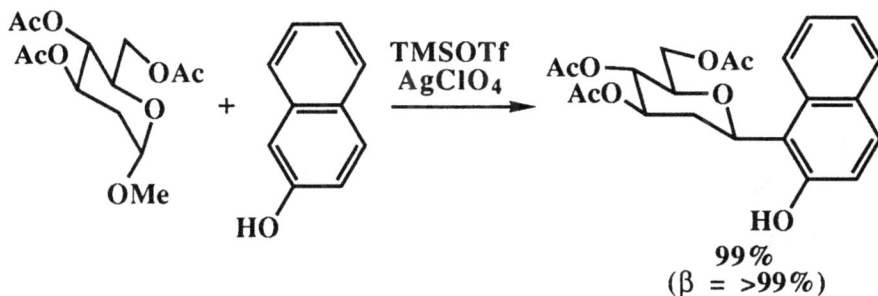

99%
(β = >99%)

---

**I.F.1-13** C. Booma and K.K. Balasubramanian, *Tetrahedron Lett.*, **33**, 3049.

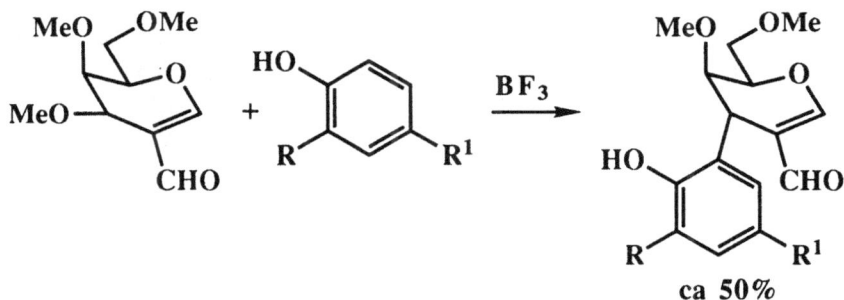

ca 50%

---

**I.F.1-14** L.F. Tietze and J. Wichmann, *Angew. Chem., Int. Ed. Engl.*, **31**, 1079; and *Liebigs Ann. Chem.*, 1063 and see also: *Chem. Ber.*, **125**, 2571.

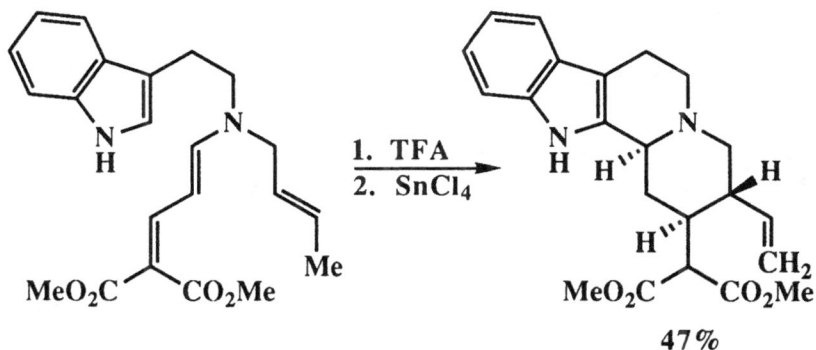

47%

**I.F.1-14**5M. Natsume et al., *Chem. Pharm. Bull.*, **40**, 2338, 2344, 2353, 2358.

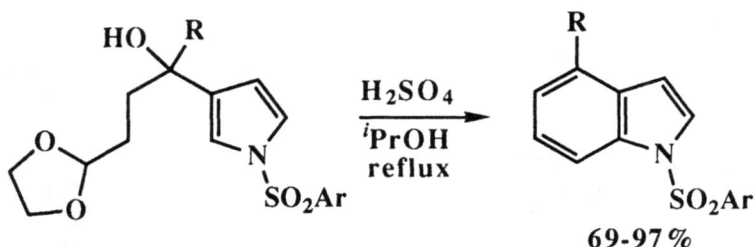

69-97%

---

**I.F.1-16** M. Nakazawa, T. Hino et al., *Heterocycles,* **33**, 801.

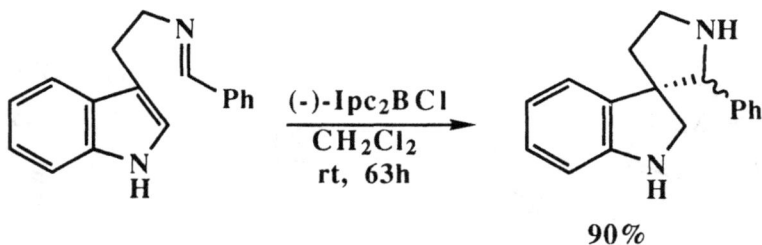

90%

## I.F.2. Coupling Reactions to Form an Aromatic Carbon-Carbon Bond

**I.F.2-1** G. Sartori et al., *Tetrahedron,* **48**, 9483; see also: M. Hovorka, J. Zavada et al., *ibid.,* **48**, 9503.

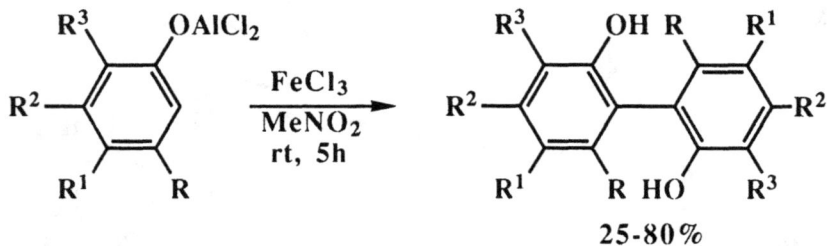

25-80%

**I.F.2-2**  Y. Landais and J.-P. Robin, *Tetrahedron*, **48**, 7185, 819; T. Wakamatsu et al., *Tetrahedron Lett.*, **33**, 4161, 4165; A.S. Kraus and W.C. Taylor, *Aust. J. Chem.*, **45**, 925.

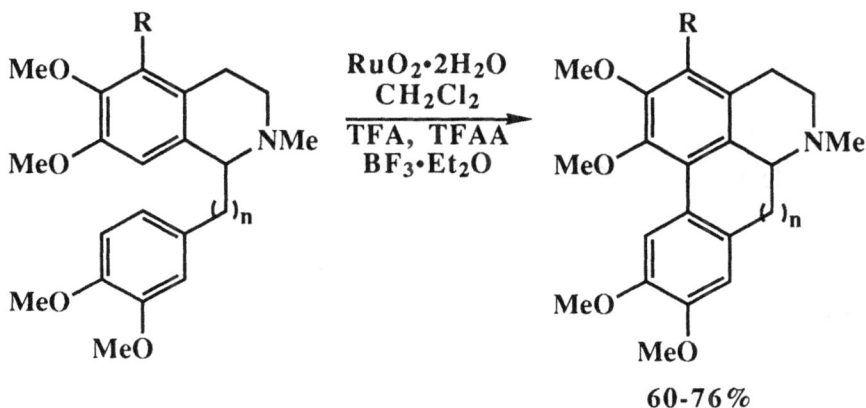

60-76%

**I.F.2-3**  D.L. Comins et al., *J. Am. Chem. Soc.*, **114**, 10971; C.F. Nutaitis and S.R. Marsh, *J. Heterocyclic Chem.*, **29**, 971.

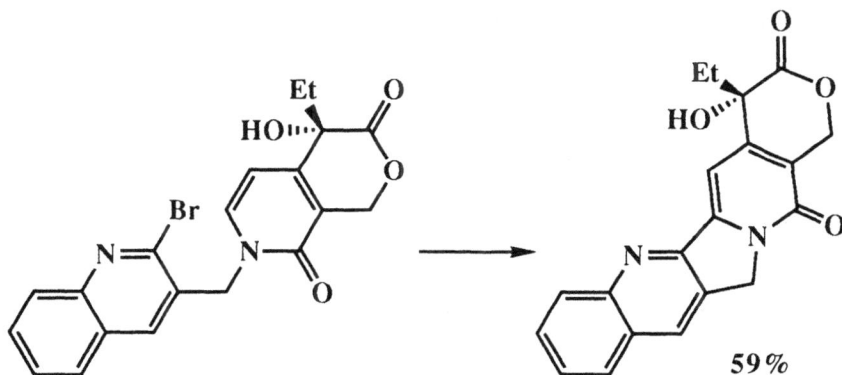

59%

**I.F.2-4**  H. Yamanaka et al., *Tetrahedron Lett.*, **33**, 5373; R. Rossi et al., *ibid.*, **33**, 4495; J.M. Kaufman et al., *J. Heterocyclic Chem.*, **29**, 1245.

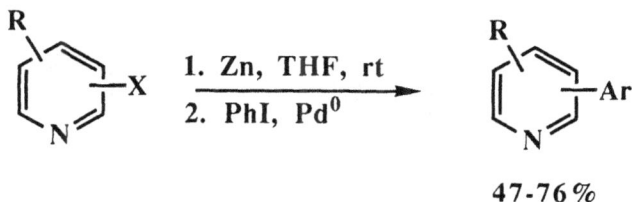

47-76%

**I.F.2-5** A.I. Meyers et al., *Tetrahedron Lett.*, **33**, 853; R. van Asselt and C.J. Elsevier, *Organometallics,* **11**, 1999.

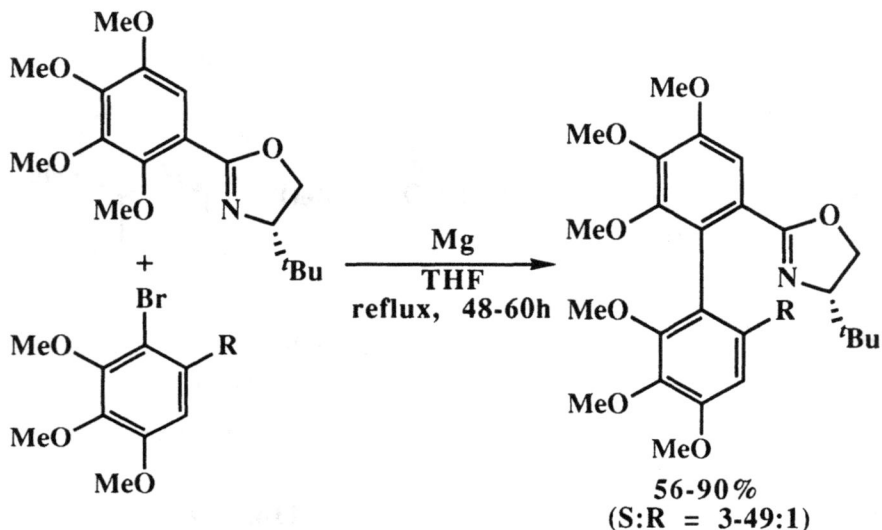

56-90%
(S:R = 3-49:1)

**I.F.2-86** K. Tomioka et al., *J. Am. Chem. Soc.*, **114**, 8733; L.H. Klemm et al., *J. Heterocyclic Chem.*, **29**, 1673.

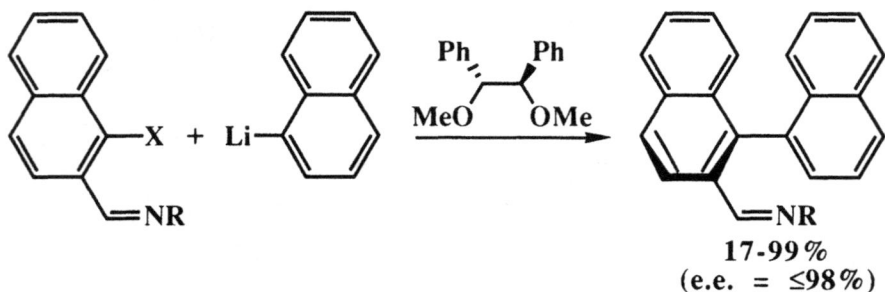

17-99%
(e.e. = ≤98%)

**I.F.2-7** R.A. Olofson et al., *Synth. Commun.*, **22**, 1807.

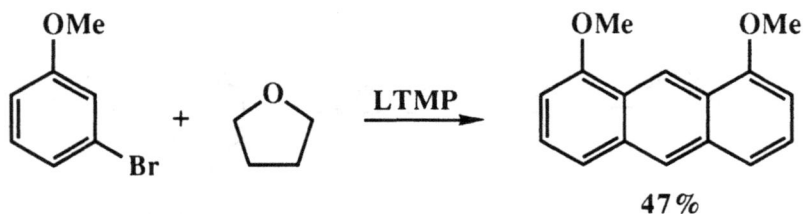

47%

**I.F.2-8** G. Dyker, *Angew. Chem., Int. Ed. Engl.*, **31**, 1023.

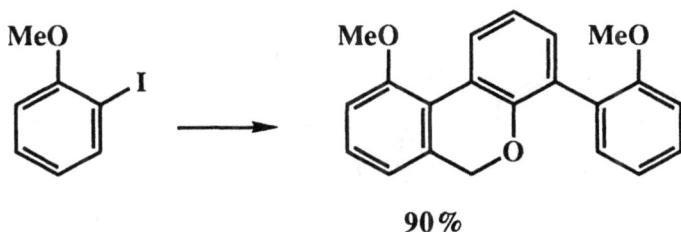

**90%**

---

**I.F.2-9** M.L. Scarpati et al., *Org. Prep. Proced. Int.*, **24**, 532; T. Iwasaki et al., *Heterocycles*, **34**, 2061; G.W. Ebert et al., *Organo-metallics*, **11**, 1560.

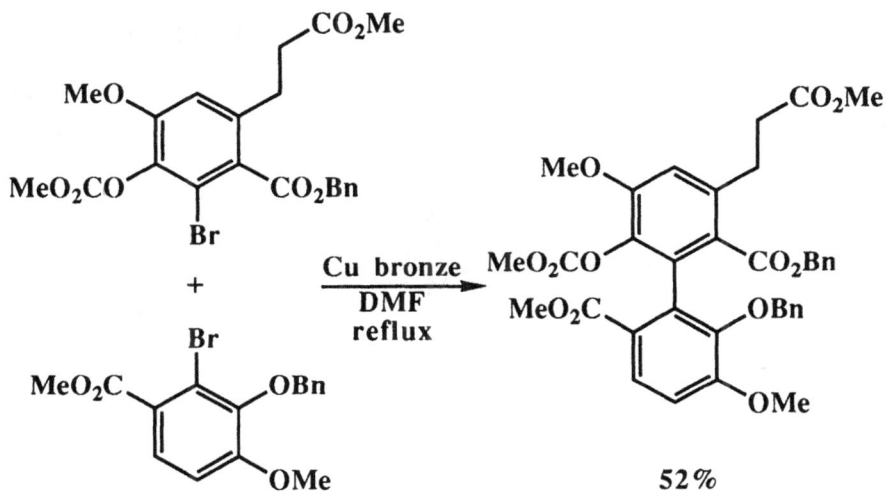

**52%**

---

**I.F.2-10** J.E. Rice and Z. Cai, *Tetrahedron Lett.*, **33**, 1675; W. Cabri et al., *J. Org. Chem.*, **57**, 1481; F. Ozawa and T. Hayashi, *J. Organmet. Chem.*, **428**, 267.

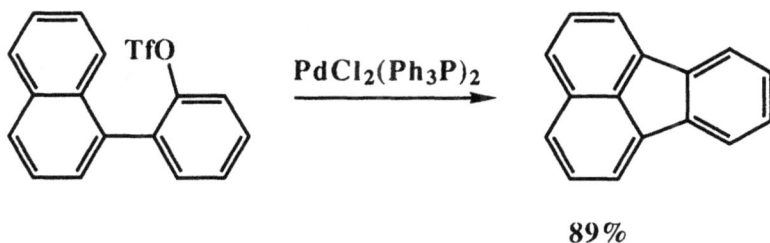

**89%**

**I.F.2-11** T. Yamato et al., *J. Chem. Res. (S)*, 420.

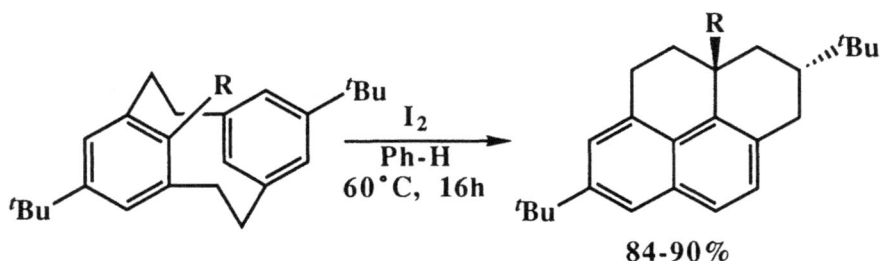

**84-90%**

## I.F.3. Other Aromatic Substitutions and Preparations

**I.F.3-1** D. St.C. Black et al., *Tetrahedron*, **48**, 7601; D. Milstein et al., *Organometallics*, **11**, 1995; C.F. Bigge et al., *J. Med. Chem.*, **35**, 1371; W.G. Rajeswaran and P.C. Srinivasan, *Synthesis*, 835; see also: L. Fillippini et al., *Tetrahedron Lett.*, **33**, 1755.

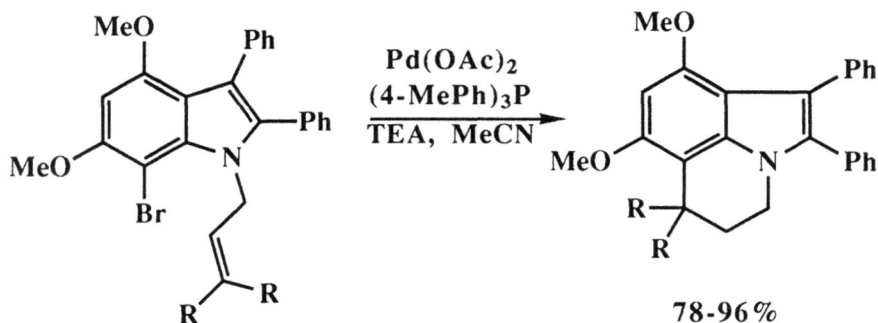

**78-96%**

**I.F.3-2** T. Ishiyama et al., *Tetrahedron Lett.*, **33**, 4465.

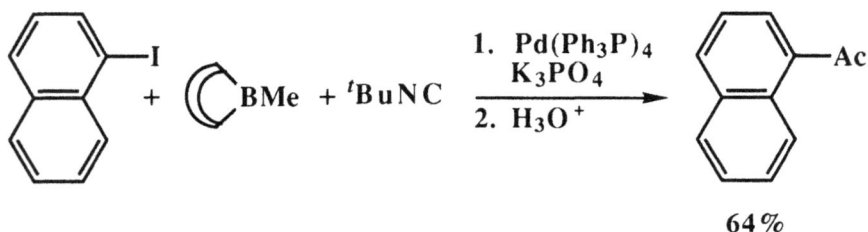

**64%**

**I.F.3-3** H. Yamanaka et al., *Chem. Pharm. Bull.*, **40**, 1136 and *Heterocycles*, **34**, 2379; D. Badone et al., *J. Org. Chem.*, 57, 6321; P. Quayle et al., *Tetrahedron Lett.*, 33, 409, 413; G.J. Hollingworth and J.B. Sweeney, *ibid.*, 33, 7049; B.C. Pearce, *Synth. Commun.*, 22, 1627; Y. Yang and A.R. Martin, *ibid.*, 22, 1757.

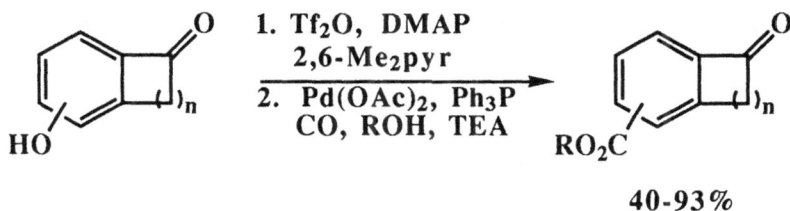

1. Tf$_2$O, DMAP
   2,6-Me$_2$pyr
2. Pd(OAc)$_2$, Ph$_3$P
   CO, ROH, TEA

**40-93%**

---

**I.F.3-4** U.T.. Bhalerao et al., *Chem. Commun.*, 1176.

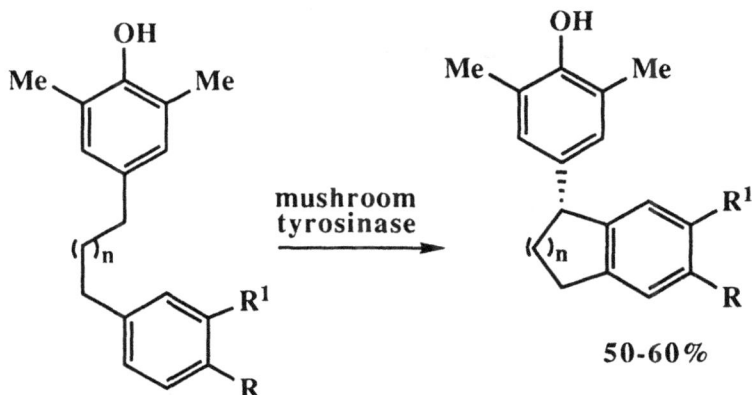

mushroom tyrosinase

**50-60%**

---

**I.F.3-5** A. Ashimori and L.E. Overman, *J. Org. Chem.*, **57**, 4571; L.E. Overman et al., *Pure Appl. Chem.*, **64**, 1813; H. Yamanaka et al., *Tetrahedron Lett.*, 33, 6845.

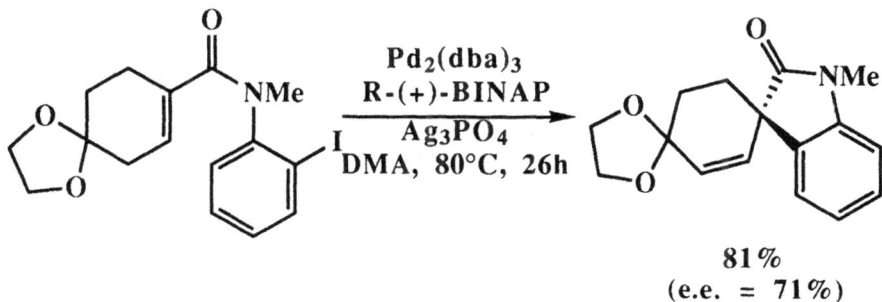

Pd$_2$(dba)$_3$
R-(+)-BINAP
Ag$_3$PO$_4$
DMA, 80°C, 26h

**81%**
**(e.e. = 71%)**

**I.F.3-6** W. Shieh and J.A. Carlson, *J. Org. Chem.*, **57**, 379; V. Snieckus et al., *Tetrahedron Lett.*, **33**, 2253; S.M. Marcuccio et al., *ibid.*, **33**, 6679; see also: K.C. Santhosh and K.K. Balasubramanian, *Chem. Commun.*, 224.

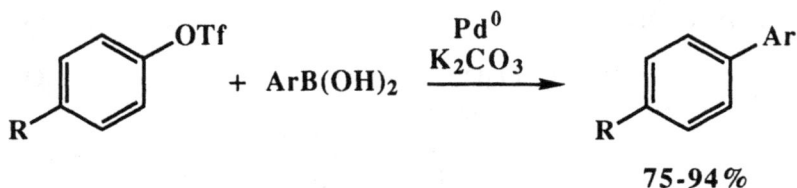

75-94%

**I.F.3-47** U. Gerlach and T. Wollmann, *Tetrahedron Lett.*, **33**, 5499.

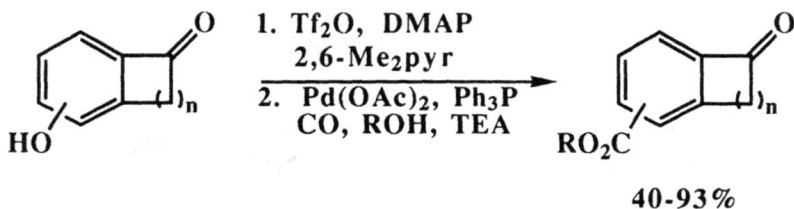

40-93%

**I.F.3-8** Y. Kita et al., *Chem. Commun.*, 429.

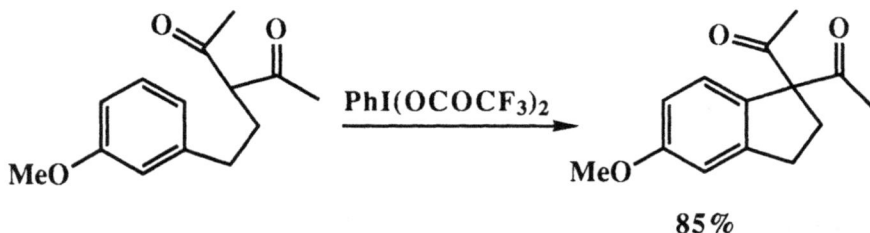

85%

**I.F.3-9** J.R. Norton et al., *J. Org. Chem.*, **57**, 6496.

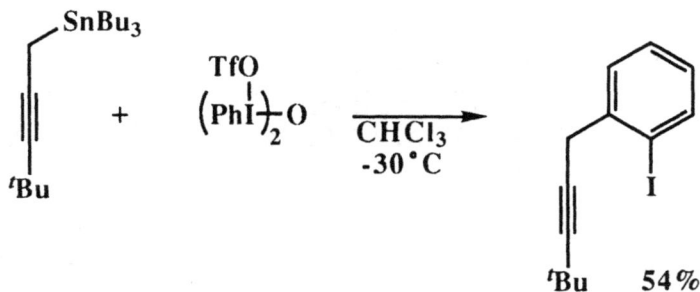

$^t$Bu    54%

**I.F.3-10** C.-P. Chuang, *Tetrahedron Lett.*, **33**, 6311.

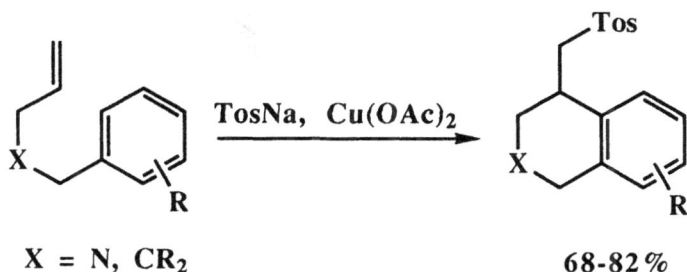

X = N, CR$_2$

TosNa, Cu(OAc)$_2$

68-82%

---

**I.F.3-11** F. Fontana et al., *Gazz. Chim. Ital.*, **122**, 167.

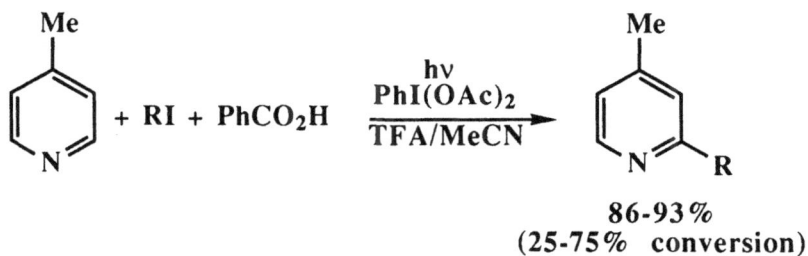

+ RI + PhCO$_2$H

$\xrightarrow[\text{TFA/MeCN}]{\substack{h\nu \\ \text{PhI(OAc)}_2}}$

86-93%
(25-75% conversion)

---

**I.F.3-12** F. Minisa et al., *Tetrahedron Lett.*, **33**, 3201.

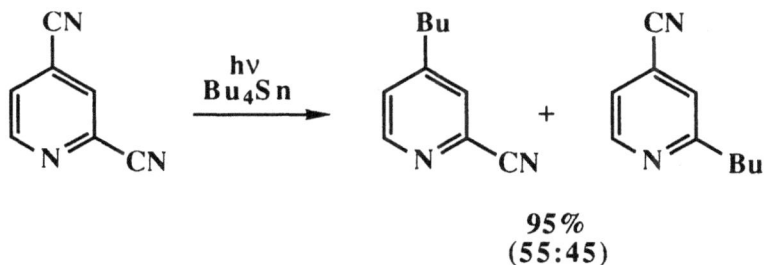

$\xrightarrow[\text{Bu}_4\text{Sn}]{h\nu}$

+

95%
(55:45)

---

**I.F.3-14** M. Novi et al., *Tetrahedron*, **48**, 325.

ArN=NS$^t$Bu +

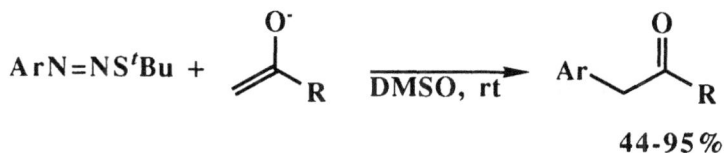

$\xrightarrow{\text{DMSO, rt}}$

44-95%

**I.F.3-14** A.F. Venkov et al., *Synth. Commun.*, **22**, 225.

$$\xrightarrow[\text{2. ArH}]{\text{1. RCOCl}}$$

**13-90%**

---

**I.F.3-15** S. Inoue et al., *Heterocycles,* **34**, 1017

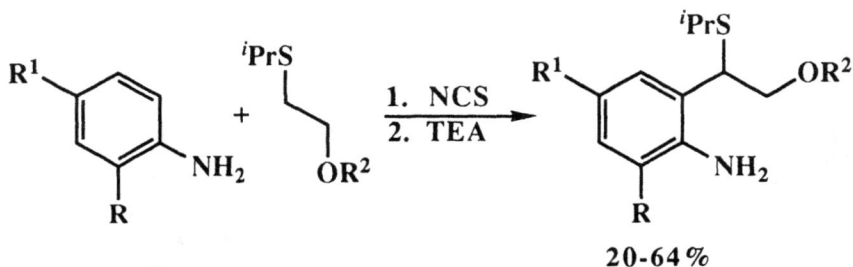

$$\xrightarrow[\text{2. TEA}]{\text{1. NCS}}$$

**20-64%**

---

**II.F.3-16** J.-C. Jacquesy, *Tetrahedron Lett.*, **33**, 8085.

$$\text{Ar-H} + \text{RCH}_2\text{NO}_2 \xrightarrow[\text{70-80°C}]{\text{TFSA}}$$

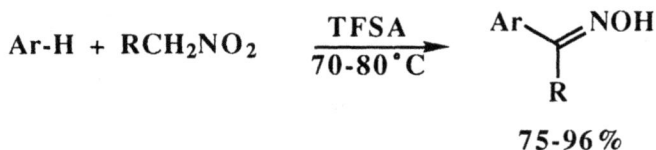

**75-96%**

---

**I.F.3-17** B. Liedholm, *J. Chem. Soc., Perkin Trans.1*, 2235.

$$\xrightarrow[\text{DMSO/aq HCl}]{\text{CuCl}_2}$$

**84-98%**

**I.F.3-18** S. Ikegami et al., *Chem. Commun.*, 1508.

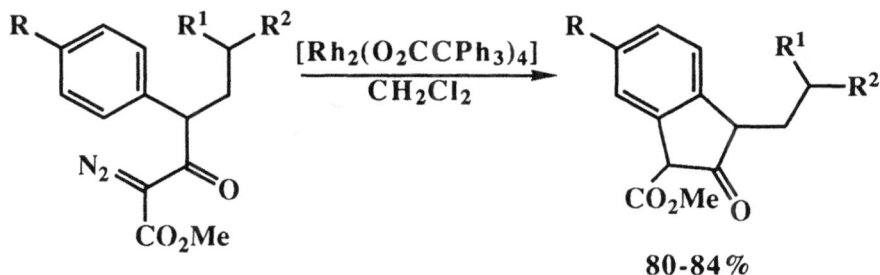

80-84%

---

**I.F.3-19** M. Ikeda et al., *Chem. Pharm. Bull.*, **39**, 3163 (1991).

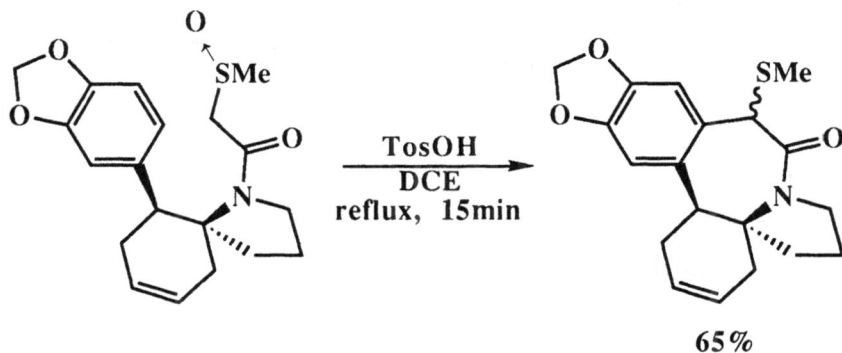

65%

---

**I.F.3-20** L.E. Fisher et al., *J. Org. Chem.*, **57**, 2700; J.C. Carretero, J.L. Garcia-Ruano and M. Visioso, *Tetrahedron*, **48**, 7373; M. Iwao and T. Kuraishi, *Heterocycles*, **34**, 1031; G. Quegniner et al., *J. Heterocyclic Chem.*, **29**, 699; V. Snieckus et al., *Tetrahedron Lett.*, **33**, 2625; V. Snieckus and F. Beaulieu, *Synthesis*, 113.

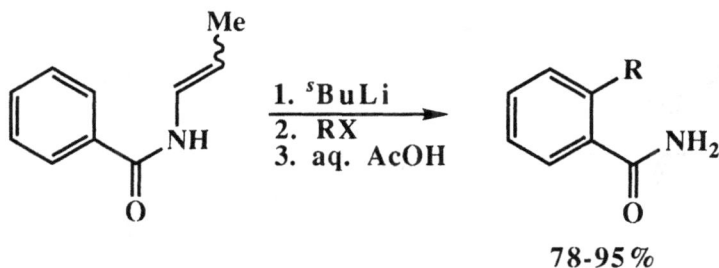

78-95%

**I.F.3-21** J. Barluenga et al., *Tetrahedron Lett.*, **33**, 6183.

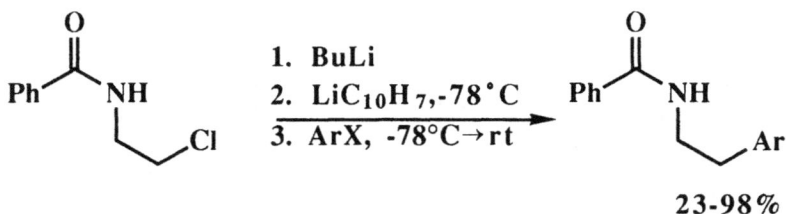

1. BuLi
2. LiC$_{10}$H$_7$,-78°C
3. ArX, -78°C→rt

23-98%

**I.F.3-22** J. Epsztajn et al., *Monatsh. Chem.*, **123**, 1125.

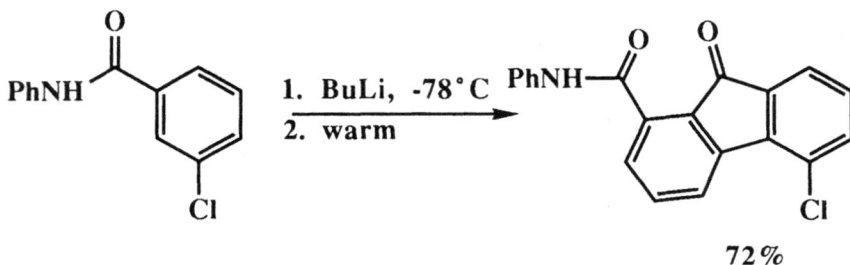

1. BuLi, -78°C
2. warm

72%

**I.F.3-23** S.-I. Murahashi et al., *Synlett,* 835.

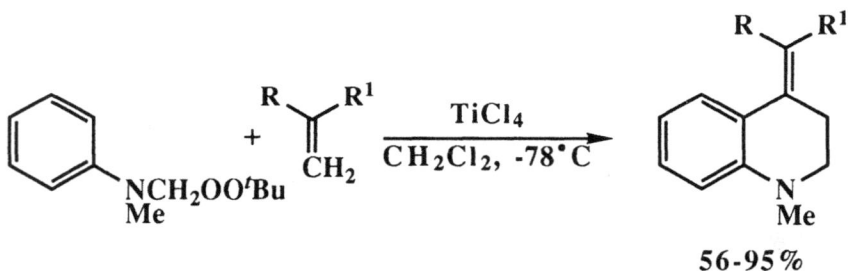

TiCl$_4$
CH$_2$Cl$_2$, -78°C

56-95%

**I.F.3-24** K.I. Booker-Milburn, *Synlett,* 327; B.L. Finkelstein, *J. Org. Chem.*, **57**, 5538.

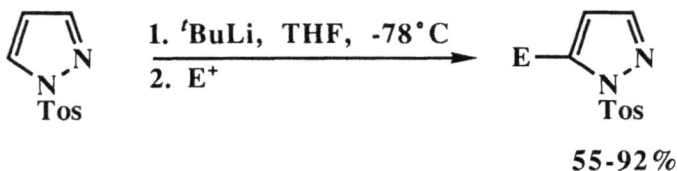

1. $^t$BuLi, THF, -78°C
2. E$^+$

55-92%

**I.F.3-25** H. Suginome et al., *Chem. Commun.*, 780.

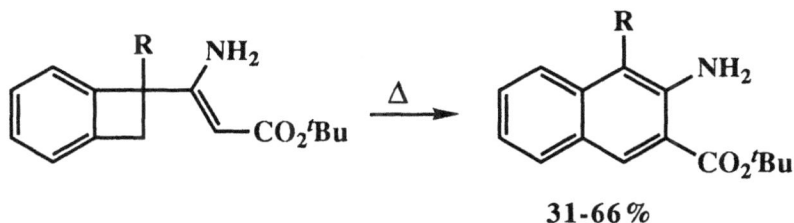

31-66%

---

**I.F.3-26** L.S. Liebeskind et al., *J. Am. Chem. Soc.*, **114**, 1412.

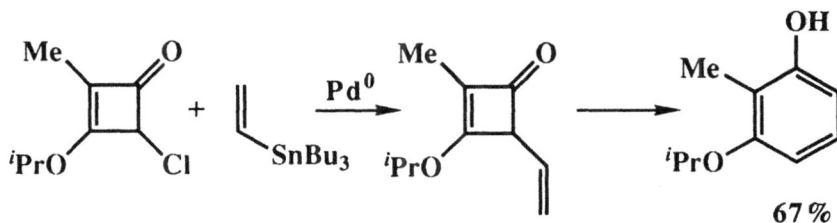

67%

---

**I.F.3-27** Y. Ishii and M. Hidai, *J. Organmet. Chem.*, **428**, 279.

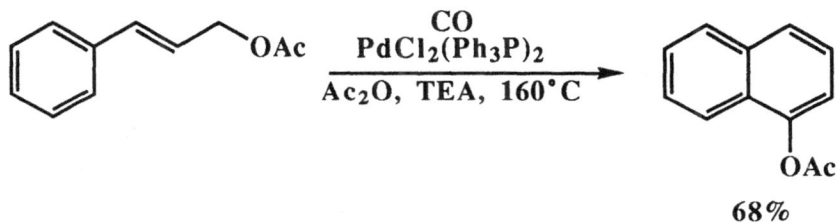

68%

---

**I.B.3-28** P. Bhatarah and E.H. Smith, *J. Chem. Soc., Perkin Trans.1*, 2163.

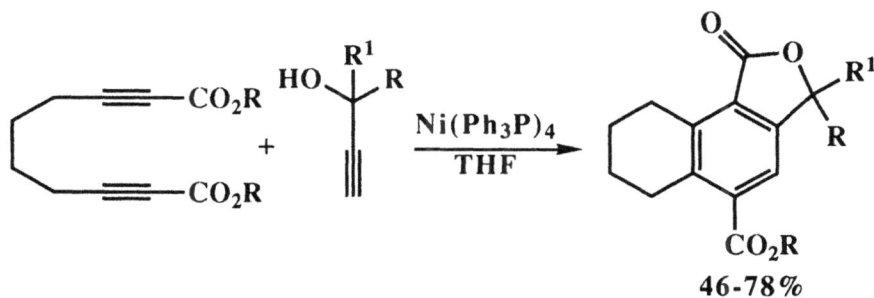

46-78%

**I.F.3-28** 9L.S. Liebeskind et al., *Organometallics*, **11**, 255.

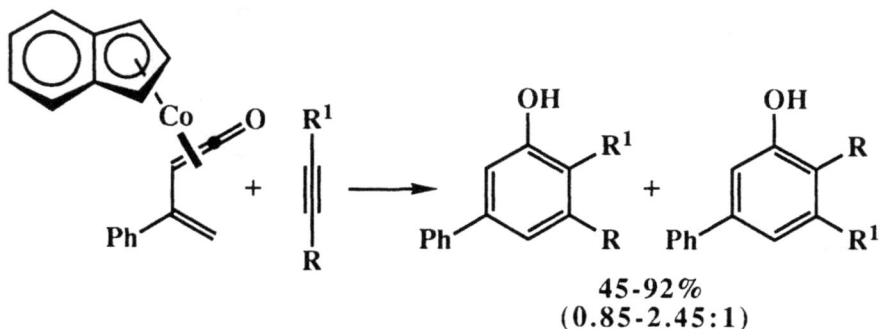

45-92%
(0.85-2.45:1)

---

**I.F.3-30** E. Nigishi et al., *Tetrahedron Lett.*, **33**, 3253.

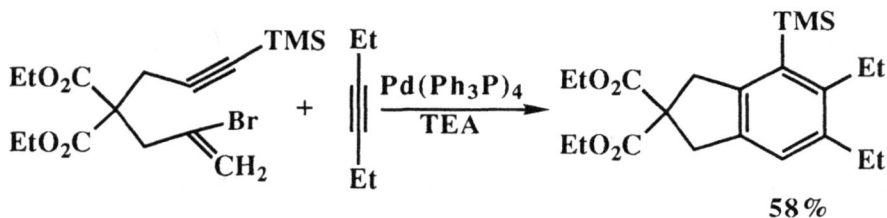

58%

---

**I.F.3-31** G. Desimoni et al., *Gazz. Chim. Ital.*, **121**, 483 (1991).

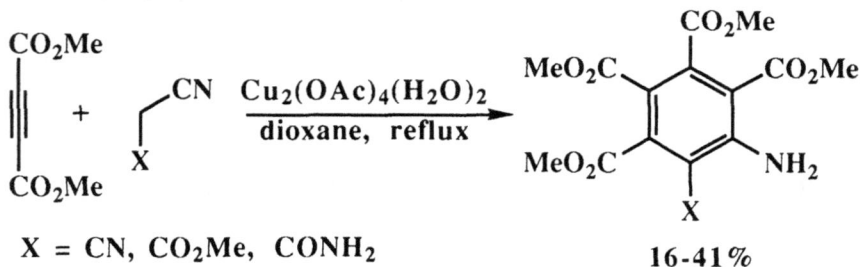

X = CN, CO₂Me, CONH₂

16-41%

## I.G.    Synthesis via Organometallics

## I.G.1   Synthesis via Organoboranes

**I.G.1-1**  H. C. Brown et al., *J. Org. Chem.*, **1992**, *57*, 3767.

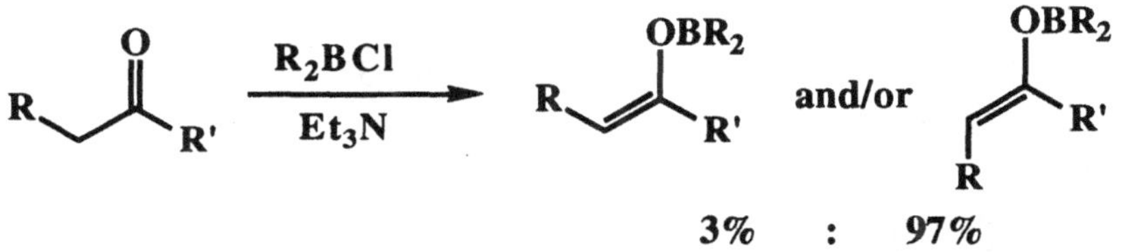

$$ \text{3\%} \quad : \quad \text{97\%} $$

**R₂BCl  =  bis(bicyclo[2,2,2]octyl)chloroborane**

---

**I.G.1-2**  M. Z. Deng*, N. S. Li, Y. Z. Huang, *J. Org. Chem.*, **1992**, *57*, 4017.

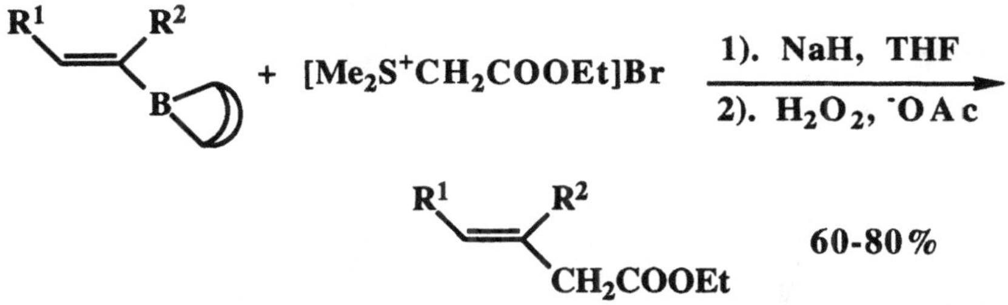

**60-80%**

## I.G.2. Carbonylation Reactions

**I.G.2-1**  H. Alper*, J. Q. Zhou, *J. Org. Chem.*, **1992**, *57*, 3729.

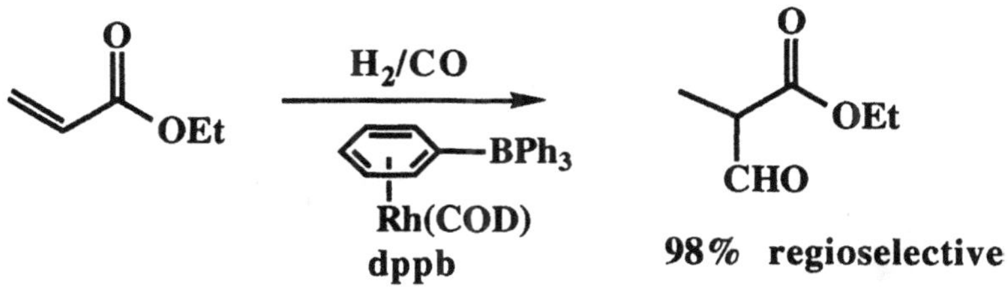

**98% regioselective**

**I.G.2-2**  S. Cacchi and A. Lupi. *Tetrahedron Lett.*, **1992**, *33*, 3939.

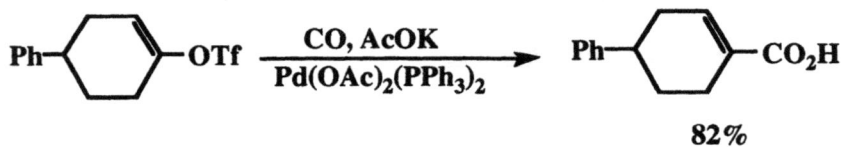

Ph—⟨cyclohexene⟩—OTf  $\xrightarrow[\text{Pd(OAc)}_2\text{(PPh}_3)_2]{\text{CO, AcOK}}$  Ph—⟨cyclohexene⟩—CO$_2$H

**82%**

---

**I.G.2-3**  R. Keese et al., *Tetrahedron Lett.*, **1992**, *33*, 1207.

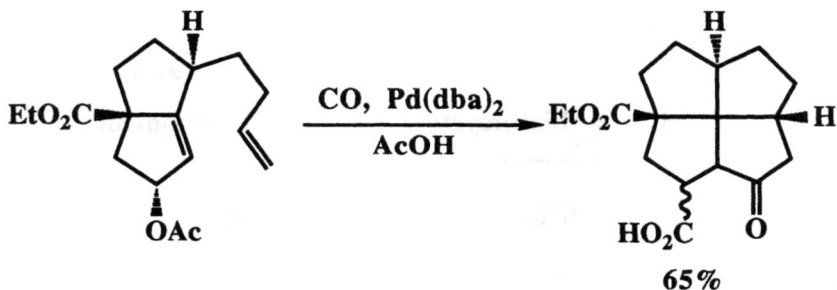

EtO$_2$C—⟨structure with H and OAc⟩  $\xrightarrow[\text{AcOH}]{\text{CO,  Pd(dba)}_2}$  EtO$_2$C—⟨structure with H, HO$_2$C, O⟩——H

**65%**

---

**I.G.2-4**  B. E. Eaton et al., *J. Am. Chem. Soc.*, **1992**, *114*, 6245.

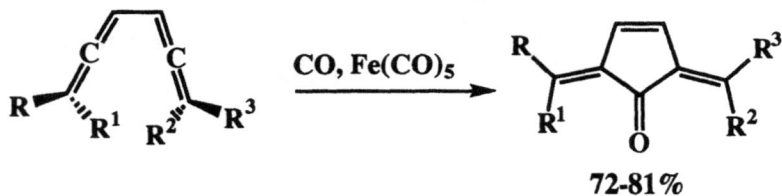

⟨diene structure: R, R$^1$, R$^2$, R$^3$⟩  $\xrightarrow{\text{CO, Fe(CO)}_5}$  ⟨cyclopentanone structure: R, R$^1$, R$^2$, R$^3$, O⟩

**72-81%**

**I.G.2-5** E. Broncato et al., *Tetrahedron Lett.*, **1992**, *33*, 7437.

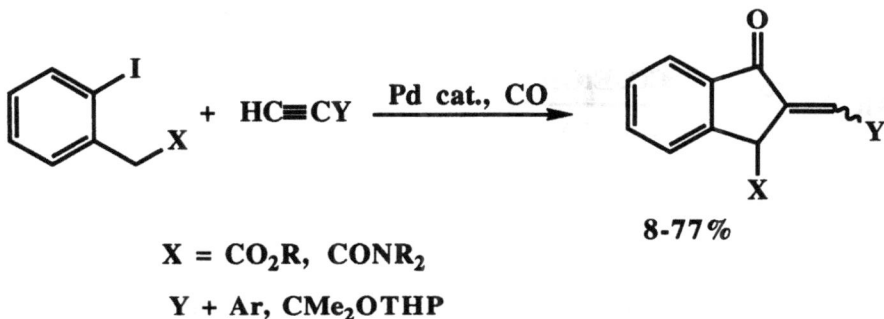

$$X = CO_2R, \quad CONR_2$$

$$Y + Ar, \quad CMe_2OTHP$$

---

**I.G.2-6** W. A. Smit*, R. Caple* et al., *J. Am. Chem. Soc.*, **1992**, *114*, 5555; K. H. Dotz and J. Christoffers *J. Organometl. Chem.* **1992**, *426*, C58; M. Hanaoka et al., *J. Chem. Soc., Chem. Commun.*, **1992**, 1014; E. G. Rowley, and N. E. Shore*, *J. Org. Chem.*, **1992**, *57*, 6853.

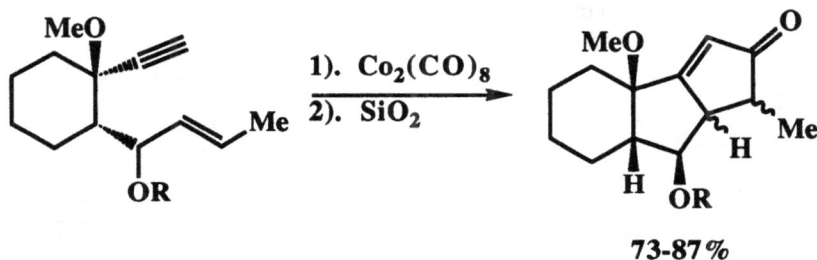

---

**I.G.2-7** M. E. Krafft et al., *Tetrahedron Lett.*, **1992**, *33*, 3829; *J. Org. Chem.*, **1992**, *57*, 5706.

**Acceleration of the Thermal Pauson-Khand Reaction by Coordination Ligands**

**I.G.2-8** R. Takeuchi et al., *J. Org. Chem.*, **1992**, *57*, 4189.

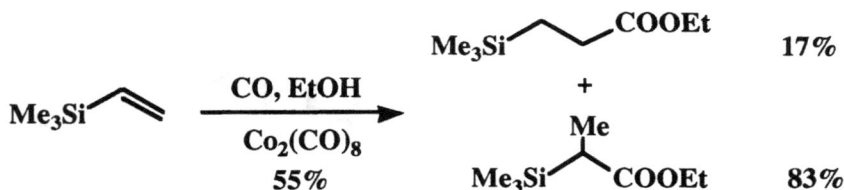

Me₃Si $\diagup\!\!\!\diagup$   $\xrightarrow[\substack{Co_2(CO)_8 \\ 55\%}]{CO, \text{ EtOH}}$   Me₃Si$\diagdown\diagup\diagdown$COOEt   17%

+

Me$_3$Si$\diagup$C(Me)COOEt   83%

---

**I.G.2-9** C.S. N. Prasad, S. R. Adapa, *Ind.J. Chem.* **1991**, *30B*, 1067.

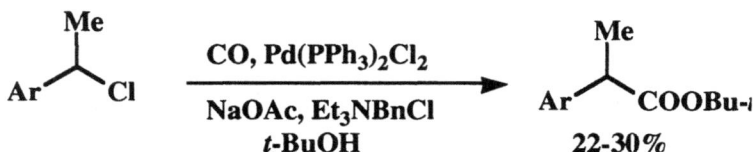

Ar–CH(Me)–Cl   $\xrightarrow[\substack{NaOAc, Et_3NBnCl \\ t\text{-BuOH}}]{CO, \text{ Pd(PPh}_3)_2Cl_2}$   Ar–CH(Me)–COOBu-*i*   22-30%

---

**I.G.2-10** K. Yamamoto et al., *Tetrahedron*, **1992**, *48*, 2333.

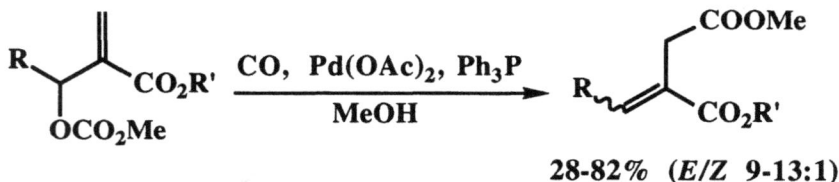

R–CH(OCO₂Me)–C(=CH₂)–CO₂R'   $\xrightarrow[\text{MeOH}]{CO, \text{ Pd(OAc)}_2, \text{ Ph}_3P}$   R–CH=C(CH₂COOMe)–CO₂R'

28-82%  (*E/Z* 9-13:1)

---

**I.G.2-11** S. Nakatani, J. Yoshida* and S. Isoe* *J. Chem. Soc., Chem. Commun.*, **1992**, 880.

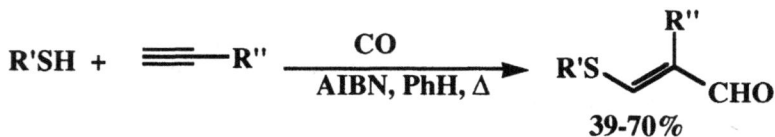

R'SH + ≡–R''   $\xrightarrow[\text{AIBN, PhH, }\Delta]{CO}$   R'S–CH=C(R'')–CHO

39-70%

**I.G.2-12** M. Miura et al., *Tetrahedron Lett.*, **1992**, *33*, 5369.

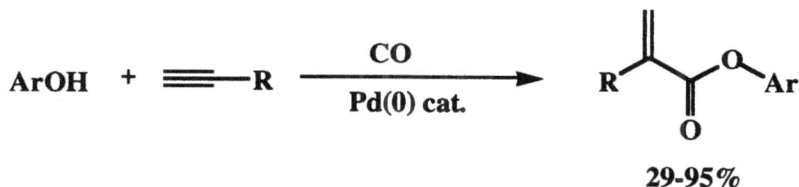

$$ArOH \quad + \quad \equiv\!\!-R \xrightarrow[\text{Pd(0) cat.}]{\text{CO}} $$

29-95%

---

**I.G.2-13** T. Okano*, N. Okabe, J. Kiji, *Bull. Chem. Soc. Jpn.*, **1992**, *65*, 2589.

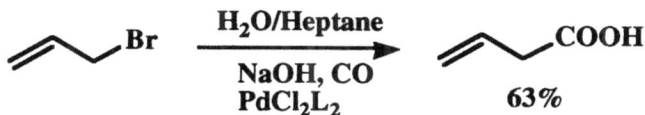

$$\xrightarrow[\substack{\text{NaOH, CO}\\ \text{PdCl}_2\text{L}_2}]{\text{H}_2\text{O/Heptane}}$$

Br → COOH

63%

---

**I.G.2-14** K. Uneyama et al., *Tetrahedron Lett.*, **1992**, *33*, 4333.

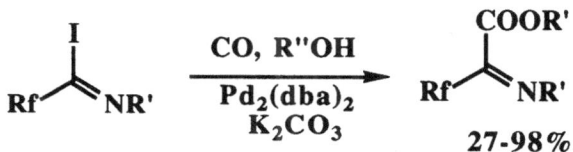

$$\xrightarrow[\substack{\text{Pd}_2\text{(dba)}_2\\ \text{K}_2\text{CO}_3}]{\text{CO, R''OH}}$$

COOR'
Rf NR'

27-98%

---

**I.G.2-15** D. H. R. Barton et al., *Tetrahedron Lett.*, **1992**, *33*, 4389.

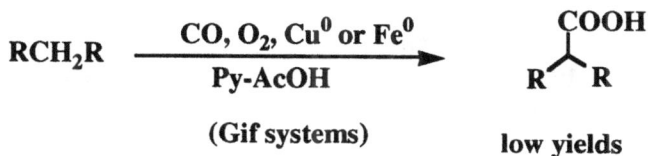

$$RCH_2R \xrightarrow[\text{Py-AcOH}]{\text{CO, O}_2\text{, Cu}^0 \text{ or Fe}^0}$$

(Gif systems)

COOH
R    R

low yields

---

**I.G.2-16** A. A. Kelkar et al., *J. Organometal. Chem.* **1992**, *430*, 111.

**Carbonylation of Methanol to Acetic Acid Using Homogeneous Ru Complex Catalyst**

**I.G.2-17**  M. Sierra et al., *Organometallics*, **1992**, *11*, 1979.

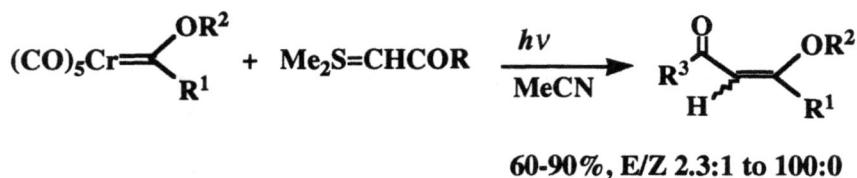

$$(CO)_5Cr = \underset{R^1}{\overset{OR^2}{<}} \quad + \quad Me_2S=CHCOR \quad \xrightarrow[\text{MeCN}]{h\nu} \quad R^3 \underset{H}{\overset{O}{\diagup}} \underset{R^1}{\overset{OR^2}{\diagdown}}$$

**60-90%, E/Z 2.3:1 to 100:0**

---

**I.G.2-18**  P. J. Stang et al., *Organometallics*, **1992**, *11*, 1017 and 1026.

**Carbonylation of Vinyl electrophiles**

---

**I.G.2-19**  A. Cabrera et al., *Bull. Soc. Chem. Belg.*, **1992**, *101*, 173.

**Hydroformylation of Alkenes with $Sn[Co(CO)_4]_4$ in Homogeneous Phase**

---

**I.G.2- 20**  P. Kalck et al., *J. Organometal. Chem.* **1992**. *426*, C16,

**Dirhodium Complexes for Low Pressure Hydroformylation**

---

**I.G.2-21**  C. Claver*, S. Castillon*, J. C. Bayon* et al., *Organometallics*, **1992**, *11*, 3525; H. Takaya et al., *Tetrahedron: Asymmetry*, **1992**, *3*, 583.

**Hydroformylation of Allyl and Vinyl Ethers/enantioselective.**

**I.G.2-22**  S. Murai et al., *Organometallics*, **1992**, *11*, 3494.

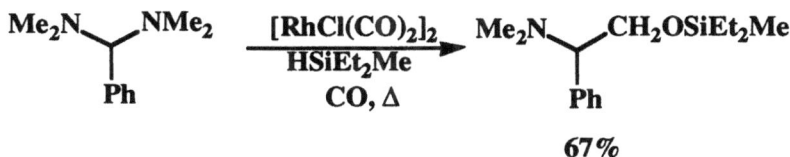

$$Me_2N \overset{\displaystyle}{\underset{Ph}{\bigvee}} NMe_2 \quad \xrightarrow[\substack{HSiEt_2Me \\ CO,\ \Delta}]{[RhCl(CO)_2]_2} \quad Me_2N \overset{\displaystyle}{\underset{Ph}{\bigvee}} CH_2OSiEt_2Me$$

**67%**

---

**I.G.2-23**  S. Murai et al., *J. Org. Chem.*, **1992**, *57*, 2; *J. Am. Chem. Soc.*, **1992**, *114*, 9710.

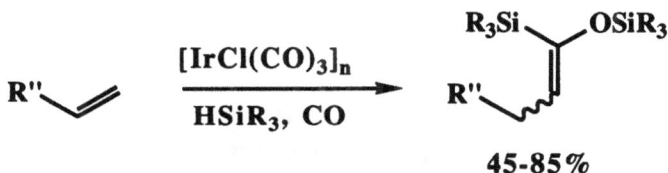

$$R'' \diagdown\!\!\diagup \quad \xrightarrow[\substack{HSiR_3,\ CO}]{[IrCl(CO)_3]_n} \quad R''$$

$R_3Si \diagup OSiR_3$

**45-85%**

---

**I.G.2-24**  Y. Tsuji and T. Ishii, *J. Organometal. Chem.* **1992**. *425*, 41.

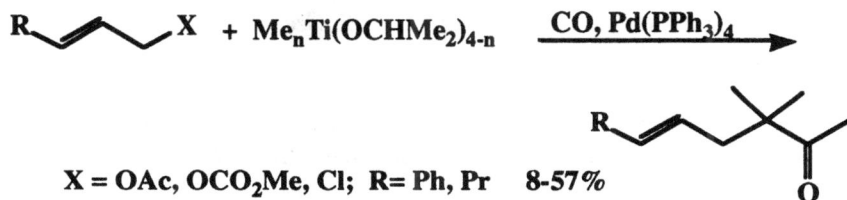

$$R \diagdown\!\!\diagup\!\!\diagdown X \ +\ Me_nTi(OCHMe_2)_{4-n} \quad \xrightarrow{CO,\ Pd(PPh_3)_4}$$

X = OAc, OCO$_2$Me, Cl;  R= Ph, Pr   **8-57%**

---

**I.G.2-25**  A. Llebaria*, F. Camps, J. M. Moreto, *Tetrahedron Lett.*, **1992**, *33*, 3683.

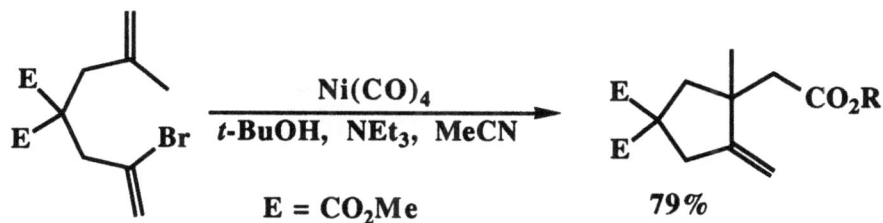

$$\xrightarrow[\substack{t\text{-BuOH, NEt}_3,\ MeCN}]{Ni(CO)_4}$$

E = CO$_2$Me                    **79%**

**I.G.2-26** S. Derien et al., *J. Organometal. Chem.*, **1992**. *424*, 213.

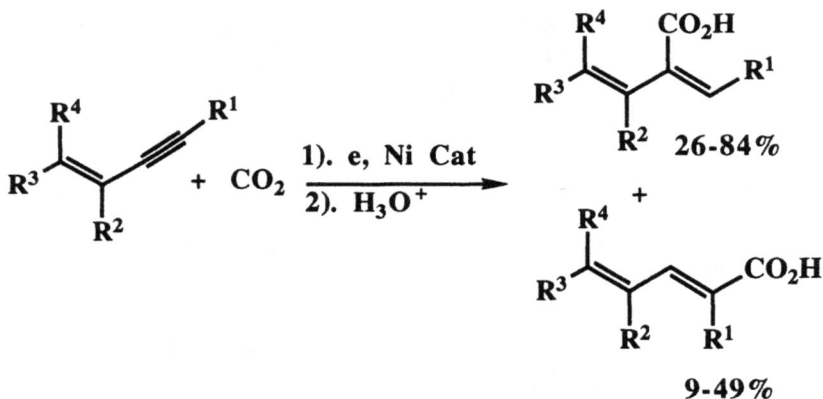

**I.G.2-27** G. Consiglio et al., *Organometallics*, **1992**, *11*, 20.

**Selectivity of the Carbonylation of Styrene by
Cationic Palladium Complexes**

**I.G.2-28** C. B. Anderson and R. Markovic, *Coll. Czech. Chem. Comm.*
**1992**, 57, 2374.

**Novel Ring Closure Carbonylation Reactions of 1,5-
Cyclooctadiene in the Presence of Pd(II)-Catalyst**

**I.G.2-29** R. B. Brossman and S. L. Buchwald, *J. Org. Chem.*, **1992**, *57*,
5803.

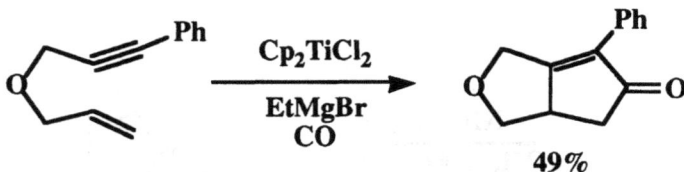

**I.G.2-30** M. E. Krafft*, C. Wright, *Tetrahedron Lett.*, **1992**, *33*, 151.

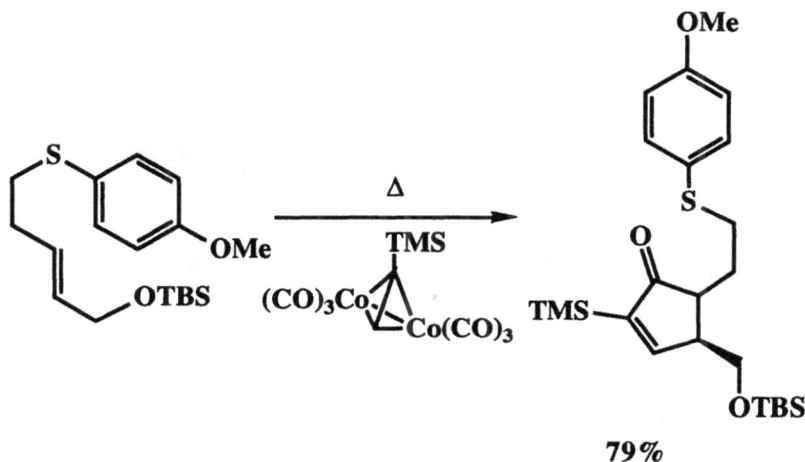

79%

---

**I.G.2-31** X. Verdaguer et al;, *J. Organometal. Chem.*, *433*, 305.

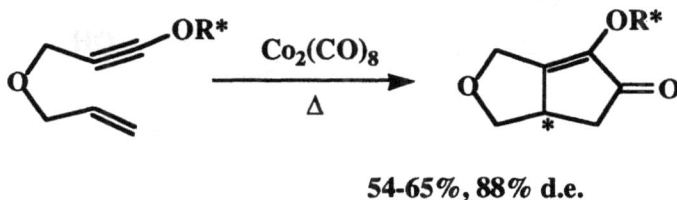

54-65%, 88% d.e.

---

**I.G.2-32** M. Tanaka et al., *J. Org. Chem.*, **1992**, *57*, 2677.

**Formylation of Aromatic Compounds with CO in
$HSO_3F$-$SbF_5$ under Atmospheric Pressure**

---

**I.G.2-33** M. Yamashita et al., *Bull. Chem. Soc. Jpn.*, **1992**, *65*, 1257.

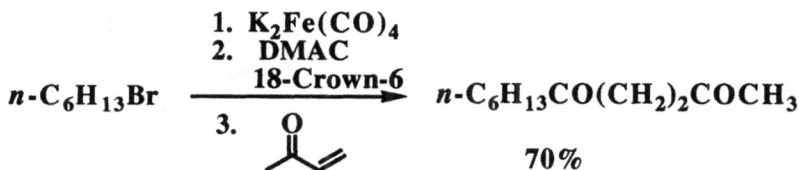

$n$-$C_6H_{13}Br$ →

1. $K_2Fe(CO)_4$
2. DMAC
   18-Crown-6
3.

$n$-$C_6H_{13}CO(CH_2)_2COCH_3$

70%

**I.G.2-34** K. Yasuda and K. Shinoda, *Bull. Chem. Soc. Jpn.*, **1992**, *65*, 289.

### Vapor Phase Carbonylation of Organic Halo Compounds

---

**I.G.2-35** T. Okano*, N. Harada and J. Kiji*, *Bull. Chem. Soc. Jpn.*, **1992**, *65*, 1741.

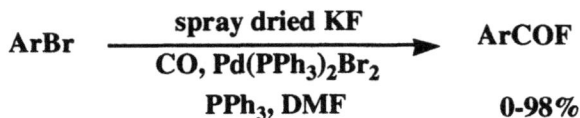

$$\text{ArBr} \xrightarrow[\substack{\text{CO, Pd(PPh}_3)_2\text{Br}_2 \\ \text{PPh}_3\text{, DMF}}]{\text{spray dried KF}} \text{ArCOF}$$

0-98%

---

**I.G.2-36** H. Arzoumanian et al., *Organometallics*, **1992**, *11*, 493.

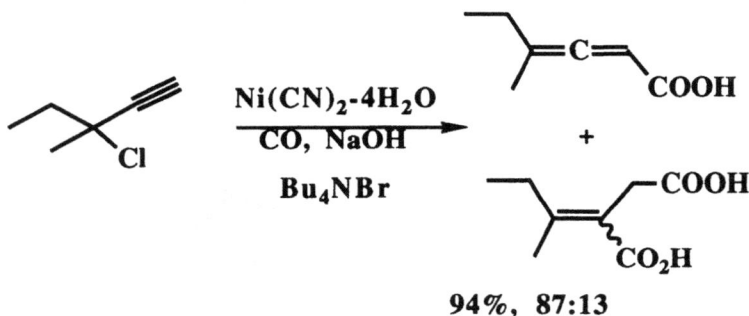

94%, 87:13

---

**I.G.2-37** C. Amatore*, A. Jutand* et al., *J. Am. Chem. Soc.*, **1992**, *114*, 7076.

### Carbon Dioxide as C1 Building Block using Pd-Catalyst and Aromatic Halides

---

**I.G.2-38** Y. Tamaru et al., *Angew. Chem., Int. Ed. Engl.* **1992**, *31*, 645.

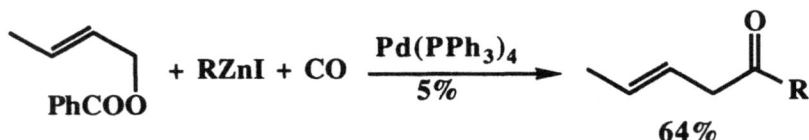

64%

**I.G.2-39** T. Gallagher et al., *J. Chem. Soc., Perkin. Trans. I*, **1992**, 433.

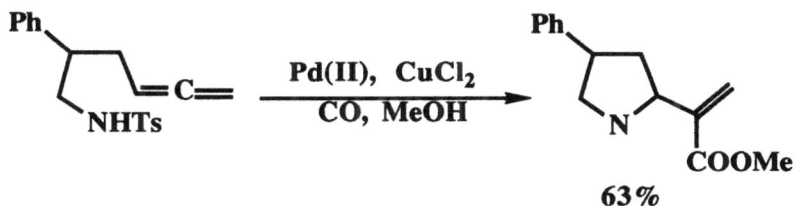

**I.G.2-40** C. A. Merlic et al., *J. Am. Chem. Soc.*, **1992**, *114*, 8722.

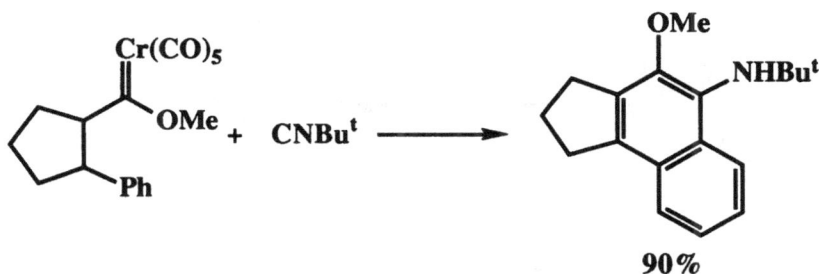

**I.G.2-41** E. P. Kundig et al., *Angew. Chem., Int. Ed. Engl.* **1992**, *31*, 1071.

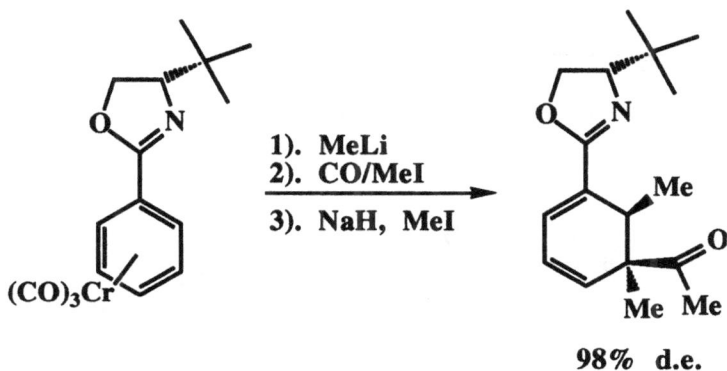

1). MeLi
2). CO/MeI
3). NaH, MeI

98% d.e.

**I.G.2-42** G. P. Roth and J. A. Thomas, *Tetrahedron Lett.*, **1992**, *33*, 1959.

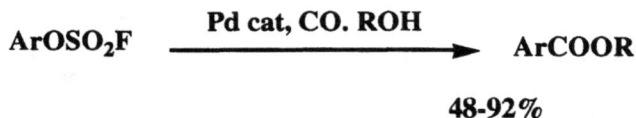

$$ArOSO_2F \xrightarrow{\text{Pd cat, CO. ROH}} ArCOOR$$

**48-92%**

---

**I.G.2-43** C. Claver*, S. Castillon* et al., *J. Chem. Soc., Chem. Commun.*, **1992**,, 639.

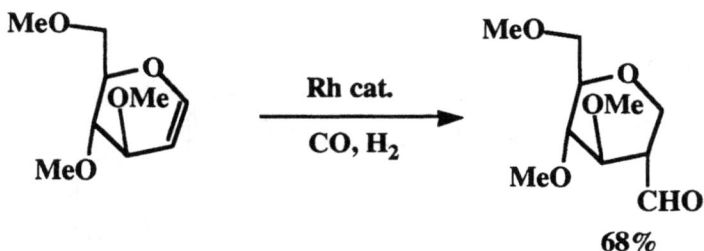

**68%**

---

**I.G.2-44** K. Yoshida et al., *J. Chem. Soc., Perkin. Trans. I*, **1992**, 1127.

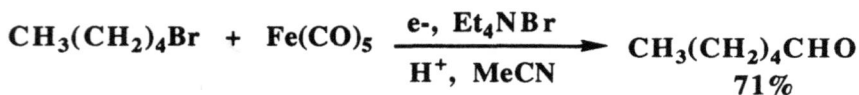

$$CH_3(CH_2)_4Br \;+\; Fe(CO)_5 \xrightarrow[\text{H}^+,\;\text{MeCN}]{\text{e-, Et}_4\text{NBr}} CH_3(CH_2)_4CHO$$

**71%**

---

**I.G.2-45** K. Yamamoto et al., *SynLett.*, **1992**, 323.

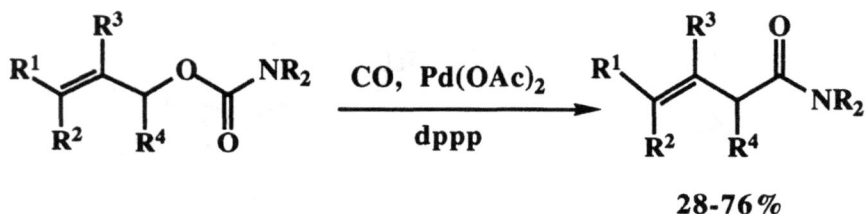

**28-76%**

**I.G.2-46** W. R. Jackson et al., *Aust. J. Chem.*, **1992**, *45*, 823.

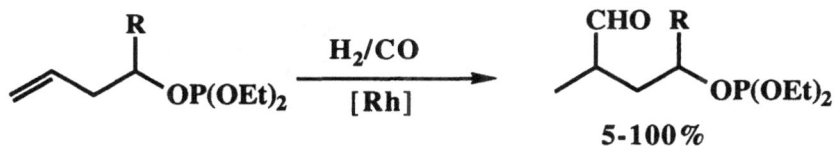

**I.G.2-47** M. Franck-Newmann et al., *Tetrahedron Lett.*, **1992**, *33*, 7361, 7365.

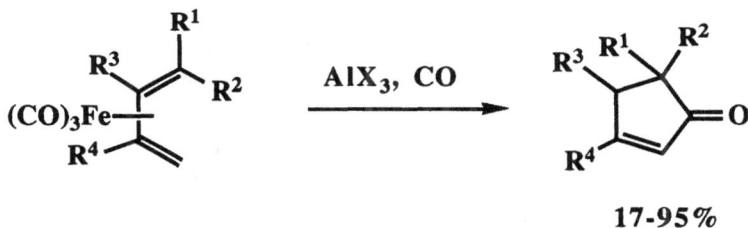

**I.G.2-48** Y. Yamaru et al., *Tetrahedron Lett.*, **1992**, *33*, 631.

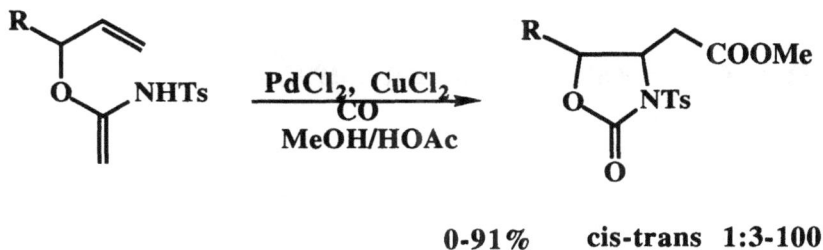

0-91%     cis-trans   1:3-100

**I.G.2-49** P. V. K. Raju and S. R. Adapa, *Ind. J. Chem.*, **1992**, *31B*, 363.

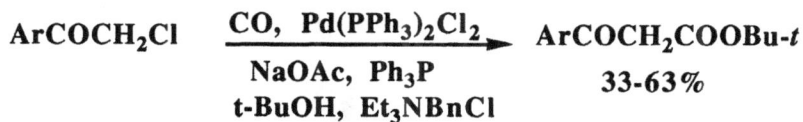

$$ArCOCH_2Cl \xrightarrow[\substack{NaOAc, \ Ph_3P \\ t\text{-}BuOH, \ Et_3NBnCl}]{CO, \ Pd(PPh_3)_2Cl_2} ArCOCH_2COOBu\text{-}t$$

33-63%

**I.G.2-50** J. M. Moreto et al., *Tetrahedron Lett.*, **1992**, *33*, 109, 113.

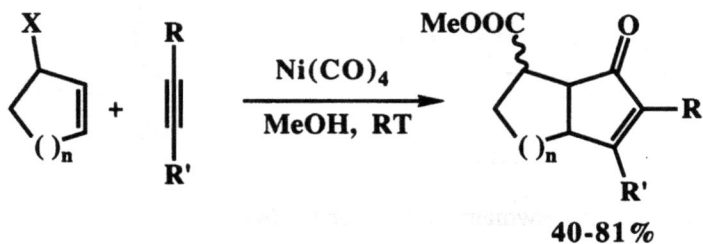

40-81%

---

**I.G.2-51** H. Tsuruta\*, G. A. King III\* et al., *Tetrahedron*, **1992**, *48*, 3473.

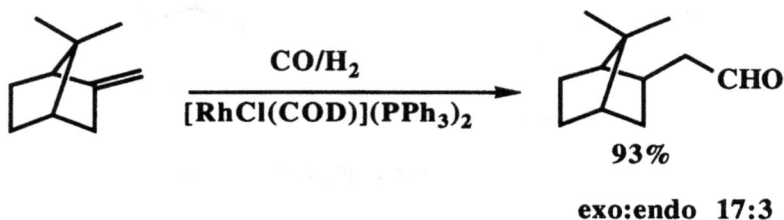

$$CO/H_2$$
$$[RhCl(COD)](PPh_3)_2$$

93%

exo:endo  17:3

---

**I.G.2-52** M. G. Finn et al., *J. Am. Chem. Soc.*, **1992**, *114*, 8735.

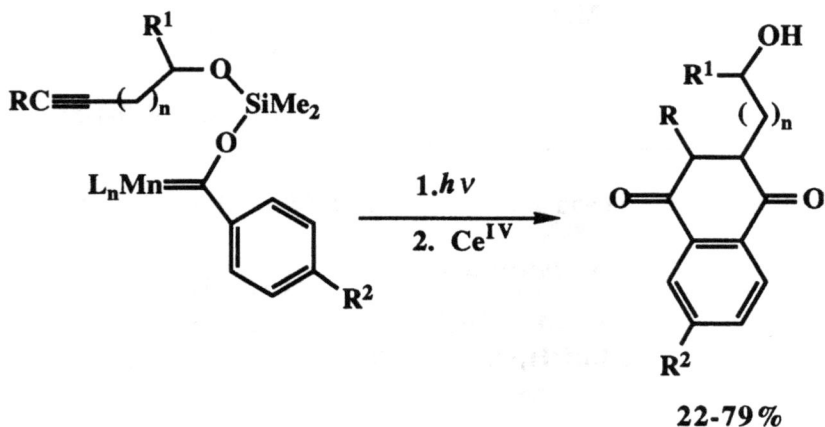

22-79%

**I.G.2-53** E. Negishi et al, *Tetrahedron Lett.,* **1992**, *33*, 1543

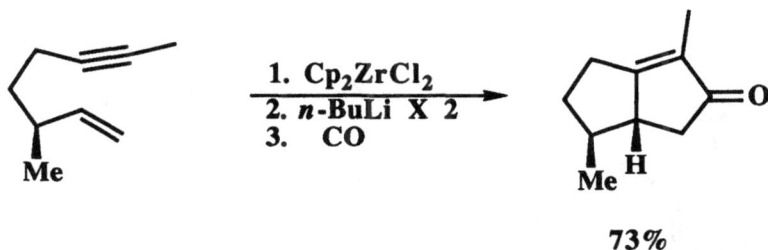

1. Cp$_2$ZrCl$_2$
2. *n*-BuLi X 2
3. CO

73%

## I.G.3. Other Synthesis via Organometallics

**I.G.3-1** T. R. Hoye and J. A. Suriano, *Organometallics*, **1992**, *11*, 2044.

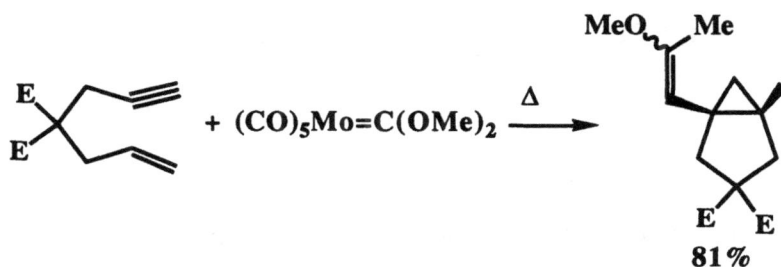

+ (CO)$_5$Mo=C(OMe)$_2$ $\xrightarrow{\Delta}$

81%

**I.G.3-2** W. D. Wulff et al., *J. Am. Chem. Soc.*, **1992**, *114*, 10785.

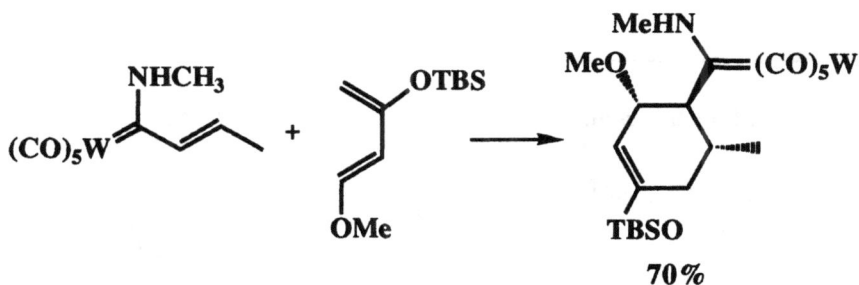

+

70%

**I.G.3-3** A. I. Myers, T. G. Gant, *J. Org. Chem.*, **1992**, *57*, 4225.

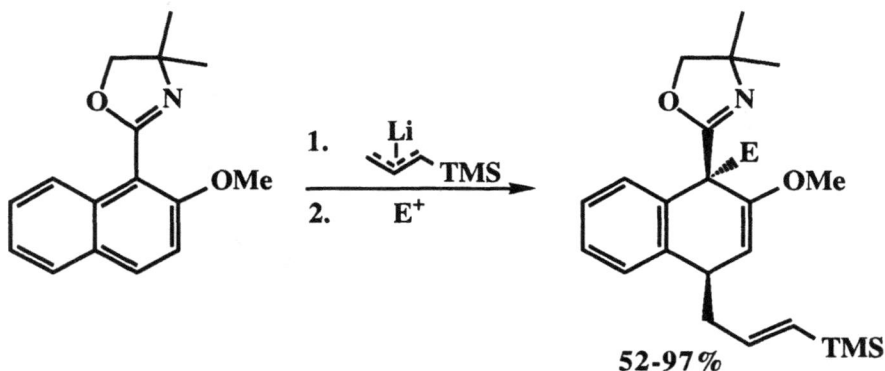

52-97%

---

**I.G.3-4** W. Cabri et al., *J. Org. Chem.*, **1992**, *57*, 3558.

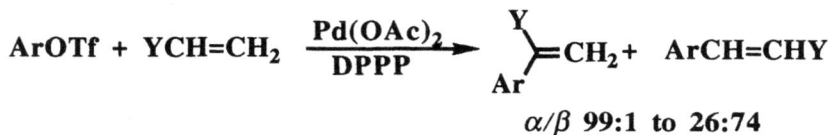

$$ArOTf + YCH=CH_2 \xrightarrow[\text{DPPP}]{\text{Pd(OAc)}_2} \underset{Ar}{\overset{Y}{\diagup}}{=}CH_2 + ArCH=CHY$$

$\alpha/\beta$ **99:1 to 26:74**

---

**I.G.3-5** F. Petit et al., *Tetrahedron Lett.*, **1992**, *33*, 2001.

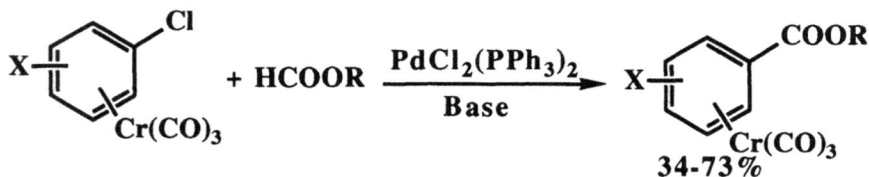

$$+ HCOOR \xrightarrow[\text{Base}]{\text{PdCl}_2\text{(PPh}_3\text{)}_2}$$

34-73%

---

**I.G.3-6** S. Ikegami et al., *Tetrahedron Lett.*, **1992**, *33*, 2709.

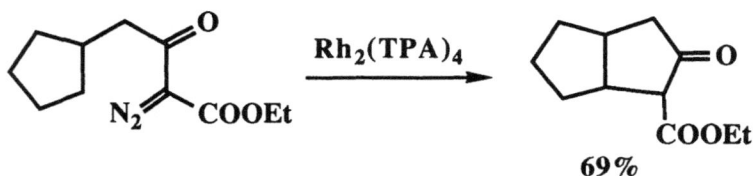

$$\xrightarrow{\text{Rh}_2\text{(TPA)}_4}$$

69%

**I.G.3-7** J. W. Herndon et al., *J. Am. Chem. Soc.*, **1992**, *114*, 8394.

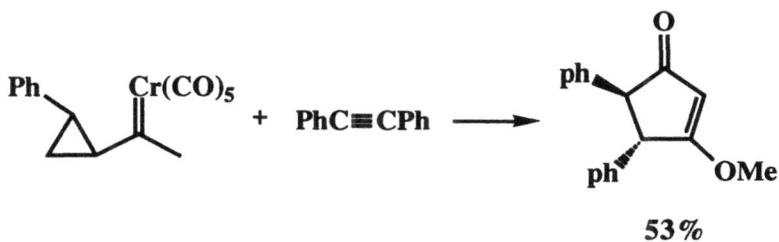

53%

---

**I.G.3-8** W. D. Wulff et al., *J. Am. Chem. Soc.*, **1992**, *114*, 10665.

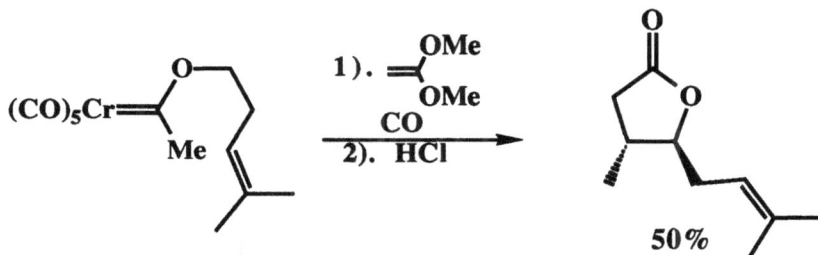

50%

---

**I.G.3-9** S. Chamberlin and W. D. . Wulff*, *J. Am. Chem. Soc.*, **1992**, *114*, 10667.

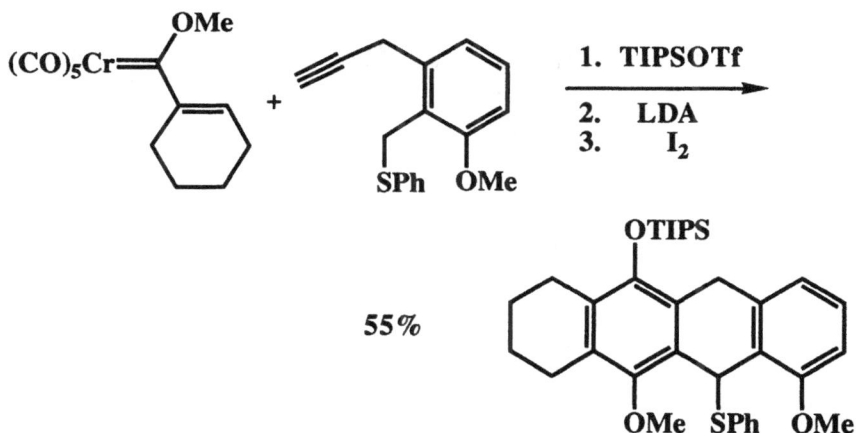

55%

**I.G.3-10** A. Padwa et al., *J. Org. Chem.*, **1992**, *57*, 1331.

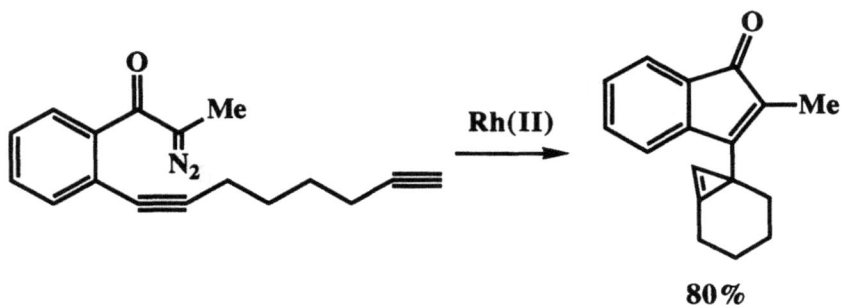

**80%**

---

**I.G.3-11** B. M. Trost et al., *J. Am. Chem. Soc.*, **1992**, *114*, 9837.

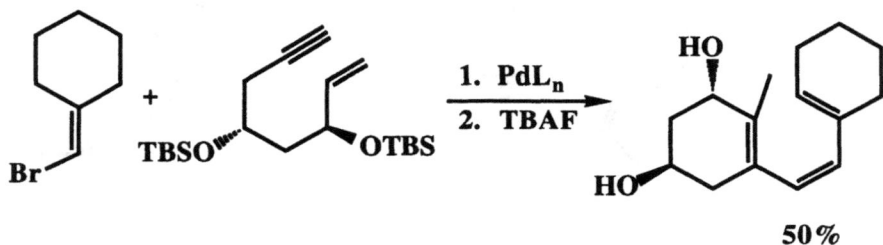

**50%**

---

**I.G.3-12** T. Jeffery, *Tetrahedron Lett.*, **1992**, *33*, 1989.

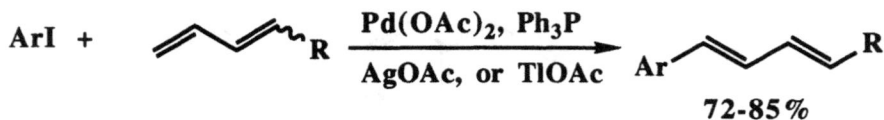

$$ArI + \text{(butadiene)}R \xrightarrow[\text{AgOAc, or TlOAc}]{Pd(OAc)_2, Ph_3P} Ar\text{(butadiene)}R$$

**72-85%**

**_I.G.3-13** T. Hudlicky and H. F. Olivo, *J. Am. Chem. Soc.*, **1992**, *114*, 9695.

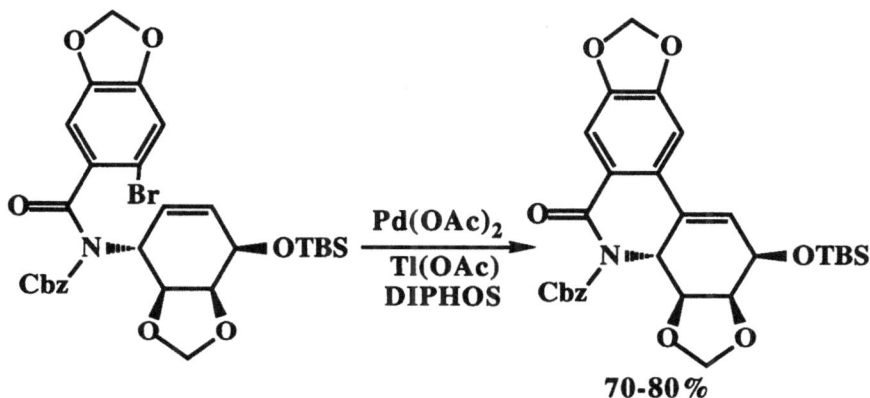

**70-80%**

**I.G.3-14** G. W. Gribble ans S. C. Conway, *Syn. Commun.*, **1992**, *22*, 2129.

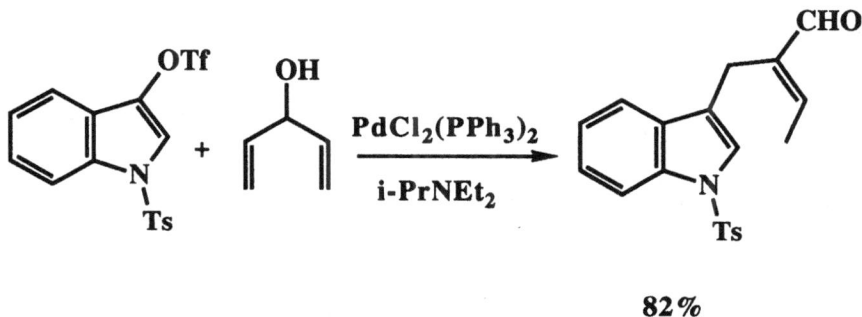

**82%**

**I.G.3-15** Z. Zhang et al., *Syn. Commun.*, **1992**, *22*, 2019.

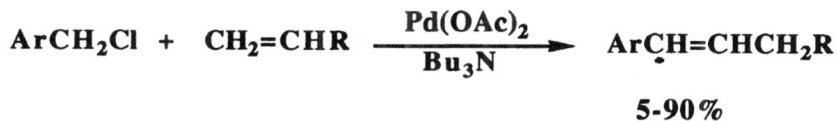

$$ArCH_2Cl + CH_2=CHR \xrightarrow[Bu_3N]{Pd(OAc)_2} ArCH=CHCH_2R$$

**5-90%**

**I.G.3-16** S. Cacchi et al., *Tetrahedron Lett.*, **1992**, *33*, 3073.

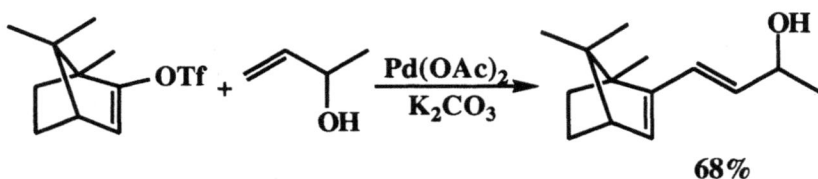

**68%**

---

**I.G.3-17** M. Shibasaki et al., *Tetrahedron Lett.*, **1992**, *33*, 2589.

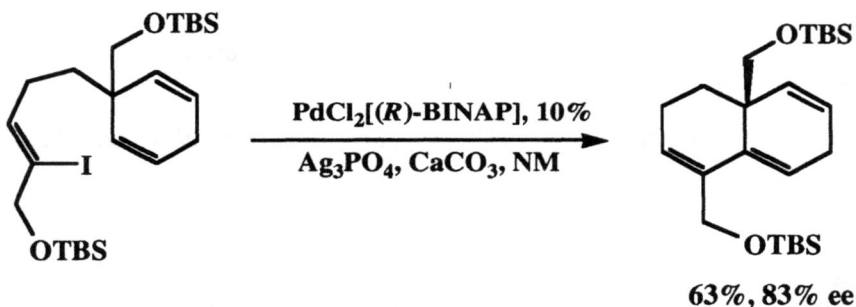

$PdCl_2[(R)\text{-}BINAP]$, 10%

$Ag_3PO_4$, $CaCO_3$, NM

**63%, 83% ee**

---

**I.G.3-18** Y. Butsugan et al., *Tetrahedron Lett.*, **1992**, *33*, 2581.

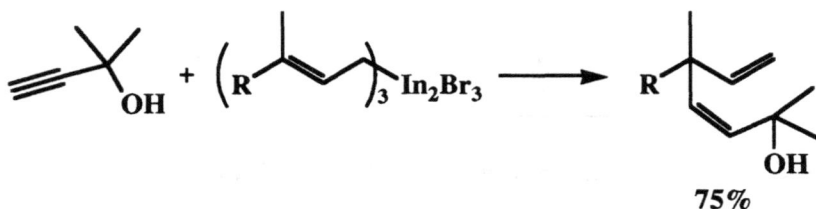

**75%**

---

**I.G.3-19** D. Seebach et al., *Angew. Chem., Int. Ed. Engl.* **1992**, *31*, 1587.

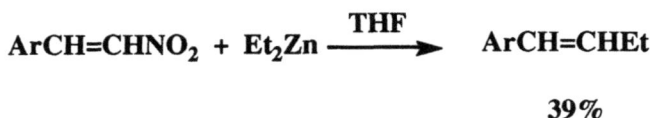

$$ArCH=CHNO_2 + Et_2Zn \xrightarrow{\ \ THF\ \ } ArCH=CHEt$$

**39%**

**I.G.3-20** G. A. Potter and R. McCague *J. Chem. Soc., Chem. Commun.*, **1992**, 635.

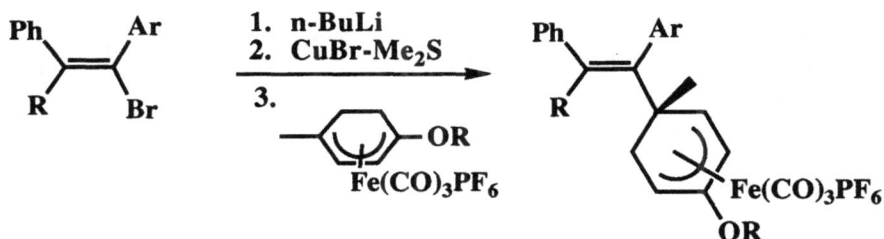

**I.G.3-21** A. G. Myers, et al., *J. Am. Chem. Soc.*, **1992**, *114*, 9369.

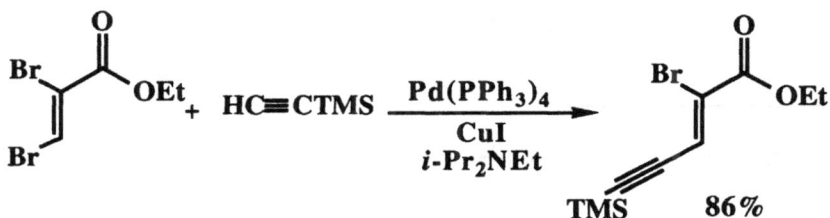

**I.G.3-22** D. Bouyssi et al., *Tetrahedron Lett.*, **1992**, *33*, 2811.

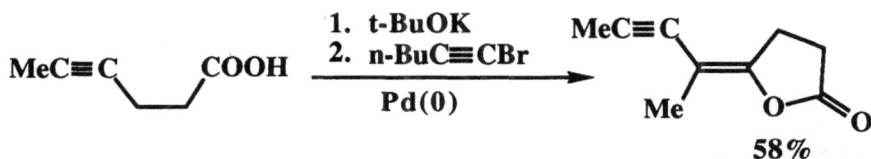

**I.G.3-23** B. M. Trost et al., *J. Am. Chem. Soc.*, **1992**, *114*, 9327.

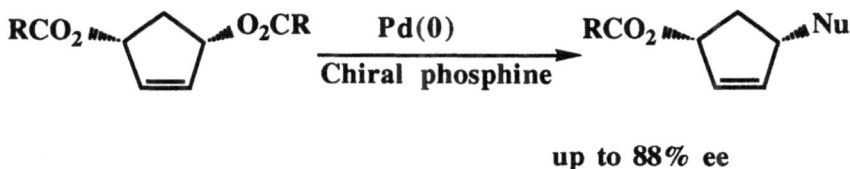

up to **88% ee**

**I.G.3-24** H. Yamanaka et al., *Synthesis*, **1992**, 552.

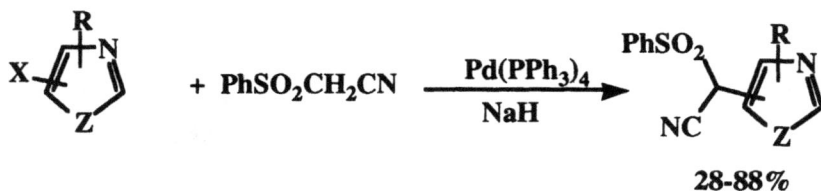

28-88%

**I.G.3-25** X. Lu et al.,*Tetrahedron Lett.*, **1992**, *33*, 2535.

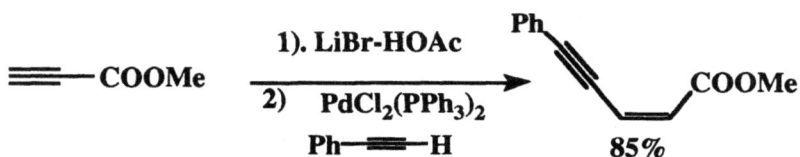

85%

**I.G.3-26** I. Marek et al., *SynLett.*, **1992**, 633,

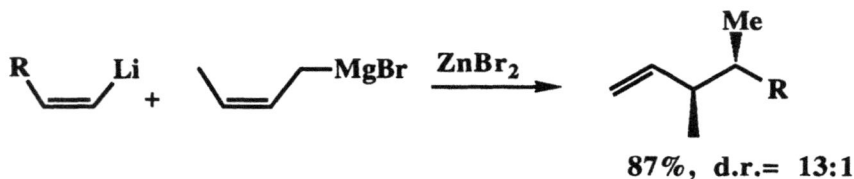

87%,  d.r.= 13:1

**I.G.3-27** R. D. Rieke* and H. Xiong, *J. Org. Chem.*, **1992**, *57*, 6560.

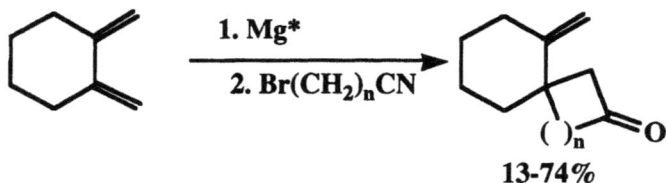

13-74%

**I.G.3-28**  T. Kauffmann* and H. Kallweit, *Chem. Ber.* **1992**, *125*, 143, 149.

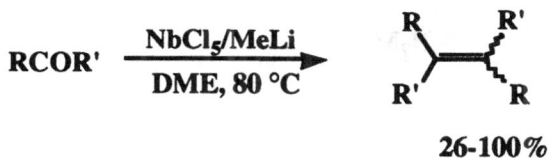

$$RCOR' \xrightarrow[\text{DME, 80 °C}]{\text{NbCl}_5/\text{MeLi}}$$

26-100%

---

**I.G.3-29**  J. Barluenga et al., *Tetrahedron Lett.*, **1992**, *33*, 7579.

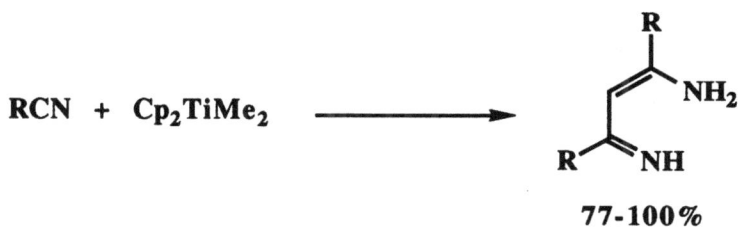

$$RCN + Cp_2TiMe_2 \longrightarrow$$

77-100%

---

**I.G.3-30**  T. Takahashi et al., *Chem. Lett.*, **1992**, 1693.

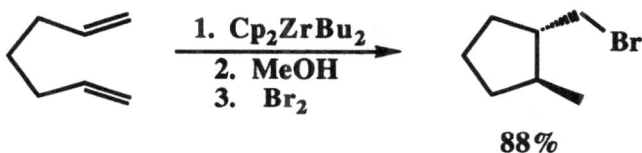

1. $Cp_2ZrBu_2$
2. MeOH
3. $Br_2$

88%

---

**I.G.3-31**  J. M. Moreto et al., *J. Am. Chem. Soc.*, **1992**, *114*, 10449.

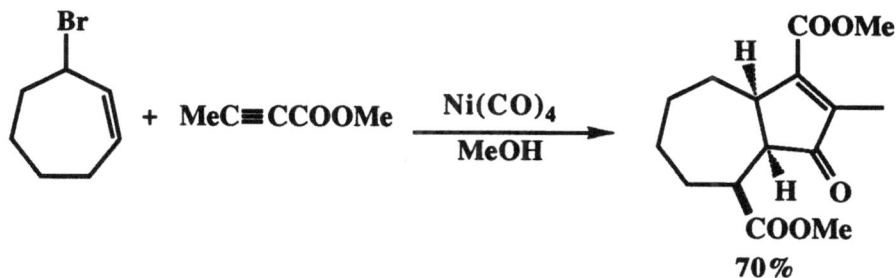

$$+ \ MeC\equiv CCOOMe \xrightarrow[\text{MeOH}]{\text{Ni(CO)}_4}$$

70%

**I.G.3-32** E. Negishi et al., *Tetrahedron Lett.*, **1992**, *33*, 1965.

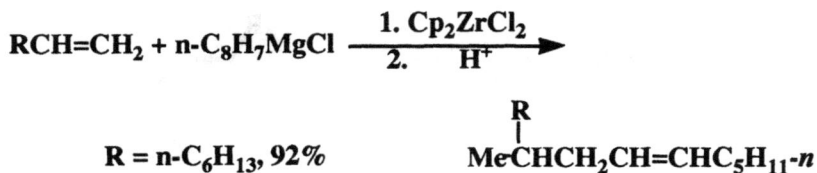

$$RCH=CH_2 + \text{n-}C_8H_7MgCl \xrightarrow[\text{2. } H^+]{\text{1. } Cp_2ZrCl_2}$$

$$R = \text{n-}C_6H_{13}, 92\%$$

$$\underset{\mid}{\overset{R}{MeCHCH_2CH=CHC_5H_{11}}}\text{-}n$$

---

**I.G.3-33** G. W. Klumpp et al., *Tetrahedron*, **1992**, *48*, 6087, 6105.

---

**I.G.3-34** F. J. Pulido et al., *J. Chem. Soc., Perkin. Trans. I*, **1992**, 327.

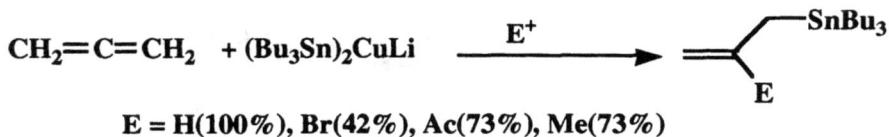

$$CH_2=C=CH_2 + (Bu_3Sn)_2CuLi \xrightarrow{E^+}$$

$$E = H(100\%), Br(42\%), Ac(73\%), Me(73\%)$$

## I.H.    Rearrangements

## I.H.1. Claisen, Cope and Similar Processes

**I.H.1-1**  S. Janardhanam and K. Rajagopalan* *J. Chem. Soc., Perkin. Trans. I,* **1992,** 2727.

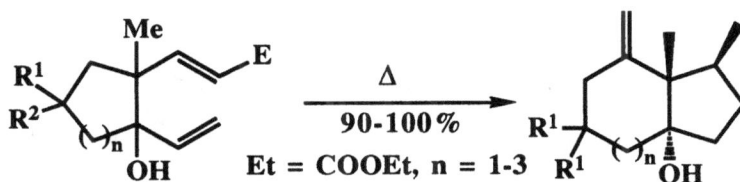

Et = COOEt, n = 1-3

**I.H.1-2**  A. Ahmed, *Ind. J. Chem.* **1992,** *31B*, 63.

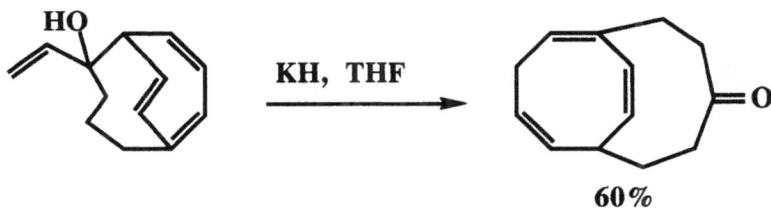

KH, THF

60%

**I.H.1-3**  D. W. Knight et al., *J. Chem. Soc., Perkin. Trans. I,* **1992,** 553.

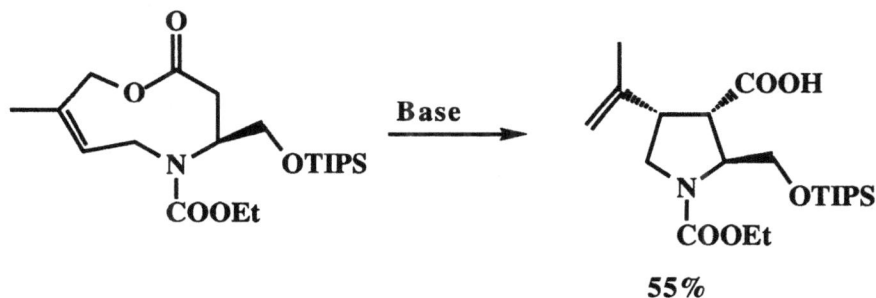

Base

55%

**I.H.1-4**  J. S. Panek* and T. D. Clark, *J. Org. Chem.*, **1992**, *57*, 4323.

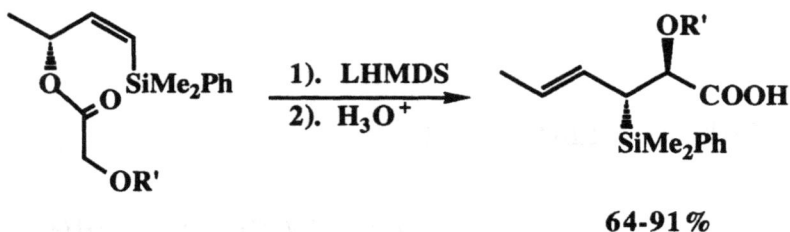

$$\text{1). LHMDS} \qquad \text{2). } H_3O^+$$

64-91%

---

**I.H.1-5**  J. R. Stille et al., *J. Org. Chem.*, **1992**, *57*, 461.

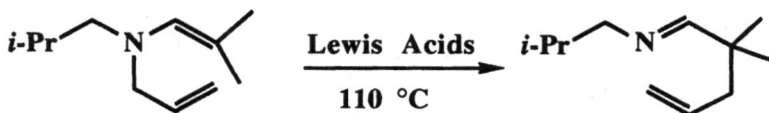

Lewis Acids
110 °C

---

**I.H.1-6**  T. Tsunoda et al., *Tetrahedron Lett.*, **1992**, *33*, 1651.

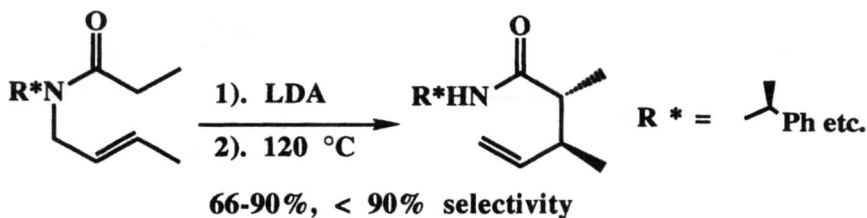

1). LDA
2). 120 °C

$$R* = \text{etc.}$$

66-90%, < 90% selectivity

---

**I.H.1-7**  A. Srikrishna and S. Nagaraju. *J. Chem. Soc., Perkin. Trans. I*,
**1992**, 311; R. S. Huber, G. B. Jones *J. Org. Chem.*, **1992**, *57*, 5778.

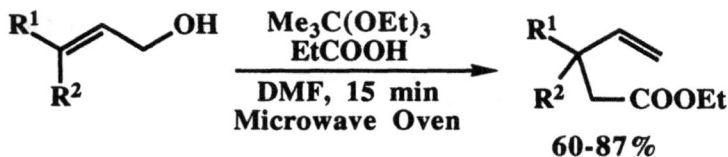

$Me_3C(OEt)_3$
EtCOOH

DMF, 15 min
Microwave Oven

60-87%

**I.H.1-8** S. Inoue et al., *J. Org. Chem.*, **1992**, *57*, 428.

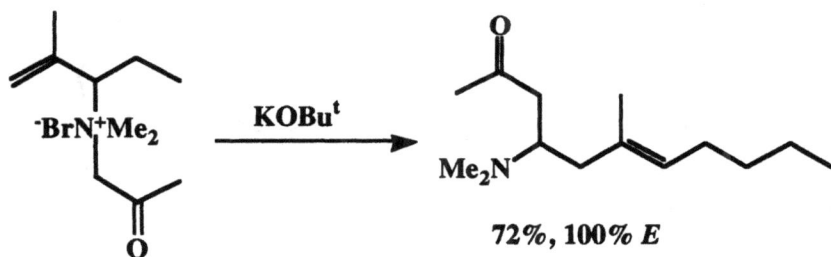

72%, 100% E

**I.H.1-9** P. J. Parsons et al., *J. Chem. Soc., Chem. Commun.*, **1992**, 350.

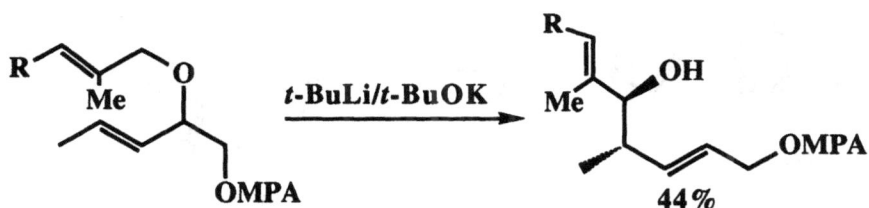

44%

**I.H.1-10** R. C. Hartley, S. Warren*, *Tetrahedron Lett.*, **1992**, *33*, 8155.

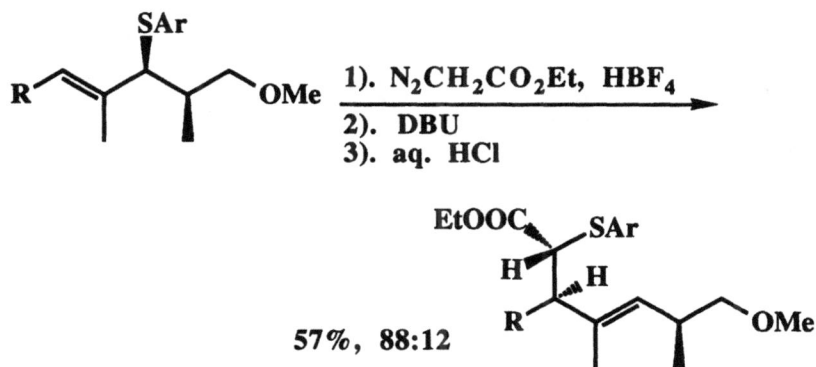

57%, 88:12

**I.H.1-11** K. V. Reedy, and S. Rajappa*, *Tetrahedron Lett.*, **1992**, *33*, 7957.

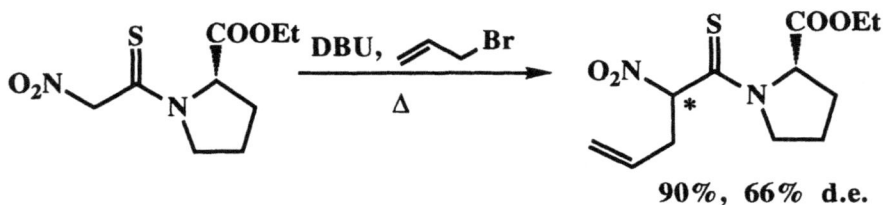

90%, 66% d.e.

---

**I.H.1-12** D. S. Brown, L.A. Paquette*, *J. Org. Chem.*, **1992**, *57*, 4512.

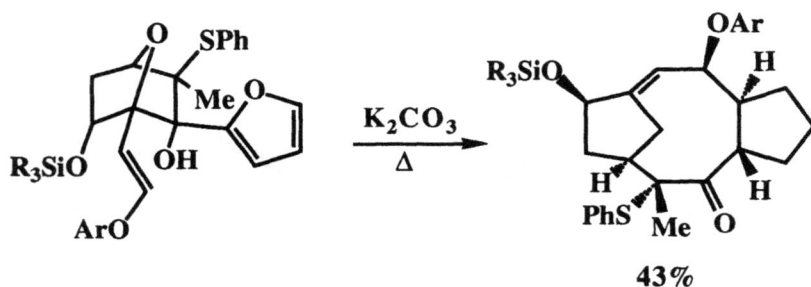

43%

---

**I.H.1-13** P. Metzner et al., *Tetrahedron*, **1992**, *48*, 10315, 10327.

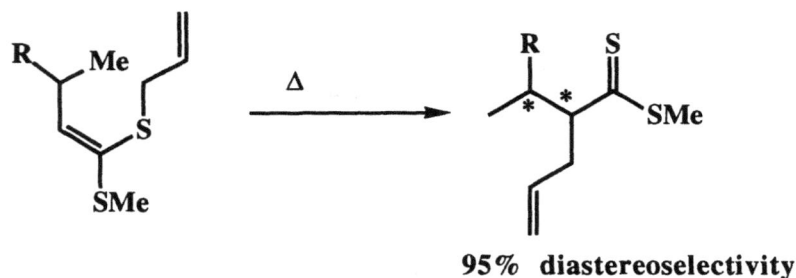

95% diastereoselectivity

**I.H.1-14** J. B. Baudin et al., *SynLett.*, **1992**, 909.

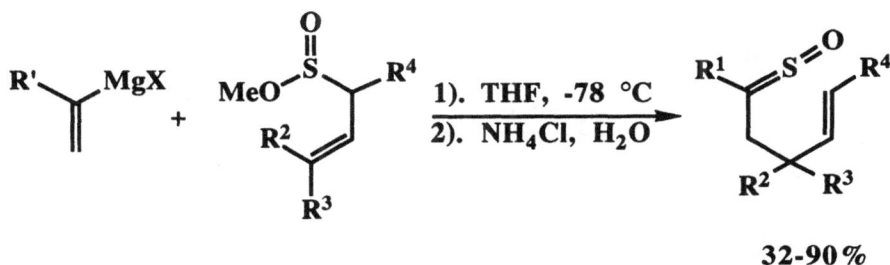

**32-90%**

---

**I.H.1-15** P. A. Grieco et al., *Tetrahedron Lett.*, **1992**, *33*, 4735.

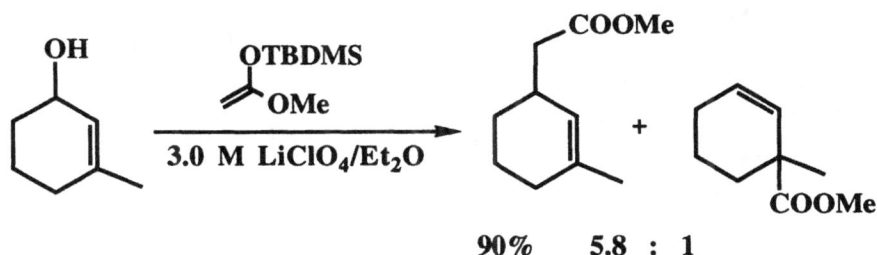

**90%**     **5.8 : 1**

---

**I.H.1-16** R. Hoffmann and R. Brukner *Chem.Ber.*, **1992**, 1471.

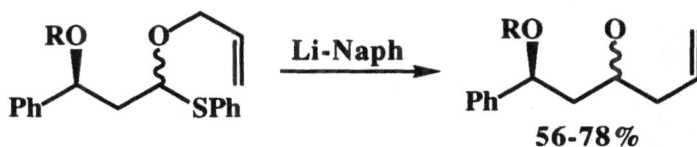

**56-78%**

---

**I.H.1-17** Y. Tamaru et al., *Tetrahedron Lett.*, **1992**, *33*, 789.

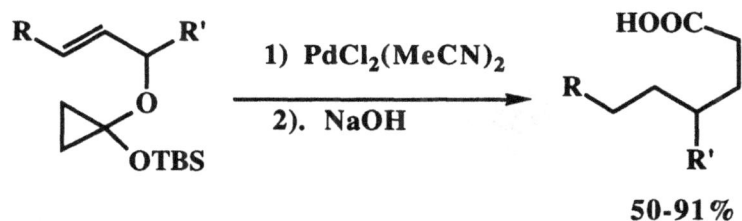

**50-91%**

**I.H.1-18**  L. A. Paquette et al., *Can. J. Chem.,* **1992**, *70,,* 1356.

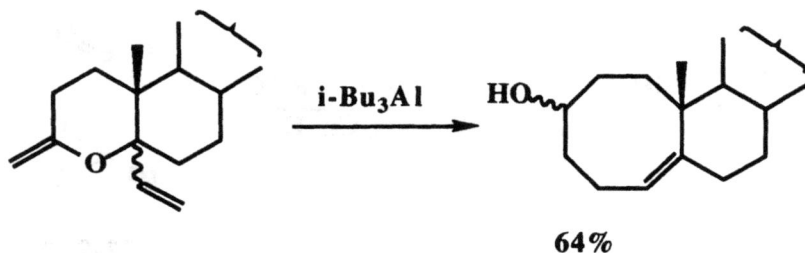

i-Bu₃Al → 64%

---

**I.H.1-19**  L. A. Paquette et al., *Tetrahedron Lett.,* **1992**, *33,* 923.

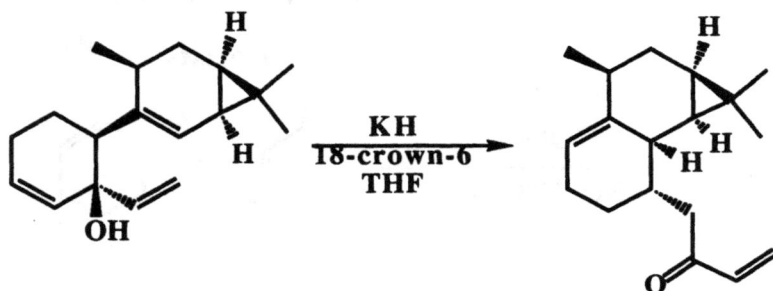

KH
18-crown-6
THF

---

**I.H.1-20**  T. Nakai et al., *Tetrahedron,* **1992**, *48,*  4087.

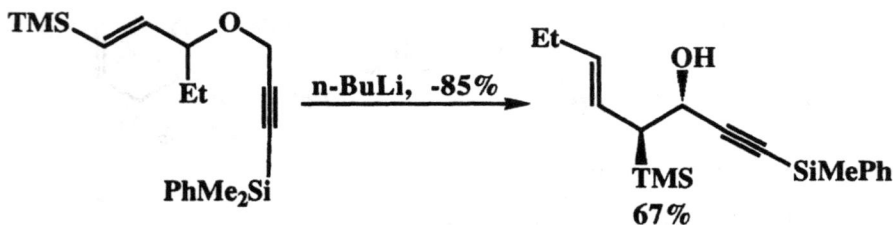

n-BuLi, -85% → 67%

**I.H.1-21**  J. L. van der Baan et al., *Tetrahedron Lett.*, **1992**, *33*, 1377.

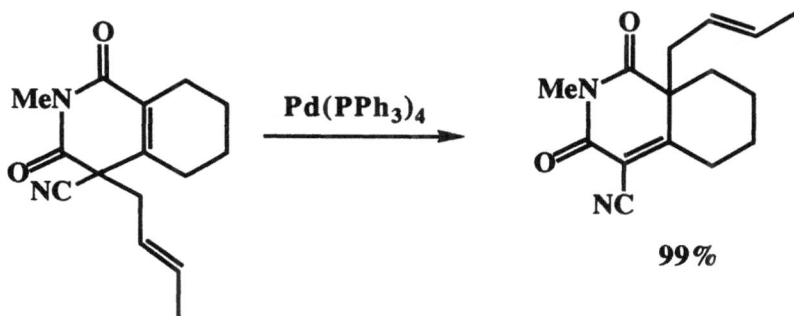

**99%**

---

**I.H.1-22**  R. Bao, S. Valverde* and B. Herradon, *SynLett.*, **1992**, 217.

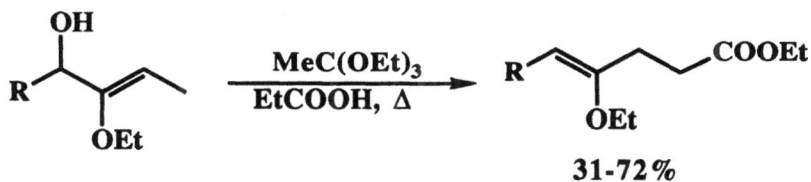

**31-72%**

---

**I.H.1-23**  P. J. Jacobi et al., *J. Org. Chem.*, **1992**, *57*, 6305,

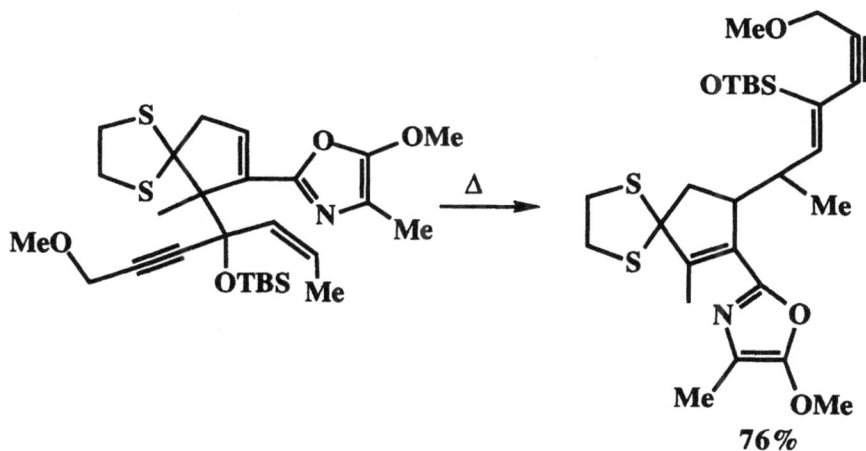

**76%**

**I.H.1-24** D. Desmaele and N. Champion, *Tetrahedron Lett.*, **1992**, *33*, 4447.

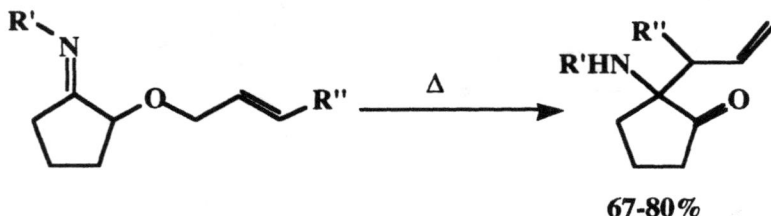

67-80%

---

**I.H.1-25** T. Troll and J. Wiedemann, *Tetrahedron Lett.*, **1992**, *33*, 3847.

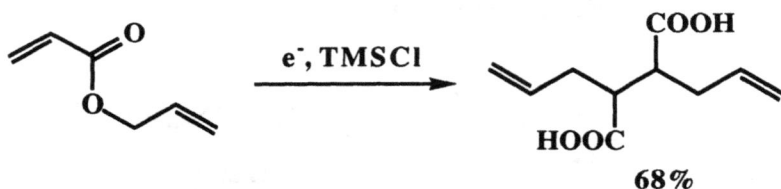

68%

---

**I.H.1-26** F. M. D. Ismail et al., *Tetrahedron Lett.*, **1992**, *33*, 3795.

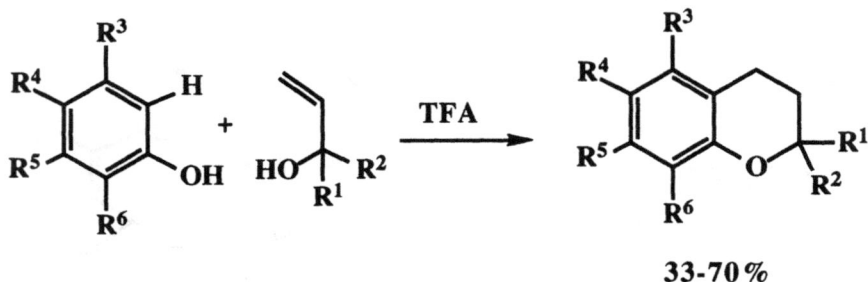

33-70%

---

**I.H.1-27** S. C. Joshi and K. N. Trivedi, *Tetrahedron*, **1992**, *48*, 563.

**A study of Claisen Rearrangements of 91,2-Dimethyl-3-prop-2-ynyloxy)-[4H]-1-benzopyran-4-one Derivatives**

**I.H.1-28** J. A. Marshall and X. Wang, *J. Org. Chem.*, **1992**, *57*, 2747;

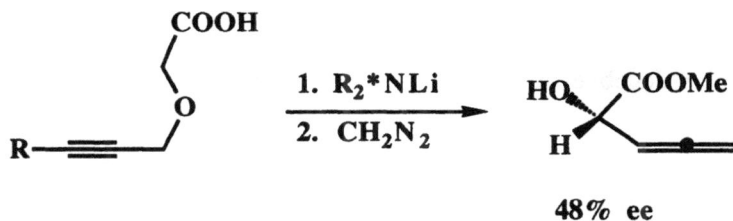

**48% ee**

**I.H.1-29** D. S. Grierson et al., *Tetrahedron Lett.*, **1992**, *33*, 4563.

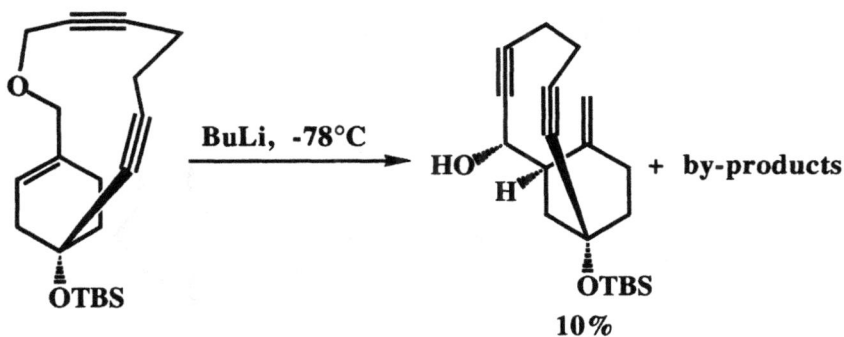

+ by-products

**10%**

**I.H.1-30** H. Yamamoto et al., *Bull. Chem. Soc. Jpn.*, **1992**, *65*, 541.

**On the Mechanism of Organoaluminum-Promoted Claisen Rearrangement of Allylic Vinyl Ethers**

## I.H.2.    Other Rearrangemnets

**I.H.2-1**  P.A. Jacobi et al., *Tetrahedron Lett.*, **1992**, *33*, 2265.

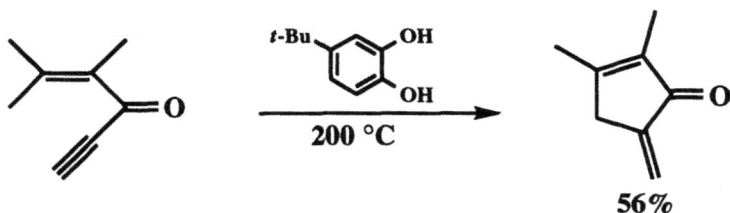

56%

---

**I.H.2-2**  T. Takeda et al., *Tetrahedron Lett.*, **1992**, *33*, 2583.

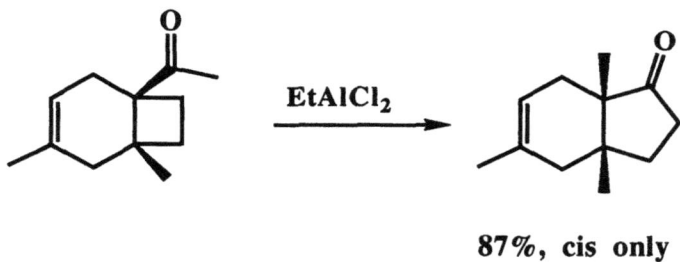

87%,  cis  only

---

**I.H.2-3**  K. Mizuno et al., *Tetrahedron Lett.*, **1992**, *33*, 2539.

50%

**I.H.2-4**  J. Nakayama et al, *Bull. Chem. Soc. Jpn.*, **1992**, *65*, 3343.

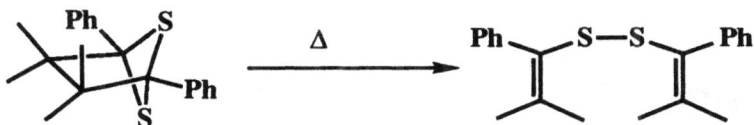

**I.H.2-5**  H. Xia, H. W. Moore*, *J. Org. Chem.*, **1992**, *57*, 3765.

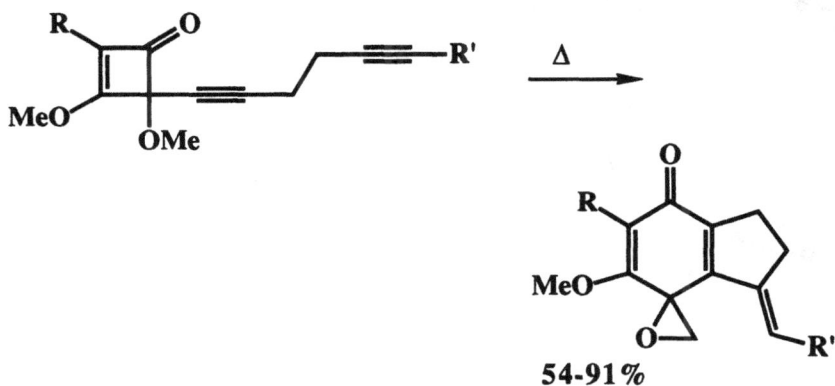

54-91%

**I.H.2-6**  N. Iwasawa et al., *Chem. Lett.*, **1992**, 473.

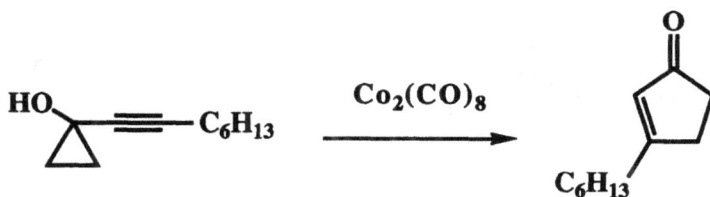

**I.H.2-7** M. L. Davies*, B. Hu *J. Org. Chem.*, **1992**, *57*, 4309.

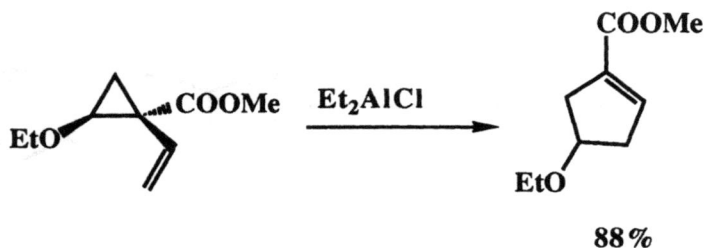

88%

---

**I.H.2-8** M. R. Uskokouic et al., *Tetrahedron Lett.*, **1992**, *33*, 7701.

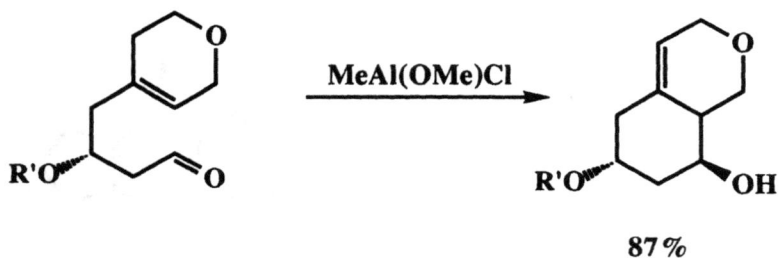

87%

---

**I.H.2-9** P. Beak and J. E. Resek *J. Org. Chem.*, **1992**, *57*, 944.

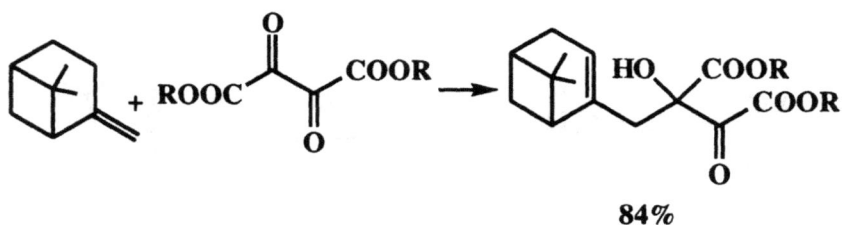

84%

**I.H.2-10**  J. R. Hwu and J. M. Wetzel, *J. Org. Chem.*, **1992**, *57*, 922.

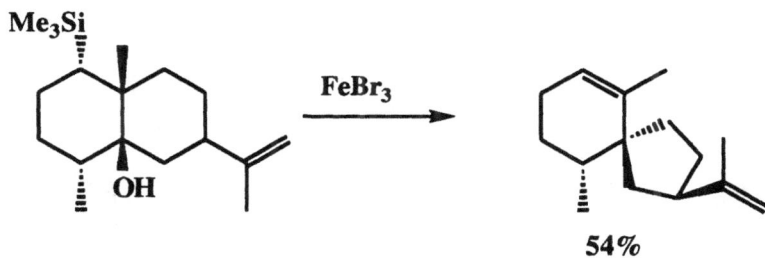

54%

**I.H.2-11**  G. Mehta and K. Venkatesan et al., *Ind. J. Chem.* **1992**, *31B*, 473.

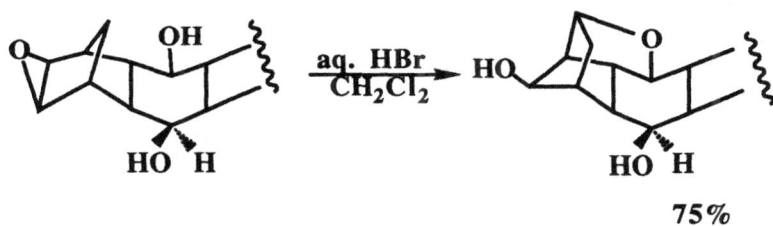

75%

**I.H.2-12** S. Ghosh et al., *Tetrahedron Lett.*, **1992**, *33*, 2363.

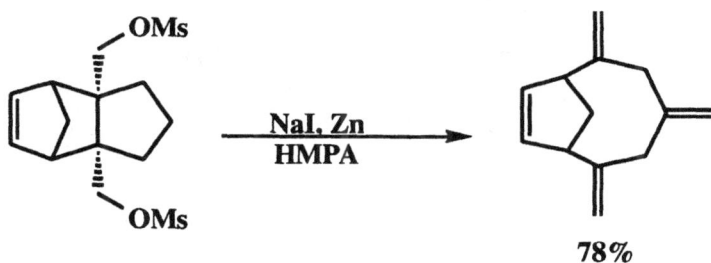

78%

**I.H.2-13**  J. W. Grissom et al.,  *Tetrahedron Lett.*, **1992**, *33*, 2315.

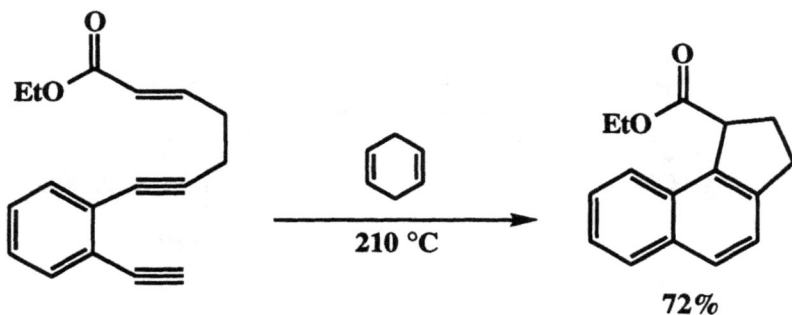

72%

---

**I.H.2-14**  H. G. Viehe et al., *Tetrahedron Lett.*, **1992**, *33*, 2511.

60%

---

**I.H.2-15**  U. K. Pandit et al., *Tetrahedron Lett.*, **1992**, *33*, 2179.

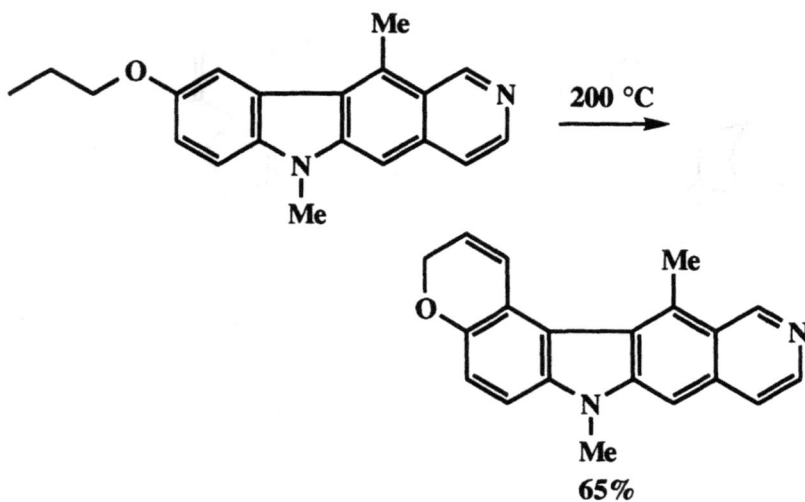

65%

**I.H.2-16**  T. W. Kwon et al., *Syn. Commun.*, **1992**, 22,  2273.

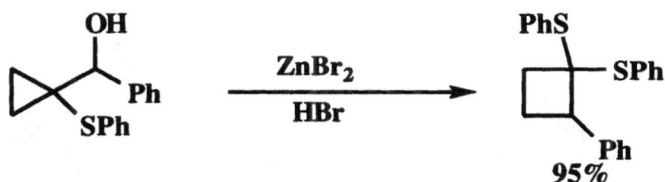

---

**I.H.2-17**  H. Moskowitz et al., *Syn. Commun.*, **1992**, 22,  1403.

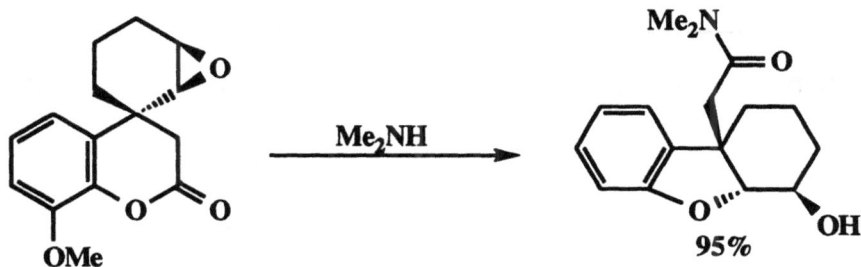

---

**I.H.2-18**  J. Aube et al., *J. Org. Chem.*, **1992**, 57, 1635.

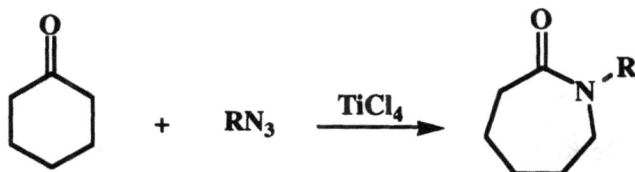

**I.H.2-19**  J. Kokosi et al., *Tetrahedron Lett.*, **1992**, *33*, 2995.

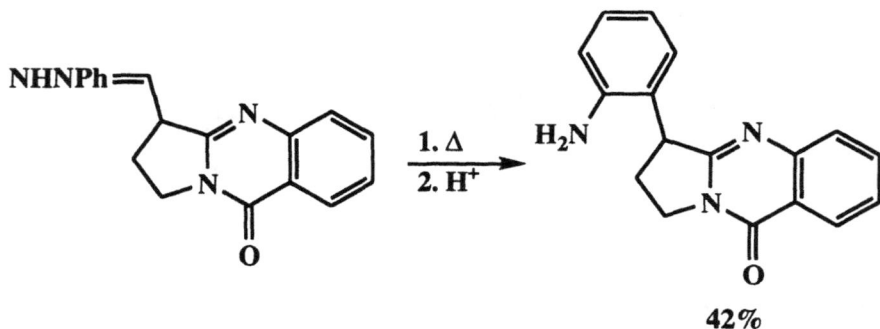

1. Δ
2. H⁺

42%

---

**I.H.2-20**  T. Sugumura, A. Tai et al., *J. Chem. Soc., Chem. Commun.*, **1992**, 324.

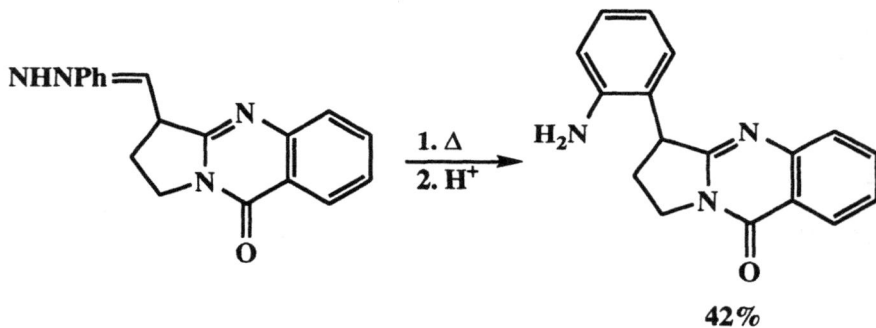

1. Δ
2. H⁺

42%

---

**I.H.2-21**  D. De Keukeleire et al., *J. Chem. Soc., Chem. Commun.*, **1992**, 419.

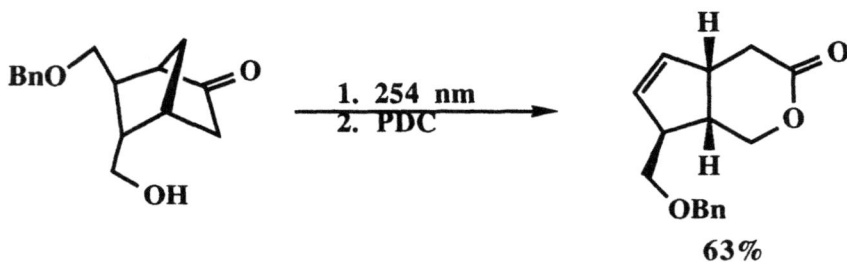

1. 254 nm
2. PDC

63%

**I.H.2-22**  T. Harada and T. Mukaiyama *Chem. Lett.*, **1992**, 81.

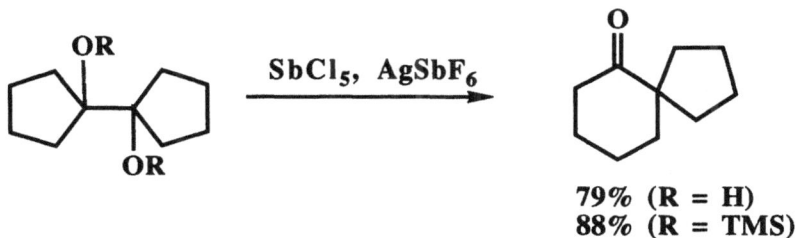

SbCl$_5$, AgSbF$_6$

79% (R = H)
88% (R = TMS)

---

**I.H.2-23**  H. R. Sonawane et al., *Tetrahedron Lett.*, **1992**, *33*, 1645.

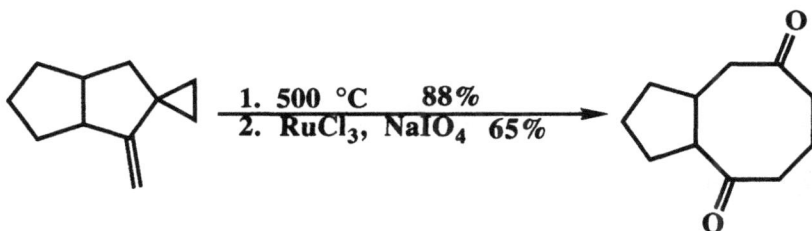

1. 500 °C    88%
2. RuCl$_3$, NaIO$_4$   65%

---

**I.H.2-24**  S. Murthy and C. N. Pillani *Tetrahedron*, **1992**, *48*, 5331.

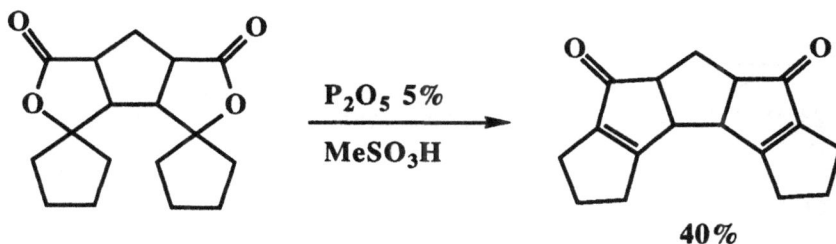

P$_2$O$_5$ 5%

MeSO$_3$H

40%

**I.H.2-25** T. Takeda et al., *Chem. Lett.*, **1992**, 1631.

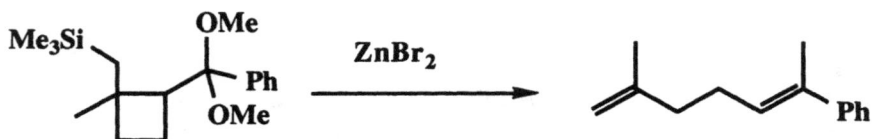

**I.H.2-26** K. M. Bol and R. M. J. Liskamp *Tetrahedron*, **1992**, *48*, 6425.

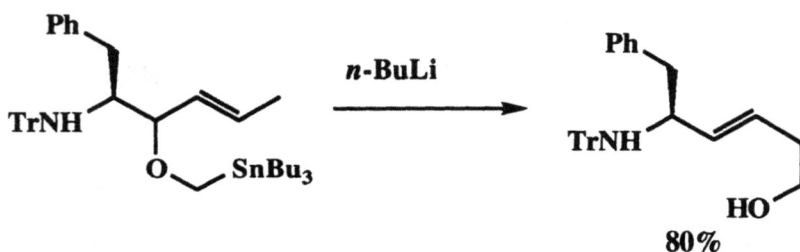

80%

**I.H.2-27** G. Ladouceur and L. A. Paquette, *Synthesis*, **1992**, 185.

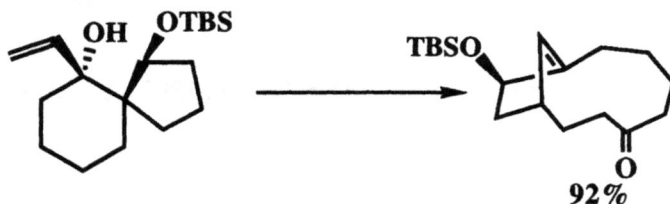

92%

**I.H.2-28** R. L. Baxter et al., *J. Chem. Soc., Perkin. Trans. I*, **1992**, 25.

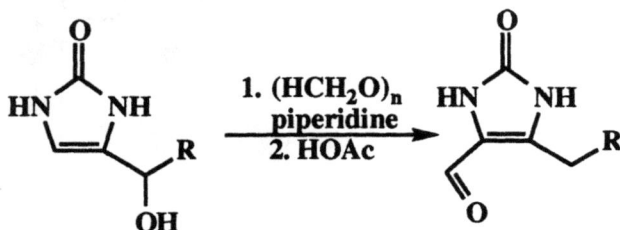

**I.H.2-29**  T. Momose et al., *J. Chem. Soc., Perkin. Trans. I*, **1992**, 517.

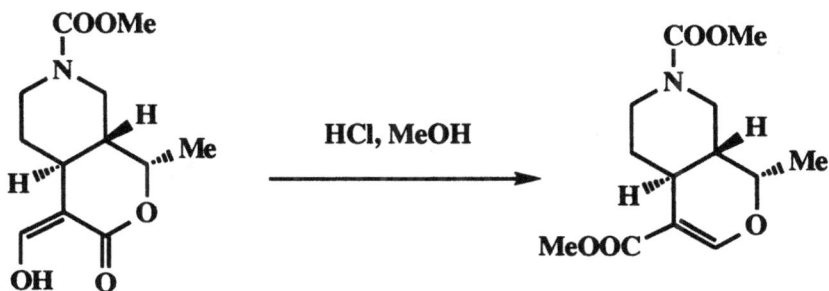

**I.H.2-30**  J. W. Frost et al., *J. Am. Chem. Soc.*, **1992**, *114*, 9725.

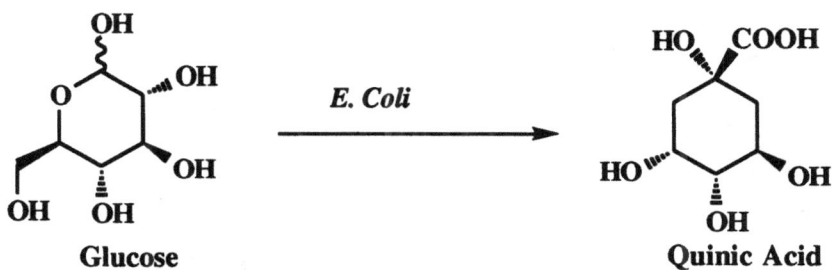

Glucose                                                Quinic Acid

**I.H.2-31**  K. Hiroi et al., *Chem. Lett.*, **1992**, 2329.

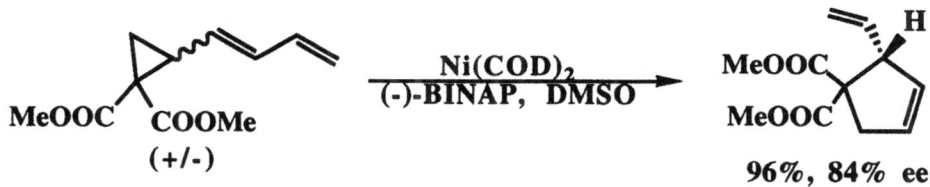

96%, 84% ee

**I.H.2-32**  F. Bickelhaupt et al., *J. Am. Chem. Soc.*, **1992**, *114*, 9191.

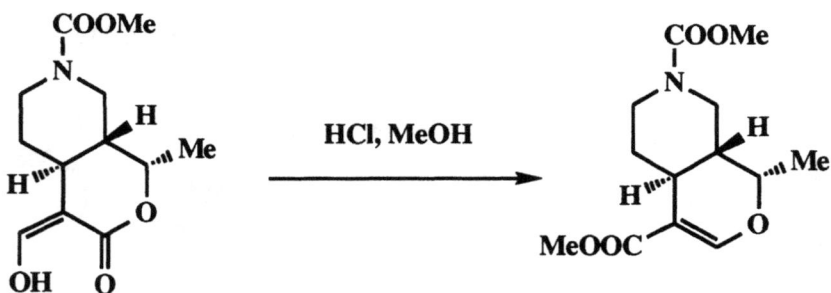

**I.H.2-33**  A. M. Seldes et al., *J. Chem. Soc., Perkin. Trans. I*, **1992**, 453.

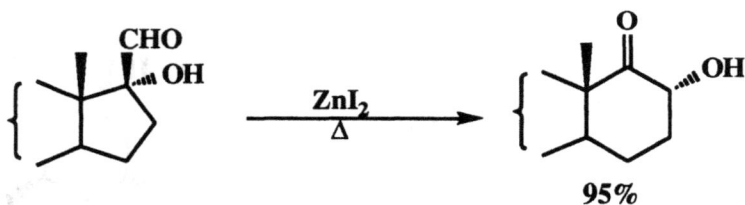

**I.H.2-34**  M. Asaoka et al., *Bull. Chem. Soc. Jpn.*, **1992**, *65*, 3206.

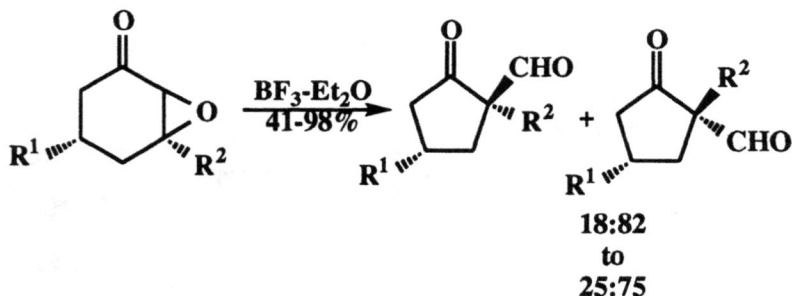

**I.H.2-35** C. Narayana et al., *J. Chem. Soc., Chem. Commun.*, **1992**, 1624.

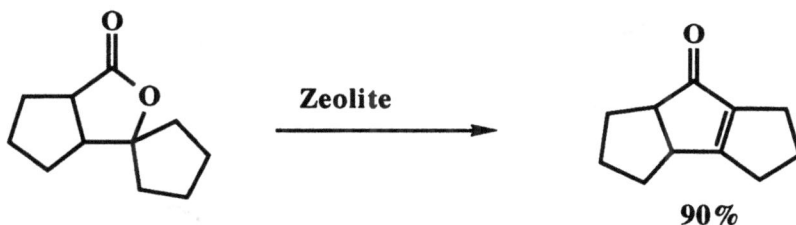

**Zeolite**

90%

---

**I.H.2-36** E. J. Verner, T. Cohen, *J. Am. Chem. Soc.*, **1992**, *114*, 375.

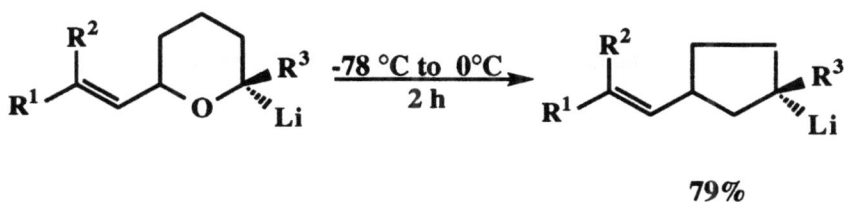

$R^2$ $R^3$ -78 °C to 0°C, 2 h $R^2$ $R^3$

$R^1$ O Li → $R^1$ Li

79%

---

**I.H.2-37** H. von der Emde, R. Bruckner, *Tetrahedron Lett.*, **1992**, *33*, 7323.

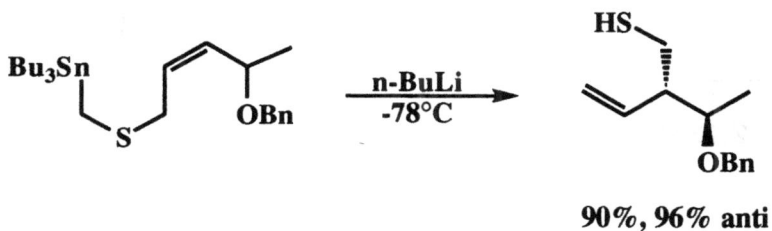

$Bu_3Sn$ OBn  n-BuLi -78°C → HS OBn

90%, 96% anti

**I.H.2-38** J. H. Babler and S. A. Schlidt, *Tetrahedron Lett.*, **1992**, *33*, 7697.

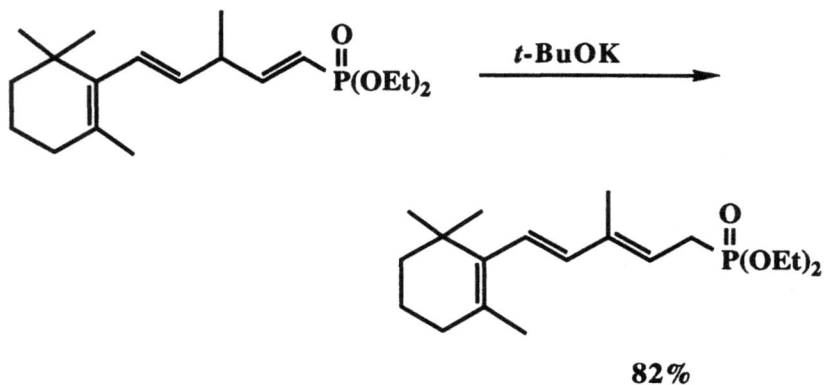

82%

---

**I.H.2-39** K. Yamakawa et al., *Tetrahedron Lett.*, **1992**, *33*, 7181, 7543.

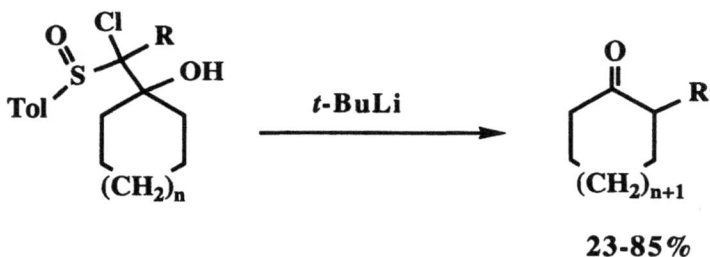

23-85%

---

**I.H.2-40** S. Z. Zard et al., *Tetrahedron Lett.*, **1992**, *33*, 7849.

59-67%

**I.H.2-41**  K. Yamakawa et al., *Tetrahedron Lett.*, **1992**, *33*, 7181, 7543.

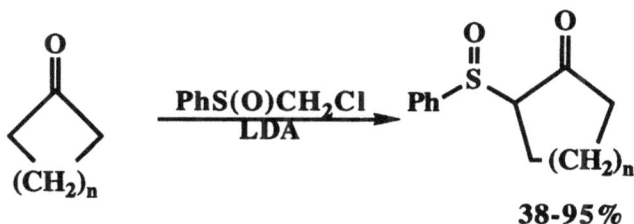

38-95%

---

**I.H.2-42**  M. G. Banwell et al., *J. Chem. Soc., Perkin Trans. I*, **1992**, 1329.

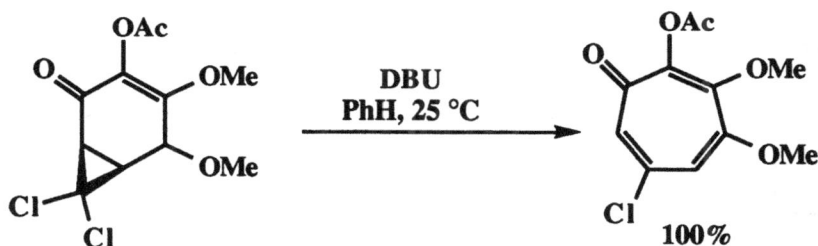

100%

---

**I.H.2-43**  K. Sakai et al., *J. Chem. Soc., Chem. Commun.*, **1992**, 1482.

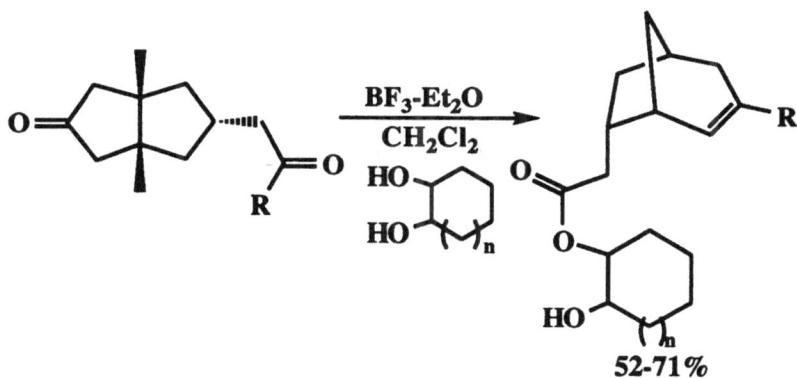

52-71%

**I.H.2-44** L. S. Liebeskind et al., *J. Org. Chem.*, **1992**, *57*, 4345.

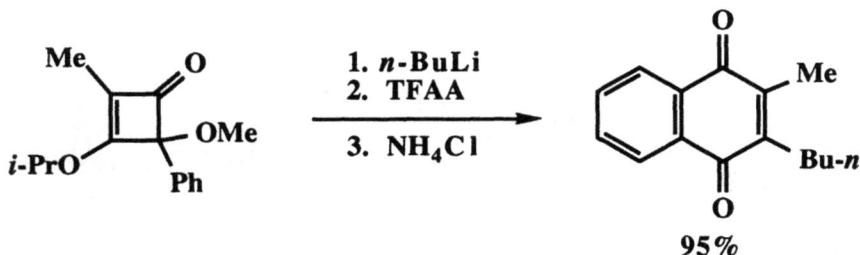

1. *n*-BuLi
2. TFAA
3. NH₄Cl

95%

Similarly for other highly substituted quinones

**I.H.2-45** R. W. Frieser et al., *Tetrahedron Lett.*, **1992**, *33*, 6715.

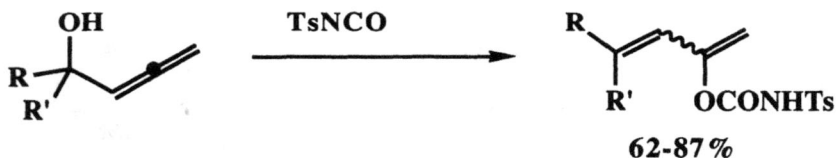

TsNCO

62-87%

**I.H.2-46** B.A. Trofimov et al., *J. Org. Chem. USSR*, **1992**, *28*, 167, 347, 490.

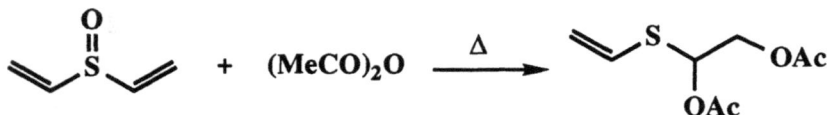

Pummerer type reaction

**I.H.2-47** T. Katsuki et al., *Bull. Chem. Soc. Jpn.*, **1992**, *65*, 1841.

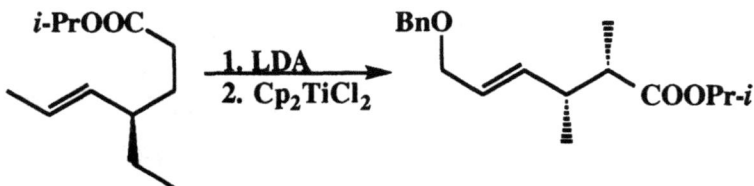

1. LDA
2. Cp₂TiCl₂

**I.H.2-48** D. M. X. Donnelly et al., *J. Chem. Soc., Perkin. Trans. I*, **1992**, 1365.

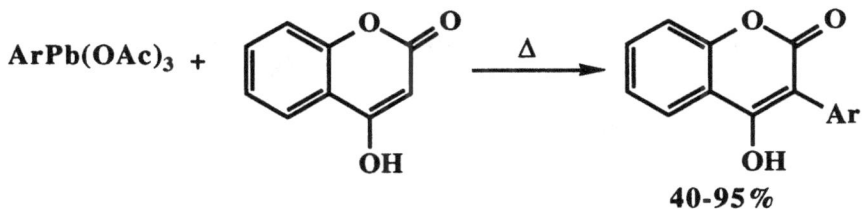

ArPb(OAc)$_3$ +

40-95%

---

**I.H.2-49** Y. Fehikawa et al., *J. Chem. Soc., Perkin. Trans. I*, **1992**, 1497.

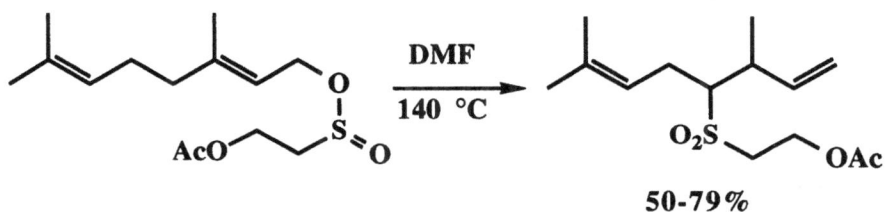

DMF

140 °C

50-79%

**[2,3]-sigmatropic rearrangemnet of allylic sulfinate to allylic sulfone**

---

**I.H.2-50** R. M. Coates et al., *J. Org. Chem.*, **1992**, *57*, 4327.

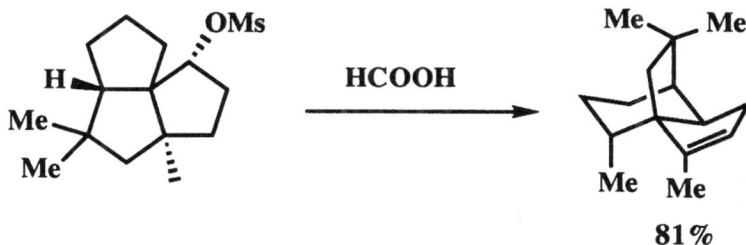

HCOOH

81%

**I.H.2-51**  H.-J. Liu et al., *Can. J. Chem.*, **1992**, *70*, 1375.

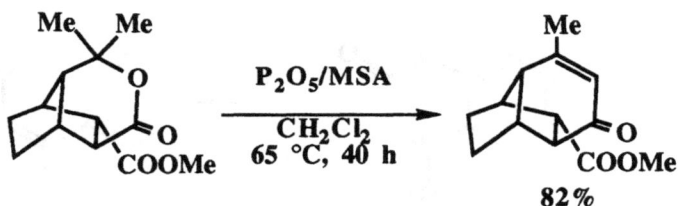

Me   Me

P₂O₅/MSA
———————→
CH₂Cl₂
65 °C, 40 h

COOMe

Me

O
COOMe

**82%**

---

**I.H.2-52**  L.A. Paquette et al., *Tetrahedron Lett.*, **1992**, *33*, 7311.

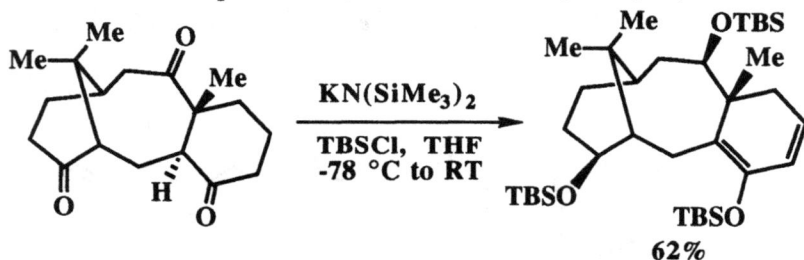

KN(SiMe₃)₂
———————————→
TBSCl, THF
-78 °C to RT

**62%**

---

**I.H.2-53**  A. Padwa et al., *Tetrahedron Lett.*, **1992**, *33*, 7303.

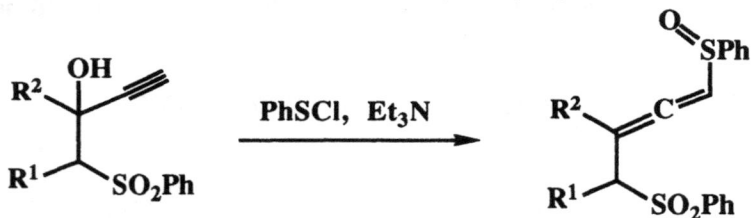

R²  OH

R¹  SO₂Ph

PhSCl, Et₃N
———————————→

O
‖
S Ph

R²  C

R¹  SO₂Ph

---

**I.H.2-54**  P. Kocovsky and J. Srogl, *J. Org. Chem.*, **1992**, *57*, 4563.

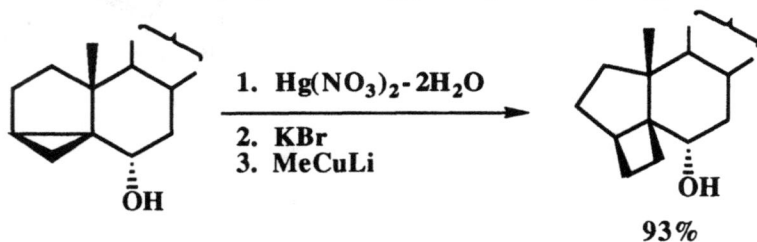

1. Hg(NO₃)₂·2H₂O
———————————————→
2. KBr
3. MeCuLi

ŌH

ŌH

**93%**

**I.H.2-55**  U. Schollkppf et al.,  *Libigs Ann Chem.*, **1992**, 1179, and 199.

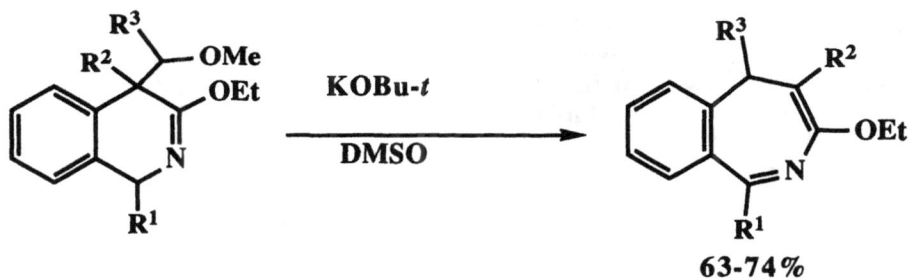

63-74%

---

**I.H.2-56**  T. Sano, Y. Tsuda et al., *Chem. Pharm. Bull.*,**1992**, *40*, 873.

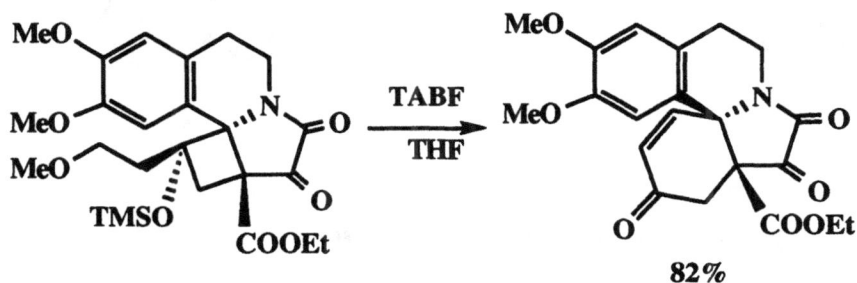

82%

---

**I.H.2-57**  R. Hoffmann and R. Bruckner, *Chem. Ber*, **1992**, *125*, 1957.

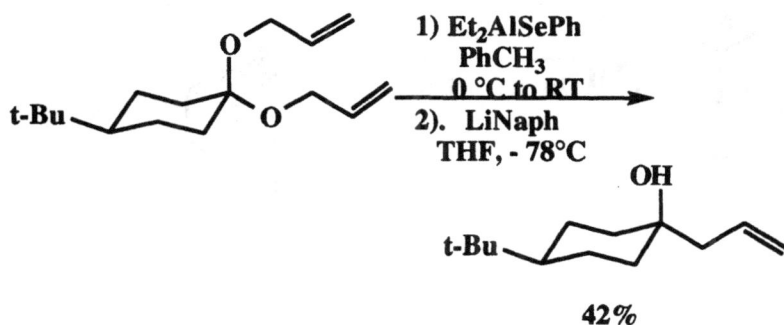

42%

**I.H.2-58**  X. Tong and J. Kallmertan, *SynLett*, **1992**, 845.

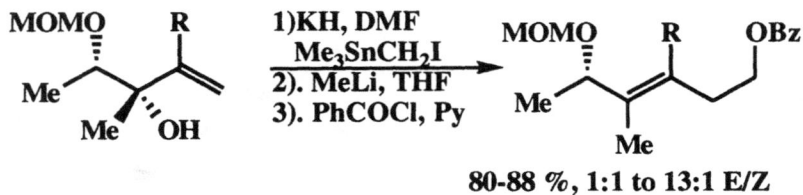

MOMO—Me—Me—OH, R (vinyl)

1) KH, DMF
   Me$_3$SnCH$_2$I
2). MeLi, THF
3). PhCOCl, Py

→

MOMO—Me—Me, R, OBz

**80-88 %, 1:1 to 13:1 E/Z**

---

**I.H.2-59**  P. Dowd and W. Zhang, *J. Org. Chem.*, **1992**, *57*, 7163.

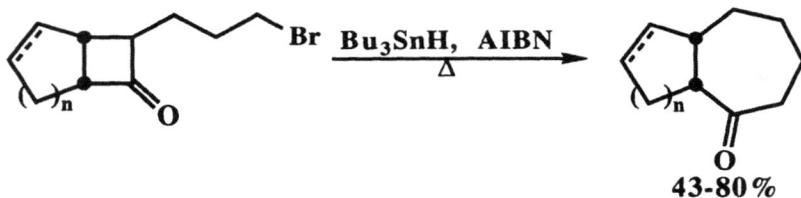

(   )$_n$ ... Br    $\underrightarrow{\text{Bu}_3\text{SnH, AIBN}}_{\Delta}$    (   )$_n$ ... O

**43-80 %**

---

**I.H.2-60**  J. R. Hanson et al., *J. Chem. Res. (S)* **1992**, 374.

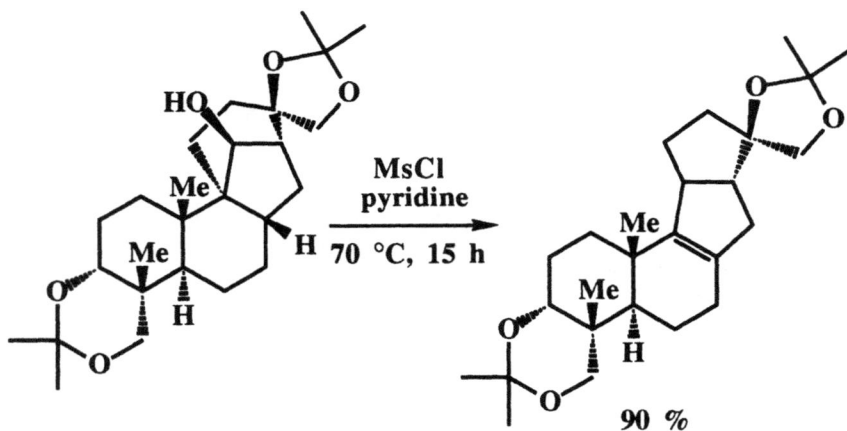

MsCl
pyridine
70 °C, 15 h

**90 %**

**I.H.2-61** J. R. Bull et al., *J. Chem. Soc., Perkin. Trans. I*, **1992**, 2545.

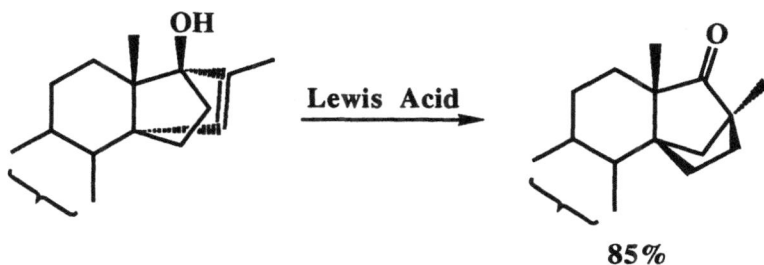

Lewis Acid

**85%**

---

**I.H.2-62** S. K. Thompson and C. H. Heathcock, *J. Org. Chem.*, **1992**, *57*, 5978.

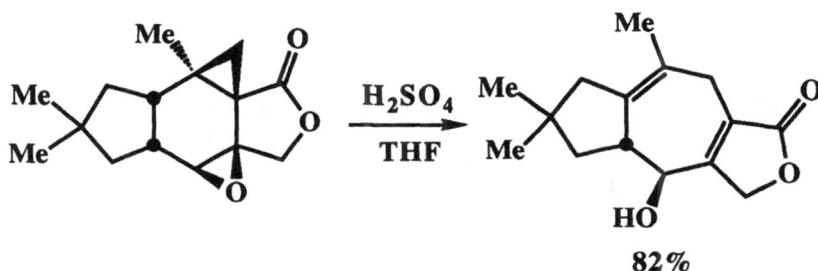

$H_2SO_4$

THF

**82%**

---

**I.H.2-63** P. E. Hudrlik et al., *J. Org. Chem.*, **1992**, *57*, 6552.

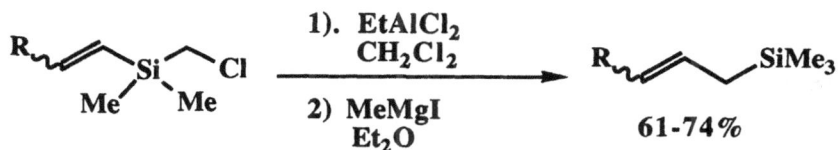

1). $EtAlCl_2$
   $CH_2Cl_2$

2) MeMgI
   $Et_2O$

**61-74%**

---

**I.H.2-64** G. Y. M. Iglesias et al., *Org. Prep. Pro. Int.*, **1992**, *24*, 690.

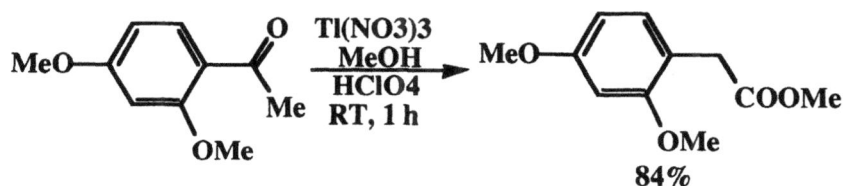

$Tl(NO3)3$
MeOH
$HClO4$
RT, 1 h

**84%**

**I.H.2-65**  M. T. Crimmins et al., *Tetrahedron Lett.*, **1992**, *33*, 181.

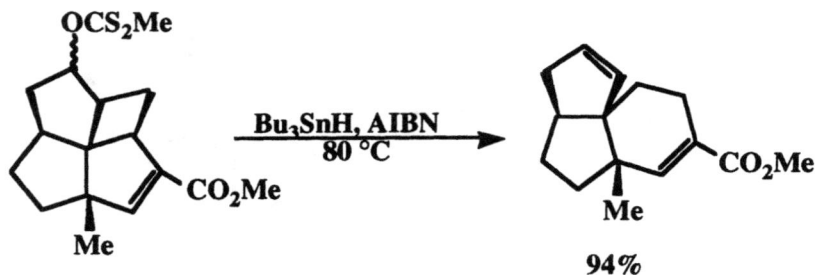

OCS$_2$Me

$$\xrightarrow[\text{80 °C}]{\text{Bu}_3\text{SnH, AIBN}}$$

CO$_2$Me

Me

CO$_2$Me

Me

94%

---

**I.H.2-66**  J. M. Takacs, Y. C. M. Young, *Tetrahedron Lett.*, **1992**, *33*, 317.

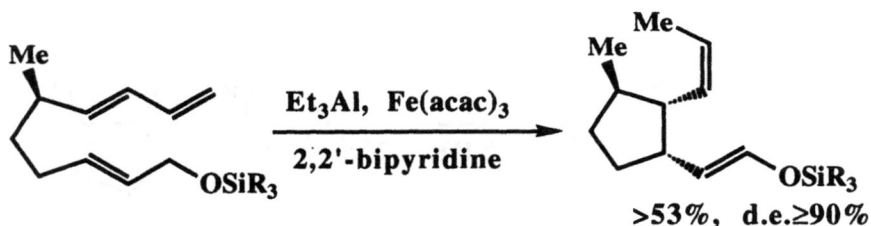

Me

OSiR$_3$

$$\xrightarrow[\text{2,2'-bipyridine}]{\text{Et}_3\text{Al, Fe(acac)}_3}$$

Me

Me

OSiR$_3$

>53%,  d.e.≥90%

---

**I.H.2-67**  T. Mayer, G. Maas, *Tetrahedron Lett.*, **1992**, *33*, 205.

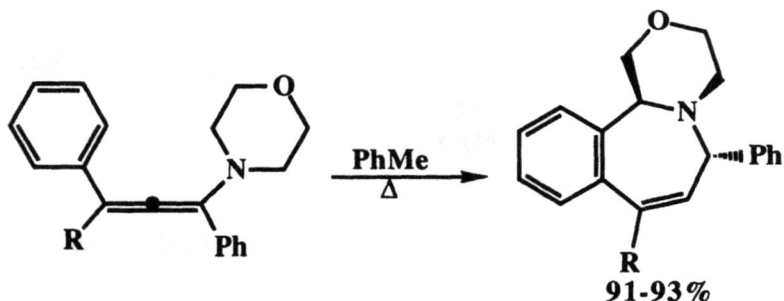

R

Ph

$$\xrightarrow[\Delta]{\text{PhMe}}$$

Ph

R

91-93%

**I.H.2-68** W. G. Dauben and R. T. Hendricks, *Tetrahedron Lett.*, **1992**, *33*, 565.

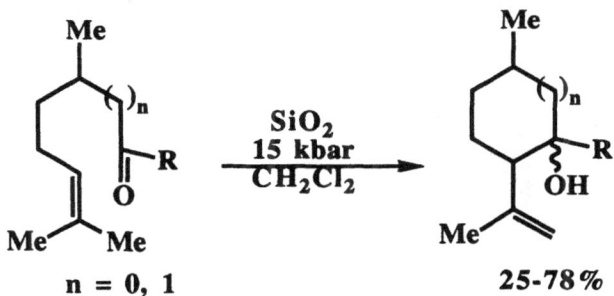

Me

SiO$_2$
15 kbar
CH$_2$Cl$_2$

R
O

Me   Me

n = 0, 1

Me

R
OH

Me

25-78%

**I.H.2-69** Y. Kita et al., *Chem. Pharm. Bull*, **1992**, *40*, 12.

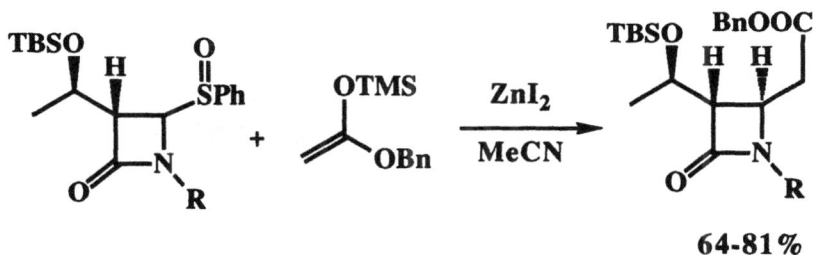

TBSO
H
O
S—Ph

N
O      R

+

OTMS

OBn

ZnI$_2$

MeCN

TBSO   BnOOC
H   H

N
O      R

64-81%

**I.H.2-70** K. Suzuki et al., *SynLett.*, **1992**, *1992*, 129.

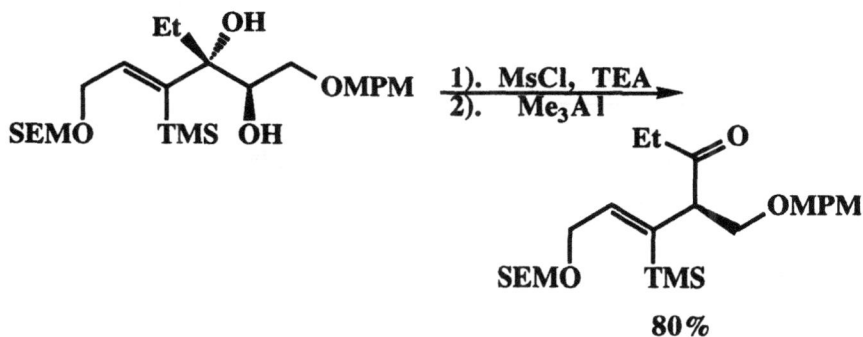

Et   OH

OMPM

SEMO   TMS   OH

1). MsCl, TEA
2).   Me$_3$Al

Et   O

OMPM

SEMO   TMS

80%

**I.H.2-71**  T. Sano et al., *Chem. Pharm. Bull.* **1992**, *40*, 36.

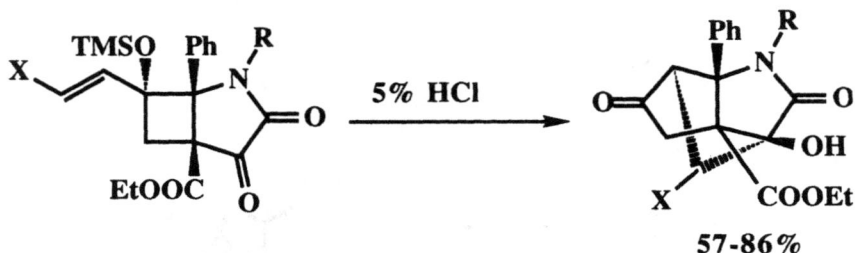

57-86%

**I.H.2-72**  A. Gambacorta et al., *Tetrahedron*, **1992**, *48*, 4459.

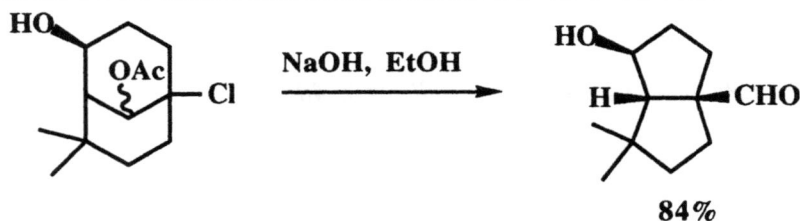

84%

**I.H.2-73**  P. Wender, T. P. Mucciaro *J. Am. Chem. Soc.*, **1992**, *114*, 5878.

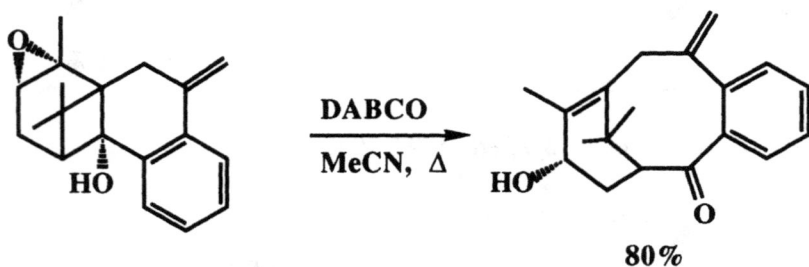

80%

**I.H.2-74**  A. Padwa and S. L. Xu, *J. Am. Chem. Soc.*, **1992**, *114*, 5881.

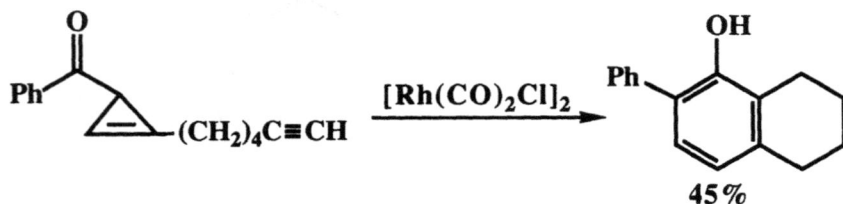

45%

**I.H.2-75** T. Yamato et al., *J. Chem. Soc., Chem. Commun.*, **1992**, 865.

**The Preparation and Novel [3,3]-Sigmatropic Rearrangement of Cyclophanes Having a Spiro Skeleton**

**I.H.2-76** S. Kim and K. H. Uh, *Tetrahedron Lett.*, **1992**, *33*, 4325.

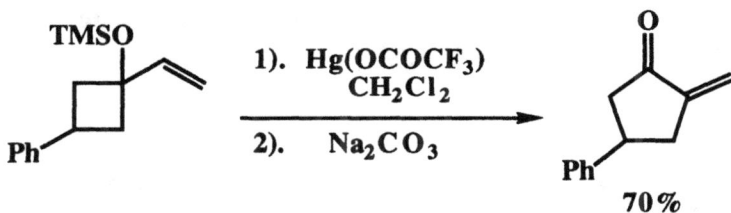

1). Hg(OCOCF$_3$)
   CH$_2$Cl$_2$

2). Na$_2$CO$_3$

70%

**I.H.2-77** P. Metze et al., *Tetrahedron*, **1992**, *48*, 1071.

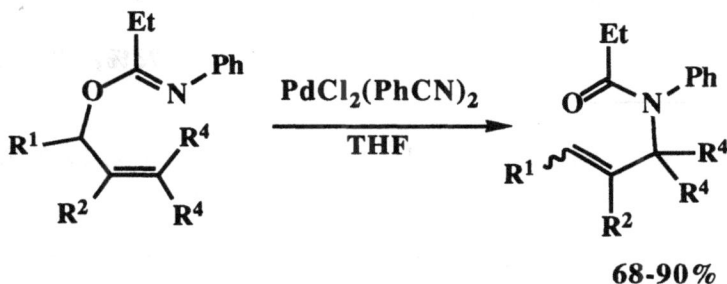

PdCl$_2$(PhCN)$_2$

THF

68-90%

**I.H.2-78** U. Chiacchio et al., *Tetrahedron*, **1992**, *48*, 123.

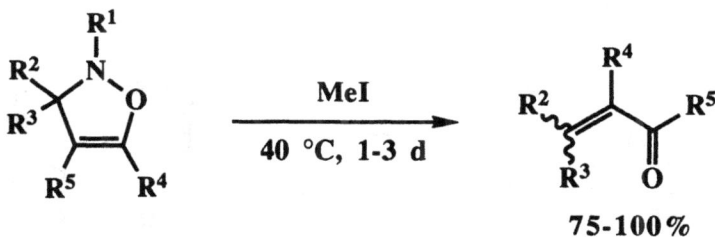

MeI

40 °C, 1-3 d

75-100%

**I.H.2-79** K. Hiroi and M. Umemura, *Tetrahedron Lett.*, **1992**, *33*, 3343.

> **Lewis acid catalyzed Intramoleculer Asymmetric Ene reactions**

---

**I.H.2-80** J. A. Marshall and M. W. Andersen, *J. Org. Chem.*, **1992**, *57*, 2766.

> **Ene-Type cyclization of unsaturated acetylenic aldehyde to macrocycles**

---

**I.H.2-81** K. Fukumoto et al., *Heterocycles*, **1992**, *33*, 549 and *J. Org. Chem.*, **1992**, *57*, 1707.

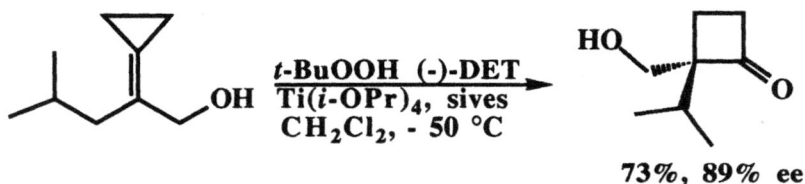

$t$-BuOOH  (-)-DET
Ti($i$-OPr)$_4$, sives
CH$_2$Cl$_2$, - 50 °C

73%, 89% ee

---

**I.H.2-82** G. D. Paderes and W. L. Jorgensen*, *J. Org. Chem.*, **1992**, *57*, 1904.

> **Computer-Assisted Mechanistc Evaluation of Organic Reactions. Ene and Retro-ene Reactions.**

---

**I.H.2-83** A. Sarkar et al., *J. Chem. Soc., Chem. Commun.*, **1992**, 793.

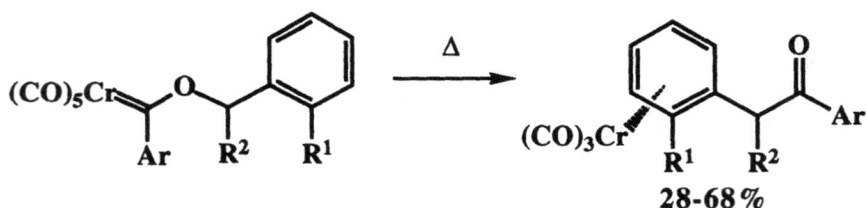

28-68%

**I.H.2-84** S. H. Gellman et al., *J. Am. Chem. Soc.*, **1992**, *114*, 6915.

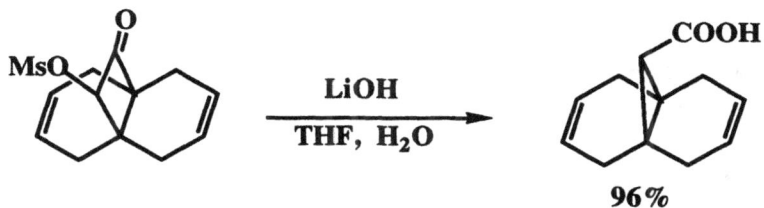

96%

---

**I.H.2-85** M. G. Banwell et al., *J. Chem. Soc., Chem. Commun.*, **1992**, 974.

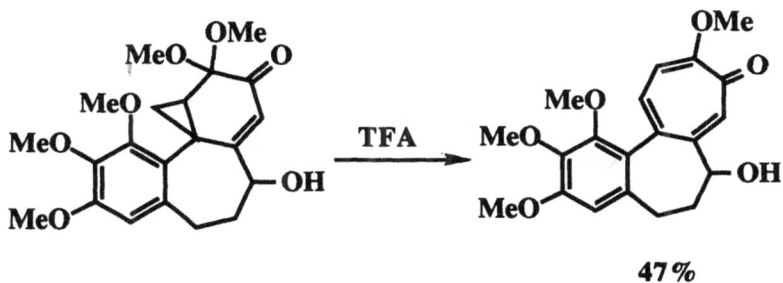

47%

# II

# OXIDATIONS

## II.A.  C-O Oxidations

### II.A.1  Alcohol → Ketone, Aldehyde

**II.A.1-1**  B. R. Chhabra et al., *SynLett* **1992**, 425.

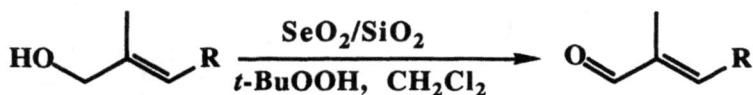

$$HO\underset{}{\diagup}\diagdown R \xrightarrow[\text{\textit{t}-BuOOH, CH}_2\text{Cl}_2]{\text{SeO}_2/\text{SiO}_2} O=\diagup\diagdown R$$

---

**II.A.1-2**  C. Guo and X. Lu*, *SynLett.*, **1992**, 405.

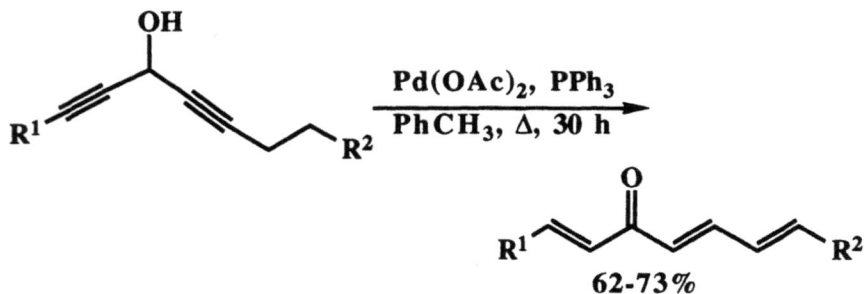

$$R^1\text{—}\underset{\text{OH}}{}\text{—}R^2 \xrightarrow[\text{PhCH}_3, \Delta, \text{ 30 h}]{\text{Pd(OAc)}_2, \text{ PPh}_3}$$

$$R^1\diagup\diagdown\underset{O}{\diagup}\diagdown\diagup R^2$$

**62-73%**

---

**II.A.1-3**  R. Somanathan et al., *Org. Prep. Pro. Int.* **1992**, *24*, 363.

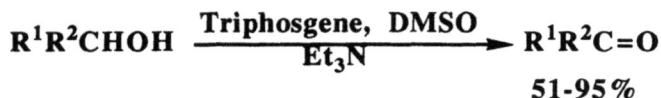

$$R^1R^2\text{CHOH} \xrightarrow[\text{Et}_3\text{N}]{\text{Triphosgene,  DMSO}} R^1R^2\text{C}=\text{O}$$

**51-95%**

**II.A.1-4** M. R. Leanna et al., *Tetrahedron Lett.*, **1992**, *33*, 5029.

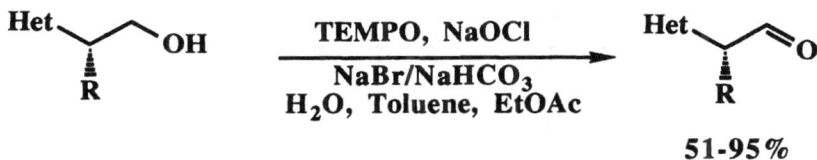

$$\text{Het} \overset{\cdot}{\underset{R}{\diagup}} \text{OH} \quad \xrightarrow[\substack{\text{NaBr/NaHCO}_3 \\ \text{H}_2\text{O, Toluene, EtOAc}}]{\text{TEMPO, NaOCl}} \quad \text{Het} \overset{\cdot}{\underset{R}{\diagup}} \diagup \text{O}$$

**51-95%**

---

**II.A.1-5** P. C. Bulman Page et al., *Tetrahedron*, **1992**, *48*, 7265.

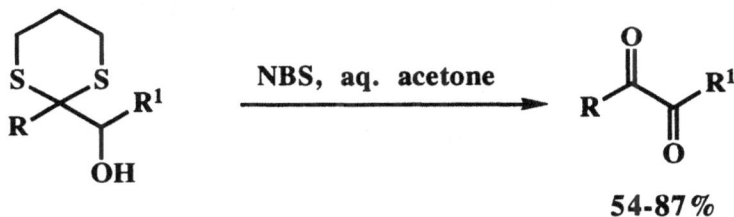

$$\xrightarrow{\text{NBS, aq. acetone}}$$

**54-87%**

---

**II.A.1-6** P. T. Meinke et al., *Tetrahedron Lett.*, **1992**, *33*, 1203.

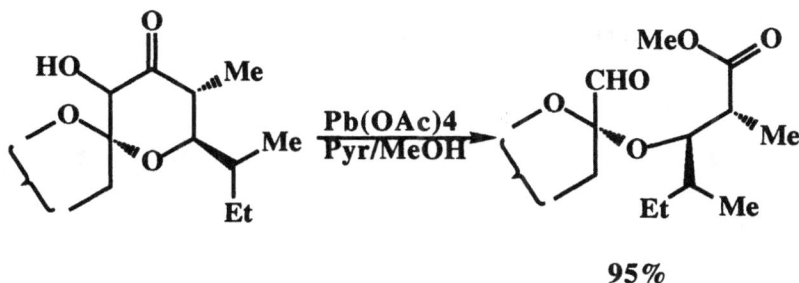

$$\xrightarrow[\text{Pyr/MeOH}]{\text{Pb(OAc)4}}$$

**95%**

---

**II.A.1-7** H. Firouzabadi and I. Mohammadpour-Baltork, *Bull. Chem. Soc. Jpn.*, **1992**, *65*, 675, 1131.

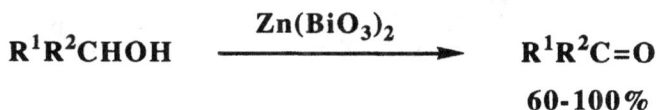

$$R^1R^2\text{CHOH} \quad \xrightarrow{\text{Zn(BiO}_3)_2} \quad R^1R^2\text{C=O}$$

**60-100%**

**II.A.1-8** H. Firouzabadi and A. Sharifi, *Synthesis*, **1992**, 999.

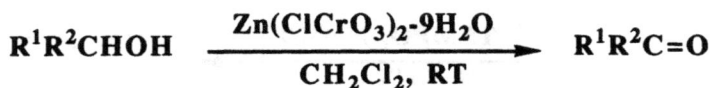

$$R^1R^2CHOH \xrightarrow[\text{CH}_2\text{Cl}_2, \text{ RT}]{\text{Zn(ClCrO}_3)_2\text{-9H}_2\text{O}} R^1R^2C=O$$

---

**II.A.1-9** T. Takeda, *J. Chem. Soc., Chem. Commun.*, **1992**, 1185.

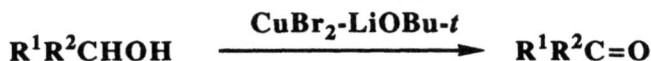

$$R^1R^2CHOH \xrightarrow{\text{CuBr}_2\text{-LiOBu-}t} R^1R^2C=O$$

---

**II.A.1-10** T. Takeda, *Chem. Lett.*, **1992**, 423,

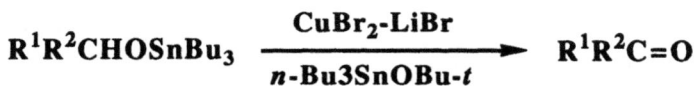

$$R^1R^2CHOSnBu_3 \xrightarrow[n\text{-Bu3SnOBu-}t]{\text{CuBr}_2\text{-LiBr}} R^1R^2C=O$$

---

**II.A.1-11** G. A, Hiegel et al., *Syn. Commun.*, **1992**, 22, 1589.

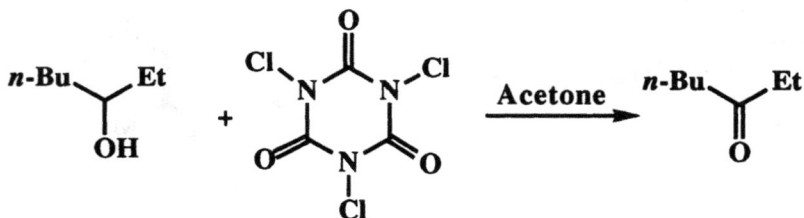

---

**II.A.1-12** J. Muzart and A. N. Ajjou *Syn. Commun.*, **1992**, 22, 1993.

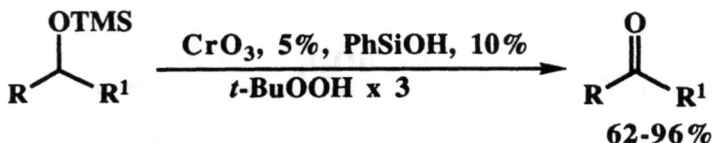

**allylic silyl ether oxidized as well.**

**II.A.1-13** S. Agarwal et al. *J. Het. Chem.*, **1992**, *29*, 257.

$$R^1R^2CHOH \xrightarrow{\left[MeN\overset{\frown}{\underset{=NH}{+}}\right]CrO_3Cl^-} R^1R^2C=O$$

---

**II.A.1-14** W. Lou and J. Lou, *Syn. Commun.*, **1992**, *22*, 767.

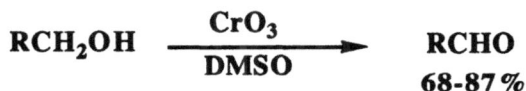

$$RCH_2OH \xrightarrow[\text{DMSO}]{CrO_3} \underset{\textbf{68-87\%}}{RCHO}$$

---

**II.A.1-14** G. Z. Wang and J. E. Backvall* *J. Chem. Soc., Chem. Commun.*, **1992**, 337.

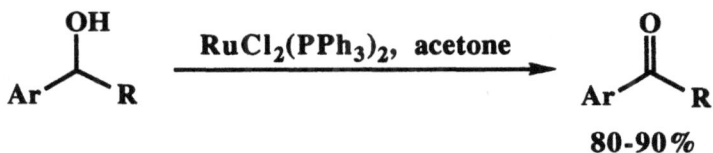

---

**II.A.1-16** O. Piva *Tetrahedron Lett.*, **1992**, *33*, 2459.

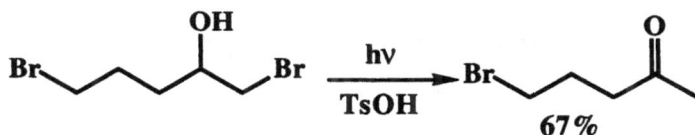

---

**II.A.1-17** Y. Hu and H. Hu *Syn. Commun.*, **1992**, *22*, 1491.

$$R^1R^2CHOH \xrightarrow{\left[MeN\overset{\frown}{\underset{=NH}{+}}\right]CrO_3Cl^-} R^1R^2C=O$$

**II.A.1-18**  E. J. Parish et al., *Syn. Commun.*, **1992**, *22*,  2839.

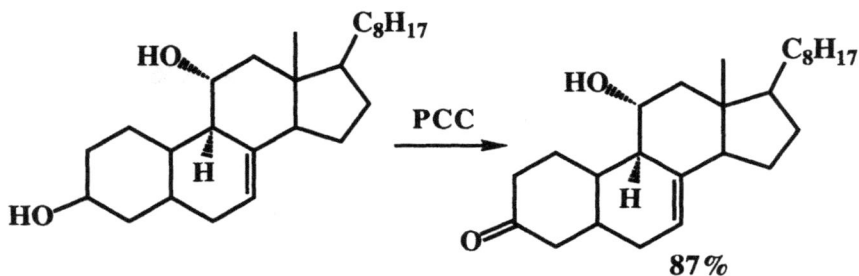

**II.A.1-19**  F. R. van Heerden et al., *Tetrahedron Lett.*, **1992**, *33*, 7399.

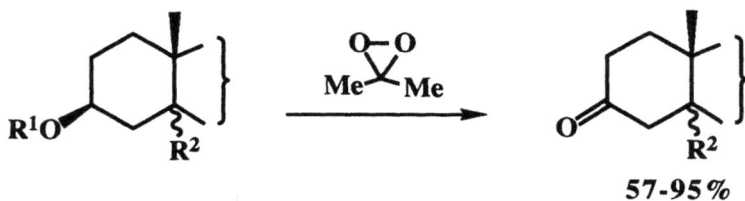

**II.A.1-20**  G. Resnati et al., *Tetrahedron Lett.*, **1992**, *33*, 7245.

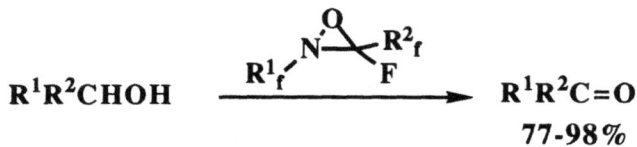

## II.B.  C-H Oxidations

### II.B.1.     C-H → to C-O

**II.B.1-1** A. B. Holmes et al., *Tetrahedron Lett.*, **1992**, *33*, 671.

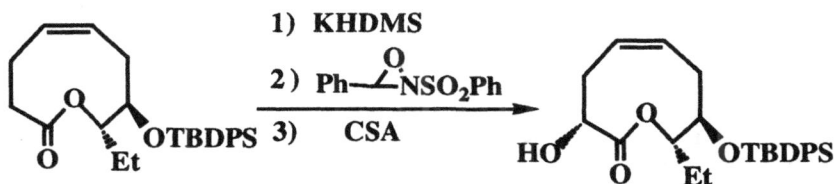

1) **KHDMS**

2) Ph—NSO$_2$Ph

3)     **CSA**

**II.B.1-2** G. J. Schroepfer Jr, et al., *J. Med. Chem.*, **1992**, *35*,, 793.

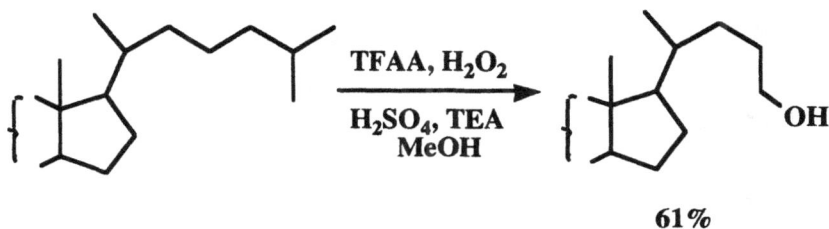

**TFAA, H$_2$O$_2$**

**H$_2$SO$_4$, TEA**
**MeOH**

**61%**

**II.B.1-3** B. M. Choudary et al., *J. Org. Chem.*, **1992**, *57*, 5841.

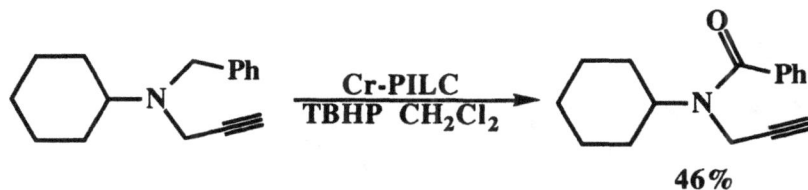

**Cr-PILC**
**TBHP  CH$_2$Cl$_2$**

**46%**

**II.B.1-4** S. M. Kerwia and C. H. Heathcock*, *J. Org. Chem.*, **1992**, *57*, 4009.

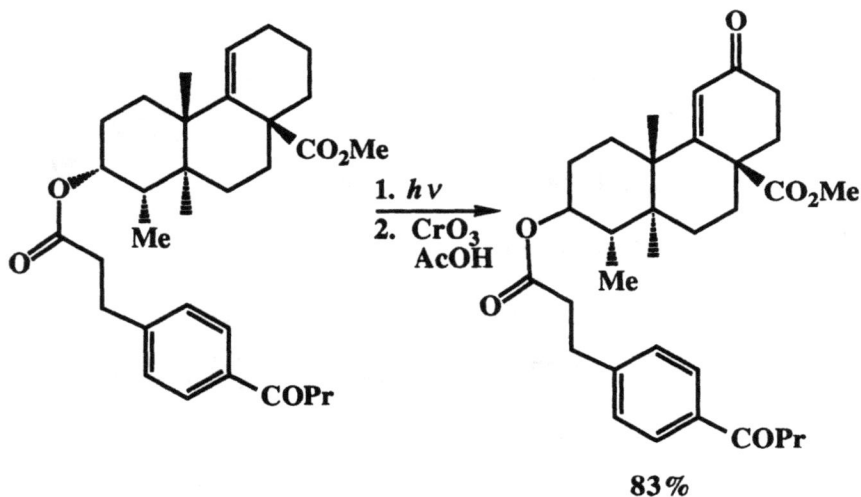

1. *hν*
2. CrO₃
   AcOH

83%

---

**II.B.1-5** B. M. Choudary et al., *Tetrahedron*, **1992**, *48*, 953.

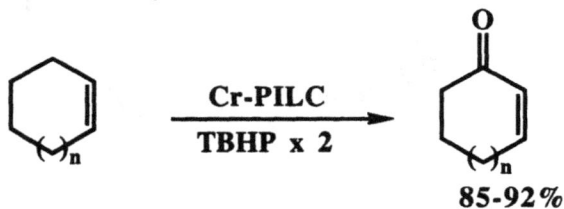

Cr-PILC

TBHP x 2

85-92%

n = 0,1,2,4
Cr-PILC = Chromia-pillared montmorillonite

---

**II.B.1-6** H. Irie et al. *Chem. Pharm. Bull.* **1991**, *39*, 3170.

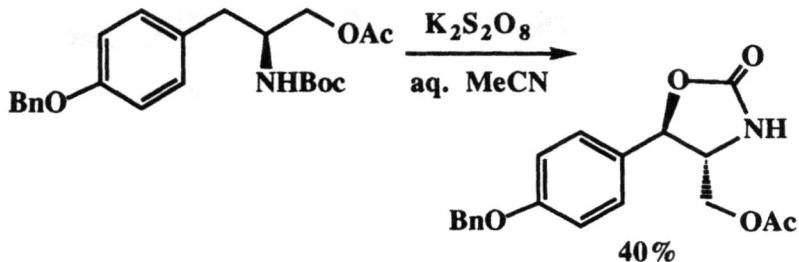

K₂S₂O₈

aq. MeCN

40%

**II.B.1-7**  T. Fujii et al., *Heterocycles*, **1992**, *34*, 1857.

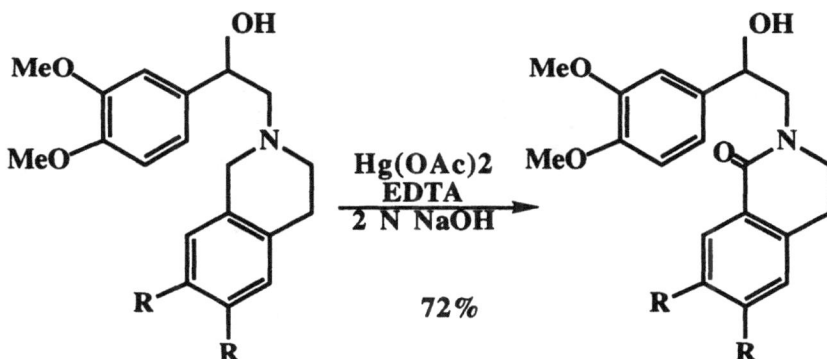

**II.B.1-8**  M. A. Tius* and M. A. Kerr , *J. Am. Chem. Soc.*, **1992**, *114*, 5959.

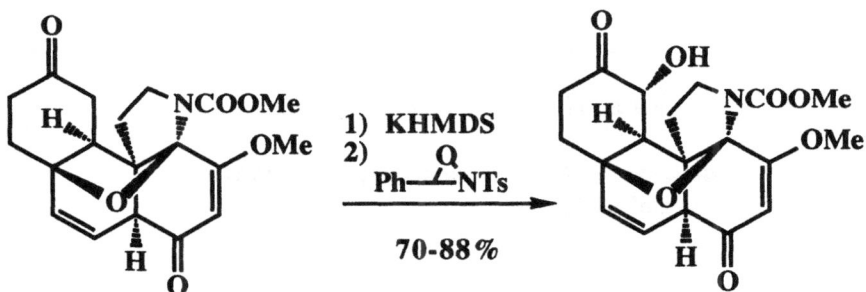

**II.B.1-9**  R. P. Kapoor et al., *Tetrahedron Lett.*, **1992**, *33*, 1495.

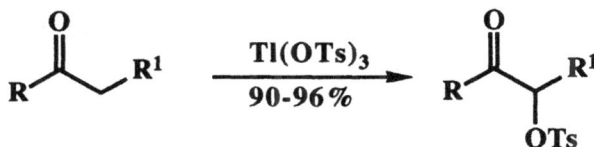

**II.B.1-10**  E. Mincione, R. Curci et al., *J. Org. Chem.*, **1992**, *57*, 5052.

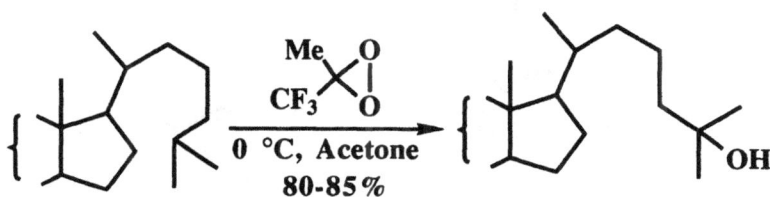

**II.B.1-11**  W. Stadbauer et al., *J. Hetero. Chem*, **1992**, 29, 1535.

51%

**II.B.1-12**  S. Murahashi et al., *J. Am. Chem. Soc.*, **1992**, *114*, 7913.

**Iron and Ruthenium-Catalyzed Oxidations of Alkanes
with Molecular Oxygen in the Presence of Aldehydes
and Acids**

**II.B.1-13**  A. F. Thomas and F. Rey *Tetrahedron*, **1992**, *48*, 1927.

**The Bayer-Villiger Reaction of Pinanones
(Bicyclo[3,1,1]heptanones)**

**II.B.1-14**  A. Nakamura et al., *Tetrahedron*, **1992**, *48*, 1557.

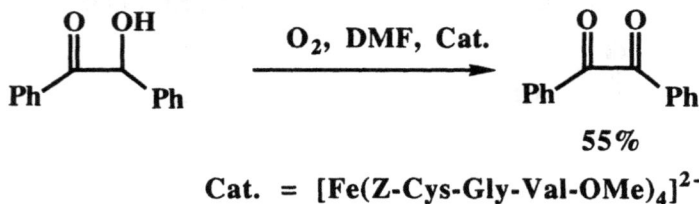

55%

Cat. = $[Fe(Z\text{-}Cys\text{-}Gly\text{-}Val\text{-}OMe)_4]^{2-}$

**II.B.1-15**  A. Marquet et al., *Syn. Commun.*, **1992**, 22, 2401.

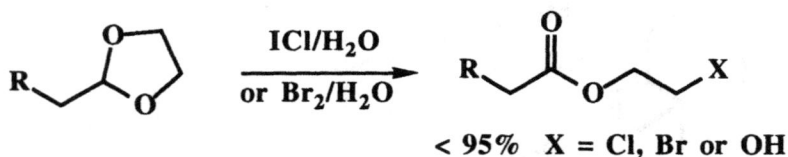

< 95%  X = Cl, Br or OH

**II.B.1-16**  S. Murahashi et al., *Chem. Lett.*, **1992**, 2237.

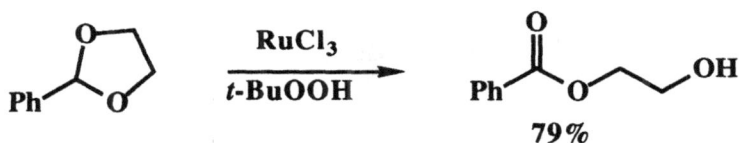

79%

---

**II.B.1-17**  Y. Matsushita et al., *Chem. Lett.*, **1992**, 2165.

1. O₂, Et₃SiH
   Co  Cat.
2. Ac₂O,  DMAP

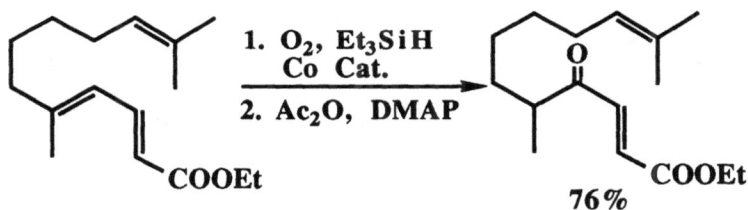

76%

---

**II.B.1-18**  E. J. Behrman *J. Chem. Soc., Perkin. Trans. I*, **1992**, 305.

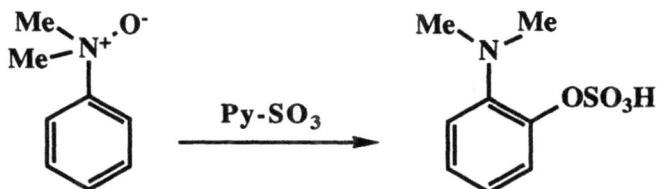

Py-SO₃

---

**II.B.1-19**  C. Venturelle et al., *Synthesis*, **1992**, 273.

1. H₂O₂/ H₂WO₄
2. O₂/Pd-C,  NaOH

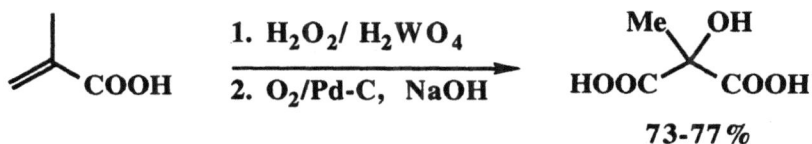

73-77%

**II.B.1-20** W. P. Griffith et al., *Syn. Commun.*, **1992**, *22*, 1967.

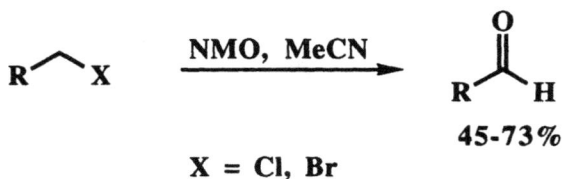

$$\text{R} \diagdown \text{X} \xrightarrow[\text{NMO, MeCN}]{} \underset{\text{R}}{\overset{\text{O}}{\|}} \text{H}$$

45-73%

X = Cl, Br

---

**II.B.1-21** T. Hosakawa et al., *Synthesis*, **1992**, 559.

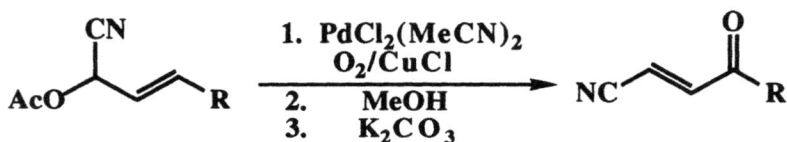

1. PdCl$_2$(MeCN)$_2$
   O$_2$/CuCl
2. MeOH
3. K$_2$CO$_3$

---

**II.B.1-22** M. Hirobe et al., *J. Am. Chem. Soc.*, **1992**, *114*, 10660.

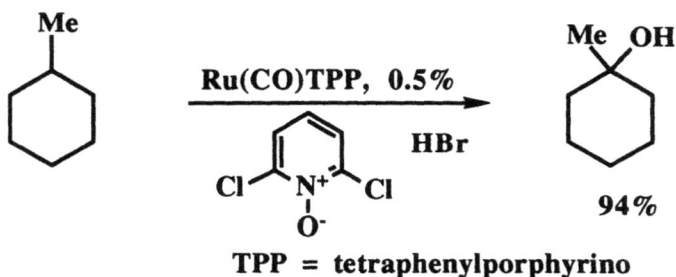

Ru(CO)TPP, 0.5%

HBr

94%

TPP = tetraphenylporphyrino

---

**II.B.1-23** I. Iovel and M. Shymanska *Syn. Commun.*, **1992**, 22, 2691.

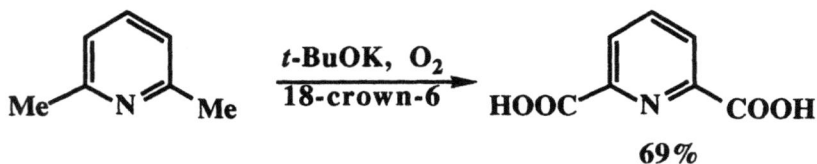

*t*-BuOK, O$_2$
18-crown-6

69%

**II.B.1-24** B. L. Feringa et al., *Tetrahedron Lett.*, **1992**, *33*, 2403.

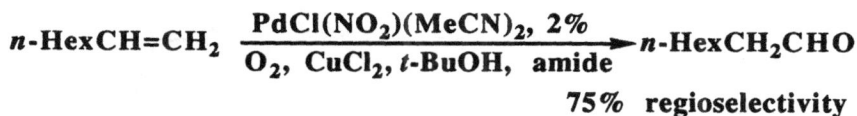

$$n\text{-HexCH=CH}_2 \xrightarrow[\text{O}_2,\ \text{CuCl}_2,\ t\text{-BuOH},\ \text{amide}]{\text{PdCl(NO}_2\text{)(MeCN)}_2,\ 2\%} n\text{-HexCH}_2\text{CHO}$$

75% regioselectivity

---

**II.B.1-25** K. Suda et al., *J. Chem. Soc., Perkin. Trans. I*, **1992**, 1283.

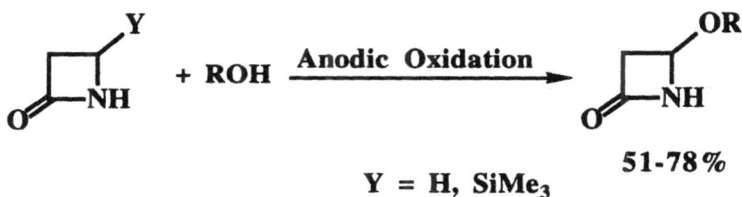

Y = H, SiMe$_3$

51-78%

---

**II.B.1-26** J. L. Courtneidge et al., *J. Chem. Soc., Perkin. Trans. I*, **1992**, 1531 and 1593.

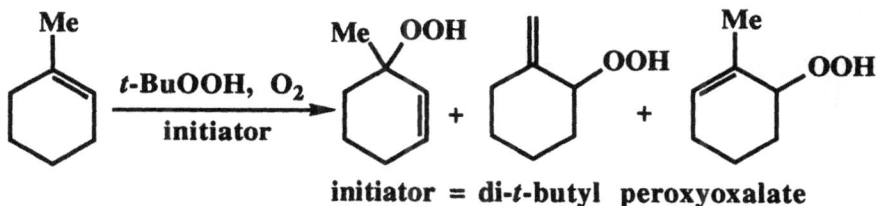

initiator = di-*t*-butyl peroxyoxalate

---

**II.B.1-27** G. Aranda et al., *Tetrahedron Lett.*, **1992**, *33*, 7845.

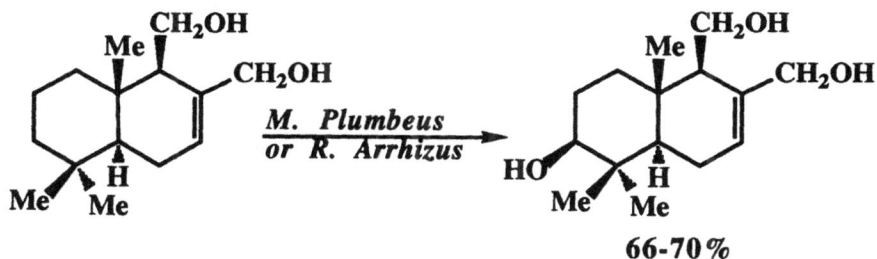

66-70%

**II.B.1-28**  S. Tang and R. M. Kennedy* *Tetrahedron Lett.*, **1992**, *33*, 7823.

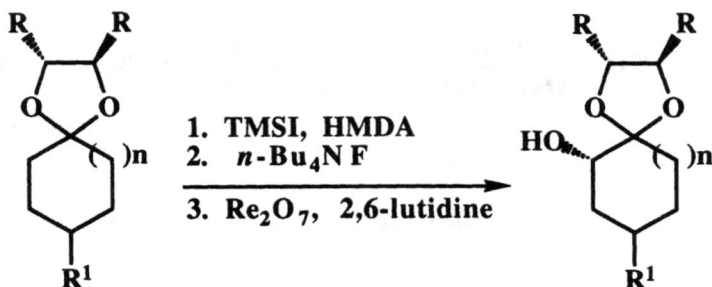

1. TMSI, HMDA
2. $n$-Bu$_4$NF
3. Re$_2$O$_7$, 2,6-lutidine

---

**II.B.1-29**  A. Tuncay et al., *Tetrahedron Lett.*, **1992**, *33*, 7647.

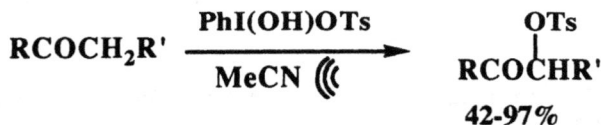

$$RCOCH_2R' \xrightarrow[\text{MeCN} \; (\!(\!(]{\text{PhI(OH)OTs}}} \quad \underset{\text{42-97\%}}{\overset{\text{OTs}}{RCOCHR'}}$$

---

**II.B.1-30**  P. Bovicelli E. Mineciene et al., *Tetrahedron Lett.*, **1992**, *33*, 7411.

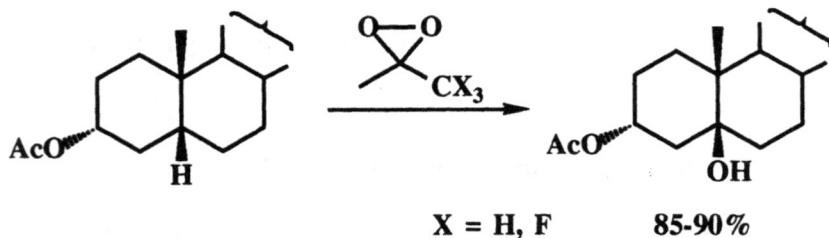

X = H, F                85-90%

**II.B.1-31**  H. Firouzabadi et al., *Bull. Chem. Soc. Jpn.*, **1992**, *65*, 2878.

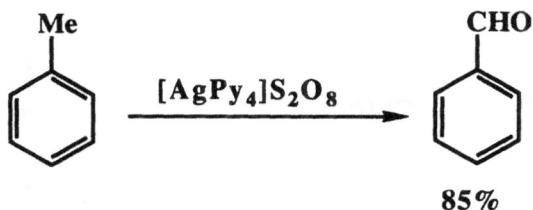

$$[AgPy_4]S_2O_8$$

85%

---

**II.B.1-32**  F. A. Davis* and A. Kumar *J. Org. Chem.*, **1992**, *57*, 3337.

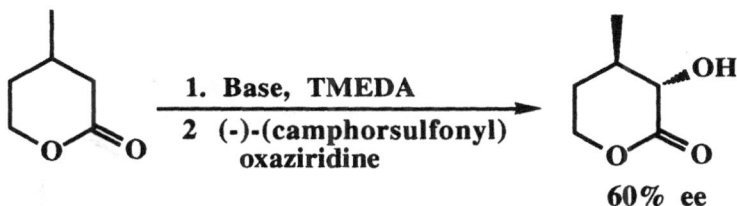

1. **Base, TMEDA**

2 **(-)-(camphorsulfonyl) oxaziridine**

60% ee

---

**II.B.1-33**  R. P. Kapoor et al., *SynLett*, **1992**, 393.

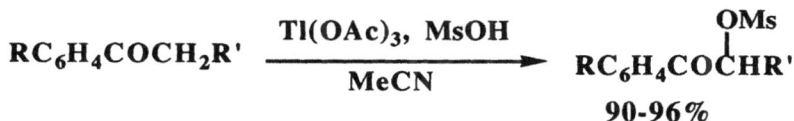

$$RC_6H_4COCH_2R' \xrightarrow[\text{MeCN}]{\text{Tl(OAc)}_3, \text{ MsOH}} RC_6H_4CO\overset{OMs}{\underset{}{\underset{}{C}}HR'}$$

90-96%

---

**II.B.1-34**  B. Waegell et al., *J. Org. Chem.*, **1992**, *57*, 5523.

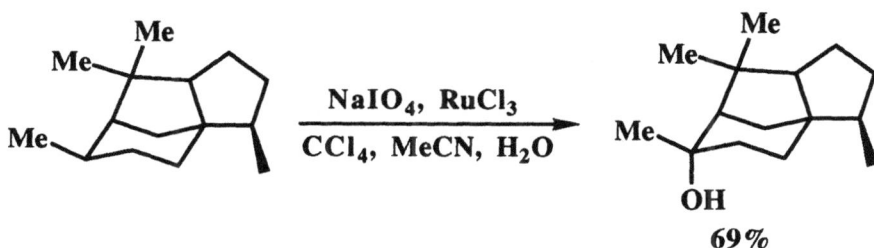

$$\xrightarrow[\text{CCl}_4, \text{ MeCN}, \text{H}_2\text{O}]{\text{NaIO}_4, \text{ RuCl}_3}$$

69%

**II.B.1-35**  H. Möhrle*, and J. Lessel,  *Chem. Ber.* **1992**, *125*, 1843.

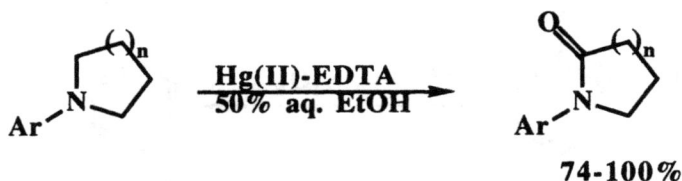

$$\xrightarrow[\text{50\% aq. EtOH}]{\text{Hg(II)-EDTA}}$$

74-100%

---

**II.B.1-36**  D. R. Williams et al., *J. Org. Chem.*, **1992**, *57*, 3730.

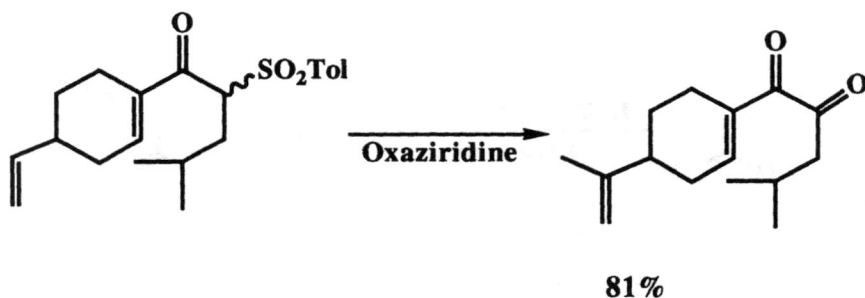

$$\xrightarrow[]{\text{Oxaziridine}}$$

81%

---

**II.B.1-37**  S. Hanessian* and R. Léger, *J. Am. Chem. Soc.*, **1992**, *114*, 3115.

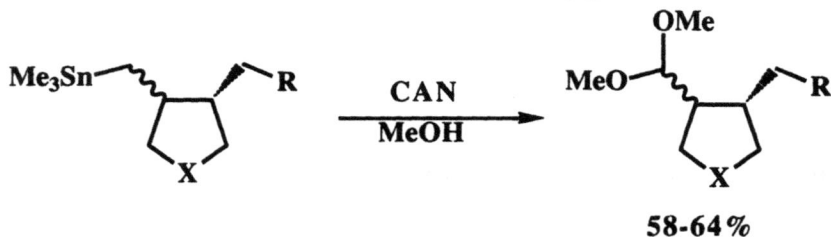

$$\xrightarrow[\text{MeOH}]{\text{CAN}}$$

58-64%

**II.B.1-38** K. Surowiec and T. Fuchigami* et al., *J. Org. Chem.*, **1992**, *57*, 5781.

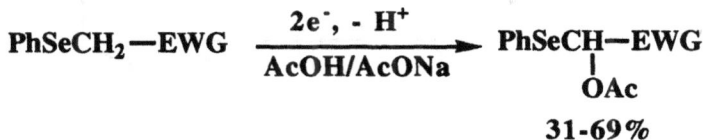

$$\text{PhSeCH}_2\text{—EWG} \quad \xrightarrow[\text{AcOH/AcONa}]{2e^-,\ -\ H^+} \quad \underset{\overset{|}{\text{OAc}}}{\text{PhSeCH—EWG}}$$

**31-69%**

---

**II.B.1-39** I. Hanna et al., *Tetrahedron Lett.*, **1992**, *33*, 5061.

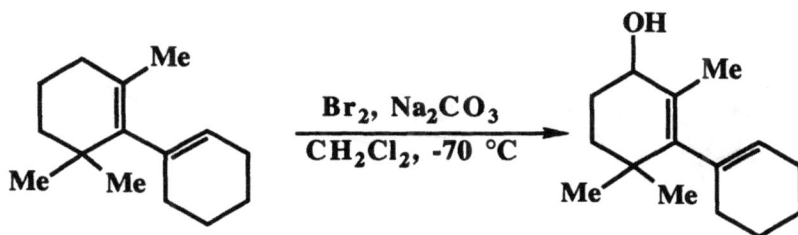

## II.B.2.          C-H → C-Hal

**II.B.2-1** V. Kumar* and M. R. Bell, *Heterocycle*, **1992**, *34*, 1289.

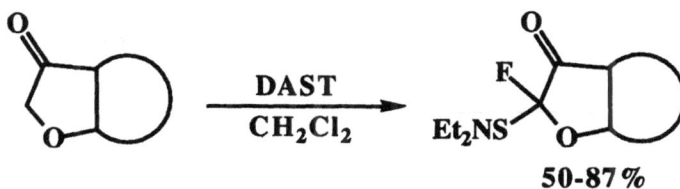

**50-87%**

**II.B.2-2**  C. R. Johnson et al., *Tetrahedron Lett.*, **1992**, *33*, 917.

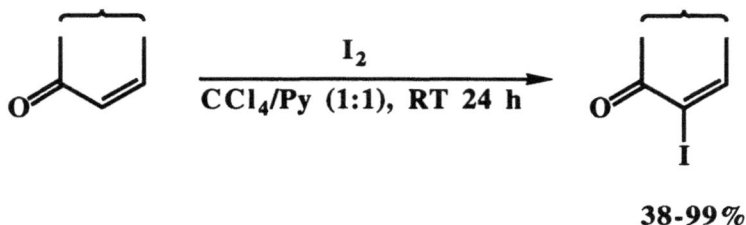

38-99%

---

**II.B.2-3**  G. Resnart, D. D. Des Marteas, *J. Org. Chem.*, **1992**, *57*, 4281.

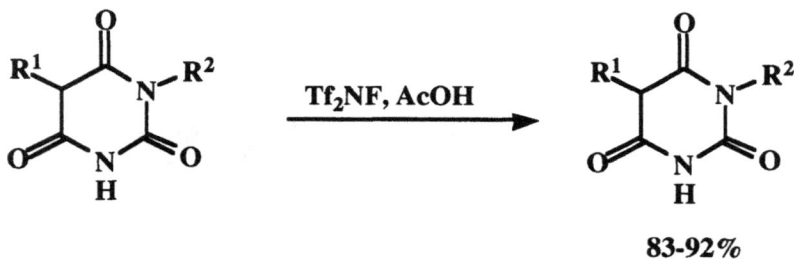

83-92%

---

**II.B.2-4**  T. Fuchigami et al., *J. Org. Chem.*, **1992**, *57*, 3753.

14-84%

---

**II.B.2-5**  F. Ghelf et al., *Tetrahedron*, **1992**, *48*, 4579.

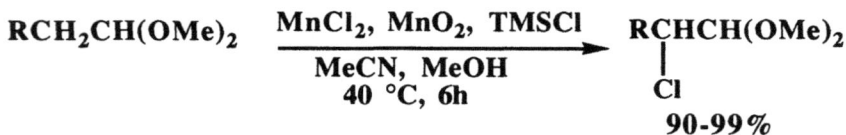

$$RCH_2CH(OMe)_2 \xrightarrow[\substack{\text{MeCN, MeOH} \\ 40\ °C,\ 6h}]{MnCl_2,\ MnO_2,\ TMSCl} \underset{\substack{| \\ Cl}}{RCHCH(OMe)_2}$$

90-99%

**II.B.2-6** T. Kurosawa*, S. Ikegawa* et al., *Steroids*, **1992**, *57*, 426.

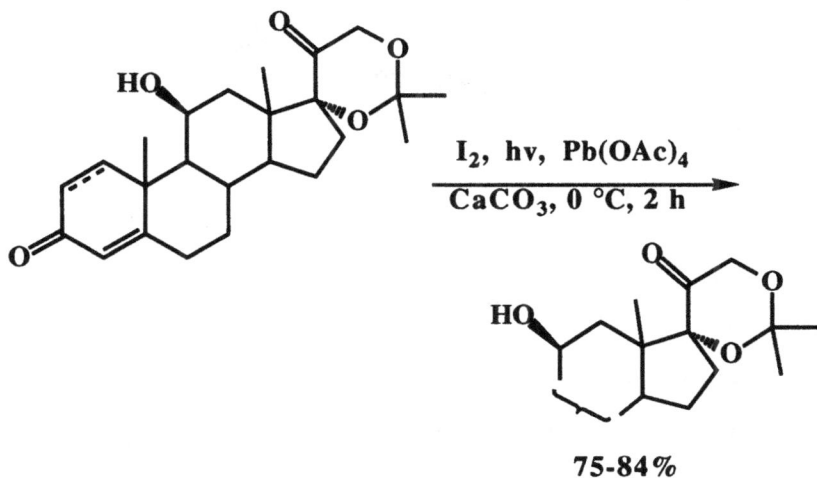

$$I_2, \ hv, \ Pb(OAc)_4$$
$$\overline{CaCO_3, \ 0 \ ^\circ C, \ 2 \ h}$$

**75-84%**

---

**II.B.2-7** G. Mignani et al., *Tetrahedron Lett.*, **1992**, *33*, 495.

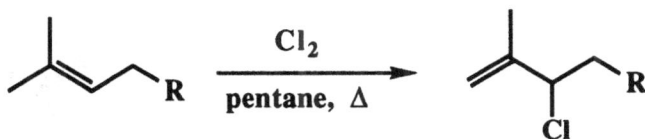

$$\xrightarrow[\text{pentane, } \Delta]{Cl_2}$$

---

**II.B.2-8** A. Sekiya et al., *Chem. Letter*, **1992**, 2183.

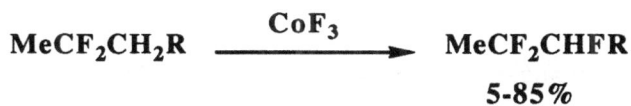

$$MeCF_2CH_2R \xrightarrow{CoF_3} MeCF_2CHFR$$
$$\textbf{5-85\%}$$

**II.B.2-9** F. A. Davis* and W. Han

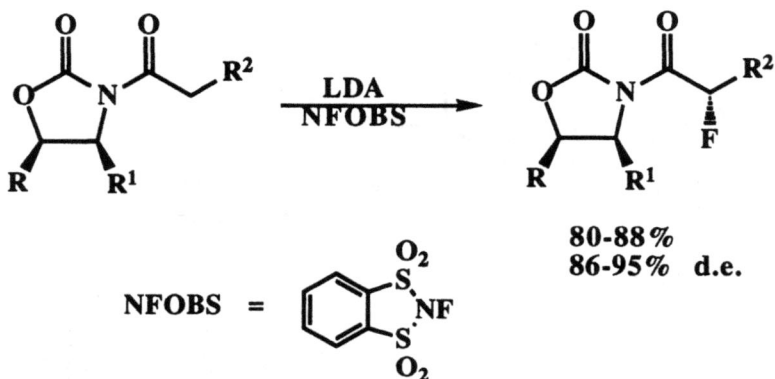

80-88%
86-95%  d.e.

NFOBS =

---

**II.B.2-10** R. V. Hoffman et al., *J. Am. Chem. Soc.*, **1992**, *114*, 6262.

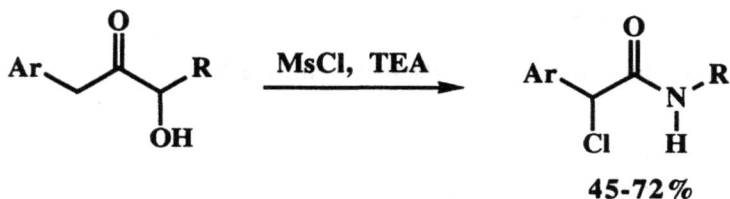

45-72%

---

**II.B.2-11** E. McNelis et al., *SynLett.*, **1992**, 131.

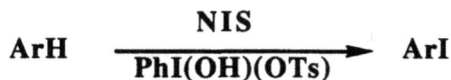

$$\text{ArH} \xrightarrow[\text{PhI(OH)(OTs)}]{\text{NIS}} \text{ArI}$$

---

**II.B.2-12** P. J. Stang et al., *Tetrahedron Lett.*, **1992**, *33*, 1419.

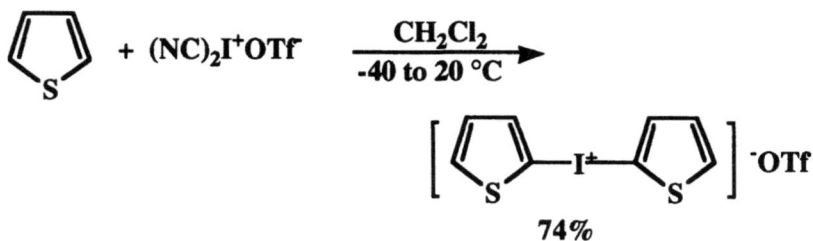

74%

**II.B.2-13** T. Kakinami*, S. Kajigaeshi* et al., *Bull. Chem. Soc. Jpn.*, **1992**, *65*, 2549.

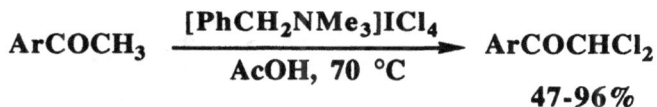

$$ArCOCH_3 \xrightarrow[\text{AcOH, 70 °C}]{[PhCH_2NMe_3]ICl_4} ArCOCHCl_2$$

**47-96%**

## II.C.    C-N Oxidations

**II.C-1** J. G. Lee et al., *Syn. Commun.*, **1992**, *22*, 2425.

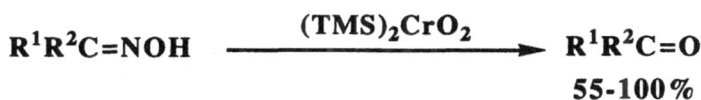

$$R^1R^2C=NOH \xrightarrow{(TMS)_2CrO_2} R^1R^2C=O$$

**55-100%**

---

**II.C-2** R. Grandi et al., *Syn. Commun.*, **1992**, *22*, 1845.

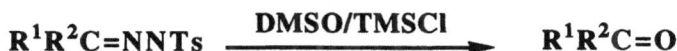

$$R^1R^2C=NNTs \xrightarrow{\text{DMSO/TMSCl}} R^1R^2C=O$$

---

**II.C-3** K. H. Park and J. B. Lee *Syn. Commun.*, **1992**, *22*, 1061.

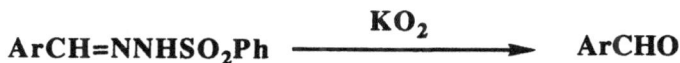

$$ArCH=NNHSO_2Ph \xrightarrow{KO_2} ArCHO$$

---

**II.C-4** M. B. Smith et al., *Org. Prep. Pro. Int.*, **1992**, *24*, 147.

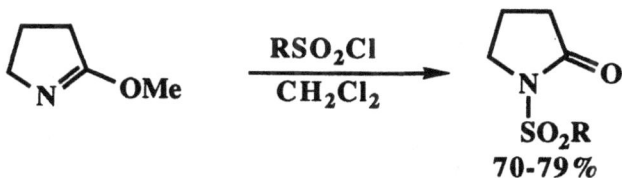

**70-79%**

**II.C-5** L. Kaczmarek et al., *Chem. Ber.* **1992**, *125*, 1965.

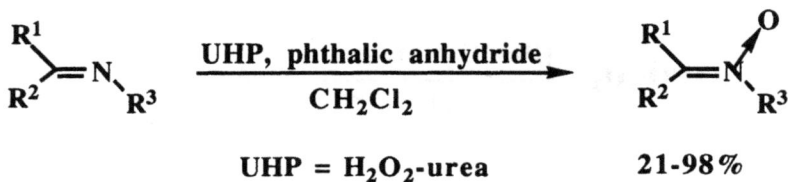

$$R^1R^2C{=}N{-}R^3 \xrightarrow[\text{CH}_2\text{Cl}_2]{\textbf{UHP, phthalic anhydride}} R^1R^2C{=}N^+(O^-){-}R^3$$

UHP = $H_2O_2$-urea          **21-98%**

---

**II.C-6** M. Petrini et al., *Tetrahedron Lett.*, **1992**, *33*, 4835.

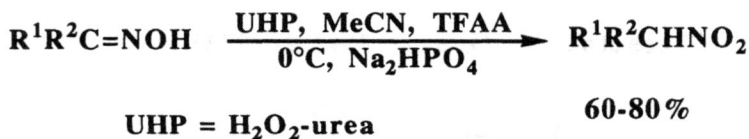

$$R^1R^2C{=}NOH \xrightarrow[\text{0°C, Na}_2\text{HPO}_4]{\textbf{UHP, MeCN, TFAA}} R^1R^2CHNO_2$$

UHP = $H_2O_2$-urea          **60-80%**

---

**II.C-7** G. A. Olah*, G. K. S. Prakash et al., *SynLett.*, **1992**,, 337.

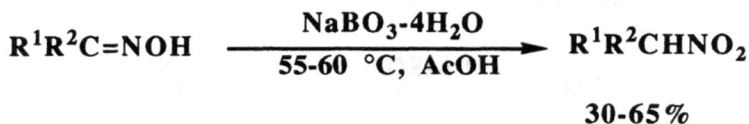

$$R^1R^2C{=}NOH \xrightarrow[\text{55-60 °C, AcOH}]{\text{NaBO}_3\text{-4H}_2\text{O}} R^1R^2CHNO_2$$

**30-65%**

## II.D. Amine Oxidations

**II.D-1** S. Rozen* and M. Kol, *J. Org. Chem.*, **1992**, *57*, 7324.

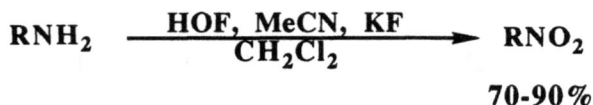

$$RNH_2 \xrightarrow[\text{CH}_2\text{Cl}_2]{\text{HOF, MeCN, KF}} RNO_2$$

**70-90%**

---

**II.D-2** A. Nishinaga*, A. Rieker et al., *Tetrahedron Lett.*, **1992**, *33*, 4425.

**Co(salen) Catalyzed Oxidation od 2,4,6-trisubstituted Anilines with tert-Butylhydroperoxide**

**II.D-3**  A. Ursini et al., *Synthesis*, **1992**, 363.

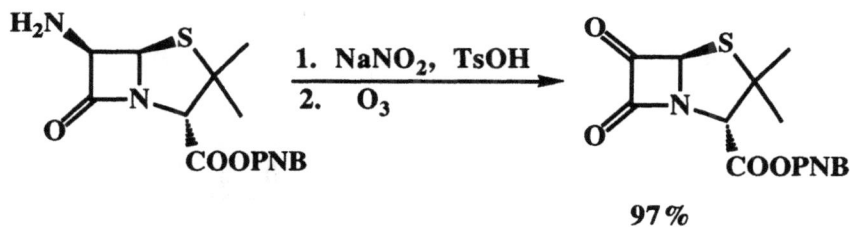

97%

---

**II.D-4**  R. J. Bergeron*, O. Phanstiel IV

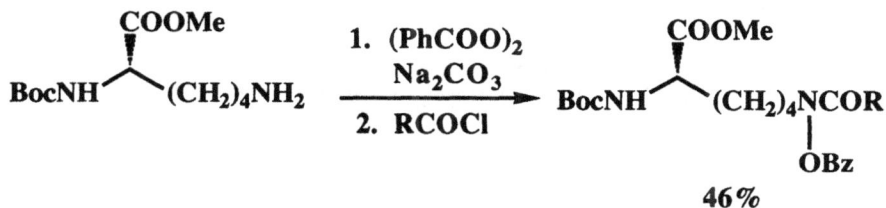

46%

---

**II.D-5**  A. S. Demir et al.,*Syn. Commun.*, **1992**, 22,  2607.

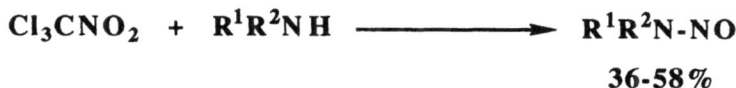

$$Cl_3CNO_2 \ + \ R^1R^2NH \longrightarrow R^1R^2N\text{-}NO$$

36-58%

---

**II.D-6**  M. Miura et al., *J. Chem. Soc., Perkin. Trans. I*, **1992,** 1387.

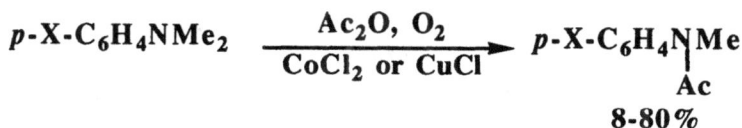

$$p\text{-}X\text{-}C_6H_4NMe_2 \xrightarrow[\text{CoCl}_2 \text{ or CuCl}]{Ac_2O, \ O_2} p\text{-}X\text{-}C_6H_4\underset{Ac}{N}Me$$

8-80%

## II.E. Sulfur Oxidation

**II.E-1**  W. Adam and L. Hadjiarapoglou, *Tetrahedron Lett.*, **1992**, *33*, 469.

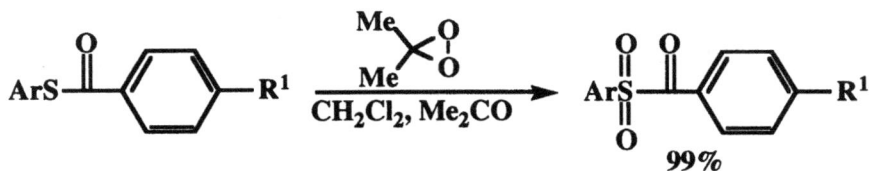

99%

**II.E-2**  V. S. Martin et al., *Tetrahedron*, **1992**, *48*,  3571.

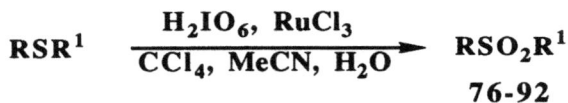

$$RSR^1 \xrightarrow[\text{CCl}_4, \text{ MeCN, } H_2O]{H_2IO_6, \text{ RuCl}_3} RSO_2R^1$$

76-92

**II.E-3**  R. P. Greenhalgh *SynLett.*, **1992**, 235.

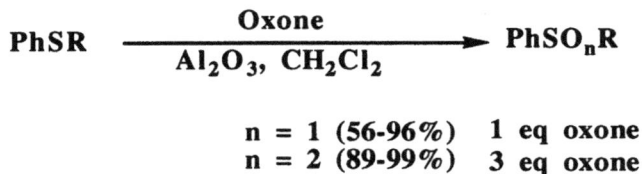

$$PhSR \xrightarrow[\text{Al}_2O_3, \text{ CH}_2Cl_2]{\text{Oxone}} PhSO_nR$$

n = 1 (56-96%)   1 eq oxone
n = 2 (89-99%)   3 eq oxone

**II.E-4**  M. F. Rahman et al., *Tetrahedron*, **1992**, *48*,  1953.

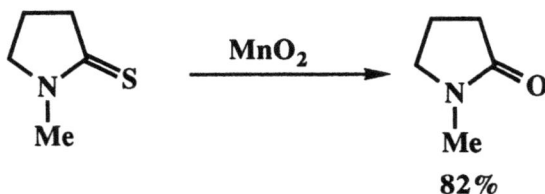

82%

**II.E-5**  F. A. Davis and R. T. Reddy, *J. Org. Chem.*, **1992**, *57*, 2599.

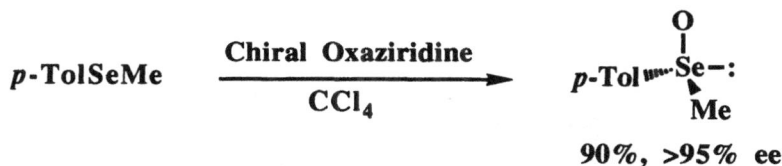

$$p\text{-TolSeMe} \xrightarrow[\text{CCl}_4]{\textbf{Chiral Oxaziridine}} p\text{-Tol}^{\text{\tiny ⋯}}\overset{\displaystyle \text{O}}{\underset{\text{Me}}{\text{Se}}}-:$$

**90%, >95% ee**

---

**II.E-6**  C. M. Rayner et al., *Tetrahedron Lett.*, **1992**, *33*, 7237 and F. A. Davis et al., *J. Am. Chem. Soc.*, **1992**, *114*, 1428.

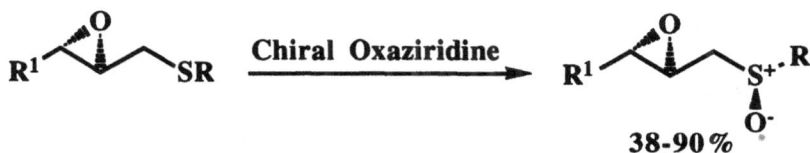

$$R^1 \overset{\text{O}}{\triangle} \diagdown \text{SR} \xrightarrow{\textbf{Chiral Oxaziridine}} R^1 \overset{\text{O}}{\triangle} \diagdown \overset{+}{\text{S}}\cdot \overset{\text{R}}{\underset{\text{O}^-}{}}$$

**38-90%**

---

**II.E-7**  T. Morimoto et al., *Bull. Chem. Soc. Jpn.*, **1992**, *65*, 1744, and A. Ohta, *Synthesis*, **1992**, 555.

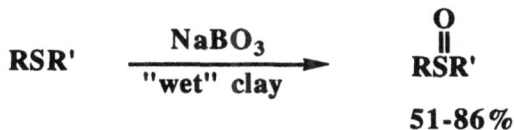

$$\text{RSR}' \xrightarrow[\text{"wet" clay}]{\text{NaBO}_3} \overset{\displaystyle \text{O}}{\overset{\|}{\text{RSR}'}}$$

**51-86%**

---

**II.E-8**  H. Suzuki et al., *Bull. Chem. Soc. Jpn.*, **1992**, *65*, 626.

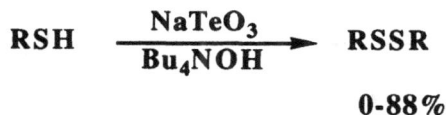

$$\text{RSH} \xrightarrow[\text{Bu}_4\text{NOH}]{\text{NaTeO}_3} \text{RSSR}$$

**0-88%**

**II.E-9** S. Colonna et al., *Tetrahedron: Asymmetry*, **1992**, *3*,, 95, and A. Fauve* and M. Madesclaire et al., *Tetrahedron: Asymmetry*, **1992**, *3*, 629.

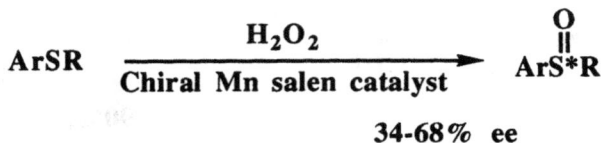

$$\text{ArSR} \xrightarrow[\text{Chiral Mn salen catalyst}]{\text{H}_2\text{O}_2} \overset{\overset{\text{O}}{\parallel}}{\text{ArS*R}}$$

**34-68% ee**

---

**II.E-10** H. E. Folsom and J. Castrillon. *Syn. Commun.*, **1992**, *22*, 1798.

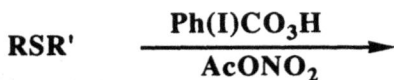

$$\text{RSR}' \xrightarrow[\text{AcONO}_2]{\text{Ph(I)CO}_3\text{H}}$$

---

**II.E-11** E. L. Clennan* and, K. Yang *J. Org. Chem.*, **1992**, *57*, 4477.

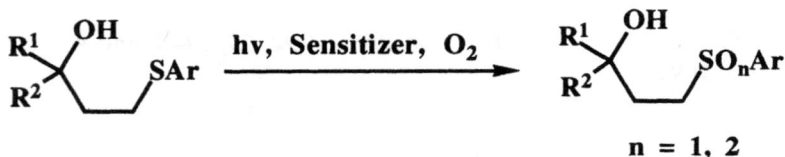

$$\underset{R^2}{\overset{R^1}{>}}\!\!\!\!\!\overset{OH}{\underset{}{\diagup}}\!\!\!\sim\!\!\text{SAr} \xrightarrow{\text{hv, Sensitizer, O}_2} \underset{R^2}{\overset{R^1}{>}}\!\!\!\!\!\overset{OH}{\underset{}{\diagup}}\!\!\!\sim\!\!\text{SO}_n\text{Ar}$$

$$n = 1, 2$$

---

**II.E-12** E. N. Jacobson et al., *Tetrahedron Lett.*, **1992**, *33*, 7111.

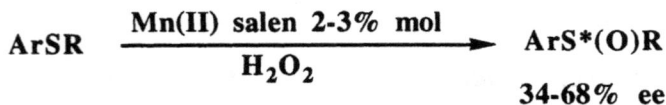

$$\text{ArSR} \xrightarrow[\text{H}_2\text{O}_2]{\text{Mn(II) salen 2-3\% mol}} \text{ArS*(O)R}$$

**34-68% ee**

---

**II.E-13** S. Uemura et al., *Tetrahedron Lett.*, **1992**, *33*, 5391.

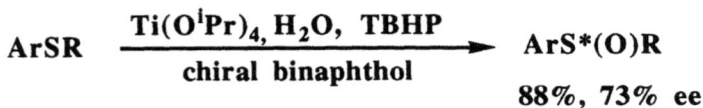

$$\text{ArSR} \xrightarrow[\text{chiral binaphthol}]{\text{Ti(O}^i\text{Pr)}_4, \text{H}_2\text{O, TBHP}} \text{ArS*(O)R}$$

**88%, 73% ee**

**II.E-14**  R. Balicki et al., *Liebigs Ann. Chem.*, **1992**, 883.

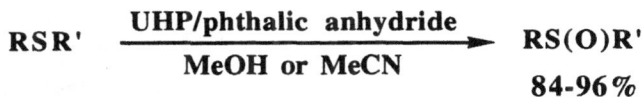

$$RSR' \xrightarrow[\text{MeOH or MeCN}]{\text{UHP/phthalic anhydride}} RS(O)R'$$

84-96%

---

**II.E-15**  W. Hanefeld et al., *Liebigs Ann. Chem.* **1992**, 337.

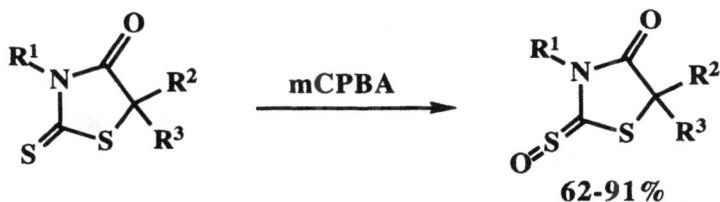

62-91%

## II.F.  Oxidative Addition to C-C Multiple Bonds

### II.F.1  Epoxidations

**II.F.1-1**  J. K. Crandall et al., *Tetrahedron*, **1992**, *48*, 1427.

**Allene Epoxidation wih dimethyldioxane**

---

**II.F.1-2**  W. Ebenezer and G. Pattenden *Tetrahedron Lett.*, **1992**, *33*, 4053.

**Trienoate Epoxidation wih dimethyldioxane**

**II.F.1-3**  W. Adam et al., *Synthesis* , **1992**,,436,  *Chem. Ber.* **1992**, *125*, 231.

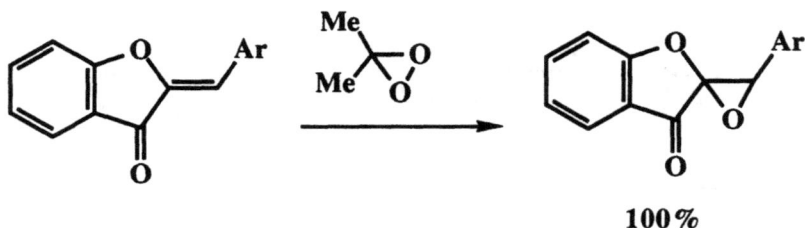

**100%**

---

**II.F.1-4**  K. A. Jorgensen et al., *J. Chem. Soc., Chem. Commun.*, **1992**, 1072.

### Regioselective Monoepoxidation of 1,3-Dienes
### Catalyzed by transition-metal Complexes

---

**II.F.1-5**  L. Wang, K. B. Sharpless et al., *J. Am. Chem. Soc.*, **1992**, *114*, 7568, 7570.

**80% ee**

---

**II.F.1-6**  S. Chandrasekaran et al., *J. Org. Chem.*, **1992**, *57*, 1928.

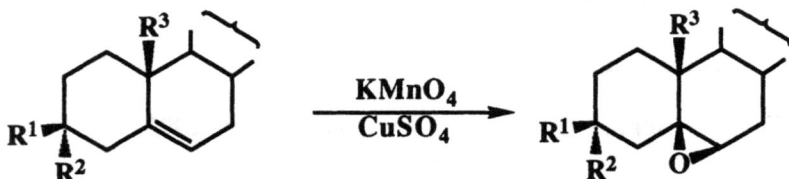

**II.F.1-7**  T. Mukaiyama et al.,  *Chem. Lett.*, **1992**, 2077, 2109,

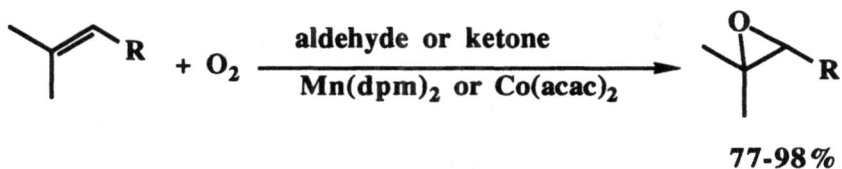

77-98%

---

**II.F.1-8**  T. Mukaiyama et al.,  *Chem. Lett.*, **1992**, 2231.

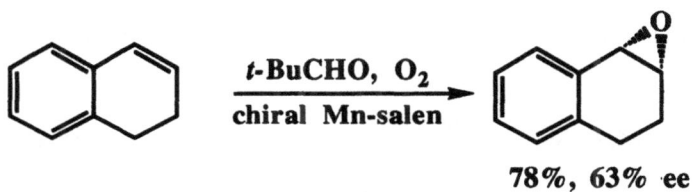

*t*-BuCHO, O$_2$
chiral Mn-salen

78%, 63% ee

---

**II.F.1-9**  K. Kaneda et al., *Tetrahedron Lett.*, **1992**, *33*, 6827.

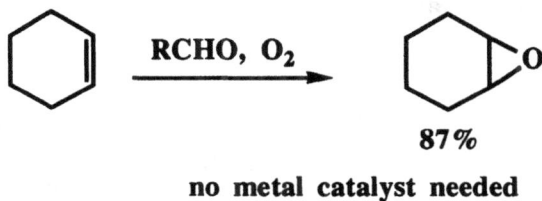

RCHO, O$_2$

87%

no metal catalyst needed

---

**II.F.1-10**  P. Caubere et al., *Tetrahedron*, **1992**, *48*, 5099.

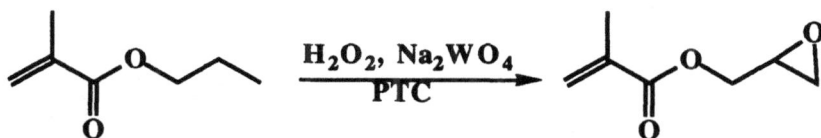

H$_2$O$_2$, Na$_2$WO$_4$
PTC

**II.F.1-11**  N. Sakai, Y. Ohfune *J. Am. Chem. Soc.*, **1992**, *114*, 998.

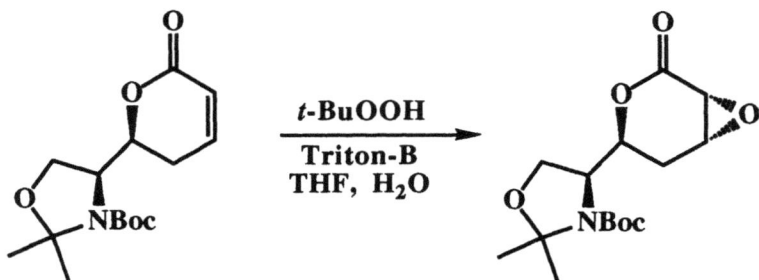

---

**II.F.1-12**  T. I. Filyakova et al., *J. Org. Chem. USSR*, **1992**, *28*, 20.

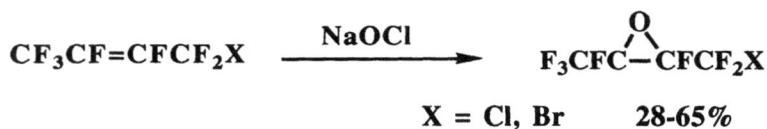

$$CF_3CF=CFCF_2X \xrightarrow{\text{NaOCl}} F_3CFC\overset{O}{\overset{\triangle}{-}}CFCF_2X$$

$$X = Cl, \ Br \qquad 28\text{-}65\%$$

---

**II.F.1-13**  S. Warren rt al, *Tetrahedron Lett.*, **1992**, *33*, 7043, and 7039.

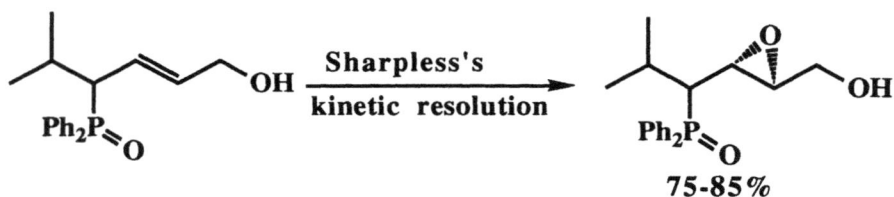

---

**II.F.1-14**  T. Katsuki et al, *Chem. Lett.*, **1992**, 1992.

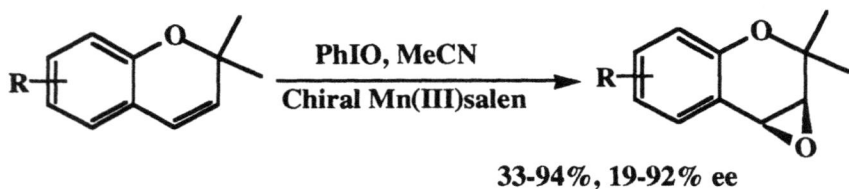

33-94%, 19-92% ee

**II.F.1-15** A. Scettri et al., *Tetrahedron Lett.*, **1992**, *33*, 5433.

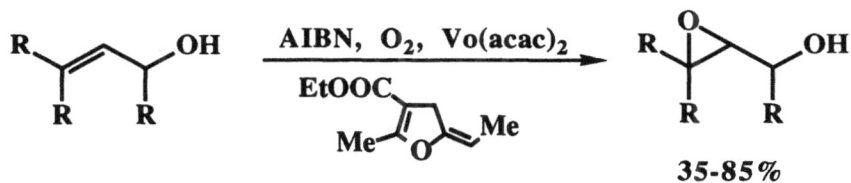

**II.F.1-16** R. F. W. Jackson et al., *Tetrahedron Lett.*, **1992**, *33*, 6197.

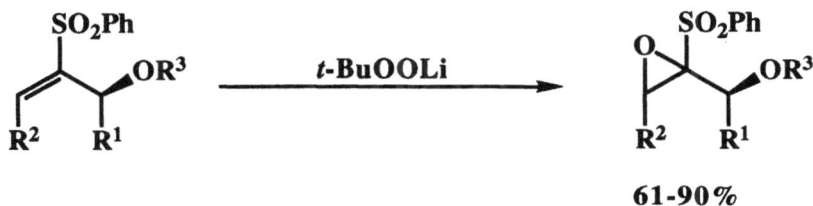

## II. F. 2. Hydroxylation

**II.F.2-1** J. S. Panek and X. Fu *J. Org. Chem.*, **1992**, *57*, 5288.

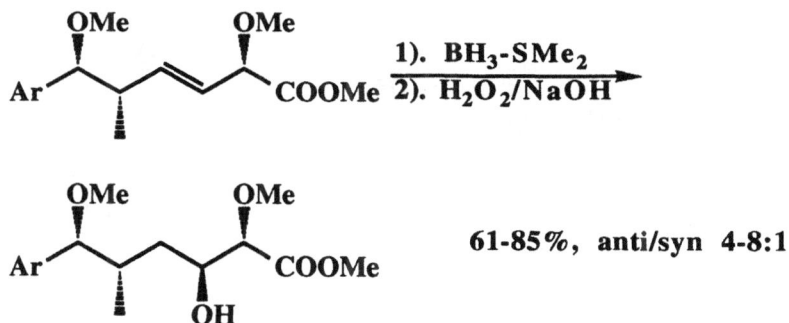

61-85%, anti/syn 4-8:1

**II.F.2-2** K. B. Sharpless et al., *Tetrahedron Lett.*, **1992**, *33*, 2095, 3833, 4273, *J. Org. Chem.*, **1992**, *57*, 2768, K. Fuji et al., *Tetrahedron Lett.*, **1992**, *33*, 4021.

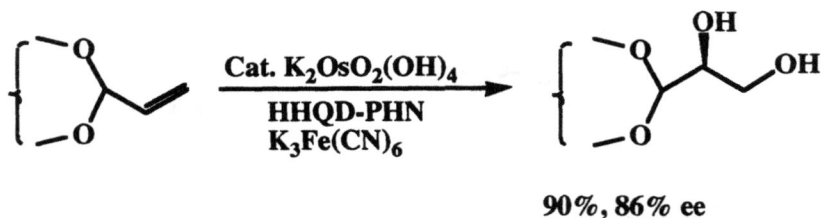

**90%, 86% ee**

---

**II.F.2-3** K. B. Sharpless, et al., *J. Org. Chem.*, **1992**, *57*, 5067.

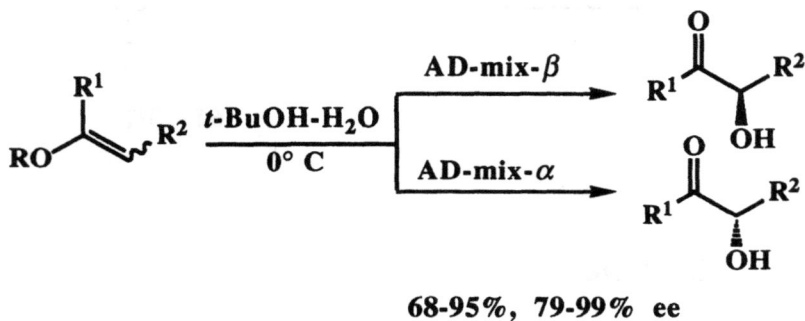

**68-95%, 79-99% ee**

---

**II.F.2-4** R. M. Moriarty et al., *Tetrahedron Lett.*, **1992**, *33*, 6065.

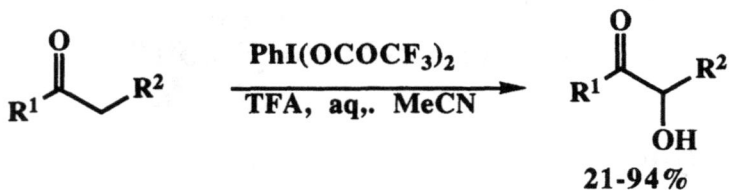

**21-94%**

**II.F.2-5**  B. B. Lohray et al., *Tetrahedron Lett.*, **1992**, *33*, 5453.

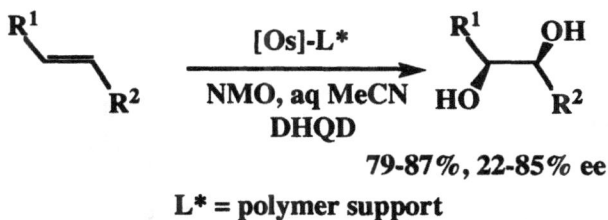

79-87%, 22-85% ee

L* = polymer support

---

**II.F.2-6**  S. I. Murahashi et al., *Tetrahedron Lett.*, **1992**, *33*, 5081.

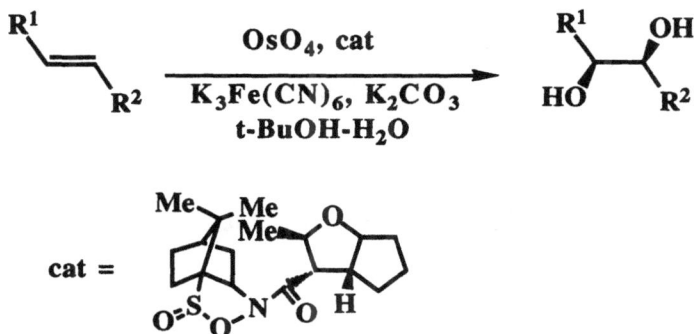

cat =

---

**II.F.2-7**  T. Hudlicky et al., *SynLett.*, **1992**, 391.

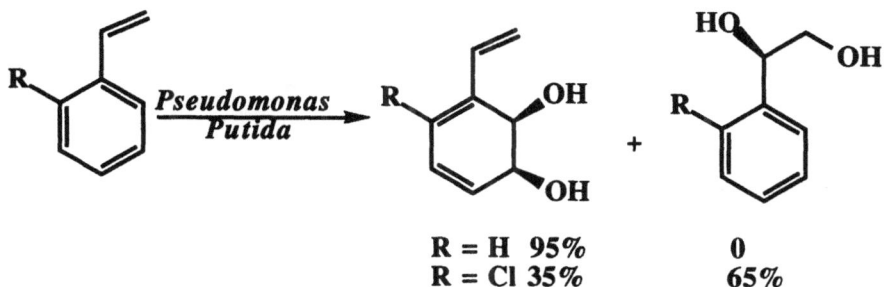

R = H  95%
R = Cl  35%

0
65%

**II.F.2-7** K. Krohn et al., *Chem. Ber.* **1992**, *125*, 2439.

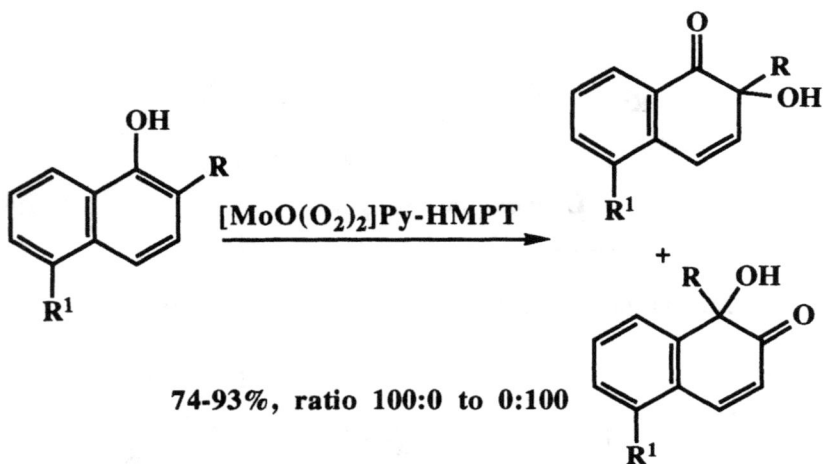

**74-93%, ratio 100:0 to 0:100**

---

**II.F.2-8** R. G. Dishi and S. J. Danishefsky *J. Am. Chem. Soc.*, **1992**, *114*, 3431.

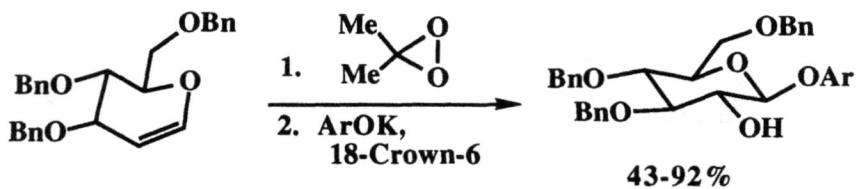

**43-92%**

**II.F.2-9**  R. L. Haltman and M. A. McEvoy *J. Am. Chem. Soc.*, **1992**, *114*, 980.

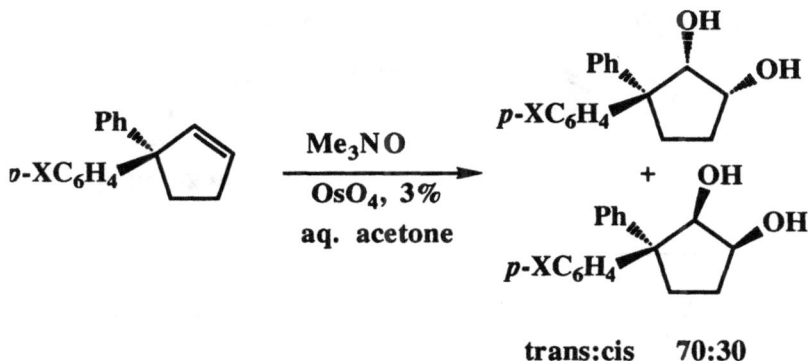

trans:cis     70:30

**II.F.2-10**  W. A Herrmann et al., *Angew. Chem., Int. Ed. Engl.* **1992**, *31*, 1345.

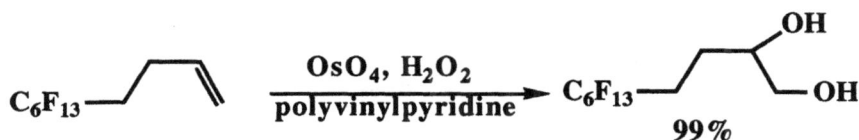

99%

**II.F.2-11**  M. R. Hale and A. H. Hoveyda *J. Org. Chem.*, **1992**, *57*, 1643.

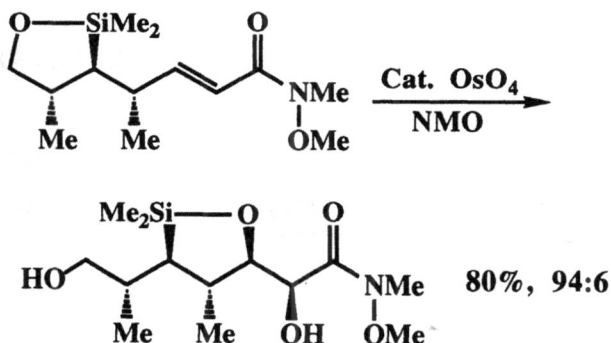

80%, 94:6

**II.F.2-12** I. Fleming et al., *J. Chem. Soc., Perkin. Trans. I*, **1992**, 3303.

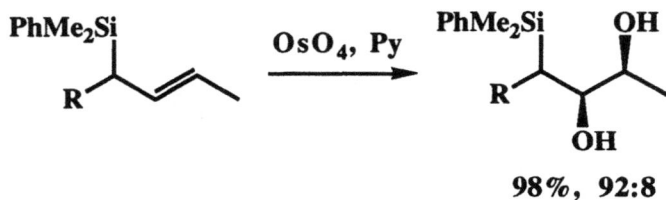

PhMe$_2$Si

R

$\xrightarrow{\text{OsO}_4,\ \text{Py}}$

PhMe$_2$Si     OH

R

ŌH

**98%, 92:8**

---

**II.F.2-13** H. B. Kagan et al., *Tetrahedron: Asymmetry*, **1992**, *3*,, 849.

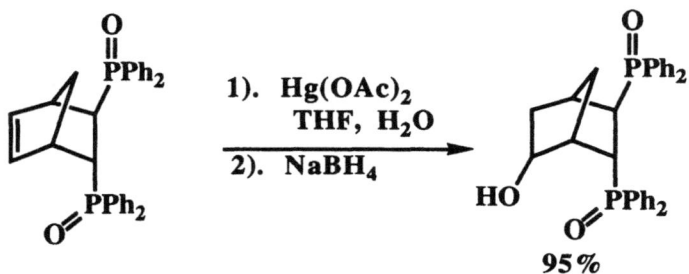

O
‖
PPh$_2$

PPh$_2$
O

$\xrightarrow{\begin{array}{l}1).\ \text{Hg(OAc)}_2\\ \ \ \ \ \text{THF, H}_2\text{O}\\ 2).\ \text{NaBH}_4\end{array}}$

O
‖
PPh$_2$

HO        PPh$_2$
O

**95%**

## II.F.3. Other Oxidative Addition to C-C Multiple Bonds

**II.F.3-1** J. P. Dulcere et al. *Bull. Soc. Chim. Fr.* , **1992**, *129*, 280.

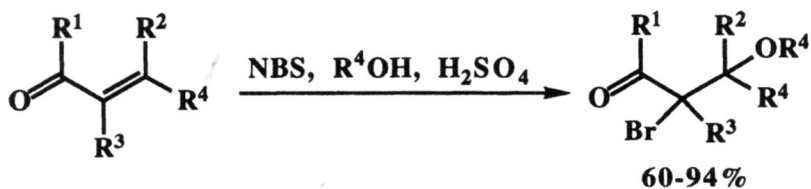

R$^1$   R$^2$

O              R$^4$

R$^3$

$\xrightarrow{\text{NBS, R}^4\text{OH, H}_2\text{SO}_4}$

R$^1$   R$^2$
            OR$^4$

O              R$^4$

Br    R$^3$

**60-94%**

**II.F.3-2** M. Tiecco, M. Tingoli et al., *J. Org. Chem.*, **1992**, *57*, 4025.

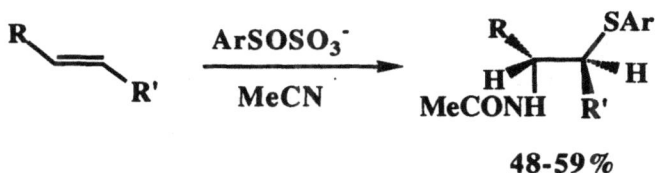

48-59%

---

**II.F.3-3** A. Veyrieres et al., *Tetrahedron Lett.*, **1992**, *33*, 221.

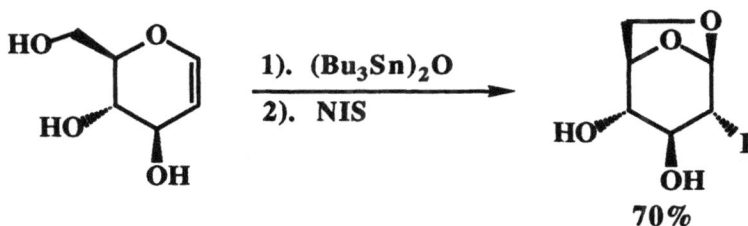

70%

---

**II.F.3-4** L. Fuentes et al., *Syn. Commun.*, **1992**, *22*, 2053.

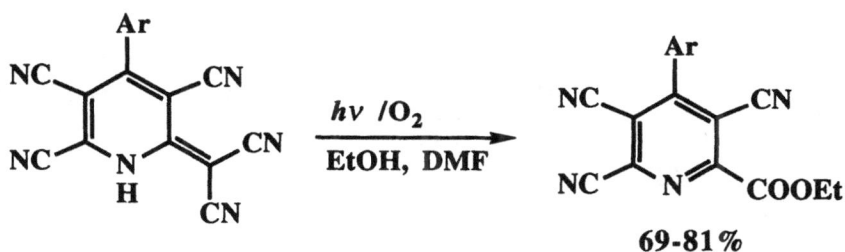

69-81%

**II.F.3-5**  K. Uchida et al., *Tetrahedron Lett.*, **1992**, *33*, 2567.

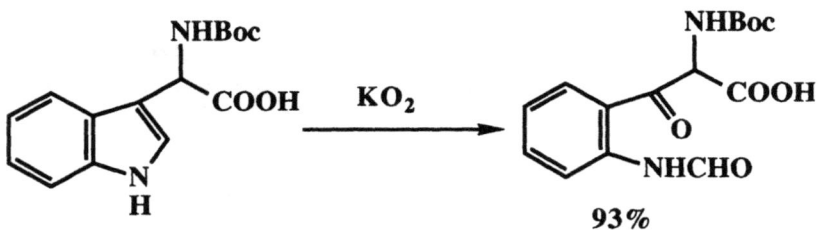

93%

**II.F.3-6**  C. R. Johnson et al., *J. Am. Chem. Soc.*, **1992**, *114*, 9414.

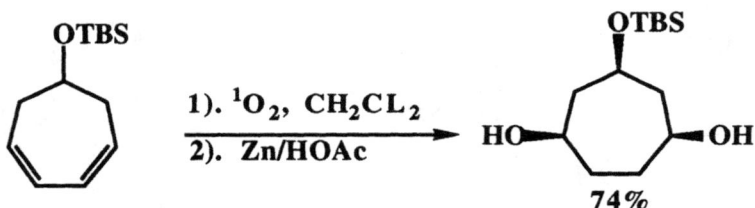

1). $^1O_2$, $CH_2CL_2$
2). Zn/HOAc

74%

**II.F.3-7**  S. Chandrasekaren et al., *J. Chem. Soc., Chem. Commun.*, **1992**, 626.

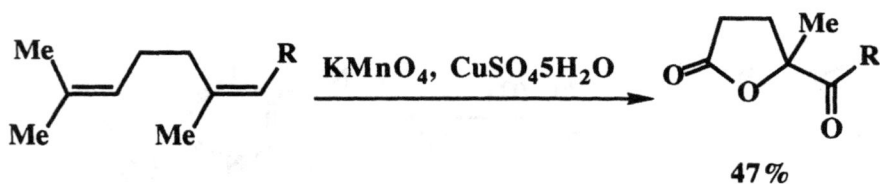

$KMnO_4$, $CuSO_45H_2O$

47%

**II.F.3-8**  R. W. Trainor et al., *Aust. J. Chem.* **1992**, *45*, 1265.

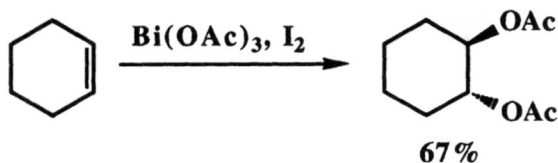

$Bi(OAc)_3$, $I_2$

67%

**II.F.3-9** J. Correia, *J. Org. Chem.*, **1992**, *57* 4555.

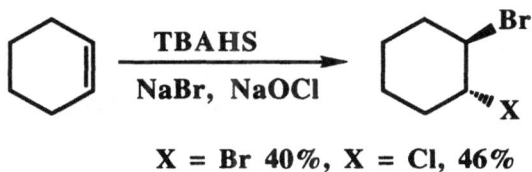

X = Br 40%, X = Cl, 46%

---

**II.F.3-10** X. Lu et al., *Tetrahedron Lett.*, **1992**, *33*, 7209.

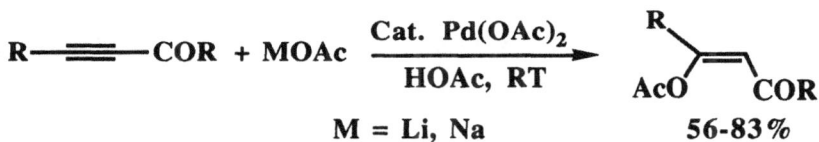

$$R\!\!-\!\!\!\equiv\!\!\!-COR + MOAc \xrightarrow[\text{HOAc, RT}]{\text{Cat. Pd(OAc)}_2}$$

M = Li, Na

56-83%

---

**II.F.3-11** T. Fujisawa et al., *Tetrahedron Lett.*, **1992**, *33*, 7003.

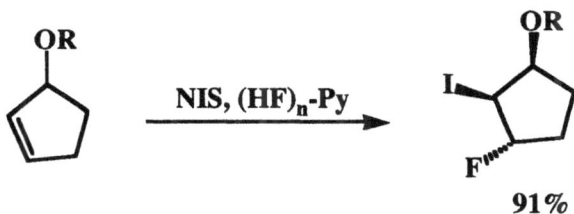

NIS, (HF)$_n$-Py

91%

---

**II.F.3-12** P. Magnus and P. Rigollier, *Tetrahedron Lett.*, **1992**, *33*, 6111.

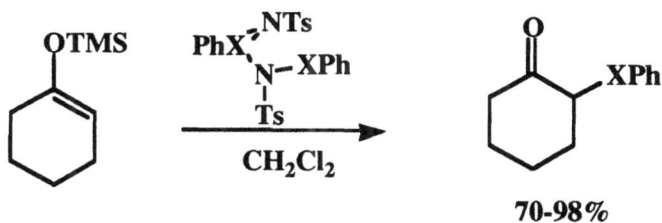

70-98%

**II.F.3-13** H. H. Wasserman and K. S. Prowse., *Tetrahedron Lett.*, **1992**, *33*, 8199.

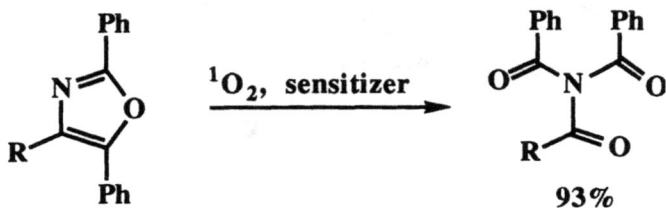

**II.F.3-14** S. C. Conway and G. W. Gribble, *Heterocycle*, **1992**, *34*, 2095.

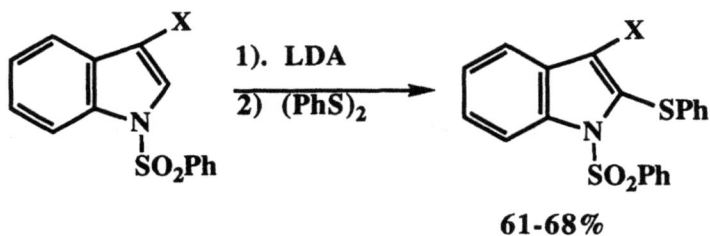

**II.F.3-15** V. Roussis* and T. D. Hubert, *Liebigs Ann. Chem*, **1992**, 539.

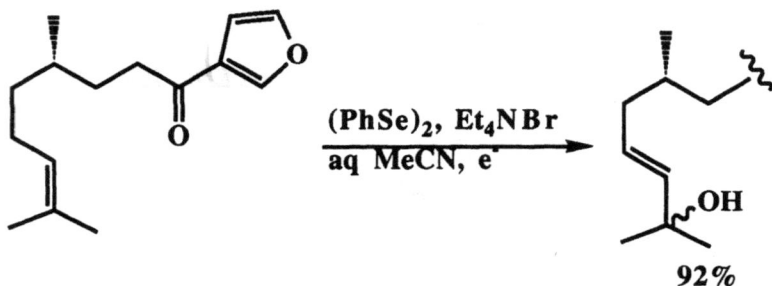

**II.F.3-16** M. T. Crimmins et al., *J. Am. Chem. Soc.*, **1992**, *114*, 5445.

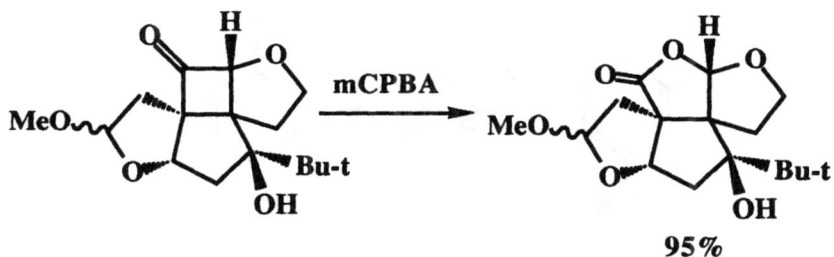

95%

---

**II.F.3-17** G. W. Kabalka et al., *Tetrahedron Lett.*, **1992**, *33*, 864.

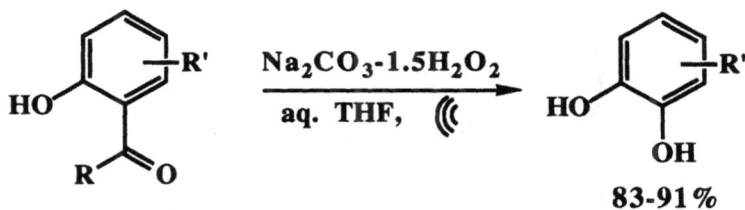

83-91%

## II.G. Phenol-Quinone Oxidations

**II.G-1** N. Adam and A. Schonberger, *Tetrahedron Lett.*, **1992**, *33*, 53.

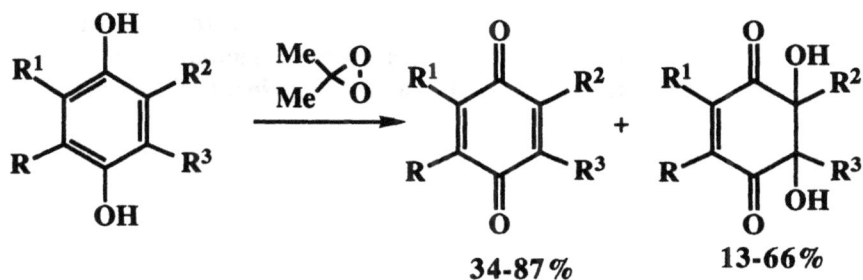

34-87%      13-66%

**II.G-2**  G. W. Morrow et al., *Syn. Commun.*, **1992**, 22,  179.

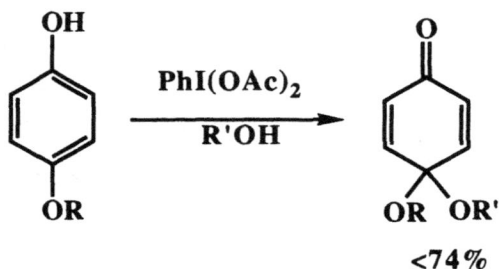

PhI(OAc)$_2$

R'OH

<74%

---

**II.G-3**  A. McKillop et al., *SynLett.*, **1992**, 201.

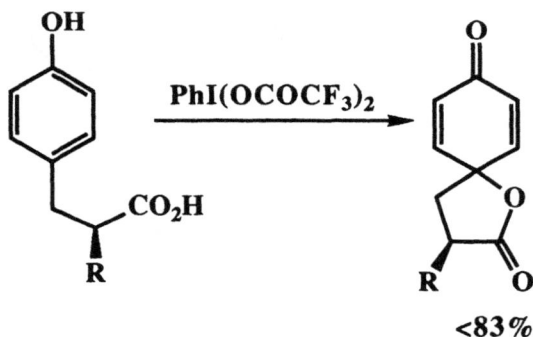

PhI(OCOCF$_3$)$_2$

<83%

---

**II.G-4**  M. Shimizu*, K. Takehira* et al., *Bull. Chem. Soc. Jpn.*, **1992**, 65, 1522.

**Synthesis of Alicyl substituted *p*-Benzoquinones from the Corresponding Phenols Using Molecular Oxygen Catalyzed by CuCl$_2$-Amine Hydrochloride Systems**

**II.G-5** H. Nishino et al., *Bull. Chem. Soc. Jpn.*, **1992**, *65*, 620,

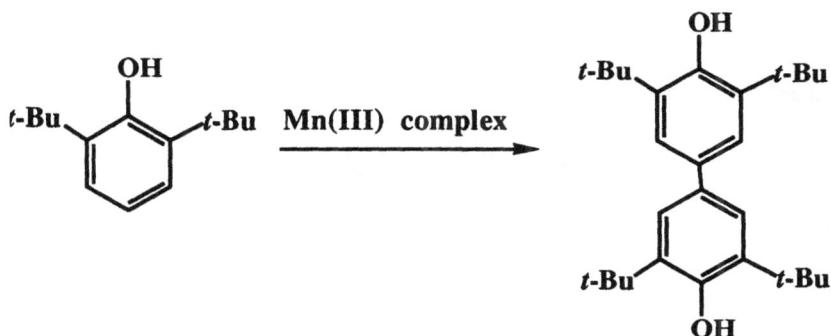

**II.G-6** M. Lissel et al., *Tetrahedron Lett.*, **1992**, *33*, 1795.

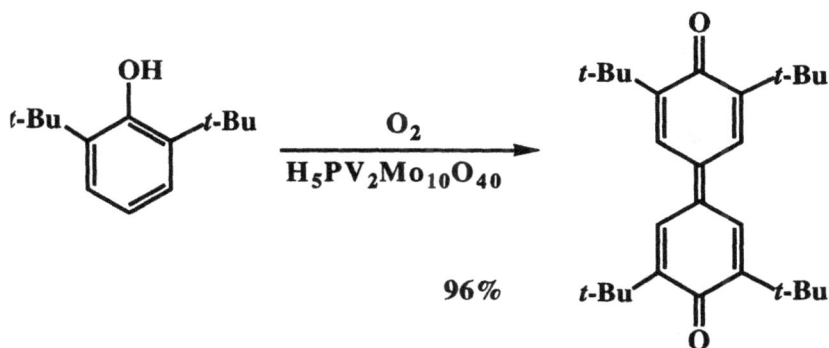

96%

**II.G-7** M. T. Maurette, et al., *Tetrahedron*, **1992**, *48*, 1869.

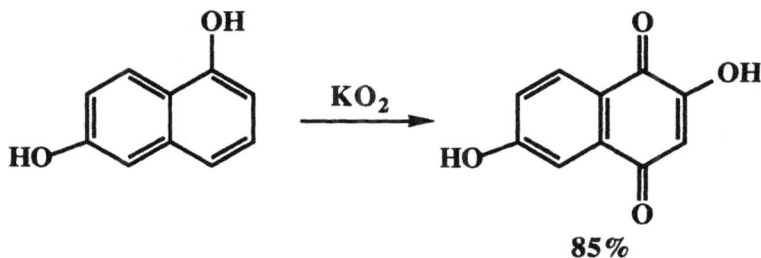

85%

**II.G-8**  U. R. Ghatak et al., *Synthesis*, **1992**, 1073.

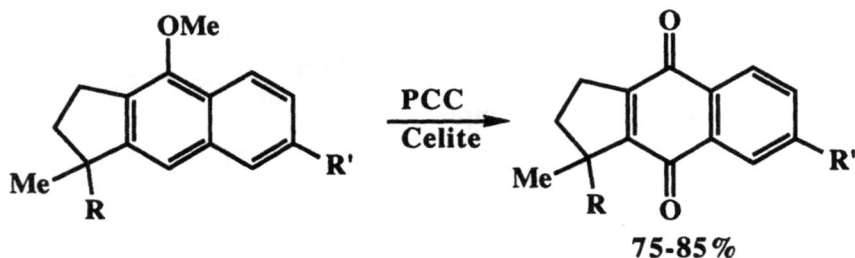

75-85%

---

**II.G-9**  J. Sykora et al., *Bull. Soc. Chem. Belg.* **1992**, *101*, 821.

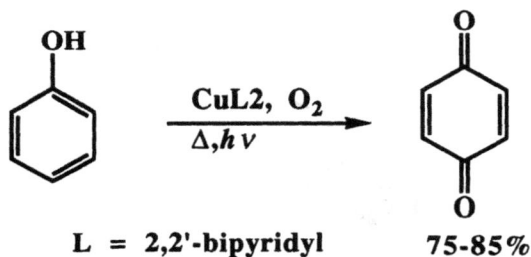

L = 2,2'-bipyridyl          75-85%

---

**II.G-10**  R. G. Srivastava and P. S. Venkataramani, *Syn. Commun.*, **1992**, 22, 35.

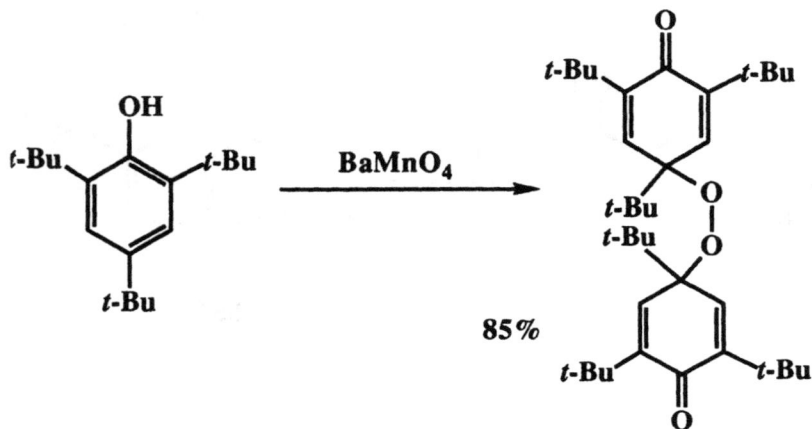

85%

**II.G-11** H. Viertler et al., *Tetrahedron Lett.*, **1992**, *33*, 8133.

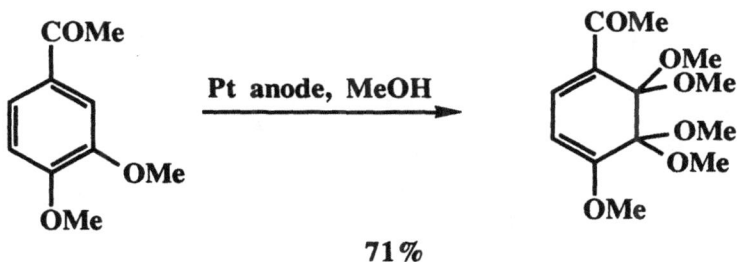

71%

## II.H. Dehydrogenation

**II.H-1** H. J. Niclas et al., *Syn. Commun.*, **1992**, *22*, 281.

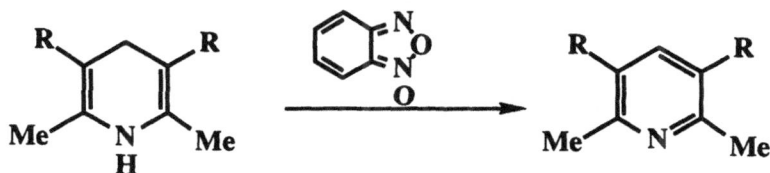

**The reagent also dehydrogenates a variety of other dihydroaromatics**

**II.H-2** J. J. Vanden Eynde et al., *Tetrahedron*, **1992**, *48*, 462.

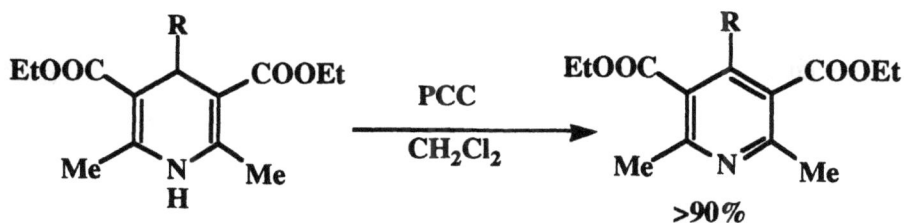

>90%

**II.H-3**  D. Seebach et al., *Synthesis*, **1992**, 39.

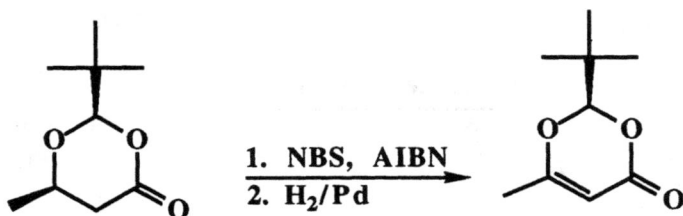

**1. NBS, AIBN**
**2. H₂/Pd**

---

**II.H-4**  P. Magnus et al., *Tetrahedron Lett.*, **1992**, *33*, 2933.

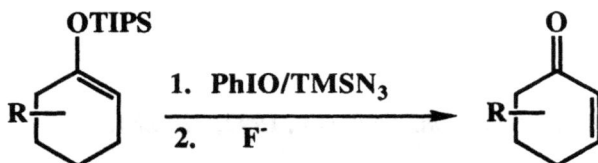

OTIPS

**1. PhIO/TMSN₃**
**2.    F⁻**

---

**II.H-5**  S. Yamazaki *Chem. Lett.*, **1992**, 823.

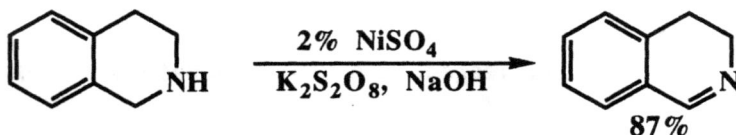

**2% NiSO₄**
**K₂S₂O₈,  NaOH**

**87%**

---

**II.H-6**  J. Yamaguchi and T. Takeda *Chem. Lett.*, **1992**, 1933.

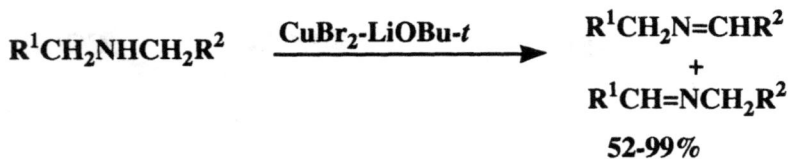

R¹CH₂NHCH₂R²     **CuBr₂-LiOBu-*t***      R¹CH₂N=CHR²
+
R¹CH=NCH₂R²

**52-99%**

**II.H-7** N.S. Simpkins *Tetrahedron Lett.*, **1992**, *33*, 8141.

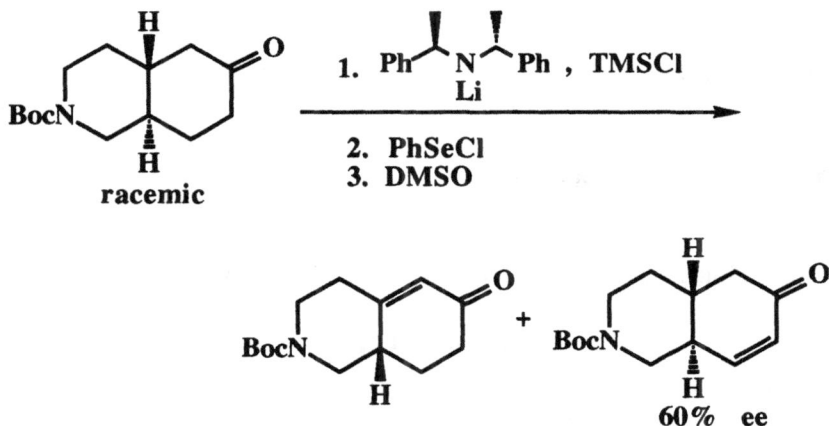

**II.H-8** J. G. Lee* and K. C. Kim *Tetrahedron Lett.*, **1992**, *33*, 6363.

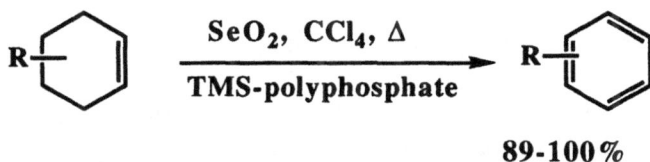

89-100%

**II.H-9** R. P. Kapoor et al., *J. Chem. Soc., Perkin. Trans. I*, **1992**, 2565, S. Kapil et al., *SynLett.*, **1992**, 751.

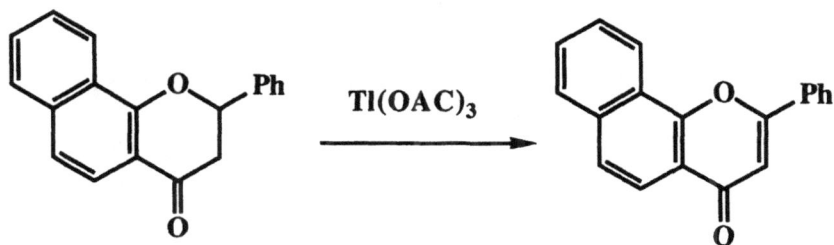

# III
# REDUCTIONS

## III.A.          C=O Reductions

**III.A-1** J. Bourguihnon et al., *Tetrahedron*, **1992**, *48*, 831.

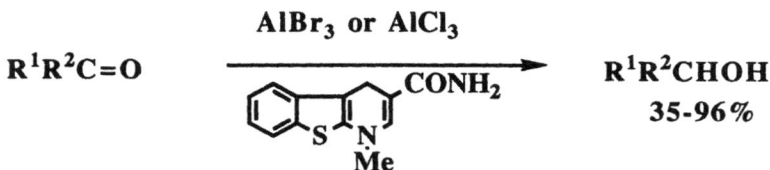

$$R^1R^2C=O \xrightarrow[\text{AlBr}_3 \text{ or AlCl}_3]{} R^1R^2CHOH$$

AlBr$_3$ or AlCl$_3$

$R^1R^2C=O$ → $R^1R^2CHOH$

**35-96%**

---

**III.A-2**  Z. Hou et al., *Organometallics* **1992**, *11*, 2711.

**Stepwise and reversible cleavage of the C-O double bond of diary ketones by lanthanide metals.**

---

**III.A-3** E. J. Corey and J. O. Link, *Tetrahedron Lett.*, **1992**, *33*, 2319, 3431, and 4141; J. Martens et al., *Tetrahedron: Asymmetry*, **1992**, *3*, 223 and 347.

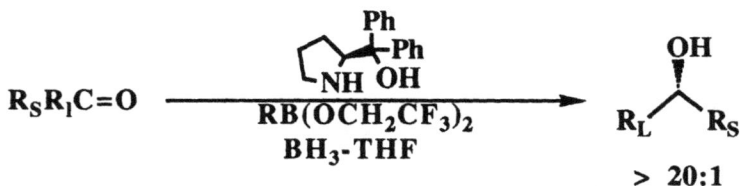

$$R_SR_1C=O \xrightarrow[\text{BH}_3\text{-THF}]{\text{RB(OCH}_2\text{CF}_3)_2} $$

$R_SR_1C=O$ →

Ph
Ph
NH OH
RB(OCH$_2$CF$_3$)$_2$
BH$_3$-THF

OH

$R_L$ $R_S$

**> 20:1**

**III.A-4** H. Brunner and M. Rotzer *J. Organometal. Chem.*, **1992**, *425*, 119.

**Optically Active Hydrido-phosphido-bridged
Catalysis for Photochemical Hydrosilylation of
PhCOMe.**

---

**III.A-5** K. Mashima et al., *J. Organometal. Chem.*, *428*, 213.

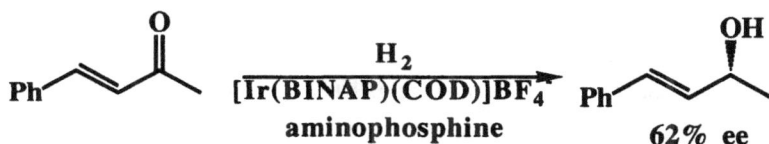

Ph (enone) →
$H_2$
[Ir(BINAP)(COD)]BF$_4$
aminophosphine
→ Ph (allylic alcohol OH) 62% ee

---

**III.A-6** R. Tacke et al., *J. Organometal. Chem.*, **1992**, *424*, 273.

**Stereochemistry of Microbial Reduction**

---

**III.A-7** K. Achiwa et al., *Tetrahedron: Asymmetry*, **1992**, *3*, 13.

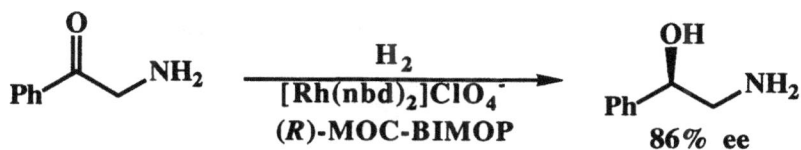

Ph—CO—CH$_2$NH$_2$ →
$H_2$
[Rh(nbd)$_2$]ClO$_4^-$
(*R*)-MOC-BIMOP
→ Ph—CH(OH)—CH$_2$NH$_2$ 86% ee

(*R*)-MOC-BIMOP =

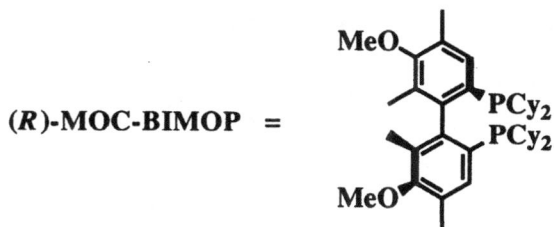

**III.A-8** S. A. King et al., *J. Org. Chem.*, **1992**, *57*, 6689.

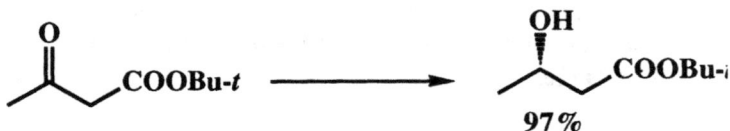

An improved produre
H$_2$ (40 psi), 0.1% HCl, MeOH,
[(R)-Ru(BINAP)Cl$_2$]$_2$TEA

---

**III.A-9** M. Periasamy et al., *Tetrahedron*, **1992**, *48*, 4622,

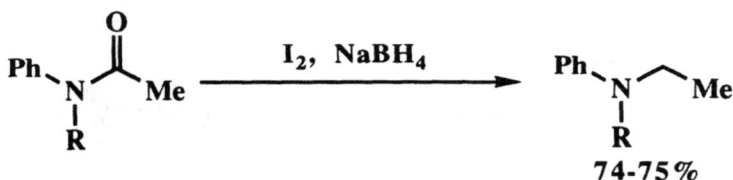

---

**III.A-10** S. C. Berk, S. L. Buchwald*, *J. Org. Chem.*, **1992**, *57*, 3751.

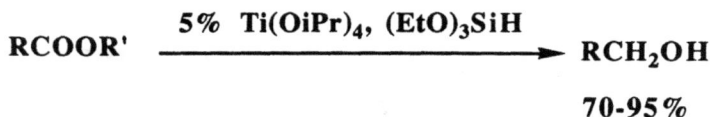

$$\text{RCOOR'} \xrightarrow{\text{5\% Ti(OiPr)}_4, \text{ (EtO)}_3\text{SiH}} \text{RCH}_2\text{OH}$$

70-95%

---

**III.A-11** H. Nishiyama et al., *J. Org. Chem.*, **1992**, *57*, 4300.

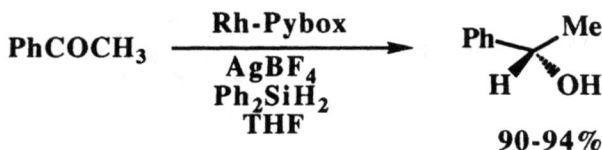

Pybox = bis(oxazolinyl)pyridine

**III.A-12**  I. Shibata et al., *J. Org. Chem.*, **1992**, *57*, 4049.

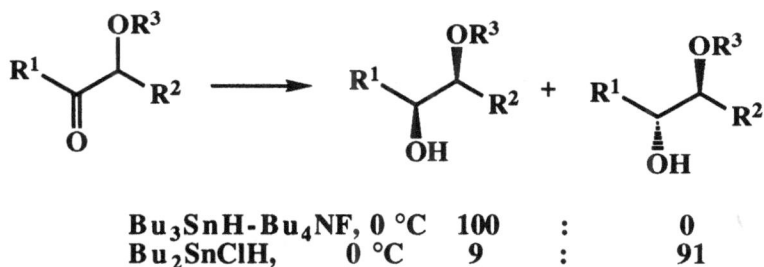

$$\mathbf{Bu_3SnH\text{-}Bu_4NF, \ 0 \ ^\circ C} \quad \mathbf{100} \quad : \quad \mathbf{0}$$
$$\mathbf{Bu_2SnClH,} \quad \mathbf{0 \ ^\circ C} \quad \mathbf{9} \quad : \quad \mathbf{91}$$

---

**III.A-13**  G. Gelbard et al., *J. Org. Chem.*, **1992**, *57*, 1789.

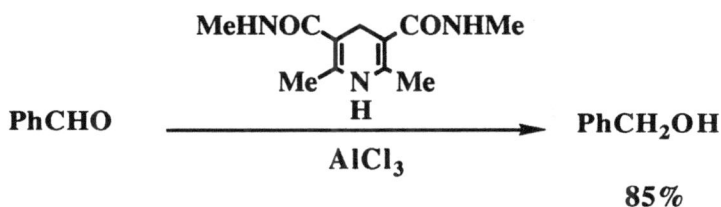

---

**III.A-14**  B. T. Cho and Y. S. Chun, *Tetrahedron: Asymmetry*, **1992**, *3*, 73.

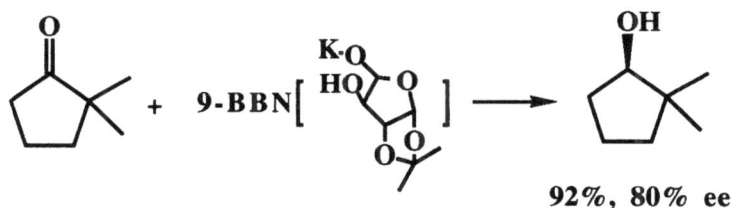

**92%, 80% ee**

---

**III.A-15**  Y. Suseela and M. Periasamy, *Tetrahedron*, **1992**, *48*, 317.

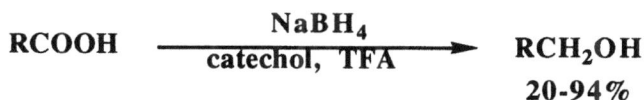

**20-94%**

**III.A-16** Y. Matsubara and I. Nishiguchi et al., *Bull. Chem. Soc. Jpn.*, **1992**, *65*, 530.

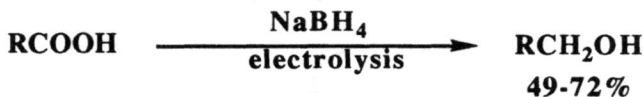

$$\text{RCOOH} \xrightarrow[\text{electrolysis}]{\text{NaBH}_4} \text{RCH}_2\text{OH}$$
$$49\text{-}72\%$$

---

**III.A-17** H. E. Hogberg et al., *J. Org. Chem.*, **1992**, *57*, 2052.

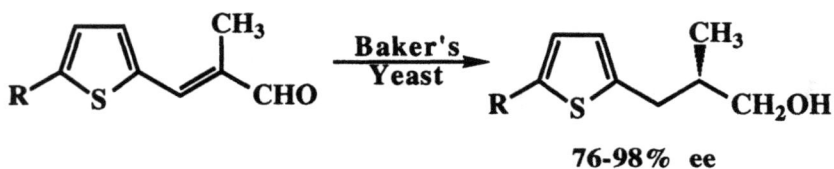

76-98% ee

---

**III.A-18** M. M. Midland and A. Kazubski *J. Org. Chem.*, **1992**, *57*, 2953.

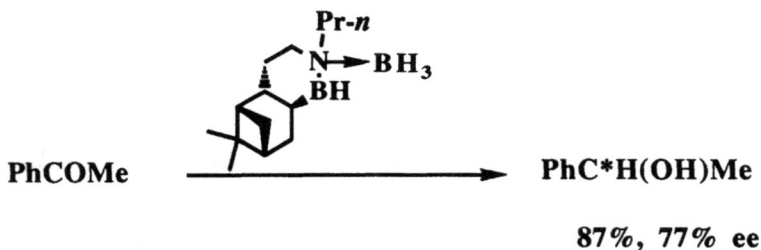

$$\text{PhCOMe} \longrightarrow \text{PhC*H(OH)Me}$$

87%, 77% ee

---

**III.A-19** H. C. Brown et al., *J. Org. Chem.*, **1992**, *57*, 2379.

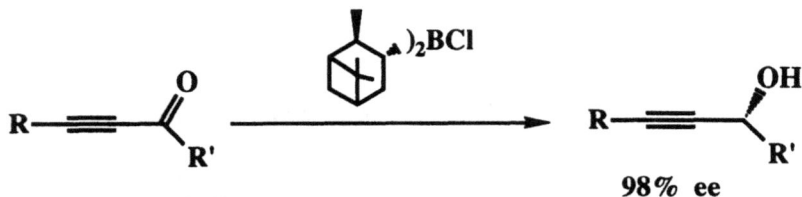

98% ee

**III.A-20**. A. Solodin, *Mon.* **1992**, *123*, 565.

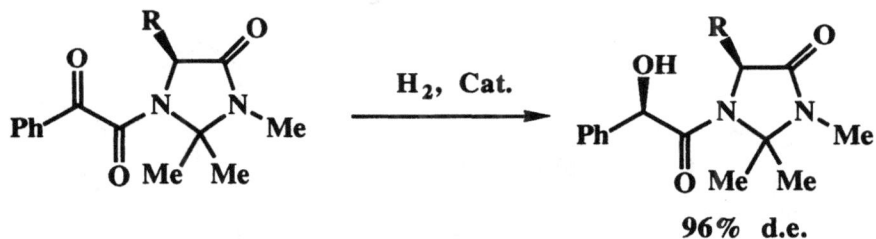

96% d.e.

---

**III.A-21** B. Singram et al., *Tetrahedron Lett.*, **1992**, *33*, 4533.

### Lithium Aminoborahydrides as Reducing Agents

---

**III.A-22** R. Ballini, *J. Chem. Soc., Perkin. Trans. I*, **1992**, 3161.

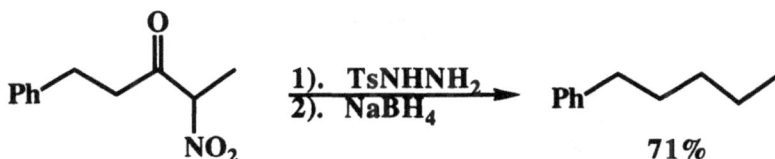

71%

---

**III.A-23** S. Sicsic et al., *J. Chem. Soc., Perkin. Trans. I*, **1992**, 3141.

### Asymmetric Microbial Reduction of Tetrolones

---

**III.A-24** G. Casy et al., *J. Chem. Soc., Chem. Commun.*, **1992**, 924.

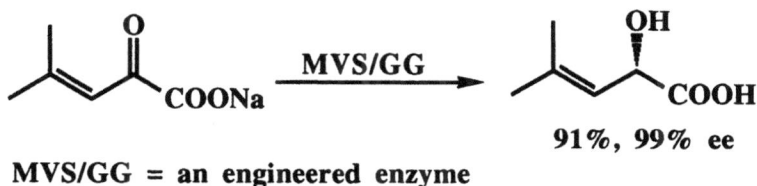

91%, 99% ee

MVS/GG = an engineered enzyme

**III.A-25**  G. Gringmann and T. Hartung, *Synthesis*, **1992**, 433, *Angew. Chem., Int. Ed. Engl.* **1992**, *31*, 761.

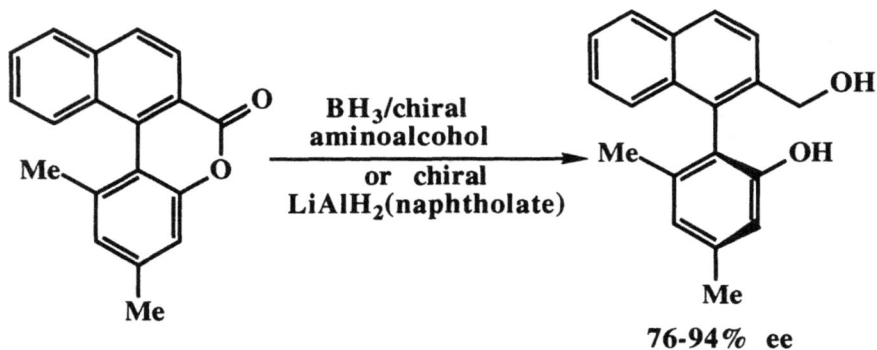

BH$_3$/chiral
aminoalcohol

or chiral
LiAlH$_2$(naphtholate)

76-94% ee

---

**III.A-26**  E. Brown et al., *Tetrahedron: Asymmetry*, **1992**, *3*, 841.

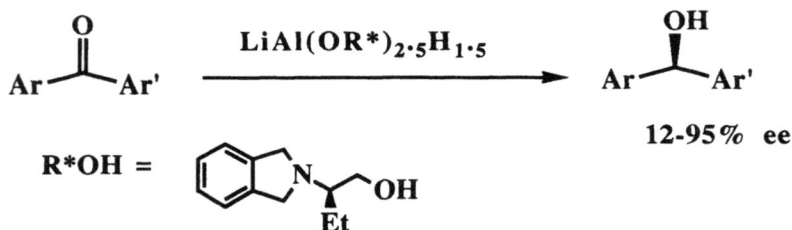

LiAl(OR*)$_{2 \cdot 5}$H$_{1 \cdot 5}$

12-95% ee

R*OH =

---

**III.A-27**  G. Solladie and N. Ghiaton, *Tetrahedron Lett.*, **1992**, *33*, 1605, 277.

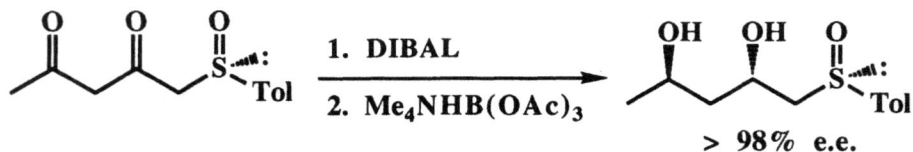

1. DIBAL

2. Me$_4$NHB(OAc)$_3$

> 98% e.e.

**III.A-28**  A. Kumar et al., *J. Chem. Soc., Chem. Commun.*, **1992**, 493.

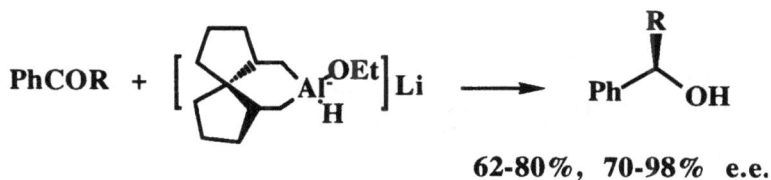

PhCOR + [ (structure) Ar OEt H ] Li ⟶ Ph R OH

**62-80%,  70-98%  e.e.**

---

**III.A-29**  S. Yanagida, *Chem. Lett.*, **1992**, 1951.

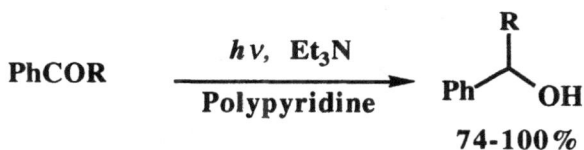

PhCOR $\xrightarrow[\text{Polypyridine}]{h\nu,\ \text{Et}_3\text{N}}$ Ph R OH

**74-100%**

**polyphenyl  affords  pinacols  only**

---

**III.A-30**  P. Salvador et al., *Tetrahedron: Asymmetry*, **1992**, *3*, 693.

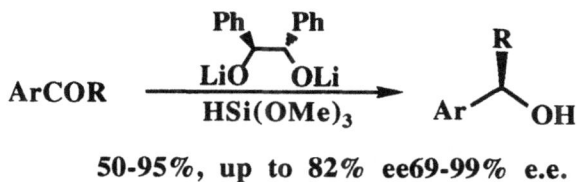

ArCOR $\xrightarrow[\text{HSi(OMe)}_3]{\begin{array}{c}\text{Ph}\quad\text{Ph}\\ \text{LiO}\quad\text{OLi}\end{array}}$ Ar R OH

**50-95%,  up to  82%  ee69-99%  e.e.**

---

**III.A-31**  D. Rawsonard and A. I Myers,  *J. Chem. Soc., Chem. Commun.*,
**1992**, 494.

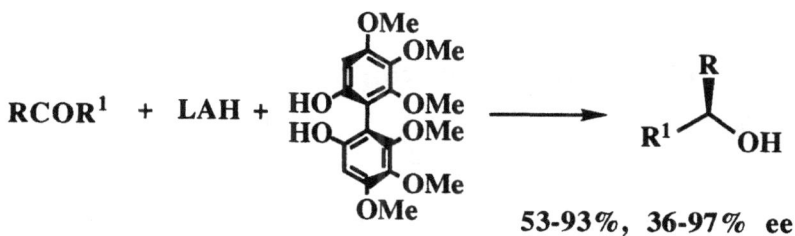

RCOR¹ + LAH + (binaphthol structure with OMe groups) ⟶ R¹ R OH

**53-93%,  36-97%  ee**

**III.A-32**  J. Markens et al., *Syn. Commun.*, **1992**, *22*, 2143, P. J. DeClercq, *Tetrahedron: Asymmetry*, **1992**, *3*, 599.

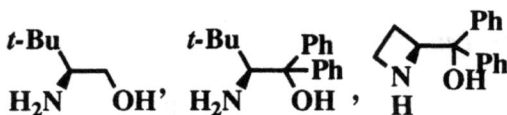

PhCOMe + BH₃  →(Chiral aminoalcohol)→  Ph—C(Me)—OH

69-99%  e.e.

t-Bu, H₂N, OH'  ;  t-Bu, H₂N, Ph, Ph, OH  ;  N-H ring, Ph, Ph, OH

---

**III.A-33**  G. Buono et al., *J. Chem. Soc., Chem. Commun.*, **1992**, 288.

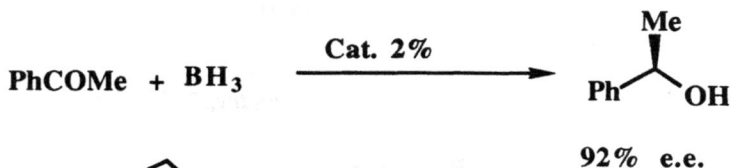

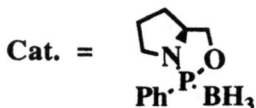

PhCOMe + BH₃  →(Cat. 2%)→  Ph—C(Me)—OH

92%  e.e.

Cat. =  (pyrrolidine) N–O, P, Ph, BH₃

---

**III.A-34**  Y. V. S. Rao and B. M. Choudary, *Syn. Commun.*, **1992**, *22*, 2711.

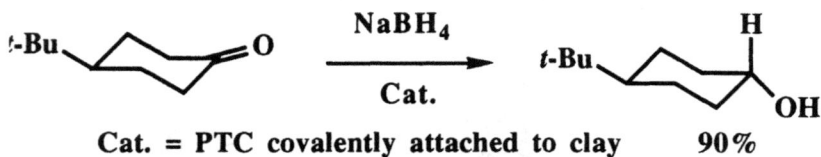

t-Bu—cyclohexanone =O  →(NaBH₄ / Cat.)→  t-Bu—cyclohexane, H, OH

Cat. = PTC covalently attached to clay        90%

**III.A-35**  S. Matsubara et al., *Chem. Lett.*, **1992**, 2173.

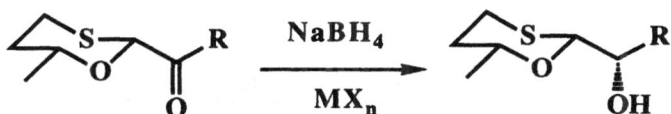

**Effects of metal salts on diastereoselectivity examined**

---

**III.A-36**  B. C. Ranu and A. R. Das, *Tetrahedron Lett.*, **1992**, *33*, 2361.

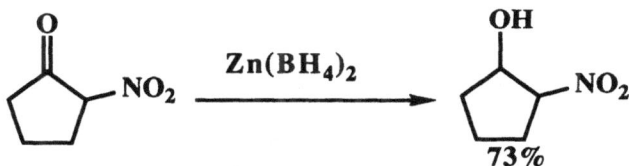

---

**III.A-37**  X. Zhou et al., *Syn. Commun.*, **1992**, 22, 1529.

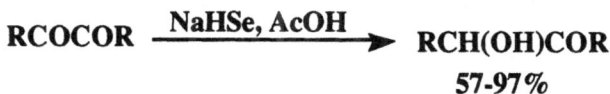

$$\text{RCOCOR} \xrightarrow{\text{NaHSe, AcOH}} \text{RCH(OH)COR}$$

**57-97%**

---

**III.A-38**  S. Spogliarich et al., *Tetrahedron: Asymmetry*, **1992**, *3*, 1001.

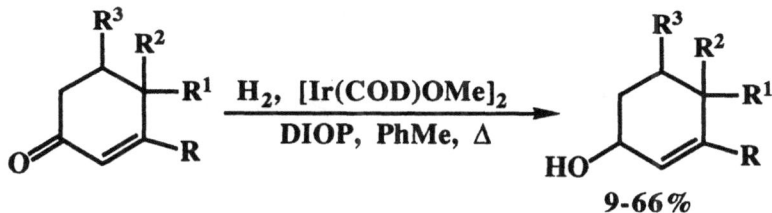

**9-66%**

**III.A-39**  C. Zhang, R.S. Phillips, *J. Chem. Soc., Perkin. Trans. I*, **1992,** 1083.

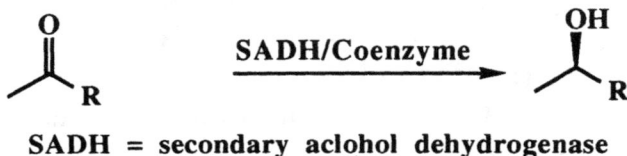

SADH = secondary aclohol dehydrogenase

---

**III.A-40**  C. H. Wong, et al., *J. Org. Chem.*, **1992,** *57*, 1532; C. Bolm et al., *Chem. Ber.* **1992,** *125*, 1169.

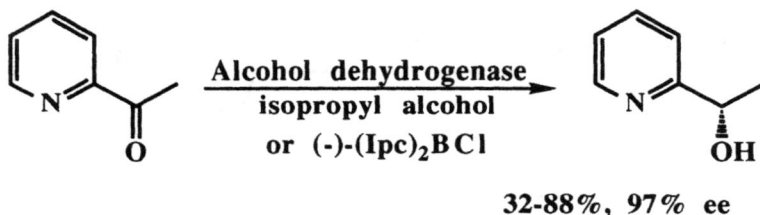

32-88%, 97% ee

---

**III.A-41**  J. Barluenga et al., *J. Org. Chem.*, **1992,** *57*, 1219.

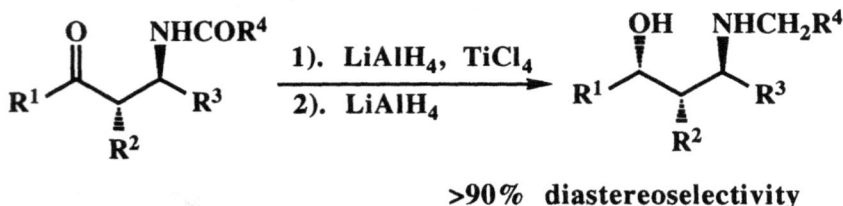

>90% diastereoselectivity

---

**III.A-42**  Y. Naoshima et al., *J. Chem. Soc., Perkin. Trans. I*, **1992,** 659.

IBY = Immobilized baker's yeast

Enantioselectivity in hexanes
was either controlled or changed by additives

**III.A-43**  E. Cesarotti et al., *Helvetica*, **1992**, *75*, 2563.

$$BnO \overset{O}{\underset{}{\bigsqcup}} OTr \xrightarrow{\text{Ru(II)BINAP, } H_2} BnO \overset{OH}{\underset{*}{\bigsqcup}} OTr$$

---

**III.A-44**  E. M. Magnus and B. Zwanebury, *Syn. Commun.*, **1992**, *22*, 783.

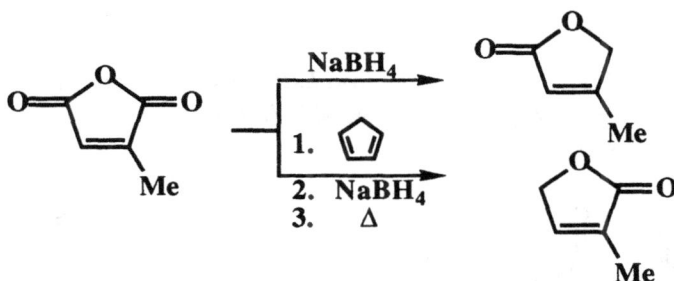

---

**III.A-45**  T. Shono et al., *J. Org. Chem.*, **1992**, *57*, 1061.

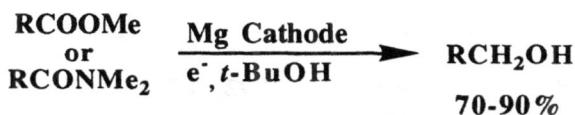

$$\begin{matrix} \text{RCOOMe} \\ \text{or} \\ \text{RCONMe}_2 \end{matrix} \xrightarrow[\text{e}^-, \text{t-BuOH}]{\text{Mg Cathode}} \begin{matrix} \text{RCH}_2\text{OH} \\ \mathbf{70\text{-}90\%} \end{matrix}$$

---

**III.A-46**  M. J. Burk and J. E. Feaster *Tetrahedron Lett.*, **1992**, *33*, 2099.

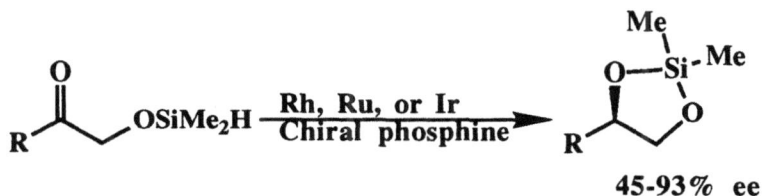

45-93% ee

**III.A-47** A. P. Davis and S. C. Hegarty, *J. Am. Chem. Soc.*, **1992**, *114*, 2745.

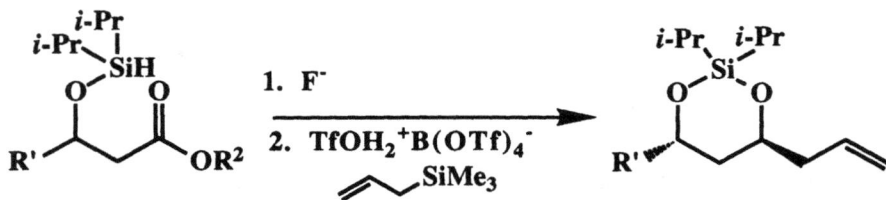

**III.A-48** T. Fujisawa et al., *Tetrahedron Lett.*, **1992**, *33*, 5567; M. Takeshita et al., *Heterocycle*, **1992**, *34*, 489; A. Kasahara et. al., *J. Chem. Soc., Perkin. Trans. I*, **1992**, 1265.

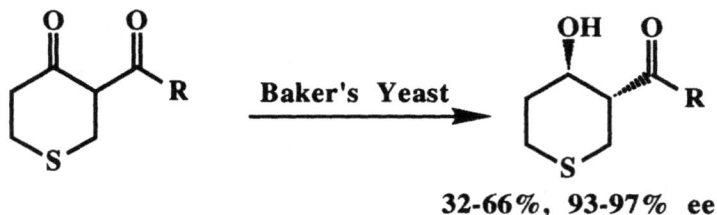

32-66%, 93-97% ee

**III.A-49** K. Oshima et al., *Chem. Lett.*, **1992**, 967; M. Koreeda et al., *Tetrahedron Lett.*, **1992**, *33*, 6599.

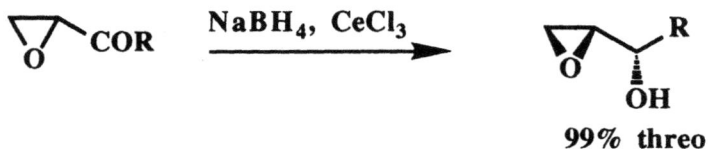

99% threo

**III.A-50** M. Caballero *Tetrahedron: Asymmetry,* **1992,** *3,* 1431.

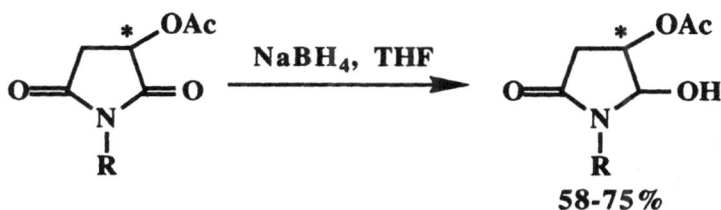

58-75%

**III.A-51** A. Kucerovy et al., *Syn. Commun.,* **1992,** *22,* 729.

**III.A-52** S. B. Mandal, et al., *Tetrahedron Lett.,* **1992,** *33,* 1647.

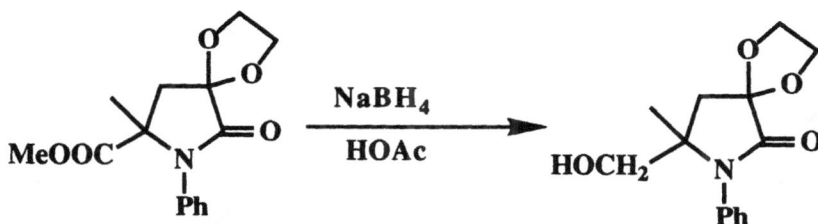

**III.A-53** R. N. Baruah *Tetrahedron Lett.,* **1992,** *33,* 5417.

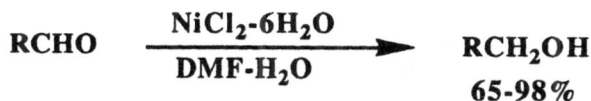

$$RCHO \xrightarrow[\text{DMF-H}_2\text{O}]{\text{NiCl}_2\text{·6H}_2\text{O}} RCH_2OH$$

65-98%

**III.A-54** H. Firouzabadi  and G. R. Afsharifar *Syn. Commun.*, **1992**, *22*, 497.

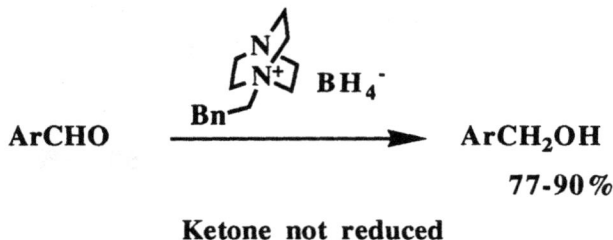

ArCHO $\xrightarrow{\text{Bn}}$ ArCH$_2$OH

77-90%

**Ketone not reduced**

---

**III.A-56** G. Kokotos et al., *Liebigs Ann. Chem,* **1992**, 961.

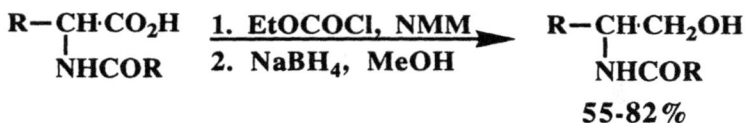

$$\underset{\text{NHCOR}}{\text{R}-\text{CH·CO}_2\text{H}} \xrightarrow[\text{2. NaBH}_4,\ \text{MeOH}]{\text{1. EtOCOCl, NMM}} \underset{\text{NHCOR}}{\text{R}-\text{CH·CH}_2\text{OH}}$$

55-82%

---

**III.A-57** K. Achina et al., *SynLett.*, **1992**, 829.

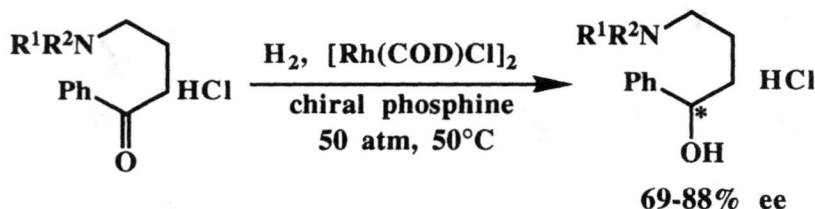

$$\xrightarrow[\substack{\text{chiral phosphine}\\ \text{50 atm, 50°C}}]{\text{H}_2,\ [\text{Rh(COD)Cl}]_2}$$

69-88% ee

---

**III.A-58** J. S. Cha et al., *Org. Prep. Pre. Int.*, **1992**, *24*, 327, 335.

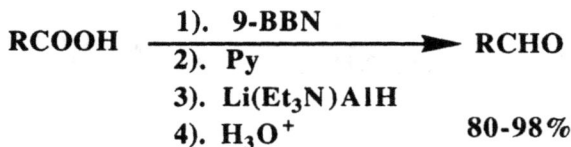

RCOOH $\xrightarrow[\substack{\text{2). Py}\\ \text{3). Li(Et}_3\text{N)AlH}\\ \text{4). H}_3\text{O}^+}]{\text{1). 9-BBN}}$ RCHO

80-98%

**III.A-59**  Y. Kamochi, T. Kudo *Bull. Chem. Soc. Jpn.*, **1992**, *65*, 3049.

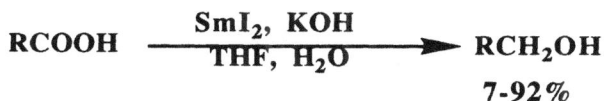

$$RCOOH \xrightarrow[\text{THF, H}_2\text{O}]{\text{SmI}_2, \text{ KOH}} RCH_2OH$$
$$\textbf{7-92\%}$$

---

**III.A-60**  C. R. Johnson and H. Sakaguchi, *SynLett.*, **1992**, 813.

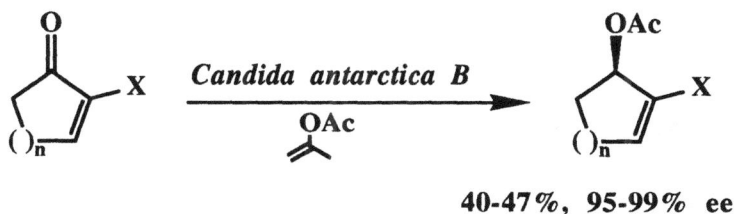

**40-47%, 95-99% ee**

---

**III.A-61**  C. M. Adams, *Syn. Commun.*, **1992**, *22*, 1385.

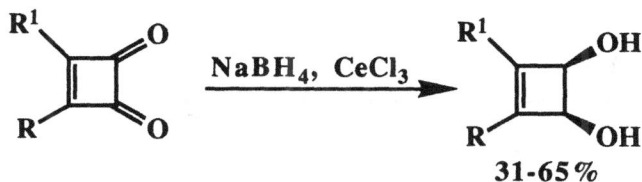

**31-65%**

---

**III.A-62**  G. R. Nakayama and P. G. Shultz, *J. Am. Chem. Soc.*, **1992**, *114*, 780.

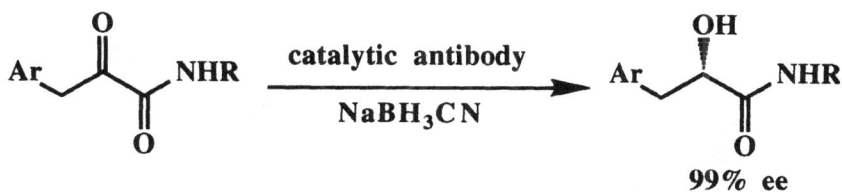

**99% ee**

**III.A-63** S. Sato, T. Moriwake et al., *SynLett.*, **1992**, 325.

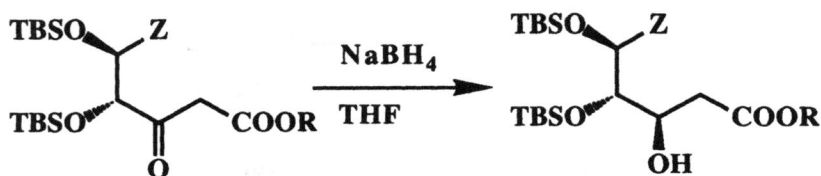

**52-85%, up to 99% de**

## III.B.    C-N Multiple Bond Reductions

### III.B.1    Imine Reductions

**III.B.1-1** A. Willoughby and S. L. Buchwald\*, *J. Am. Chem. Soc.*, **1992**, *114*, 7652.

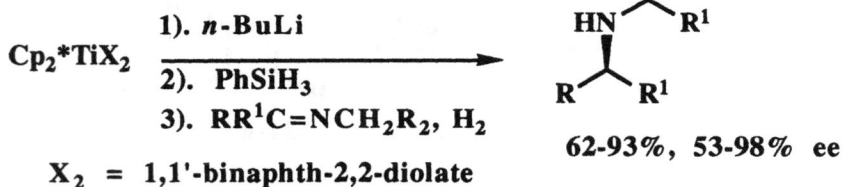

**62-93%, 53-98% ee**

**III.B.1-2** N. Hamamichi\* and T. Miyasaka et al, *Chem. Pharm. Bull.*, **1992**, *40*, 843, 2585.

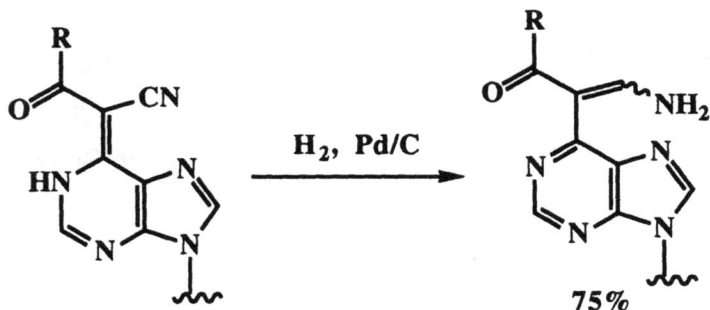

**75%**

**III.B.1-3** B. T. Cho* and Y. S. Chen *Tetrahedron: Asymmetry*, **1992**, *3*, 1583; M. Nakagawa* and T. Hino et al., *Tetrahedron: Asymmetry*, **1992**, *3*, 227.

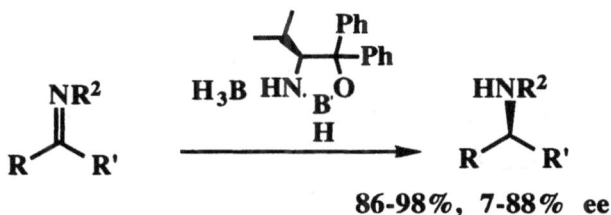

86-98%, 7-88% ee

---

**III.B.1-4** G. Z. Wang and J. E. Backvall*, *J. Chem. Soc., Chem. Commun.*, **1992**, 980.

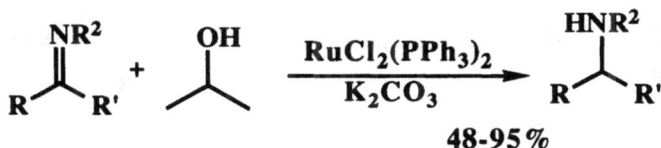

48-95%

---

**III.B.1-5** M. J. Burk* and J. E. Feaster, *J. Am. Chem. Soc.*, **1992**, *114*, 6266.

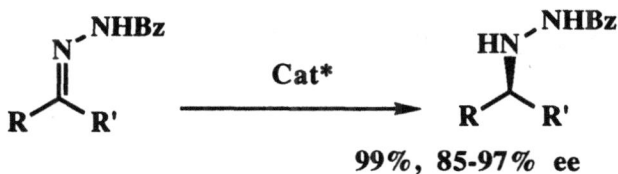

99%, 85-97% ee

cat. = [(COD)Rh((*R,R*)-Et-DuPhos)]OTf

**III.B.1-6** T. Mukaiyama et al., *Chem. Lett.*, **1992**, 181.

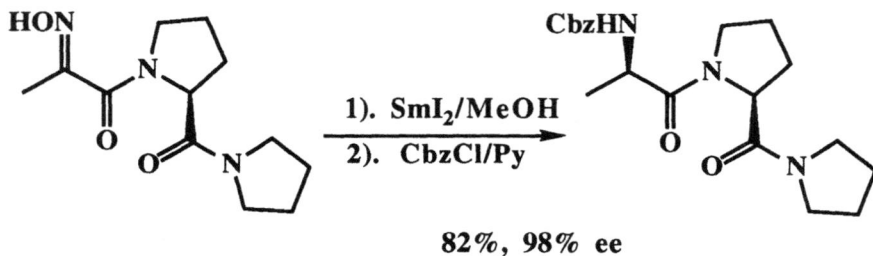

$$\xrightarrow{\begin{array}{c}1). \ \ \mathbf{SmI_2/MeOH}\\ 2). \ \ \mathbf{CbzCl/Py}\end{array}}$$

**82%, 98% ee**

---

**III.B.1-7** A. Guy and J. F. Barbetti, *Syn. Commun.*, **1992**, *22*, 853.

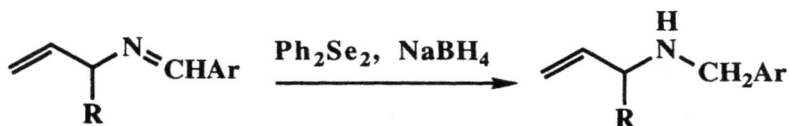

$$\xrightarrow{\mathbf{Ph_2Se_2, \ NaBH_4}}$$

---

**III.B.1-8** D. R. Williams and M. H. Osterhout, *J. Am. Chem. Soc.*, **1992**, *114*, 8751.

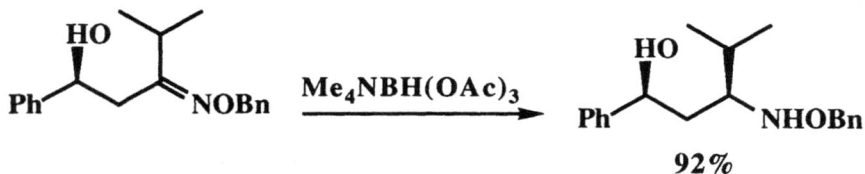

$$\xrightarrow{\mathbf{Me_4NBH(OAc)_3}}$$

**92%**

---

**III.B.1-9** C. Lensink and J. G. de Vries*, *Tetrahedron: Asymmetry*, **1992**, *3*, 235.

> **Improving Enantioselectivity by Using a Mono-sulphonated Diphosphine as Ligand for Homogeneous Imine Hydrogenation.**

**III.B.1-10** C. A. Willoughby and S. L. Buchwald*, *J. Am. Chem. Soc.*, **1992**, *114*, 7562.

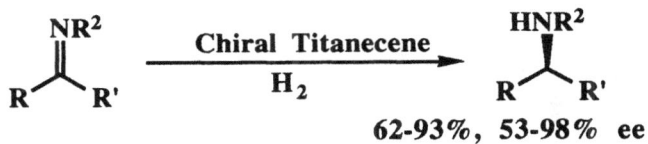

62-93%, 53-98% ee

## III.B.2    Reduction of Heterocycles

**III.B.2-1**  N. Bodor et al., *Tetrahedron*, **1992**, *48*, 4767.

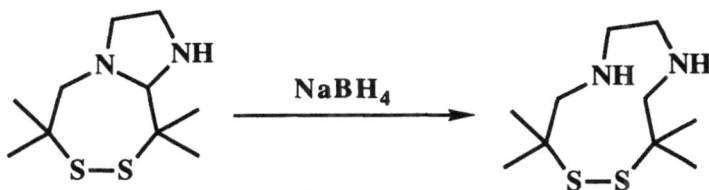

**III.B.2-2**  B. Hu et al., *Syn. Commun.*, **1992**, *22*, 1179.

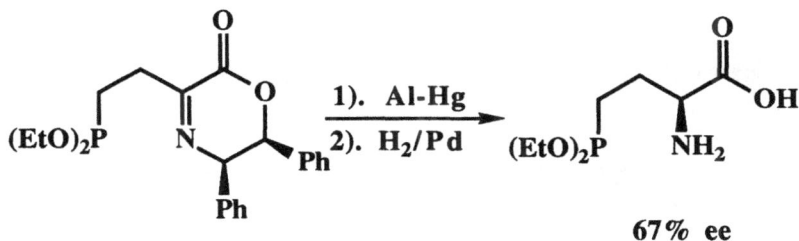

67% ee

**III.B.2-3** S. Szantay et al., *Rec. Tran. Chim.*, **1992**, *110*, 437.

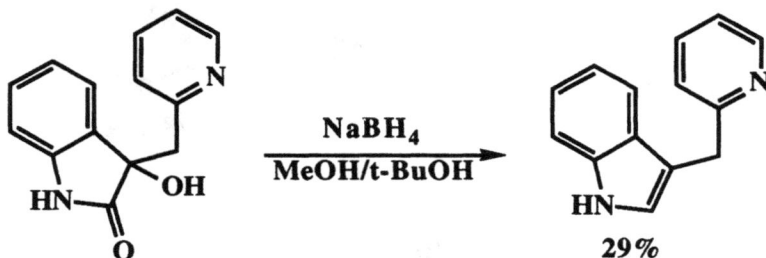

NaBH₄
MeOH/t-BuOH

29%

## III.C Reduction of Sulfur Compound

**III.C-1** J. Drabowicz et al., *SynLett.*, **1992**, 252.

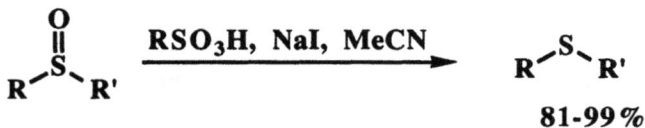

$$R-\overset{\overset{\displaystyle O}{\|}}{S}-R' \xrightarrow{\text{RSO}_3\text{H, NaI, MeCN}} R-S-R'$$

81-99%

**III.C-2** H. Bartsch*, T. Erker, *Tetrahedron Lett.*, **1992**, *33*, 199.

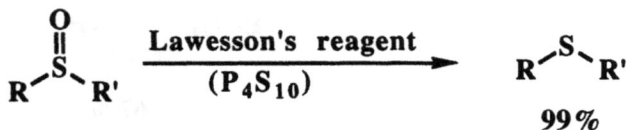

$$R-\overset{\overset{\displaystyle O}{\|}}{S}-R' \xrightarrow[\text{(P}_4\text{S}_{10})]{\text{Lawesson's reagent}} R-S-R'$$

99%

## III.D.  N-O Reductions

**III.D-1**  R. V. Hoffman et al., *J. Org. Chem.*, **1992**, *57*, 5700.

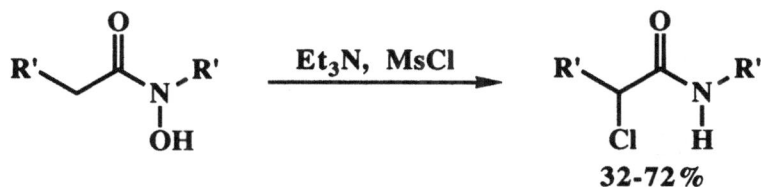

**32-72%**

---

**III.D-2**  J. R. Hwu et al., *J. Org. Chem.*, **1992**, *57*, 5254.

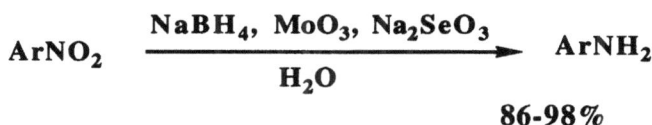

$$ArNO_2 \xrightarrow[\text{H}_2\text{O}]{\text{NaBH}_4,\ \text{MoO}_3,\ \text{Na}_2\text{SeO}_3} ArNH_2$$

**86-98%**

**Na$_2$SeO$_3$ as a promoter for the Mo catalyst**

---

**III.D-3**  J. Marquet et al., *Tetrahedron Lett.*, **1992**, *33*, 7053.

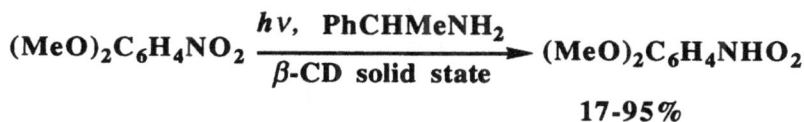

$$(MeO)_2C_6H_4NO_2 \xrightarrow[\beta\text{-CD solid state}]{h\nu,\ \text{PhCHMeNH}_2} (MeO)_2C_6H_4NHO_2$$

**17-95%**

---

**III.D-4**  H. Alper* and G. Vasapollo, *Tetrahedron Lett.*, **1992**, *33*, 7472.

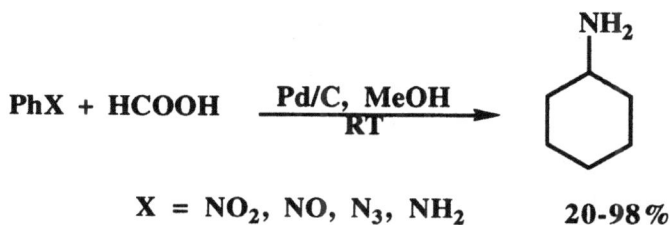

$$PhX + HCOOH \xrightarrow[\text{RT}]{\text{Pd/C, MeOH}}$$

**X = NO$_2$, NO, N$_3$, NH$_2$**          **20-98%**

**III.D-5** M. A. Schwartz et al., *Tetrahedron Lett.*, **1992**, *33*, 1687.

$$\underset{\overset{|}{OH}}{R-N-R'} \xrightarrow{\text{CS}_2, \text{ RT}} \text{RNHR}'$$

## III.E.    C-C Multiple Bond Reductions

### III.E.1    C=C Reductions

**III.E.1-1** J. M. Brown and G. C. Lloyd-Jones, *J. Chem. Soc., Chem. Commun.* **1992**, 711.

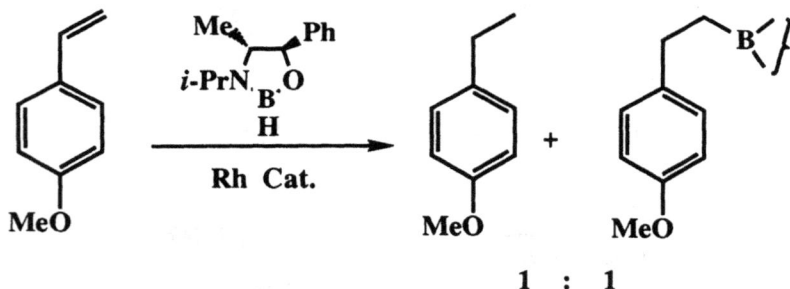

1   :   1

**III.E.1-2** E. M. Beccalli and A. Marchesini *Tetrahedron*, **1992**, *48*, 48, 5359.

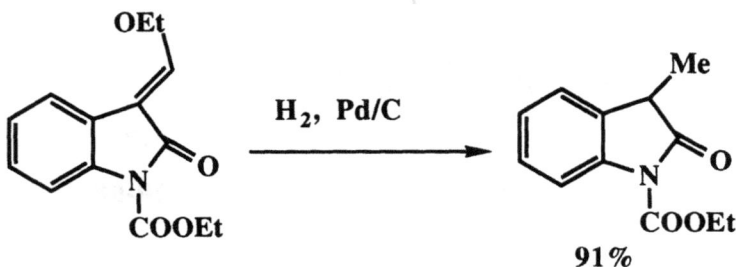

91%

**III.E.1-3** S. Ram and L. D. Spicer, *Syn. Commun.*, **1992**, *22*, 2683.

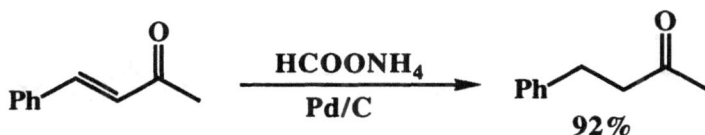

**III.E.1-4** A. Cabrera and H. Apler *Tetrahedron Lett.*, **1992**, *33*, 5007.

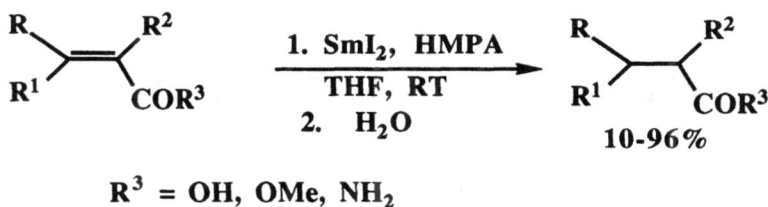

$$R^3 = OH, OMe, NH_2$$

**III.E.1-5** M. Saburi rt al, *Tetrahedron Lett.*, **1992**, *33*, 5783.

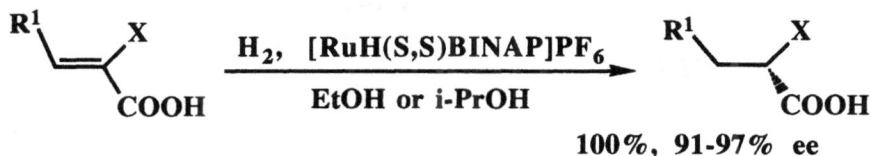

100%, 91-97% ee

**III.E.1-6** J. P. Genet et al., *Tetrahedron Lett.*, **1992**, *33*, 5343,

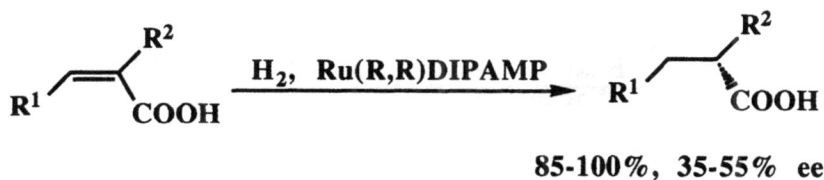

85-100%, 35-55% ee

**III.E.1-7** J. Inanaga et al., *Chem. Lett.*, **1992**, 2117.

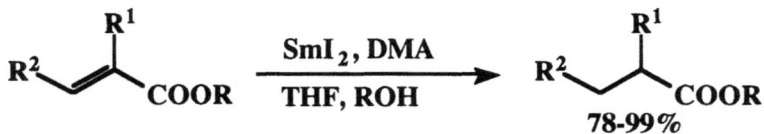

**III.E.1-8** M. Saburi *Tetrahedron Lett.*, **1992**, *33*, 7877.

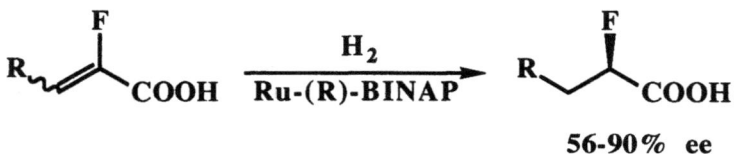

**III.E.1-9** A. J. Lauret and S. Lesniak et al., *Tetrahedron Lett.*, **1992**, *33*, 3311.

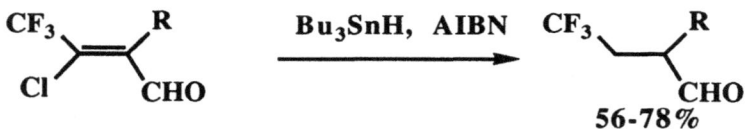

**III.E.1-10** K. Takabe et al., *Tetrahedron: Asymmetry*, **1992**, *3*, 1399.

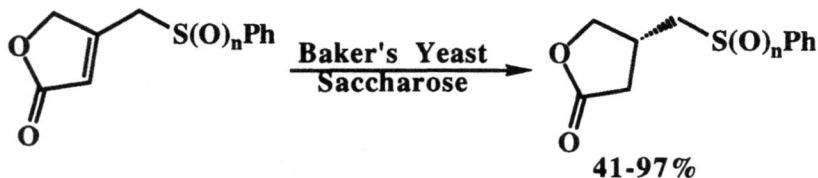

**III.E.1-11** Y. Ohishi et al., *Chem. Pharm. Bull.* **1992**, *40*, 907.

23-79%

---

**III.E.1-12** J. W. Hemdon* and J. J. Matasi*, *Tetrahedron Lett.*, **1992**, *33*, 5725.

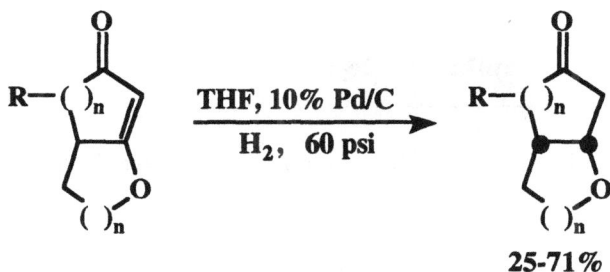

25-71%

---

**III.E.1-13** H. Takaya et al., *Tetrahedron Lett.*, **1992**, *33*, 635.

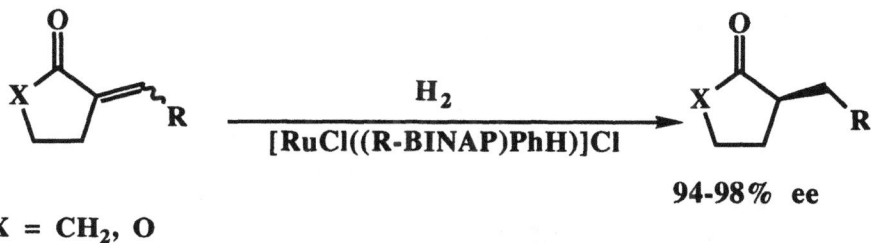

94-98% ee

X = CH$_2$, O

**III.E.1-14** T. Toyokuni et al., *J. Org. Chem.*, **1992**, *57*, 6693.

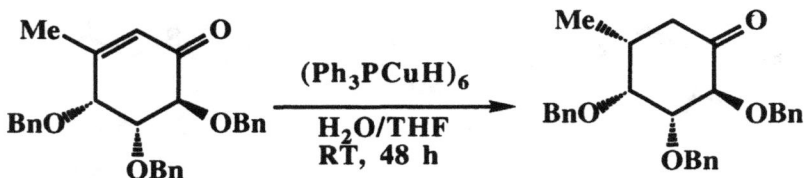

**III.E.1-15** M. Lautens et al., *Angew. Chem., Int. Ed. Engl.* **1992**, *31*, 232.

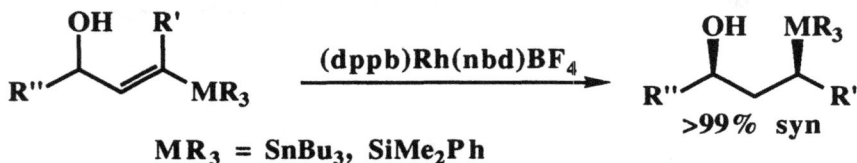

**III.E.1-16** B. C. Ranu and R. Chakraborty, *Tetrahedron*, **1992**, *48*, 5317.

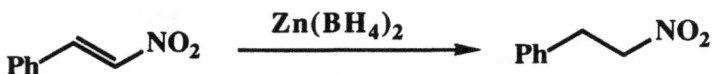

**III.E.1-17** A. Terfort, *Synthesis,* **1992**, 951.

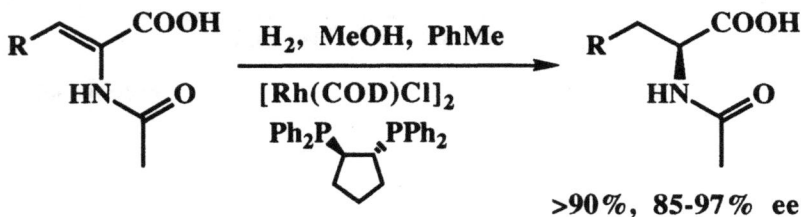

**III.E.1-18** G. A. Molander and J. O. Hoberg, *J. Org. Chem.*, **1992**, *57*, 3266.

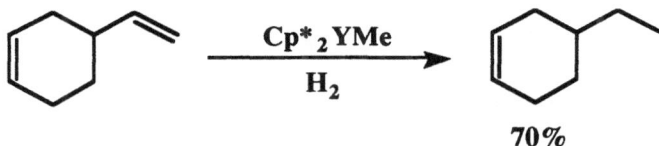

**70%**

---

**III.E.1-19** V. B. Shur et al., *J. Organometal. Chem.*, *439*, 303.

**Catalytic Activity of the Titanocene Complex with Tolane in Homogeneous Hydrogenation of Unsaturated Hydrocarbons**

---

**III.E.1-20** A. Corma et al., *J. Organometal. Chem.*, *431*, 233.

**Optically Active compounds of transition metals (Rh, Ru, Co and Ni) with 2-aminocarbonylpyrrolidine ligands. Selective catalysts for hydrogenation of prochiral olefines**

---

**III.E.1-21** H. Kuno et al., *Bull. Chem. Soc. Jpn.*, **1992**, *65*, 1240.

**Regioselective Hydrogenation of Unsaturated Compounds Using Platinum-Zeolite Coupled with Organosilicon Alkoxide by CVD Method**

---

**III.E.1-22** L. Shao et al., *J. Organometal. Chem.*, *435*, 133, 155.

**Asymmetric Hydrogenation of Prochiral Carboxylic Acids and Functionalized Carbonyl Compounds Catalyzed by Ru(II)-BINAP Complexes with Aryl Nitriles**

**III.E.1-23**  H. Krause and C. Sailer *J. Organometal. Chem.*, *423*, 271.

**Influence of β-arranged Substituents in Chiral 7 Membered Rhodium Diphosphine Rings on Asymmetric Hydrogenation of Amino acids Precursors**

## III.E.2    C-C Triple Bond Reductions

**III.E.2-1**  J. M. Tour et al, joc, 4786.

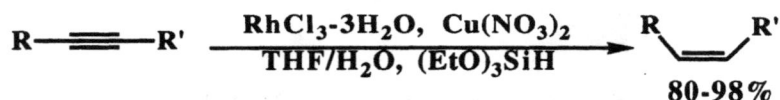

$$R\!-\!\!\equiv\!\!-R' \xrightarrow[\text{THF/H}_2\text{O, (EtO)}_3\text{SiH}]{\text{RhCl}_3\text{-3H}_2\text{O, Cu(NO}_3)_2}$$

R       R'

**80-98%**

---

**III.E.2-2**  K. Takai, K. Utimoto et al, *J. Org. Chem.*, **1992**, *57*, 1615.

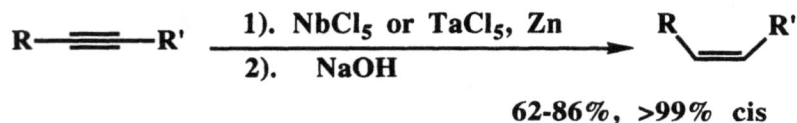

$$R\!-\!\!\equiv\!\!-R' \xrightarrow[\text{2). NaOH}]{\text{1). NbCl}_5 \text{ or TaCl}_5, \text{ Zn}}$$

R       R'

**62-86%, >99%  cis**

---

**III.E.2-3**  Y. Masude et al, *J. Chem. Soc., Perkin. Trans. I*, **1992**, 2725.

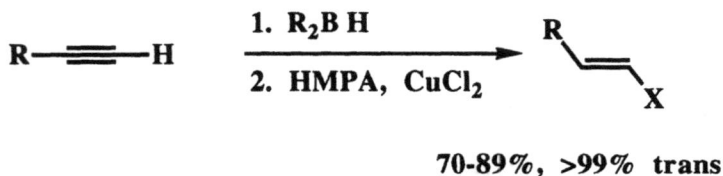

$$R\!-\!\!\equiv\!\!-H \xrightarrow[\text{2. HMPA, CuCl}_2]{\text{1. R}_2\text{B H}}$$

R

X

**70-89%, >99%  trans**

**III.E.2-4**  R. F. dela Pradilla* et al *Tetrahedron Lett.*, **1992**, *33*, 6101.

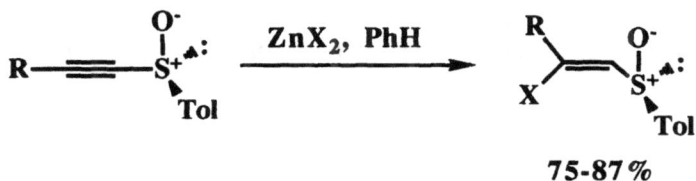

**75-87%**

---

**III.E.2-5**  C. Bianchini* and L.A. Oro* et al, *Organometallics*, **1992**, *11*, 138.

### Selective Hydrogenation of 1-Alkynes to Alkenes Catalyzed by an Iron(II) complex

---

**III.E.2-6**  A. Moyano et al, *Tetrahedron Lett.*, **1992**, *33*, 2863.

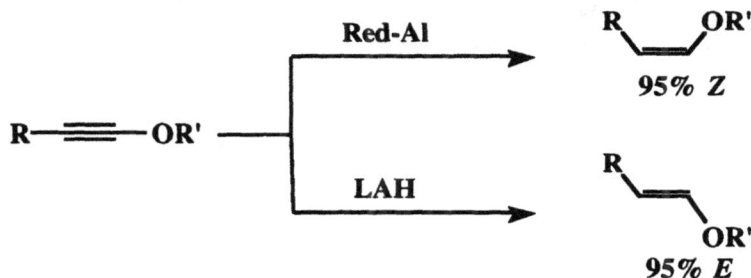

95% Z

95% E

## III.F.1.   C-O → C-H

**III.F.1-1**  T. Mandai, J. Tsuji*, et al., *J. Org. Chem.*, **1992**, *57*, 1326, 1327, 6090.

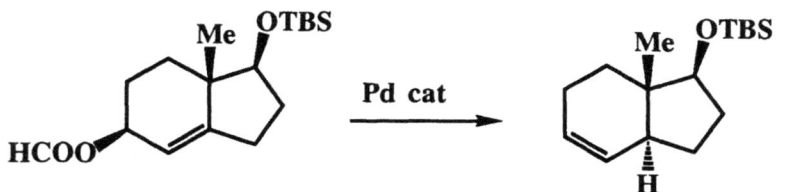

**Pd cat**

**Hydride delivered from the opposite face of the formate**

---

**III.F.1-2**  T. Mandai et al., *Tetrahedron Lett.*, **1992**, *33*, 2987.

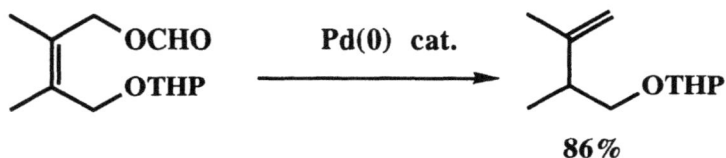

**Pd(0)  cat.**

**86%**

---

**III.F.1-3**  M. E. Jung et al., *Tetrahedron Lett.*, **1992**, *33*, 2921.

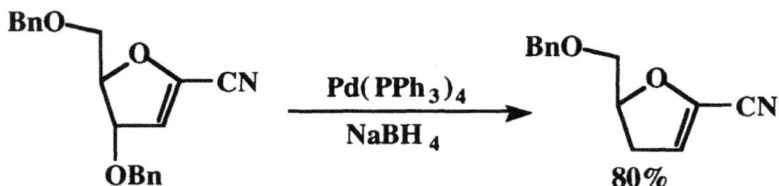

**Pd(PPh$_3$)$_4$**

**NaBH$_4$**

**80%**

**III.F.1-4**  S. Torii et al., *Chem. Lett.*, **1992**, 1895.

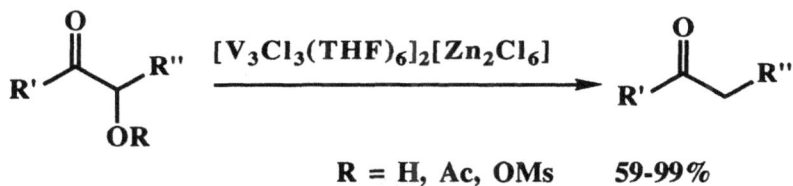

R = H, Ac, OMs     **59-99%**

---

**III.F.1-5**  S. Hanessian et al., *Tetrahedron Lett.*, **1992**, *33*, 573.

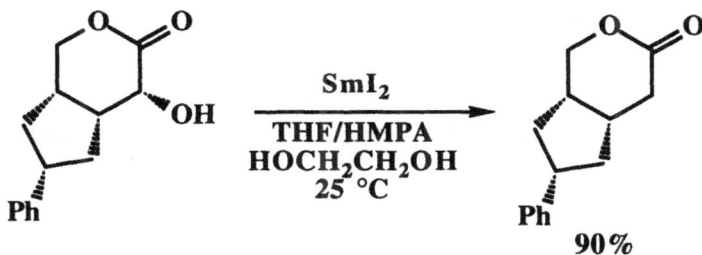

**SmI$_2$**

**THF/HMPA**
**HOCH$_2$CH$_2$OH**
**25 °C**

**90%**

---

**III.F.1-6**  H. Ohmori et al., *Tetrahedron Lett.*, **1992**, *33*, 1347.

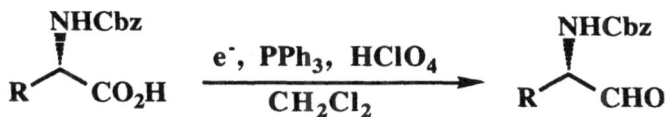

e$^-$, PPh$_3$,  HClO$_4$

CH$_2$Cl$_2$

**III.F.1-7** H.-D. Scharp et al., *Liebigs Ann. Chem.,* **1992**, 103.

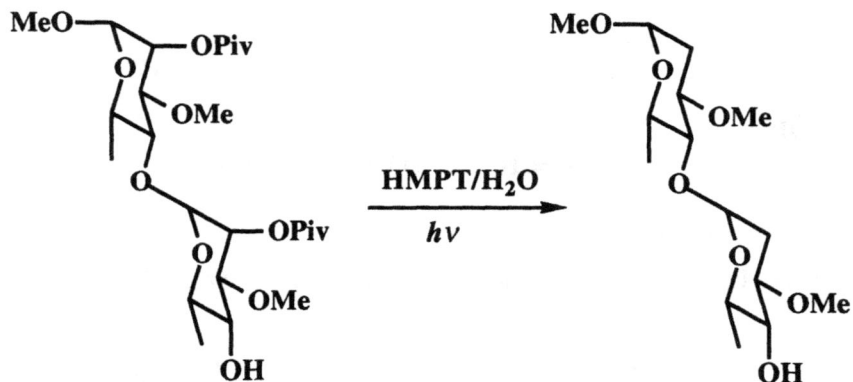

$$\text{HMPT/H}_2\text{O} \xrightarrow{h\nu}$$

**III.F.1-8** M. N. Greco and B. E. Marganoll, *Tetrahedron Lett.,* **1992**, *33*, 5009.

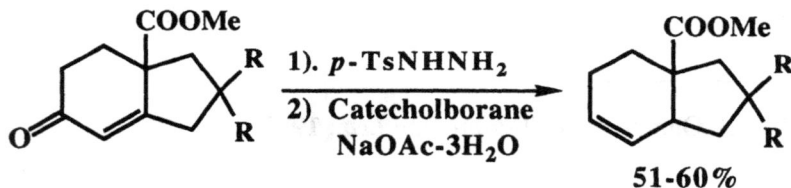

1). *p*-TsNHNH$_2$

2) Catecholborane
NaOAc-3H$_2$O

51-60%

**III.F.1-9** Y. Nishiyama, S. Hamanaka, et al., *Tetrahedron Lett.,* **1992**, *33*, 6347.

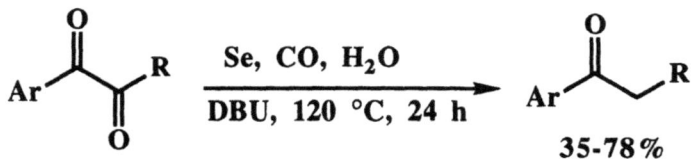

$$\xrightarrow[\text{DBU, 120 °C, 24 h}]{\text{Se, CO, H}_2\text{O}}$$

35-78%

**III.F.1-10** S. Torii et al., *Bull. Chem. Soc. Jpn.*, **1992**, *65*, 3200.

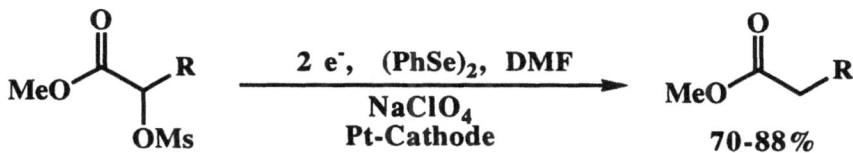

$$\text{MeO} \overset{O}{-} \underset{OMs}{\overset{|}{C}} R \xrightarrow[\substack{\text{NaClO}_4 \\ \text{Pt-Cathode}}]{2\ e^-,\ (\text{PhSe})_2,\ \text{DMF}} \text{MeO} \overset{O}{-} CH_2 R$$

70-88%

---

**III.F.1-11** M. Tada et al., *J. Chem. Soc., Perkin. Trans. I*, **1992**, 1897.

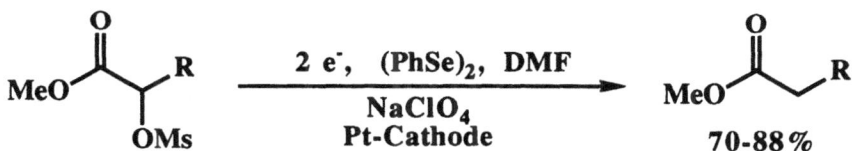

$$\text{MeO} \overset{O}{-} \underset{OMs}{\overset{|}{C}} R \xrightarrow[\substack{\text{NaClO}_4 \\ \text{Pt-Cathode}}]{2\ e^-,\ (\text{PhSe})_2,\ \text{DMF}} \text{MeO} \overset{O}{-} CH_2 R$$

70-88%

---

**III.F.1-12** J. J. Eisch et al., *J. Org. Chem.*, **1992**, *57*, 2143.

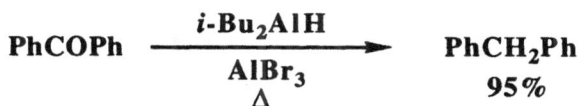

$$\text{PhCOPh} \xrightarrow[\substack{\text{AlBr}_3 \\ \Delta}]{i\text{-Bu}_2\text{AlH}} \text{PhCH}_2\text{Ph}$$

95%

---

**III.F.1-13** M. P. Sibi et al., *Syn. Commun.*, **1992**, *22*, 809.

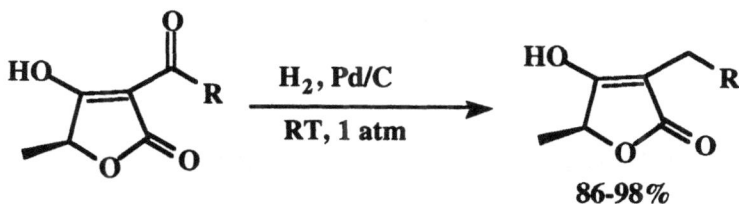

$$\xrightarrow[\text{RT, 1 atm}]{\text{H}_2,\ \text{Pd/C}}$$

86-98%

**III.F.1-14** C. Chatgilialoglu et al., *J. Org. Chem.*, **1992**, *57*, 2427; and J. C. Jaszberenyi et al., *Tetrahedron Lett.*, **1992**, *33*, 6629.

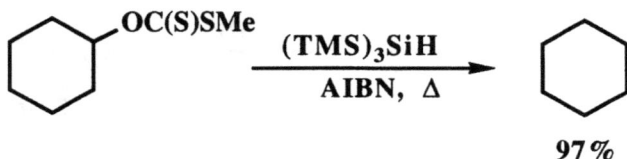

$$\text{Cyclohexyl-OC(S)SMe} \xrightarrow[\text{AIBN, } \Delta]{(TMS)_3SiH} \text{Cyclohexane}$$

**97%**

---

**III.F.1-15** D. H. R. Barton et al., *Tetrahedron Lett.*, **1992**, *33*, 2311,

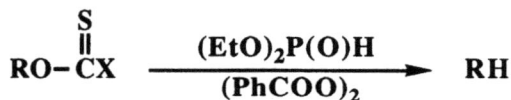

$$\underset{RO-CX}{\overset{S}{\parallel}} \xrightarrow[\text{(PhCOO)}_2]{(EtO)_2P(O)H} RH$$

---

**III.F.1-16** W. P. Newnann* and M. Petersein, *SynLett.*, **1992**, 801.

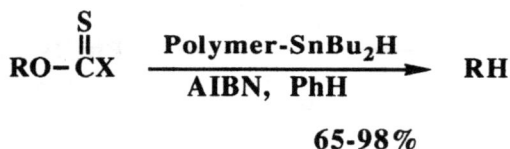

$$\underset{RO-CX}{\overset{S}{\parallel}} \xrightarrow[\text{AIBN, PhH}]{\text{Polymer-SnBu}_2H} RH$$

**65-98%**

## III.F.2    C-Hal → C-H

**III.F.2-1** M. Sakai et al., *Bull. Chem. Soc. Jpn.*, **1992**, *65*, 1739.

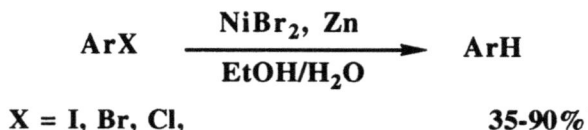

$$\text{ArX} \xrightarrow[\text{EtOH/H}_2O]{\text{NiBr}_2, \text{Zn}} \text{ArH}$$

X = I, Br, Cl,                                    **35-90%**

**III.F.2-2**  D. H. R. Barton et al., *Tetrahedron Lett.*, **1992**, *33*, 5709.

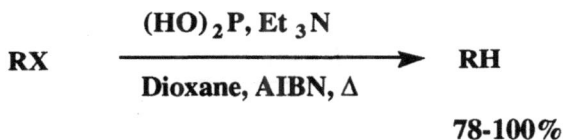

$$RX \quad \xrightarrow[\textbf{Dioxane, AIBN, } \Delta]{\textbf{(HO)}_2\textbf{P, Et}_3\textbf{N}} \quad RH$$

**78-100%**

---

**III.F.2-3**  A. de Meijere*, J. Salaun* et al., *Tetrahedron Lett.*, **1992**, *33*, 3307.

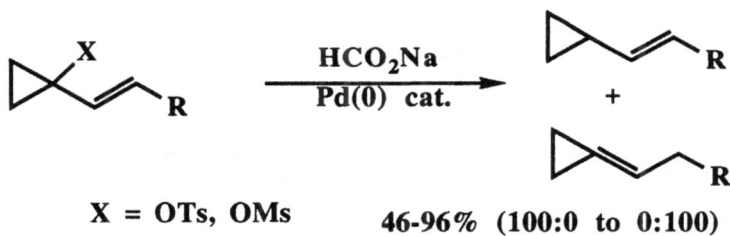

X = OTs, OMs          **46-96%  (100:0 to 0:100)**

---

**III.F.2-4**  N. DeKompe et al., *J. Org. Chem.*, **1992**, *57*, 5761.

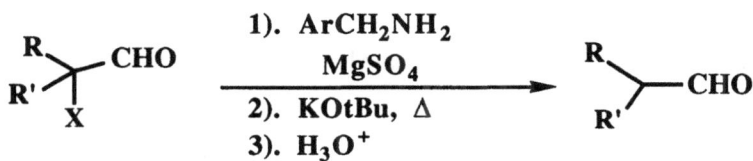

1). $ArCH_2NH_2$
   $MgSO_4$
2). KOtBu, $\Delta$
3). $H_3O^+$

---

**III.F.2-5**  S. P. Watson, *Syn. Commun.*, **1992**, *22*, 2971.

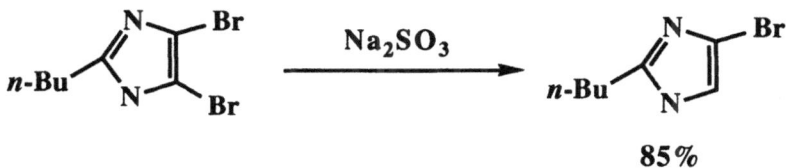

$Na_2SO_3$

**85%**

**III.F.2-6**  Y. Yishi et al., *Chem. Lett.*, **1992**, 293.

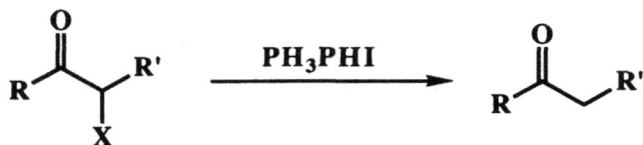

α-haloethers reduced similarly

---

**III.F.2-7**  K. Sato et al., *Chem. Lett.*, **1992**, 1469.

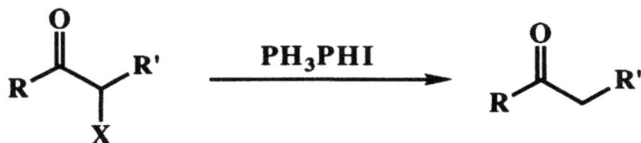

α-haloethers reduced similarly

---

**III.F.2-8**  Y. Ishii et al., *Chem. Lett.*, **1992**, 2431.

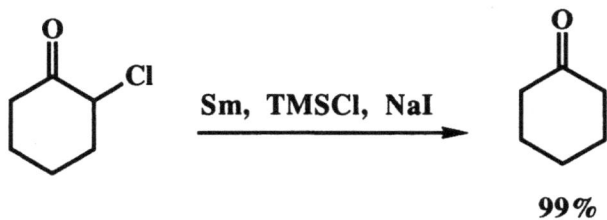

**99%**

**III.F.2-9** D. P. Curran et al., *Tetrahedron Lett.*, **1992**, *33*, 2295.

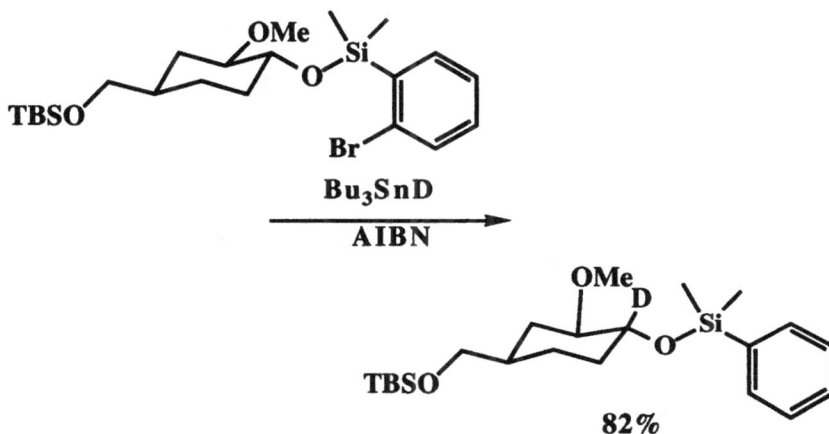

**82%**

---

**III.F.2-10** J. Hawari, *J. Organometal. Chem.*, **1992**, *427*, 91.

> **Regioselectivity of Dechlorination : Redcutive
> Dechlorination of Polychlorobiphenyl by
> Polymethylhydrosiloxane-Alkali Metal**

---

**III.F.2-11** D. T. Ferrughelli and I. T. Horvath, *J. Chem. Soc., Chem. Commun.*, **1992**, 806.

> **The Use of Bifunctional Homogeous Rhodium
> Catalyst for the Conversion of Chloroaromatics into
> Saturated Hydrocarbons**

---

**III.F.2-12** Z. N. Parnes et al., *J. Org. Chem. USSR*, **1992**, *28*, 216.

> **Effects Of Structures Of Organosilane Hydride on
> Their Reactivities in the Ionic Hydrogenolysis of Alkyl
> Halide**

## III.F.3.    C-S  →  C-H

**III.F.3-1** G. F. Cooper et al., *Tetrahedron Lett.*, **1992**, *33*, 5895.

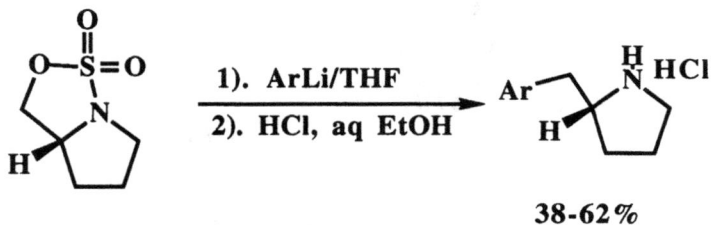

**38-62%**

---

**III.F.3-2** T. S. Chou* and S. Y. Chang, *J. Org. Chem.*, **1992**, *57*, 5015.

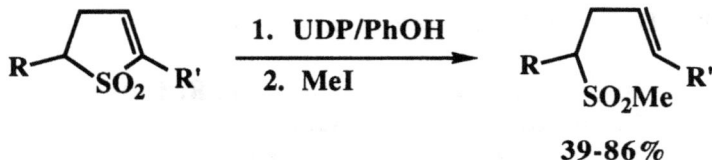

**39-86%**

---

**III.F.3-3** T. G. Back et al., *J. Org. Chem.*, **1992**, *57*, 1986.

**Desulfurization of Benzo- and Dibenzothiophenes
with Nickel Boride**

---

**III.F.3-4** J. H. Penn and W. H. Owens *Tetrahedron Lett.*, **1992**, *33*, 3737.

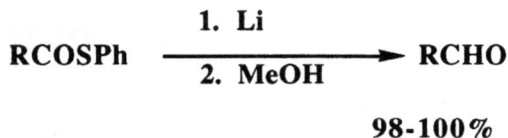

$$RCOSPh \xrightarrow[\text{2. MeOH}]{\text{1. Li}} RCHO$$

**98-100%**

**III.F.3-5**  A. Pllter*, R. S. Ward* et al., *Tetrahedron: Asymmetry*, **1992**, *3*, 239.

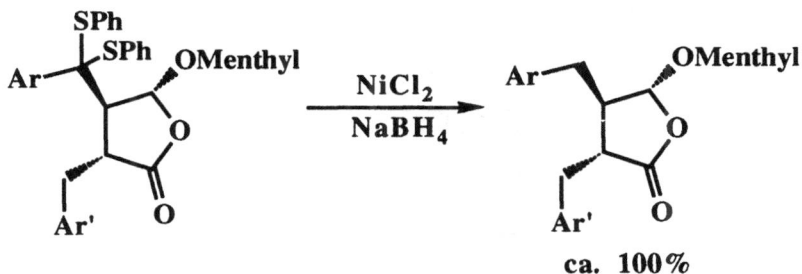

**ca.  100%**

---

**III.F.3-6**  K. K. Park et al., *J. Chem. Soc., Perkin. Trans. I*, **1992**, 601.

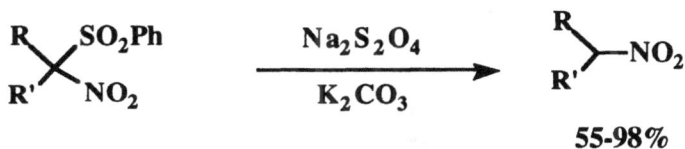

**55-98%**

---

**III.F.3-7**  S. Murumoto and I. Kuwajima, *Chem. Lett.*, **1992**, 1421.

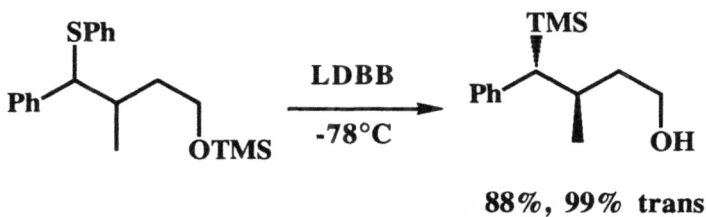

**88%,  99%  trans**

## III.F.4     C-N → C-H

**III.F.4-1** F. S. Guziec, Jr*, D. Wei. *J. Org. Chem.*, **1992**, *57*, 3772.

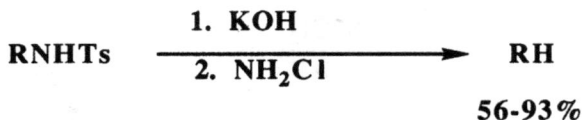

$$RNHTs \xrightarrow{\begin{array}{c}\textbf{1. KOH}\\\textbf{2. NH}_2\textbf{Cl}\end{array}} RH$$

**56-93%**

---

**III.F.4-2** M. P. Sibi and R. Sharma, *SynLett.*, **1992**, 497.

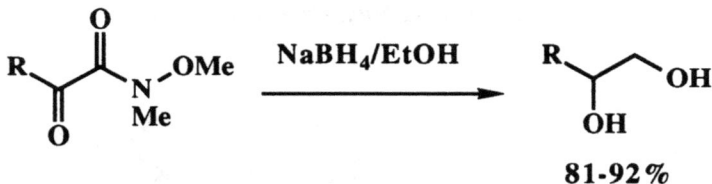

**81-92%**

---

**III.F.4-3** R. Balicki, *Gazz. Chim. Ital.*, **1992**, *122*, 133.

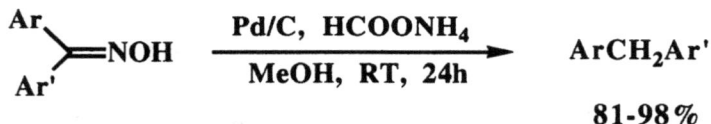

$$\xrightarrow[\textbf{MeOH, RT, 24h}]{\textbf{Pd/C, HCOONH}_4} ArCH_2Ar'$$

**81-98%**

---

**III.F.4-4** Y. Kamochi*, T. Kudo, *Tetrahedron*, **1992**, *48*, 4300.

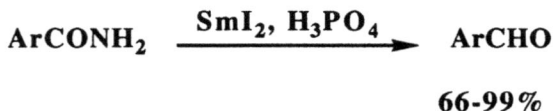

$$ArCONH_2 \xrightarrow{\textbf{SmI}_2,\ \textbf{H}_3\textbf{PO}_4} ArCHO$$

**66-99%**

## III.G.    Reductive Clevage, oxiranes

### III.G. 1    Oxiranes

**III.G.1-1** S. Torii et al., *SynLett.*, **1992**, 510.

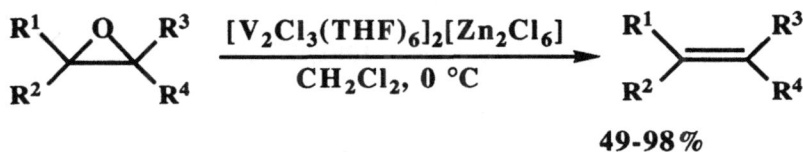

$$R^1 \quad O \quad R^3 \xrightarrow[\text{CH}_2\text{Cl}_2,\ 0\ ^\circ\text{C}]{[\text{V}_2\text{Cl}_3(\text{THF})_6]_2[\text{Zn}_2\text{Cl}_6]} R^1 \quad R^3$$

R²        R⁴                                              R²        R⁴

**49-98%**

---

**III.G.1-2** B. C. Rank* and A. R. Das, *J. Chem. Soc., Perkin. Trans. I*, **1992**, 1881.

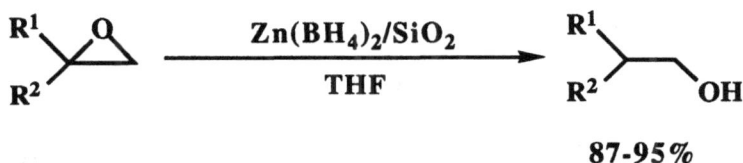

$$R^1 \quad O \xrightarrow[\text{THF}]{\text{Zn(BH}_4)_2/\text{SiO}_2} R^1$$

R²                                          R²        OH

**87-95%**

---

**III.G.1-3** G. A. Epling and Q. Wang, *J. Chem. Soc., Chem. Commun.*, **1992**, 1133.

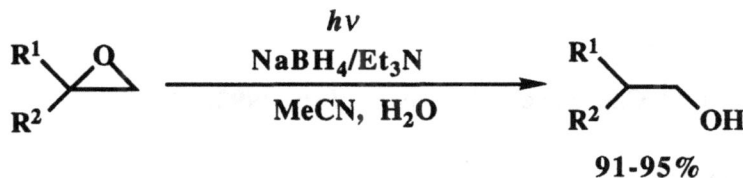

$$R^1 \quad O \xrightarrow[\text{MeCN, H}_2\text{O}]{\overset{h\nu}{\text{NaBH}_4/\text{Et}_3\text{N}}} R^1$$

R²                                          R²        OH

**91-95%**

**III.G.1-4**  E. Hasegawa et al., *J. Org. Chem.*, **1992**, *57*, 5352.

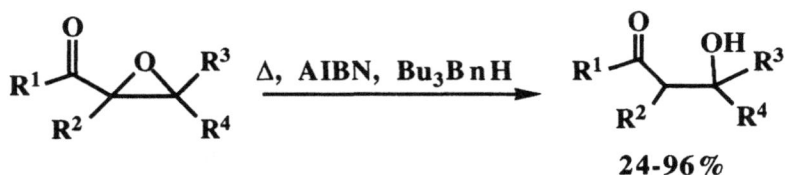

R¹—C(=O)—(epoxide with R², R³, R⁴)  →  Δ, AIBN, Bu₃BnH  →  product

**24-96%**

---

**III.G.1-5**  J. J. Eisch et al., *J. Org. Chem.*, **1992**, *57*, 1619.

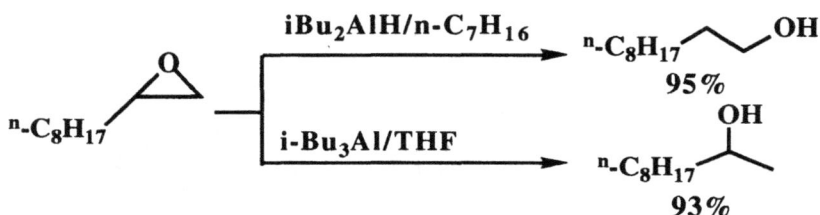

iBu₂AlH/n-C₇H₁₆ → n-C₈H₁₇CH₂CH₂OH  **95%**

i-Bu₃Al/THF → n-C₈H₁₇CH(OH)CH₃  **93%**

---

**III.G.1-6**  H. Wantanabe et al., *Steroids*, **1992**, *57*, 444.

NaBH₄ / diglyme / 80 °C, 15 h

**63%**

---

**III.G.1-7**  J. M. Chong et al., *Tetrahedron Lett.*, **1992**, *33*, 33.

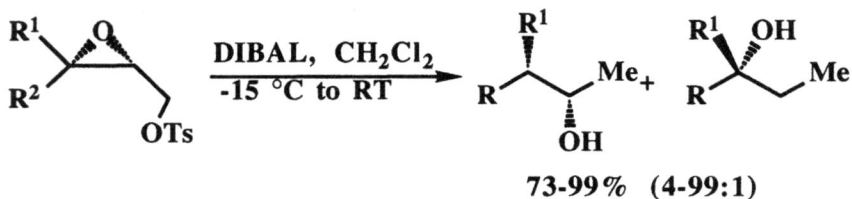

DIBAL, CH₂Cl₂ / -15 °C to RT

**73-99%  (4-99:1)**

**III.G.1-8**  L. Hansson, R, Carlson\*, *Acta. Chem. Scand.*, **1992**, *46*, 103.

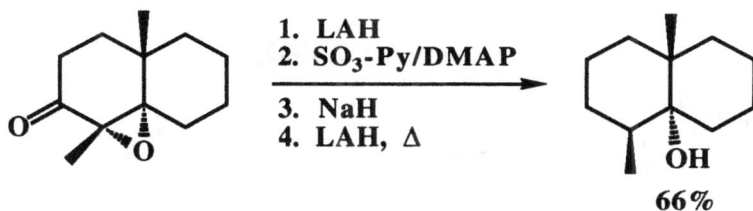

1. LAH
2. SO₃-Py/DMAP

3. NaH
4. LAH, Δ

66%

---

**III.G.1-9**  J. S. Yadav et al., *Tetrahedron Lett.*, **1992**, *33*, 7973.

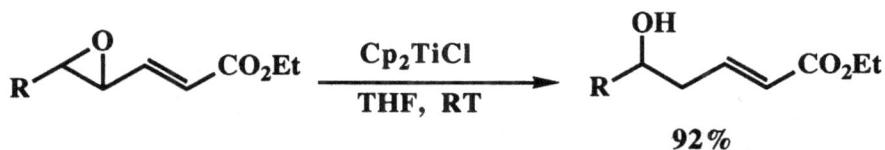

$$\text{Cp}_2\text{TiCl}$$
THF, RT

92%

## III.G.2    N-O Cleavage

R. Zimmer et al., *Liebigs Ann. Chem.*, **1992**, 709.

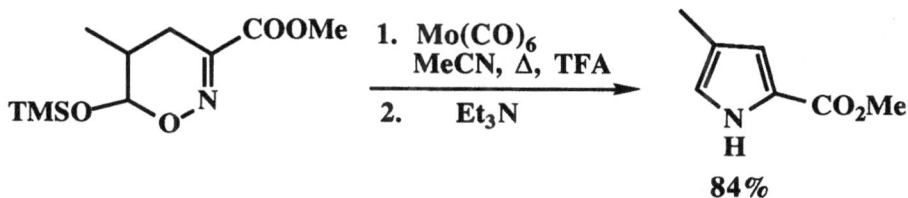

1. Mo(CO)₆
   MeCN, Δ, TFA
2. Et₃N

84%

## III. H.    Reduction of Azides

**III.H-1**  G. Kokotos, et al., *J. Chem. Res.S*, **1992**, 391.

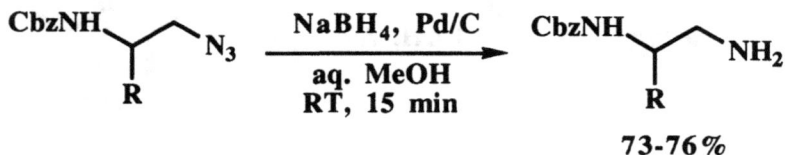

CbzNH ⟶ N₃ (R)  →[NaBH₄, Pd/C / aq. MeOH / RT, 15 min]→ CbzNH ⟶ NH₂ (R)

**73-76%**

---

**III.H-2**  M. M. Joullie et al., *Tetrahedron Lett.*, **1992**, *33*, 3595.

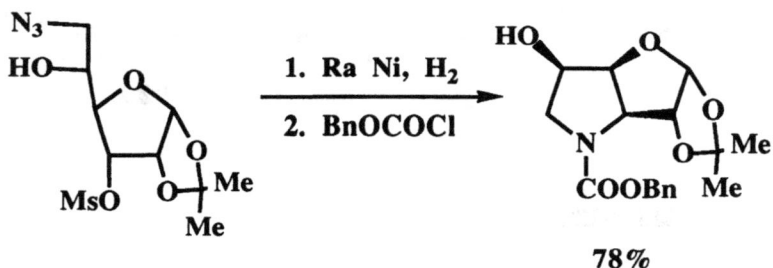

→[1. Ra Ni, H₂ / 2. BnOCOCl]→

**78%**

## III.I.    Reductive Cyclizations

**III.I-1**  M. Shibasaki et al., *J. Am. Chem. Soc.*, **1992**, *114*, 4418.

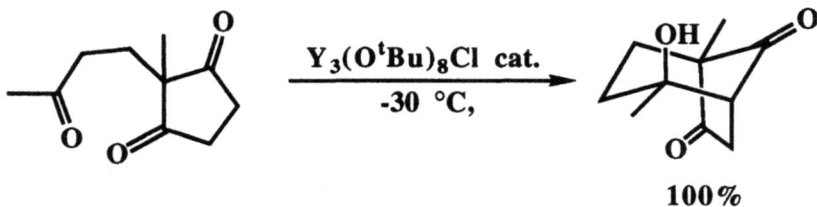

→[Y₃(OᵗBu)₈Cl cat. / -30 °C,]→

**100%**

**III.I-2** G. A. Molander and J. O. Hoberg, *J. Am. Chem. Soc.*, **1992**, *114*, 3123.

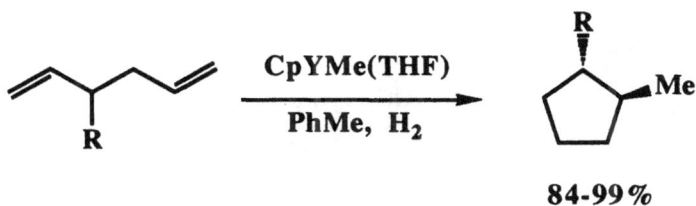

84-99%

## III.J.  Other Reductions

**III.J-1**  A. G. Godfrey, B. Ganem*, *Tetrahedron Lett.*, **1992**, *33*, 7461.

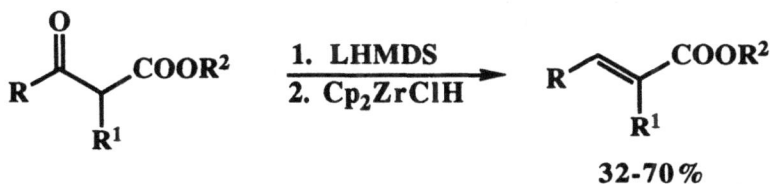

32-70%

**III.J-2**  J. M. O'Connor and J. Wa, *J. Org. Chem.*, **1992**, *57*, 5075.

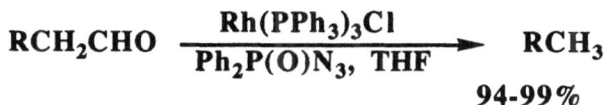

$$RCH_2CHO \xrightarrow[\text{Ph}_2\text{P(O)N}_3, \text{ THF}]{\text{Rh(PPh}_3)_3\text{Cl}} RCH_3$$

94-99%

**III.J-3**  C. Chapuis et al., *Tetrahedron Lett.*, **1992**, *33*, 6135.

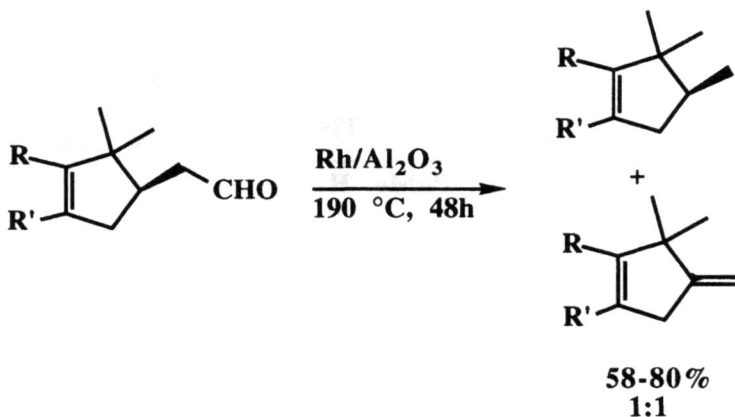

$$\text{R/Al}_2\text{O}_3 \quad 190\ °C,\ 48h$$

58-80%
1:1

---

**III.J-4**  T. Shono et al., *Tetrahedron*, **1992**, *48*, 8253.

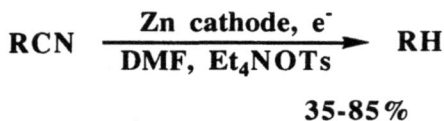

$$\text{RCN} \xrightarrow[\text{DMF, Et}_4\text{NOTs}]{\text{Zn cathode, e}^-} \text{RH}$$

35-85%

---

**III.J-5**  T. Konosu , S. Oida, *Chem. Pharm. Bull.*, **1992**, *40*, 609.

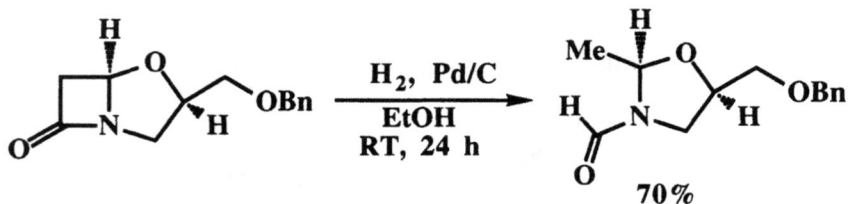

$$\xrightarrow[\substack{\text{EtOH} \\ \text{RT, 24 h}}]{\text{H}_2,\ \text{Pd/C}}$$

70%

**III.J-6**  C. Chatgilialoglu, M. Ballestri et al., *Tetrahedron Lett.*, **1992**, *33*, 1787.

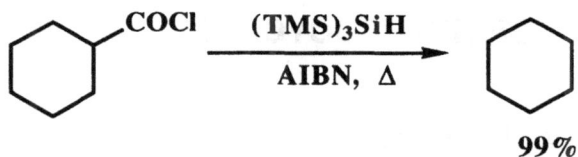

99%

**III.J-7**  S. Chandrasekaren et al., *Tetrahedron Lett.*, **1992**, *33*, 6371.

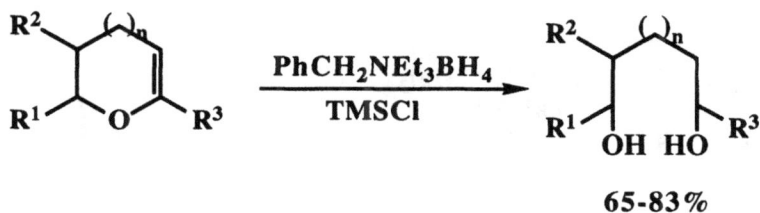

65-83%

**III.J-8**  J. Vossy et al., *Tetrahedron Lett.*, **1992**, *33*, 5045.

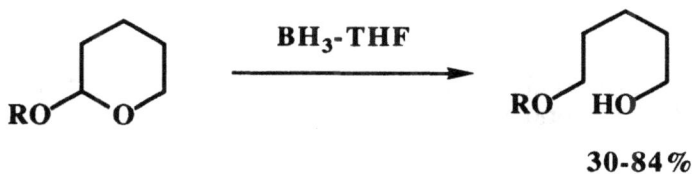

30-84%

# IV

# SYNTHESIS OF HETEROCYCLES

## IV.A. Oxiranes, Aziridines and Thiiranes

**IV.A-1** H.C. Kolb and K.B. Sharpless, *Tetrahedron*, **48**, 10515; M.P. DeNinno et al., *J. Org. Chem.*, **57**, 7115; M. Yoshita et al., *Heterocycles*, **33**, 507.

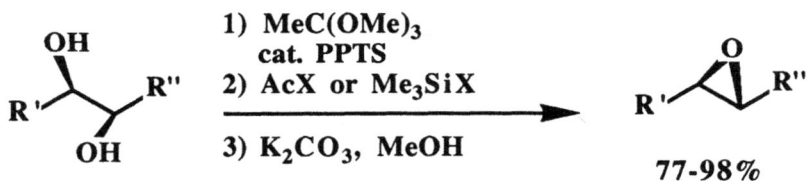

**IV.A-2** I. Shibata et al., *J. Org. Chem.*, **57**, 6909; A. Schwartz et al., *ibid.*, **57**, 851; S. Florio and L. Troisi, *Tetrahedron Lett.*, **33**, 7953.

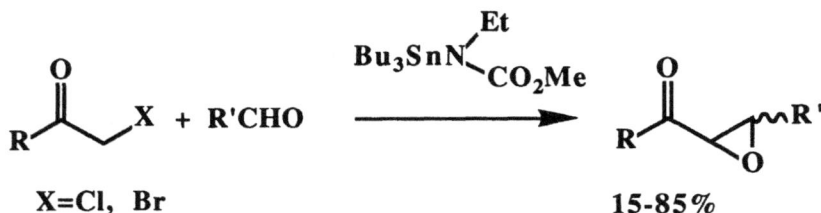

**IV.A-3**  A. Baba et al., *Synthesis*, 693.

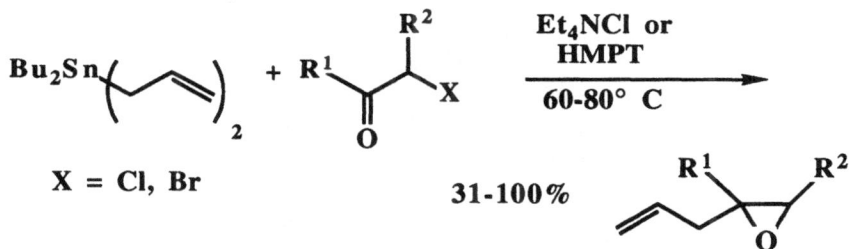

$$Bu_2Sn\left(\diagup\diagdown\right)_2 + R^1\underset{O}{\overset{R^2}{\underset{|}{C}}}\!\!-X \xrightarrow[\text{60-80° C}]{\substack{\textbf{Et}_4\textbf{NCl or}\\ \textbf{HMPT}}}$$

X = Cl, Br

31-100%

$$R^1 \overset{R^2}{\underset{O}{\triangle}}$$

---

**IV.A-4**  Y.-Z. Huang et al., *Chem. Commun.*, 986.

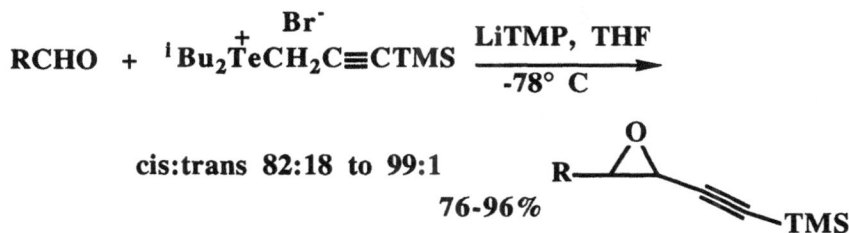

$$RCHO + {}^i Bu_2\overset{+}{Te}CH_2C{\equiv}CTMS \cdot Br^- \xrightarrow[\text{-78° C}]{\textbf{LiTMP, THF}}$$

**cis:trans 82:18 to 99:1**

76-96%

$$R\!\!-\!\!\overset{O}{\triangle}\!\!-\!\!\equiv\!\!-TMS$$

---

**IV.A-5**  M.B. Berry and D. Craig, *Synlett.*, 41; E. Kuyl-Yeheskiely et al., *Tetrahedron Lett.*, **33**, 3013.

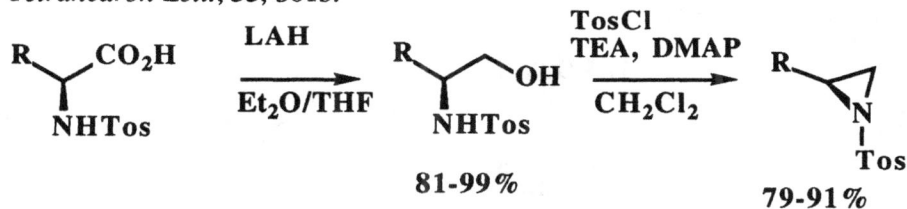

$$R\underset{NHTos}{\overset{}{\diagup}}CO_2H \xrightarrow[\textbf{Et}_2\textbf{O/THF}]{\textbf{LAH}} R\underset{NHTos}{\overset{}{\diagup}}OH \xrightarrow[\textbf{CH}_2\textbf{Cl}_2]{\substack{\textbf{TosCl}\\ \textbf{TEA, DMAP}}} R\overset{}{\underset{\underset{Tos}{|}}{\overset{}{\triangle}N}}$$

81-99%

79-91%

---

**IV.A-6**  B. Zwaneburg et al., *Rec Trav. Chim.*, **111**, 1,16, 59, 69 and 75; P. Molina et al., *Tetrahedron Lett.*, **33**, 2387.

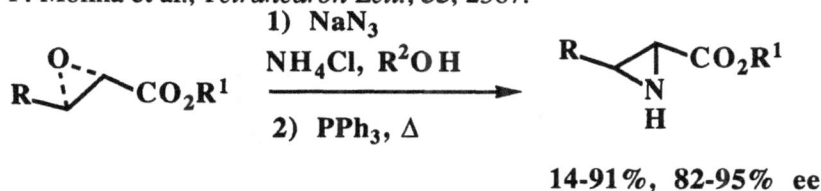

$$R\underset{}{\overset{O}{\diagdown}}\!\!-CO_2R^1 \xrightarrow[\textbf{2) PPh}_3, \Delta]{\substack{\textbf{1) NaN}_3\\ \textbf{NH}_4\textbf{Cl, R}^2\textbf{OH}}} R\overset{}{\underset{\underset{H}{|}}{\overset{}{\triangle}N}}CO_2R^1$$

14-91%, 82-95% ee

**IV.A-7**  J. Moulines et al., *Tetrahedron Lett.*, **33**, 487; M. Vaultier et al., *ibid.*, **33**, 5351.

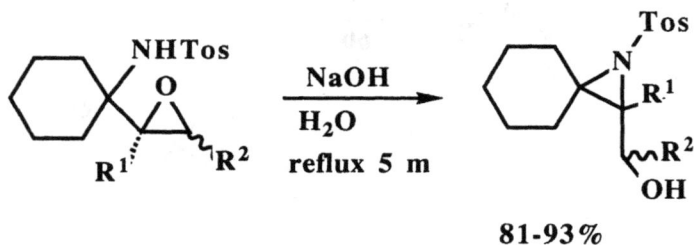

NaOH / H₂O / reflux 5 m

81-93%

---

**IV.A-8**  J. Aube et al., *J. Am. Chem. Soc.*, **114**, 5466.

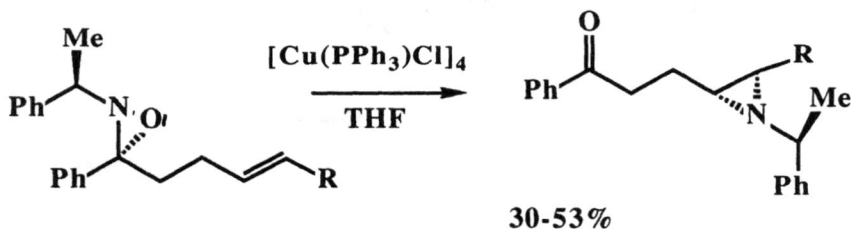

[Cu(PPh₃)Cl]₄ / THF

30-53%

---

**IV.A-9**  J.M. Vilalgordo and H. Heimgartner, *Helv. Chim. Acta*, **75**, 1866, 2270 and 2515.

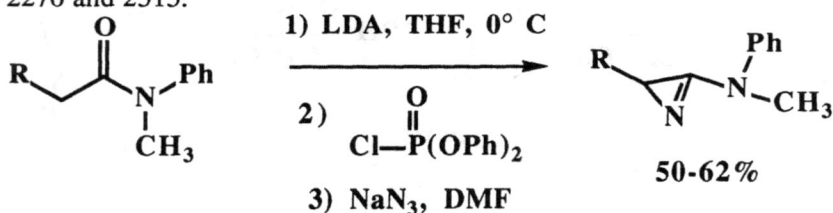

1) LDA, THF, 0° C

2)  Cl—P(OPh)₂ (O)

3) NaN₃, DMF

50-62%

---

**IV.A-10**  E. Vedejs and H. Sano, *Tetrahedron Lett.*, **33**, 3261.

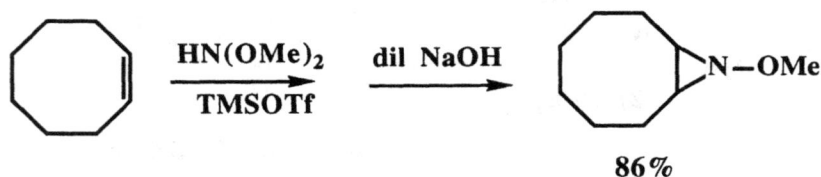

HN(OMe)₂ / TMSOTf    dil NaOH

86%

**IV.A-11**  H. Quast and T. Hergenrother, *Chem Ber.*, **125**, 2095.

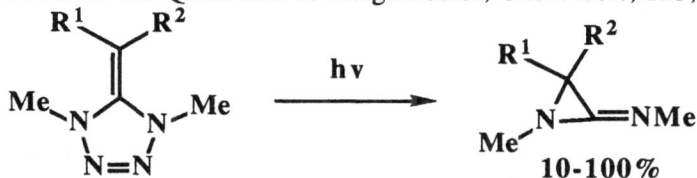

10-100%

---

**IV.A-12**  Y. Tominaga, A. Hosomi et al., *Tetrahedron Lett.*, **33**, 85.

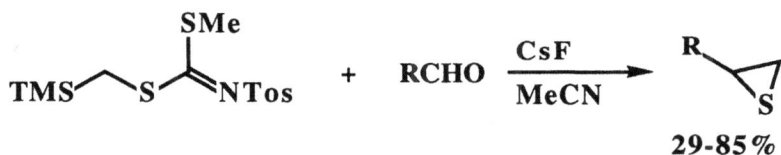

29-85%

---

**IV.A-13**  T.C. Owen and J.K. Leone, *J. Org. Chem.*, **57**, 6985.

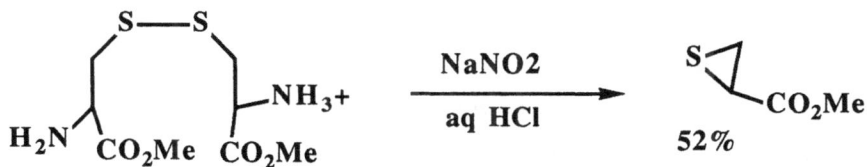

52%

## IV.B.   Oxetanes, Azetidines and Thietanes

**IV.B-1**  T. Toda et al., *Heterocycles*, **33**, 511.

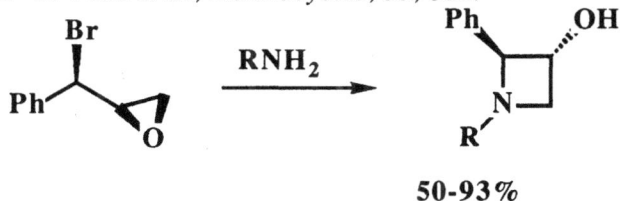

50-93%

**IV.B-2**  H.A.J. Carless and A.F.E. Halfhide, *J Chem. Soc., Perkin Trans. 1*, 1081.

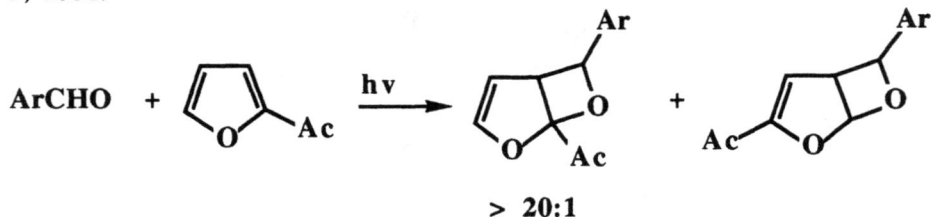

ArCHO  +  [furan with Ac]  $\xrightarrow{h\nu}$  [bicyclic product with Ar, O, Ac]  +  [bicyclic product with Ar, Ac, O]

> 20:1

---

**IV.B-3**  A. Whiting et al., *Tetrahedron*, **48**, 9553.

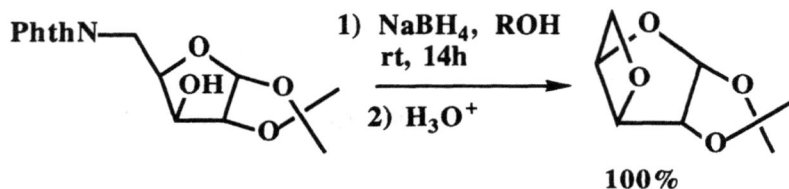

PhthN— [sugar structure with O, OH, O]  $\xrightarrow[\text{2) } H_3O^+]{\begin{array}{c}\text{1) } NaBH_4, \text{ ROH}\\ \text{rt, 14h}\end{array}}$  [bicyclic product]

100%

---

**IV.B-4**  B. Zwanenburg et al., *Tetrahedron*, **48**, 9985.

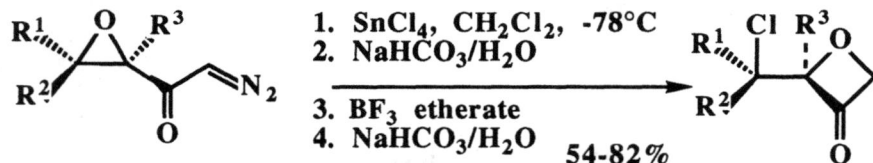

[epoxide structure with $R^1$, $R^2$, $R^3$, O, N$_2$]  $\xrightarrow[\begin{array}{l}\text{3. } BF_3 \text{ etherate}\\ \text{4. } NaHCO_3/H_2O\end{array}]{\begin{array}{l}\text{1. } SnCl_4, CH_2Cl_2, -78°C\\ \text{2. } NaHCO_3/H_2O\end{array}}$  [product with $R^1$, $R^2$, Cl, $R^3$, O]

54-82%

---

**IV.B-5**  T.Toda et al., *Chem. Lett.*, 1655.

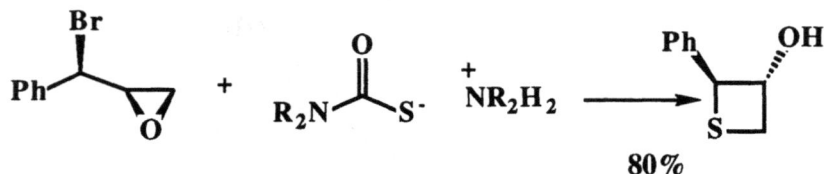

[structure with Br, Ph, O]  +  $R_2N\overset{O}{\underset{}{\parallel}}\!\!-S^-$  $\overset{+}{NR_2H_2}$  $\longrightarrow$  [product with Ph, OH, S]

80%

**IV.B-6** M. Sakamoto et al., *Chem. Commun.*, 891; **see also:** K. Oda et al., *Chem. Pharm. Bull.*, **40**, 585; H. Takechi et al., *Synthesis*, 778; E. Er and P. Margaretha, *Helv. Chim. Acta*, **75**, 2265.

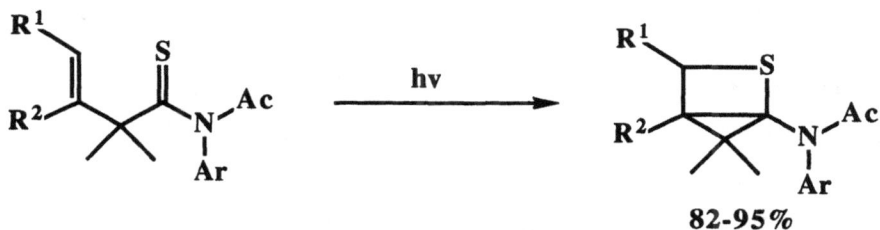

82-95%

**IV.B-7** R.J. Stoodley et al., *J. Chem. Soc., Perkin Trans. 1*, 2371.

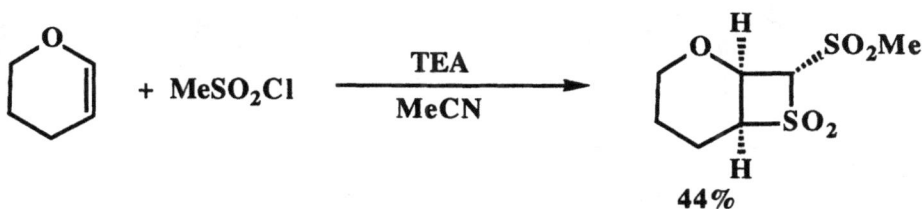

44%

## IV.C.   Lactams

**IV.C-1** C. Betschart and L.S. Hegedus, *J. Am. Chem. Soc.*, **114**, 5010; B. Alcaide, M.A. Sierra et al., *J. Org. Chem.*, **57**, 447.

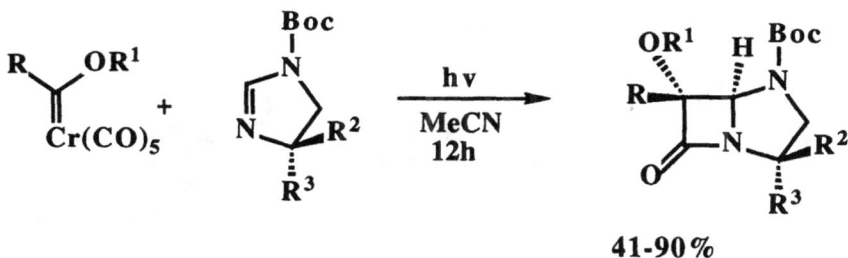

41-90%

**IV.C-2** S.T. Purrington and K.-W Sheu, *Tetrahedron Lett.*, **33**, 3289.

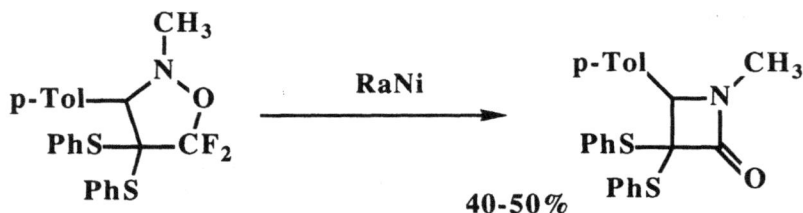

40-50%

**IV.C-3** O. Piva et al., *Tetrahedron Lett.*, **33**, 1993.

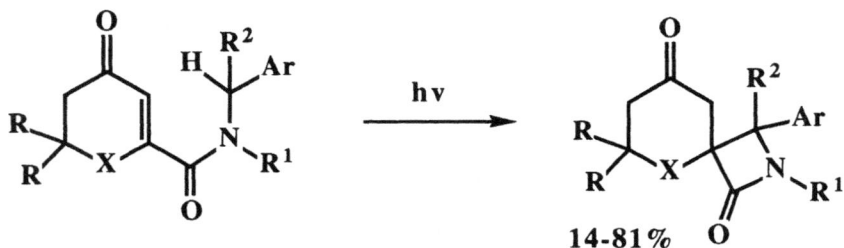

14-81%

**IV.C-4** H. Amri et al., *Tetrahedron Lett.*, **33**, 6159.

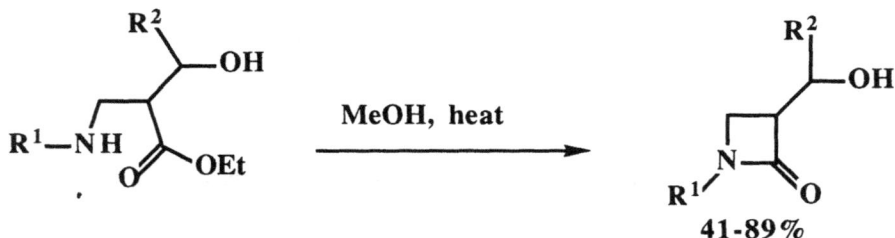

41-89%

**IV.C-5** T. Hiyama et al., *J. Org. Chem.*, **57**, 1864.

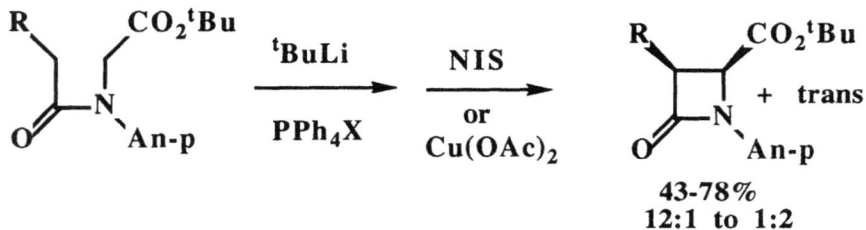

43-78%
12:1 to 1:2

**IV.C-6** M, Cinquini, F. Cozzi et al., *J. Org. Chem.*, **57**, 4155; G. van Koten et al., *ibid.*, **57**, 3906 and *Rec. Trav. Chim.*, **111**, 497.

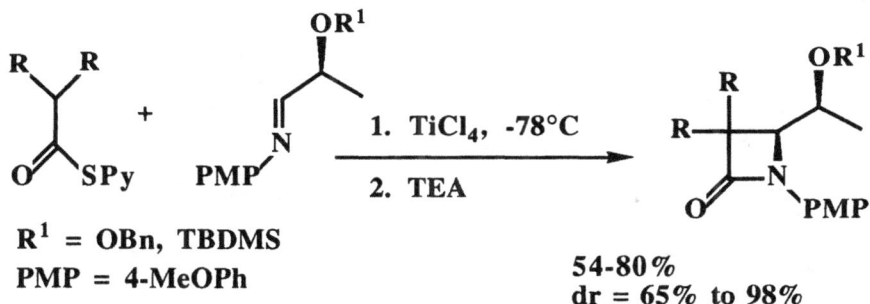

$R^1$ = OBn, TBDMS
PMP = 4-MeOPh

54-80%
dr = 65% to 98%

**IV.C-7** T. Fujisawa et al., *Tetrahedron Lett.*, **33**, 7903 and *Tetrahedron*, **48**, 5629 and *Chem. Lett.*, 1349.

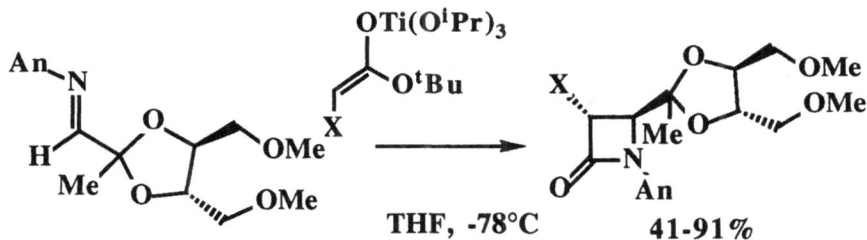

THF, -78°C          41-91%

**IV.C-8** J. Kajima Mulengi and N. Fatmi, *Bull. Soc. Chim.*, *Belg*, **101**, 257.

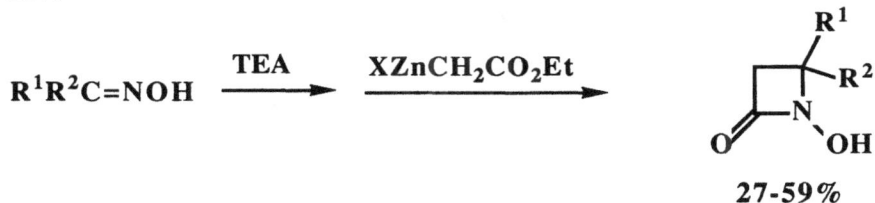

$R^1R^2C=NOH$  →[TEA]  →[$XZnCH_2CO_2Et$]

27-59%

**IV.C-9** M.S. Manhas et al., *Tetrahdron Lett.*, **33**, 3603; G.I. Georg et al., *ibid.*, **33**, 2111; S. Terashima et al., *Tetrahedron*, **48**, 1853; A.K. Bose et al., *ibid.*, **48**, 4831; K. Jahnisch et al., *Leibigs*, 781; C. Palomo et al., *J. Am. Chem. Soc.*, **114**, 9360 and *J. Org. Chem.*, **57**, 1571; B.M. Bhawal et al., *Synlett*, 749; B. Alcaide et al., *J. Org. Chem.*, **57**, 5921; K. Lempert et al., *J. Chem. Soc., Perkin Trans. 1*, 369; S.S. Bari et al., *Synthesis*, 439; R. Sunagawa et al., *Chem. Pharm. Bull.*, **40**, 1094; J.S. Sandhu et al., *J. Chem. Soc., Perkin Trans 1*, 1821.

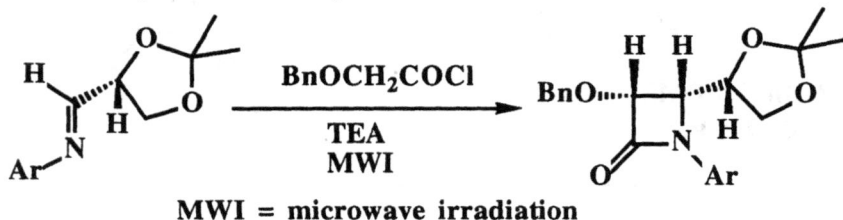

MWI = microwave irradiation

70-75%

**similar approaches to beta-lactams from ketenes**

---

**IV.C-10** J.-Q. Zhou and H. Alper, *J. Org. Chem.*, **57**, 3328; M.E. Krafft et al., *J. Am. Chem. Soc.*, **114**, 9215.

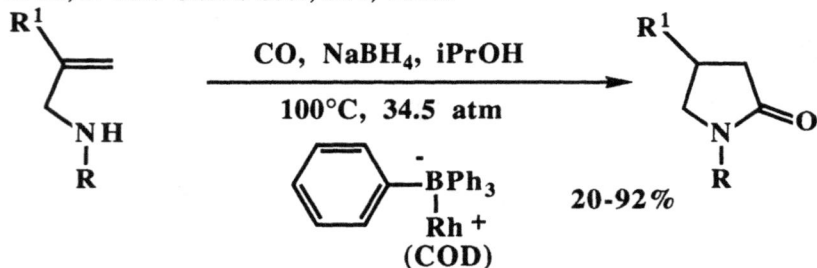

---

**IV.C-11** S. Masamune et al., *Tetrahedron Lett.*, **33**, 1937.

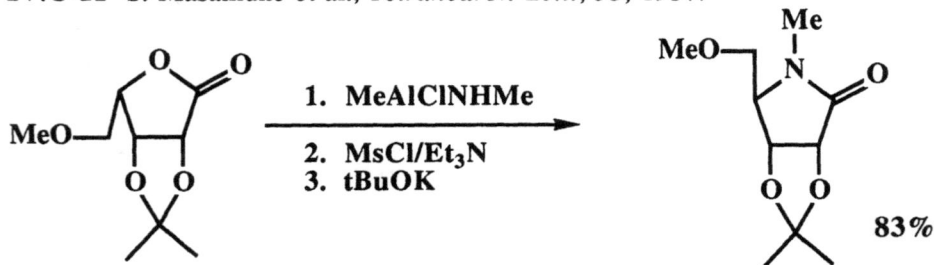

83%

**IV.C-12**  A.W. Lee et al., *J. Org. Chem.*, **57**, 4404; M.P. Doyle et al., *Tetrahedron Lett.*, **33**, 7819.

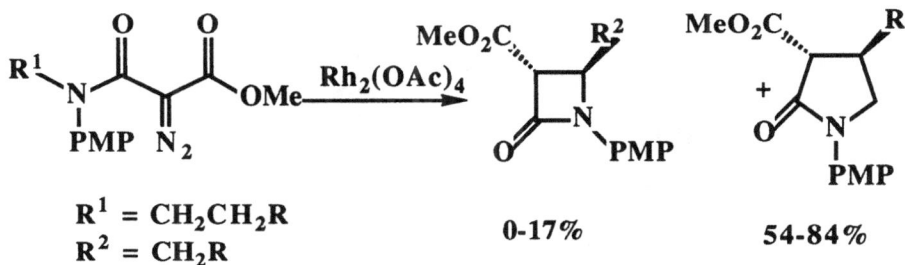

$R^1 = CH_2CH_2R$
$R^2 = CH_2R$

0-17%        54-84%

---

**IV.C-13**  J. Villieras et al., *Tetrahedron Asym.*, **3**, 351 and 511.

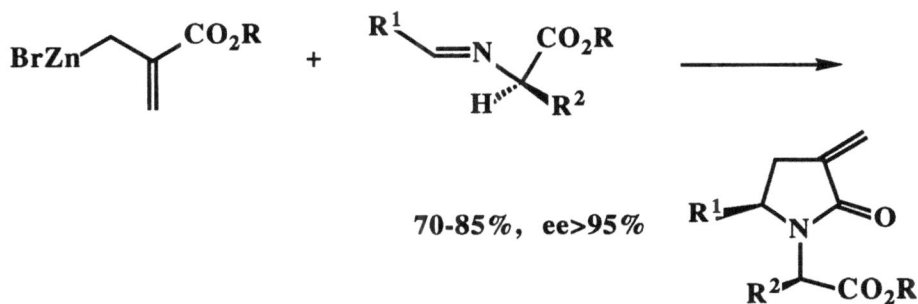

70-85%,  ee>95%

---

**IV.C-14**  K. Itoh et al., *J. Org. Chem.*, **57**, 1682.

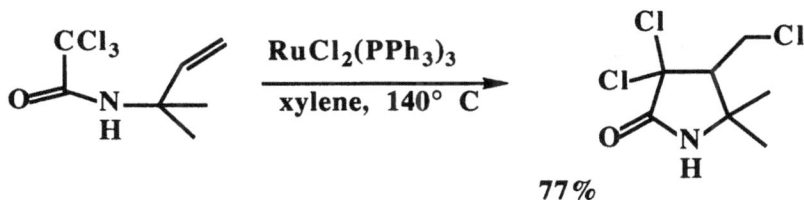

77%

**IV.C-15**  U. Groth, L. Richter and U. Schollkopf, *Liebigs Ann. Chem.*, 903; H. Amri et al., *Tetrahedron Lett.*, **33**,7345.

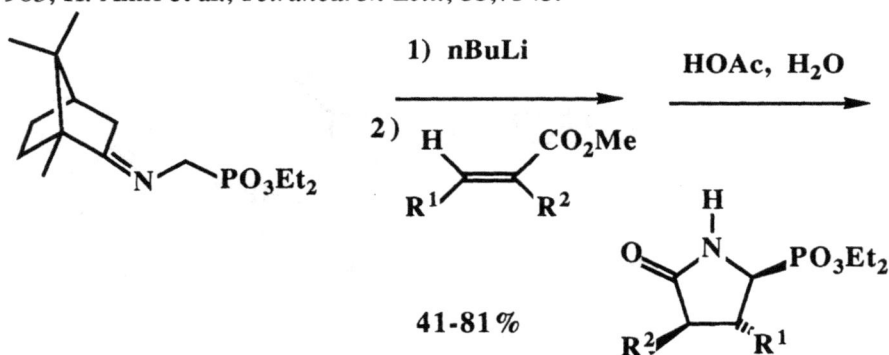

1) nBuLi

2)

HOAc, H₂O

41-81%

---

**IV.C-16**  P.A. Jacobi and S. Rajeswari, *Tetrahedron Lett.*, **33**, 6235.

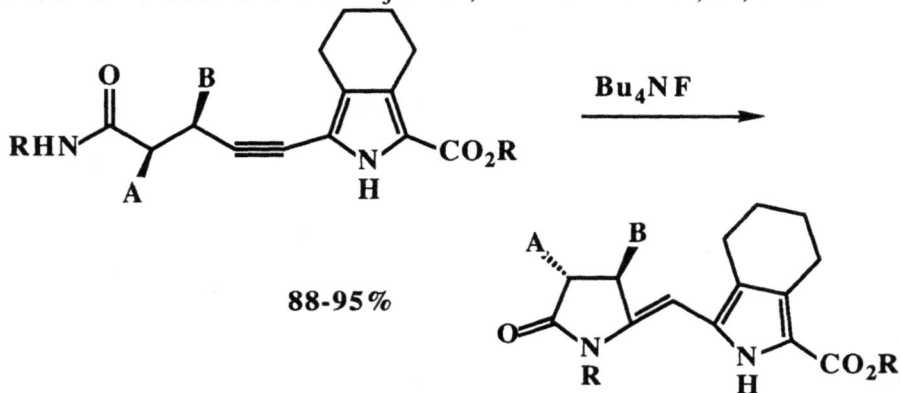

Bu₄NF

88-95%

---

**IV.C-17**  W.N. Speckamp et al., *Rec. Trav. Chim.*, **111**, 360.

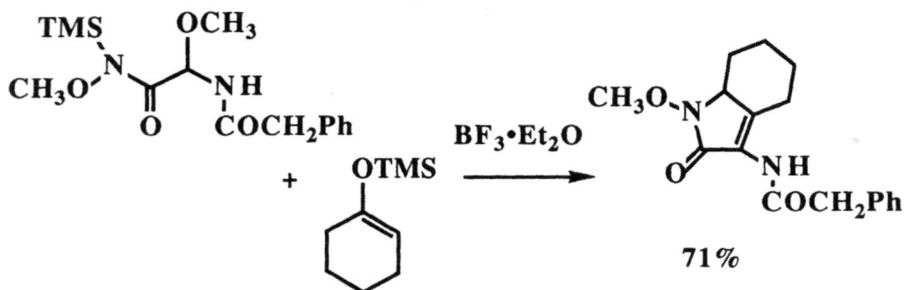

BF₃•Et₂O

71%

**IV.C-18**  M. Ikeda et al., *Heterocycles*, **33**, 139.

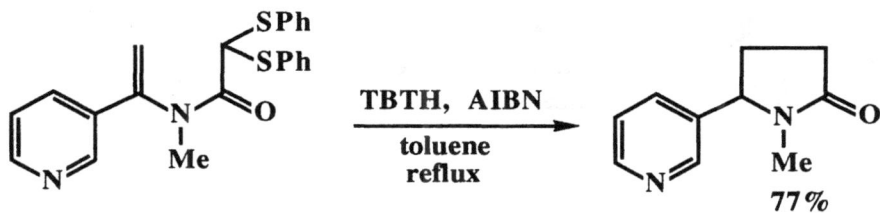

**IV.C-19**  P.G. Andersson and J.E. Backvall, *J. Am. Chem. Soc.*, **114**, 8696.

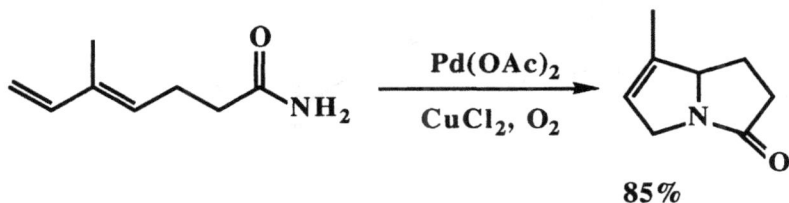

**IV.C-20**  K. Jones et al., *J. Chem. Soc., Chem. Commun.*, 1766 and 1767.

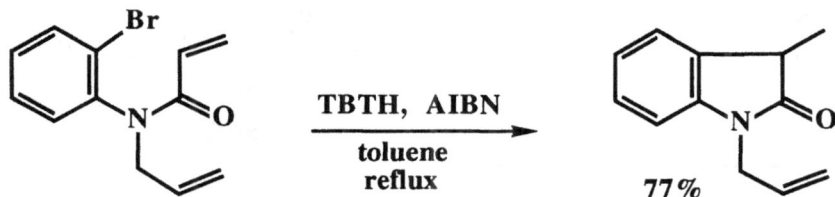

**IV.C-21**  J.A. Joule et al., *Synthesis*, 769.

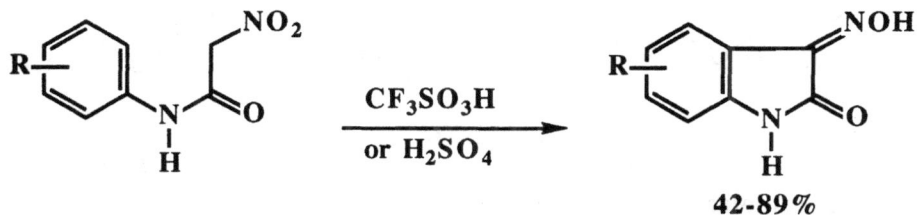

**IV.C-22**  V. Voerckel et al., *Chem. Ber.*, **125**, 2719.

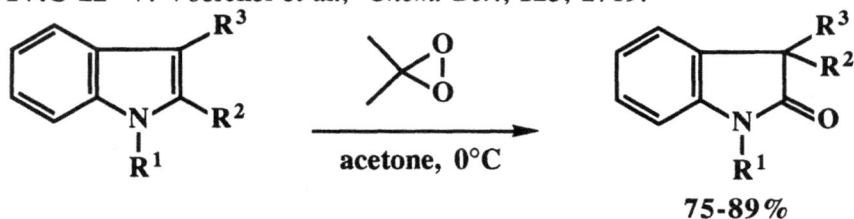

75-89%

---

**IV.C-23**  J.-M. Vierfond et al., *Heterocycles*, **34**, 911.

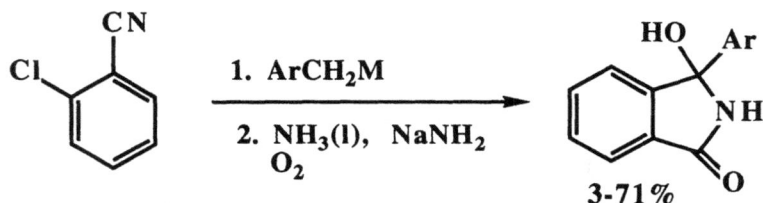

1. ArCH$_2$M

2. NH$_3$(l),  NaNH$_2$
   O$_2$

3-71%

---

**IV. C-24**  C.M. Marson et al., *J. Org. Chem.*, **57**, 5045.

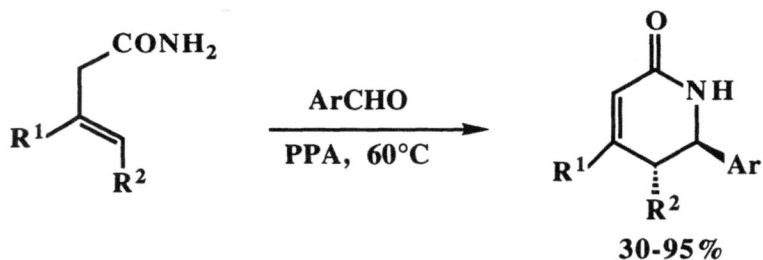

ArCHO

PPA,  60°C

30-95%

---

**IV.C-25**  Y. Leblanc et al., *Tetrahedron Lett.*, **33**, 5717.

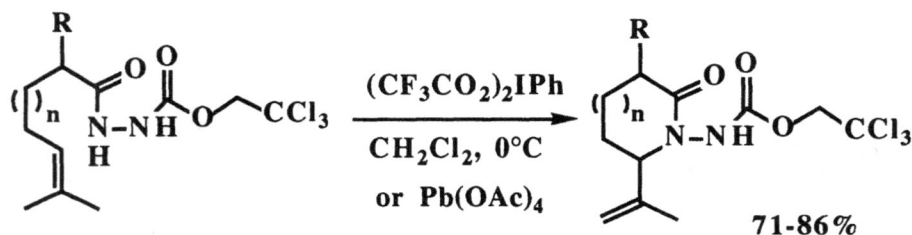

(CF$_3$CO$_2$)$_2$IPh

CH$_2$Cl$_2$, 0°C

or  Pb(OAc)$_4$

71-86%

intramolecular ene reaction

**IV.C-26**  E. Pinto de Souza and P.S. Fernandes, *Ind. J. Chem.*, **31b**, 578.

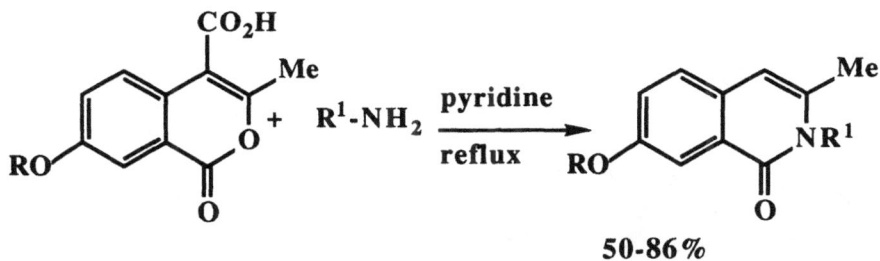

50-86%

---

**IV.C-27**  C.-q. Shin et al., *Heterocycles*, **33**, 589.

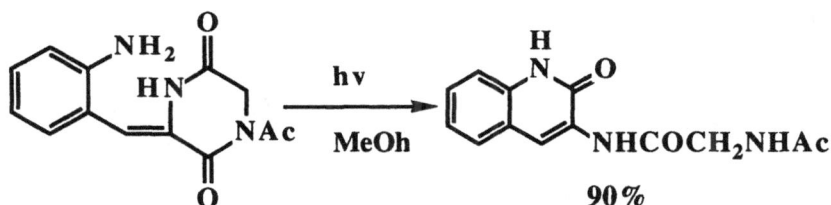

90%

---

**IV.C-28**  T. Naito et al., *Heterocycles*, **34**, 1783.

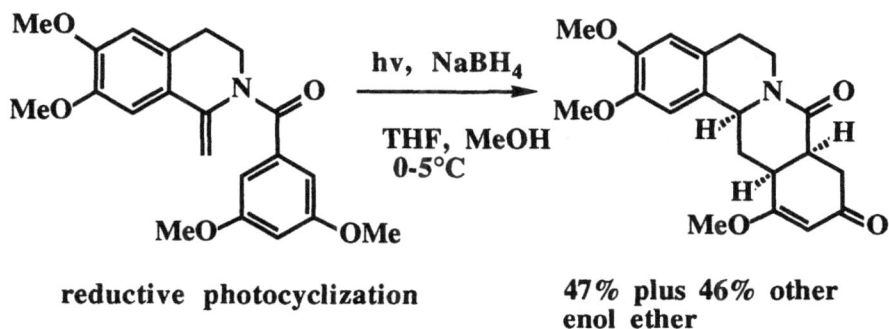

reductive  photocyclization

47% plus 46% other
enol ether

**IV.C-29**  L. Castedo et al., *J. Org. Chem.*, **57**, 5907.

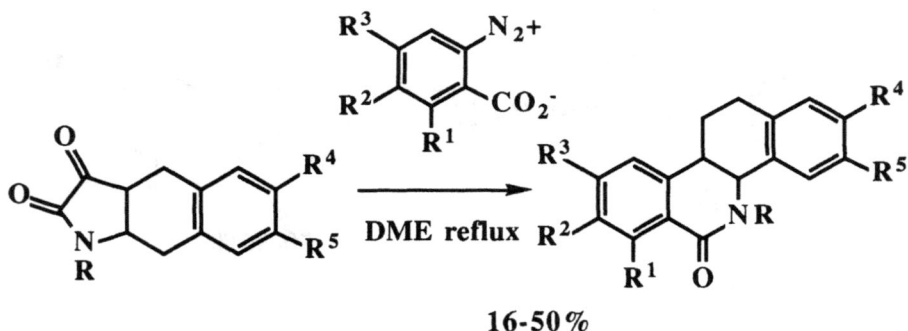

16-50%

---

**IV.C-30**  J.-C. Gramain et al., *Syn Commun.*, **22**, 189.

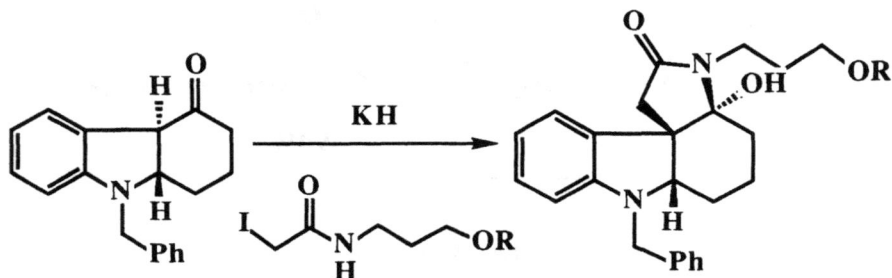

---

**IV.C-31**  C.J. Moody et al., *J. Chem. Soc., Perkin 1.*, 797 and 813 and 823.

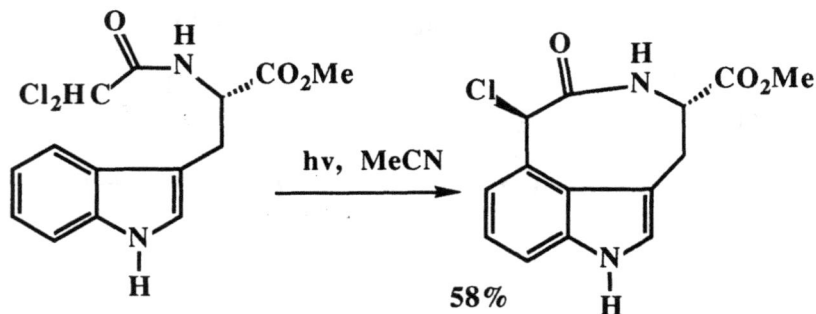

58%

**IV.C-32**  C.J. Moody et al., *J. Chem. Soc., Perkin Trans. 1*, 831.

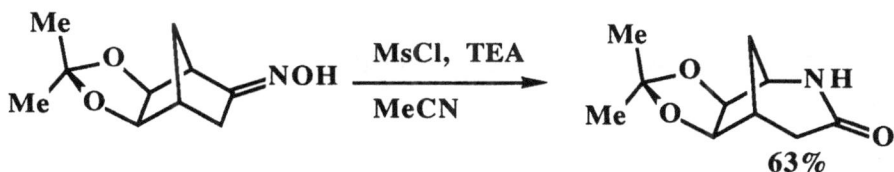

---

**IV.C-33**  S. Ogawa et al., *Chem. Commun.*, 1064.

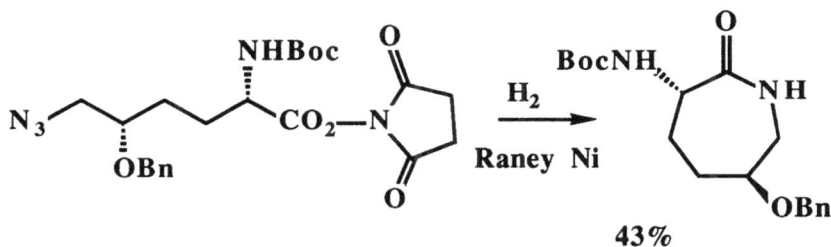

---

**IV.C-34**  P.A. Evans and A.B. Holmes, *Tetrahedron Lett.*, **33**, 6857.

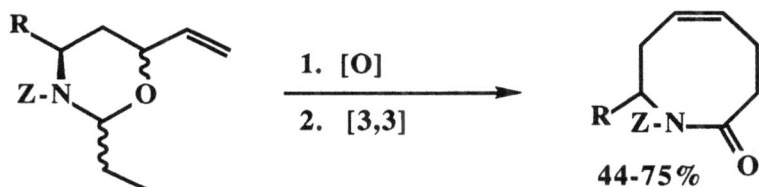

selenoxide elimination to ketene aminal followed
by Claisen rearrangement

---

**IV.C-35**  A.L. Gutman, C. Abell et al., *Tetrahedron Lett.*, **33**, 3943.

**"Enzymatic Formation of Lactams in Organic Solvents"**

## IV.D.    Lactones

**IV.D-1**  A.V. Rama Rao et al., *Tetrahedron Lett.*, **33**, 3907.

### Radical cyclization approach to chiral γ-lactone

---

**IV.D-2**  H. Yamamoto et al., *Synlett.*, 31.

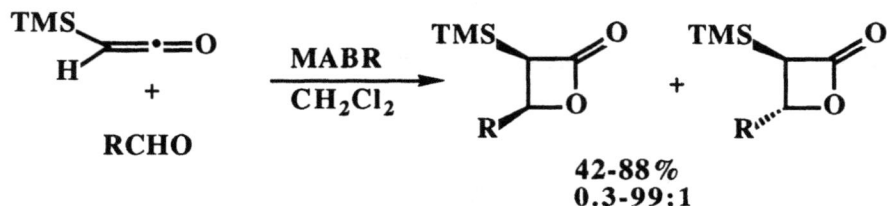

42-88%
0.3-99:1

---

**IV.D-3**  T. Joh et al., *Chem. Lett.*, 1305.

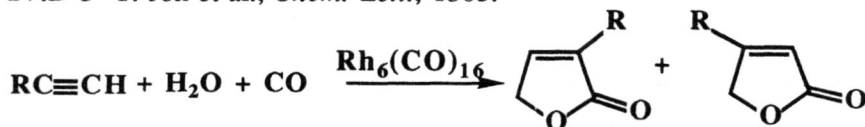

$$RC \equiv CH + H_2O + CO \xrightarrow{Rh_6(CO)_{16}}$$

---

**IV.D-4**  S.E. Denmark and D.C. Forbes, *Tetrahedron Lett.*, **33**, 5037.

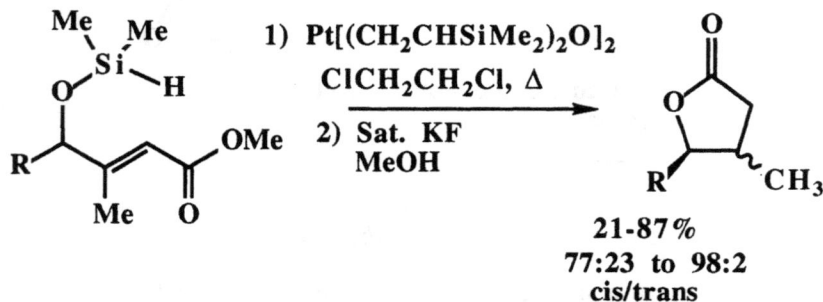

1) Pt[(CH₂CHSiMe₂)₂O]₂
   ClCH₂CH₂Cl, Δ

2) Sat. KF
   MeOH

21-87%
77:23 to 98:2
cis/trans

**IV.D-5** M.E. Krafft et al., *J. Org. Chem.*, **57**, 5277.

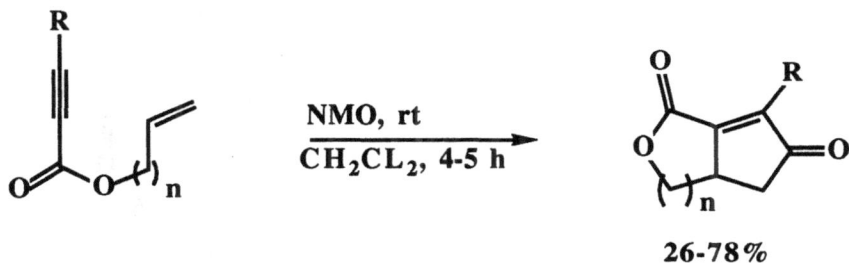

**26-78%**

**Pauson-Khand reactions carried out with corresponding dicobalthexacarbonyl complex**

---

**IV.D-6** M.D. Bachi and E. Bosch, *J. Org. Chem.*, **57**, 4696.

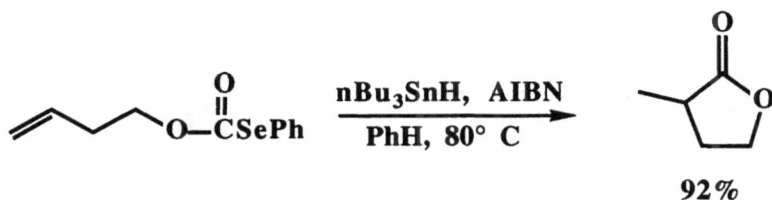

**92%**

---

**IV.D-7** I. Fleming and S.K. Ghosh, *J. Chem. Soc., Chem. Commun.*, 1777.

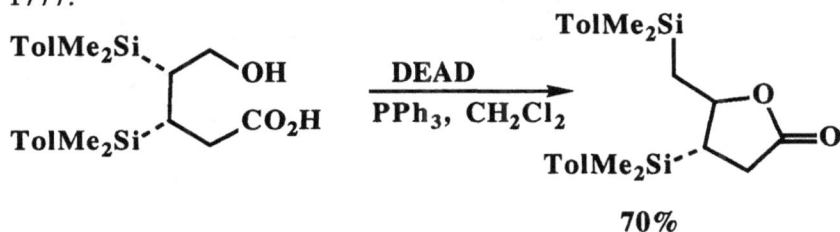

**70%**

---

**IV.D-8** Y. Lin et al., *J. Organomet Chem.*, **429**, 269.

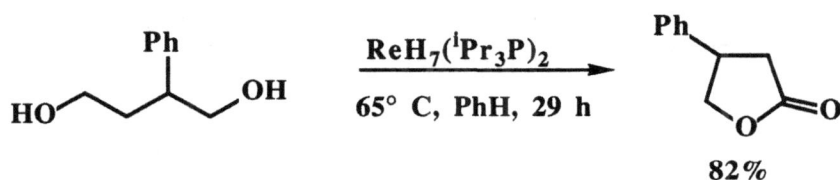

**82%**

**IV.D-9** K.B. Sharpless et al., *Tetrahedron Lett.*, **33**, 6407 and 6411.

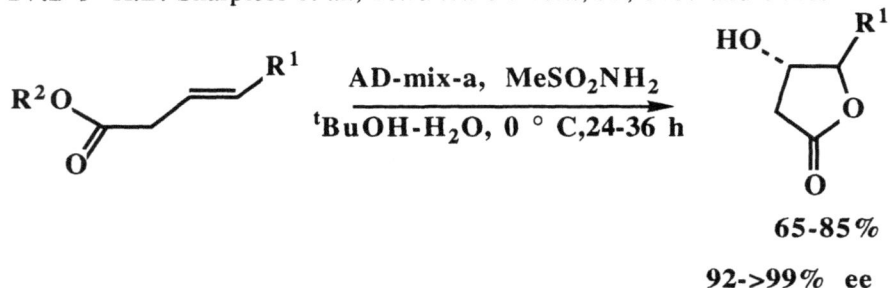

**AD = asymmetric dihydroxylation**

---

**IV.D-10** M.P. Sibi and J.A. Gaboury, *Tetrahedron Lett.*, **33**, 5681; C. Bonini et al., *Tetrahedron Asym.*, **3**, 29; P. Pedrini, M.E. Guerzoni et al., *ibid.*, **3**, 107.

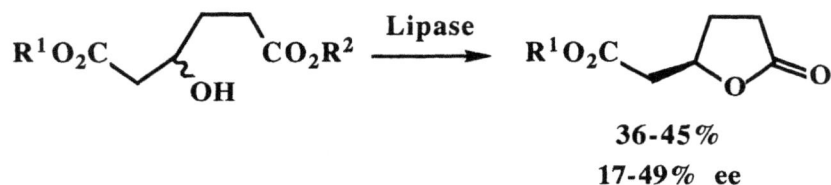

---

**IV.D-11** G. Casy, *Tetrahedron Lett.*, **33**, 8159.

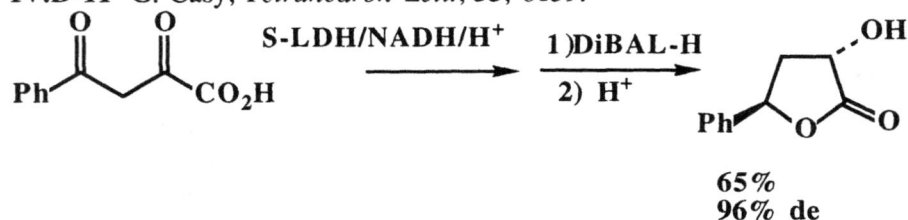

---

**IV.D-12** S. Kagabu et al., *Synthesis*, 830.

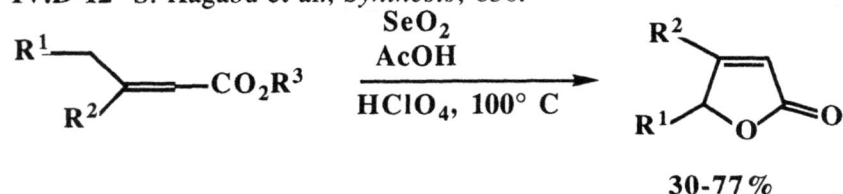

**IV.D-13**  A.R. Sidduri and P. Knochel, *J. Am. Chem. Soc.*, **114**, 7579; M.-Y. Chen and J.-M. Fang, *J. Org. Chem.*, **57**, 2937; J.I. Levin, *Synth. Commun.*, **22**,961.

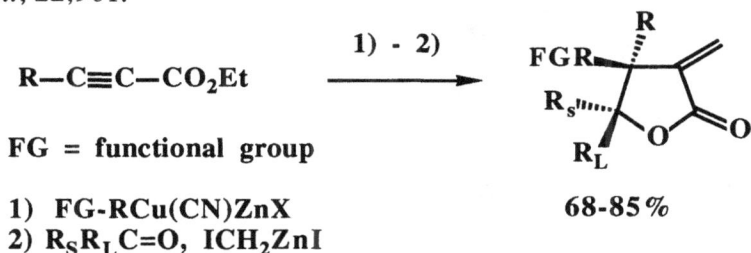

$$R—C\equiv C—CO_2Et \xrightarrow{\text{1) - 2)}}$$

FG = functional group

1)  FG-RCu(CN)ZnX
2)  $R_SR_LC=O$, $ICH_2ZnI$

68-85%

---

**IV.D-14**  O. Kitagawa, T. Taguchi et al., *Chem. Commun.*, 1005; D. Guillerm and G. Guillerm, *Tetrahedron Lett.*, **33**, 5047; S. Shibuya et al., *ibid.*, **33**, 6999; I.F. Pelyvas, J. Thiem et al., *Liegigs Ann. Chem.*, 3; A. Valla and M. Giroud, *Synthesis*, 690; H. Cerfontain et al., *Rec. Trav. Chim.*, **111**, 478; D.W. Knight et al., *Tetrahedron Lett.*, **33**, 6505,6507 and 6511.

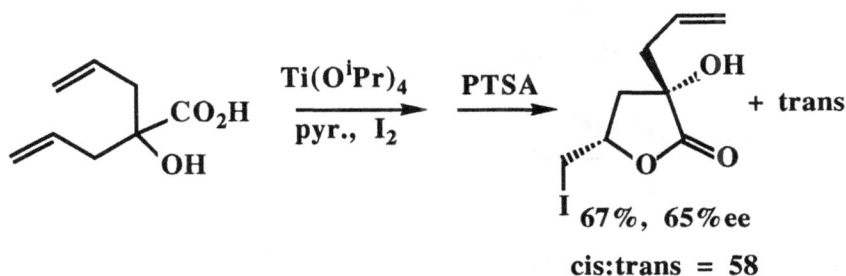

$$\xrightarrow[\text{pyr., } I_2]{\text{Ti(O}^i\text{Pr)}_4} \quad \xrightarrow{\text{PTSA}}$$

+ trans

67%, 65%ee

cis:trans = 58

---

**IV.D-15**  M. Casey et al., *Tetrahedron Lett.*, **33**, 965; T. Naito et al., *Chem. Pharm. Bull.*, **40**, 2579; I. Coldham and S. Warren, *J. Chem. Soc.,Perkin Trans 1*, 2303.

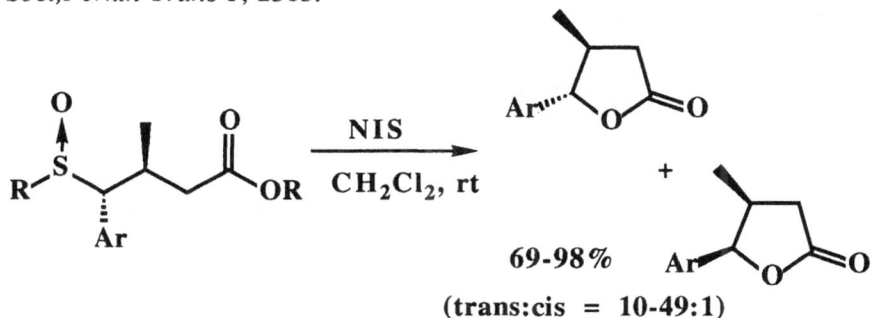

$$\xrightarrow[\text{CH}_2\text{Cl}_2, \text{ rt}]{\text{NIS}}$$

+

69-98%

(trans:cis = 10-49:1)

**IV.D-16** S. Cacchi et al., *J. Org. Chem.*, **57**, 976; T. Mandai, J. Tsuji et al., *Synlett.*, 671.

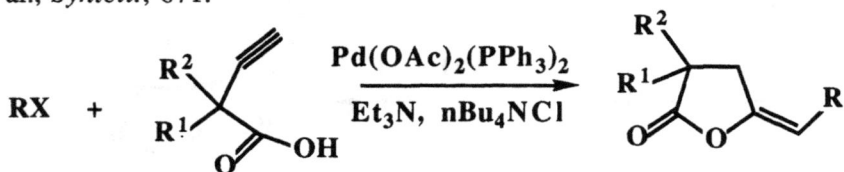

RX +

R = vinyl and aryl; X=OTf, I, Br

---

**IV.D-17** G.T. Crisp and A.G. Meyer; *J. Org. Chem.*, **57**, 6972.

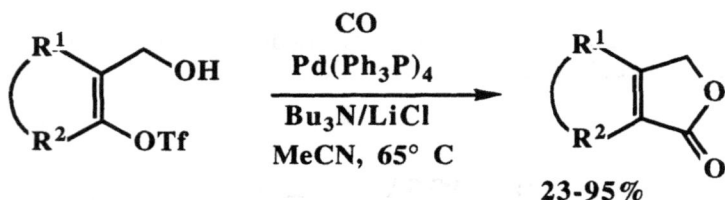

23-95%

---

**IV.D-18** R.R. Schmidt et al., *Tetrahedron Lett.*, **33**, 8035 and *Synlett.*, 429.

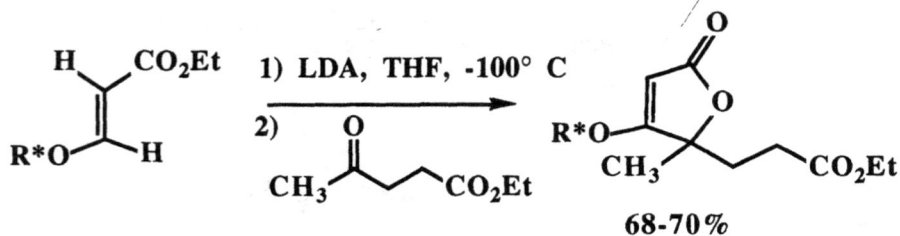

1) LDA, THF, -100° C

2)

68-70%

diastereomer ratio 1:1 to 4.5:1

**IV.D-19**  H. Xiong and R.D. Rieke, *J. Org. Chem.*, **57**, 7007.

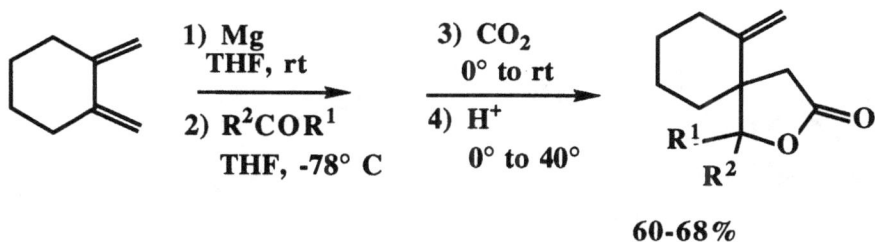

60-68%

**IV.D-20**  T. Hudlicky et al., *Synth. Commun.*, **22**, 151; K. Tadano et al., *J. Org. Chem.*, **57**, 3789.

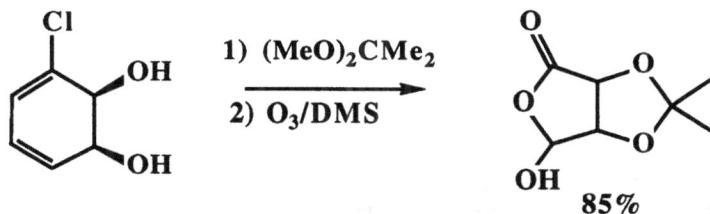

85%

**IV.D-21**  K. Tadano et al., *Tetrahedron Lett.*, **33**, 7899.

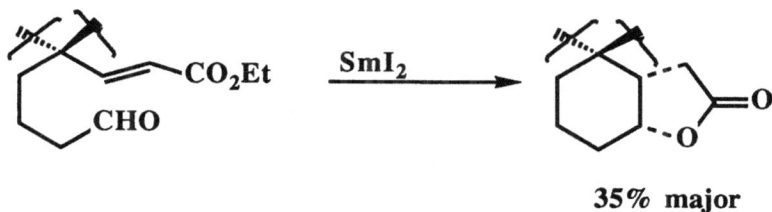

35% major

**IV.D-22**  L. Castedo et al., *J. Org. Chem.*, **57**, 2029.

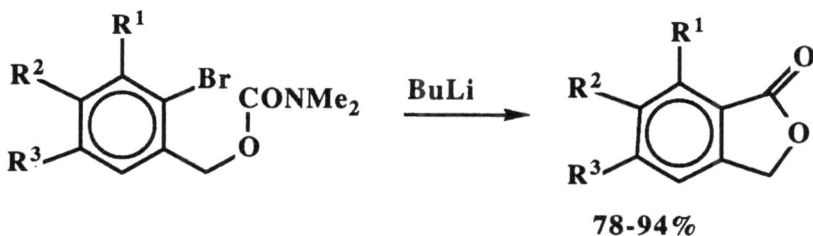

78-94%

324

**IV.D-23**  R.S. Mali et al., *Chem. Commun.*, 883.

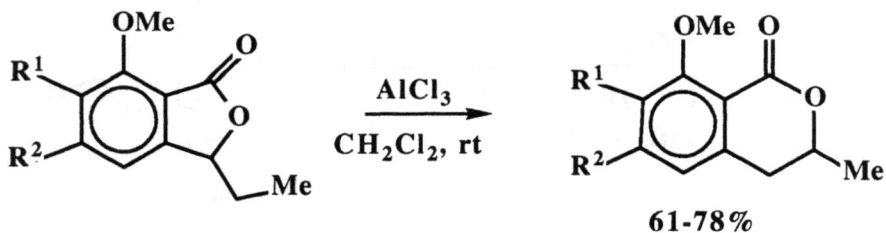

61-78%

---

**IV.D-24**  B.H. Bhide et al., *Indian J. Chem.*, **31B**, 143.

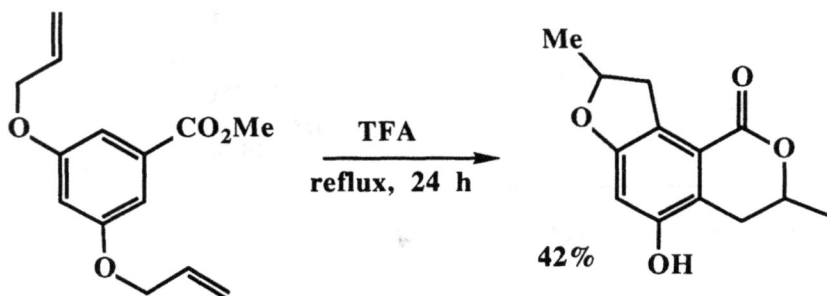

42%

---

**IV.D-25**  J.E. Pickett and P.C. Van Dort, *Tetrahedron Lett.*, **33**, 1161.

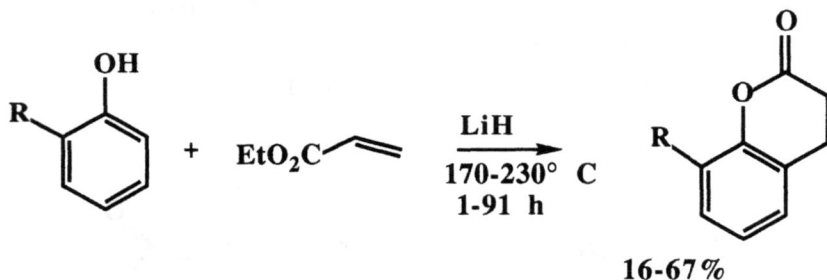

16-67%

**IV.D-26**  J. Epsztajn, M.W. Plotka et al., *Synth. Commun.*, **22**, 1239; R.A. Ward and G. Procter, *Tetrahedron Lett.*, **33**, 3359.

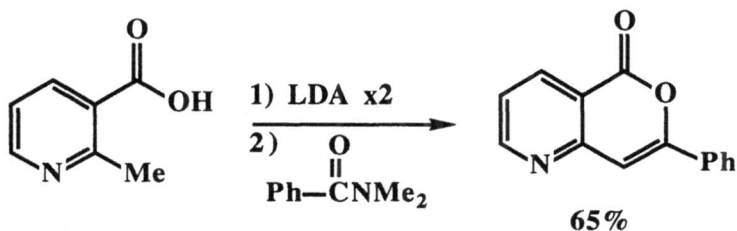

**IV.D-27**  T. Tsuda et al., *J. Organomet. Chem.*, **429**, C46.

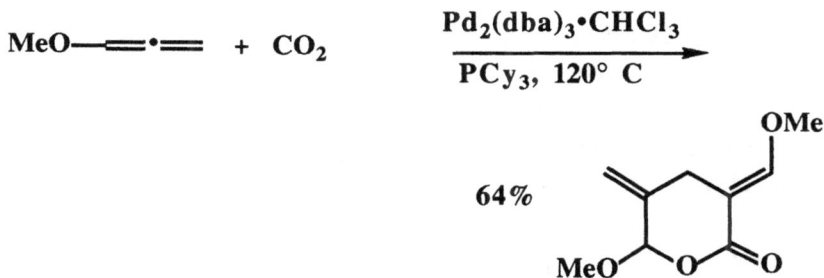

**IV.D-28**  K. Tsushima and A. Murai, *Tetrahedron Lett.*, **33**, 4345; N.E. Schore et al., *J. Organomet. Chem.*, **431**, 335.

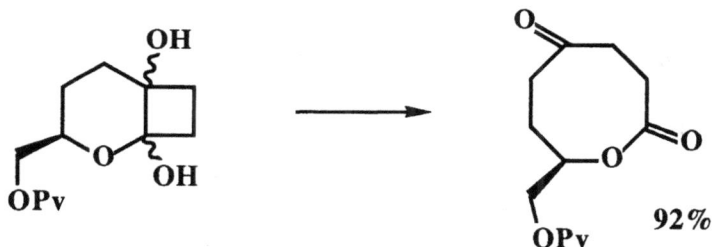

**IV.D-29** H. Hiemstra, W.N. Speckamp et al.,*Tetrahedron Lett.*, **33**, 5141;E. Lee et al., *J. Am. Chem. Soc.*, **114**, 10981.

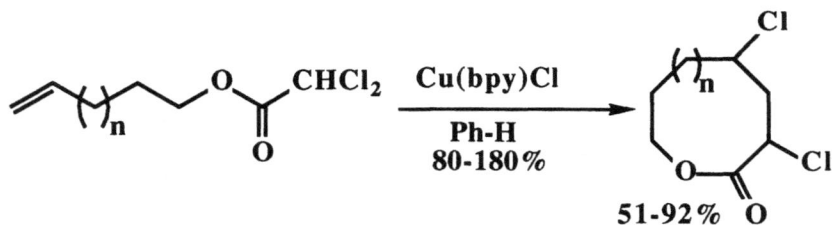

**IV.D-30** M.L. Morin-Fox and M.A. Lipton, *Tetrahedron Lett.*, **33**, 5699.

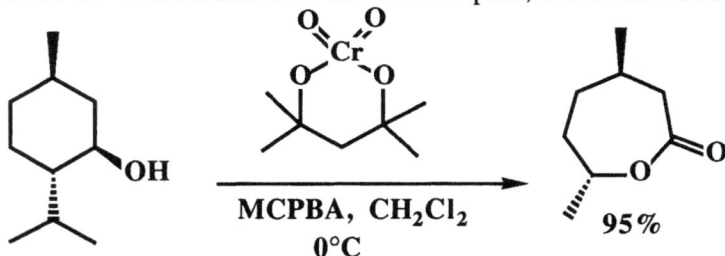

**IV.D-31** L. Weiler et al., *Can. J. Chem.*, **70**, 1427.

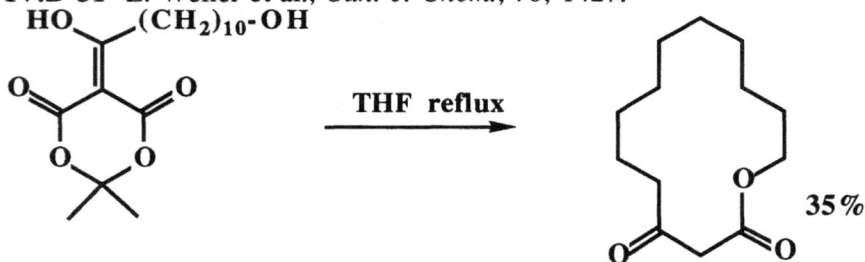

**IV.D-32** J.E. Baldwin et al., *Tetrahedron*, **48**, 2957.

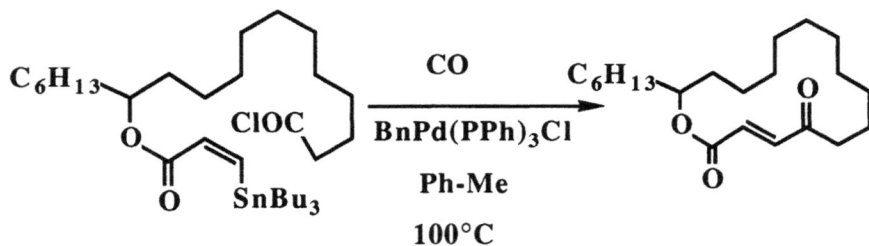

## IV.E.    Furans and Thiophenes

**IV.E-1**  J.S. Clark, *Tetrahedron Lett.*, **33**, 6193; M.C. Pirrung and J. Zhang, *ibid.*, **33**, 5987.

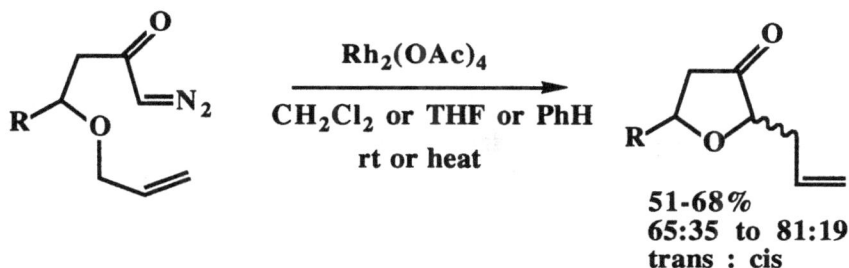

51-68%
65:35 to 81:19
trans : cis

**IV.E-2**  M. Julia et al., *Bull. Soc. Chim. Fr.*, **129**, 387; V.E. Marquez et al., *Tetrahedron Lett.*, **33**, 1539; P. Pale et al., *ibid.*, **33**, 7857 and 4905; J.S. Yadav et al., *ibid.*, **33**, 3687; Y. Chapleur et al., *J. Chem. Soc., Perkin Trans 1*, 991 and 999; A. Srikrishna et al., *Synth. Commun.*, **22**, 1221; H.M.R. Hoffmann et al., *Angew. Chem., Int. Ed. Engl.*, **31**, 910; M. Tsukazaki and V. Snieckus, *Can. J. Chem.*, **70**, 1486; K.A. Parker and D. Fokas, *J. Am. Chem. Soc.*, **114**, 9688.

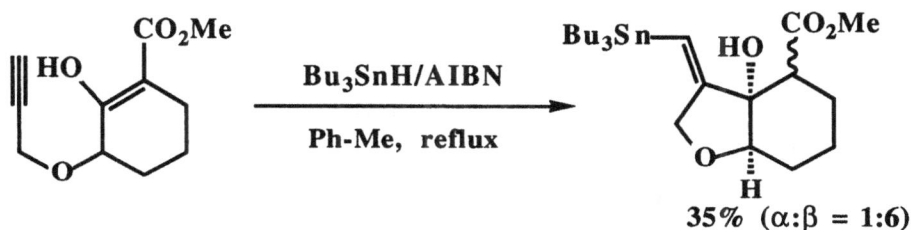

35% (α:β = 1:6)

**IV.E-3**  R.M. Kennedy and S. Tang, *Tetrahedron Lett.*, **33**, 3729, 5299 and 5303.

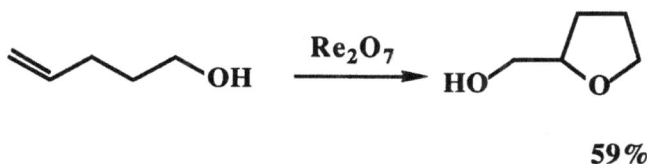

59%

**IV.E-4**  G. Boisvert and R. Giasson, *Tetrahedron Lett.*, **33**, 6587; G. Pattenden et al., *ibid.*, **33**, 2851.

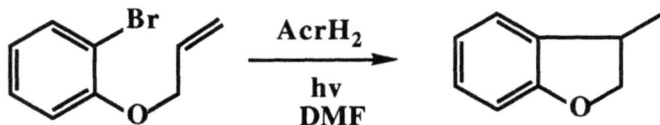

$AcrH_2$  =  **10-methyl-9,10-dihydroacridine**

---

**IV.E-5**  S. Ozaki et al., *Chem. Commun.*, 1120; S. Torii et al., *Tetrahedron Lett.*, **33**, 6495 and *Synlett*, 515.

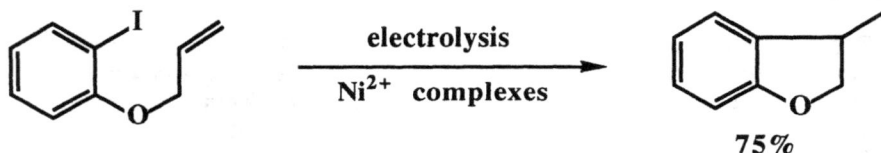

---

**IV.E-6**  J.S. Swenton et al., *J. Org. Chem.*, **57**, 2135; M. Schmittel and M. Rock, *Chem. Ber.*, **125**, 1611.

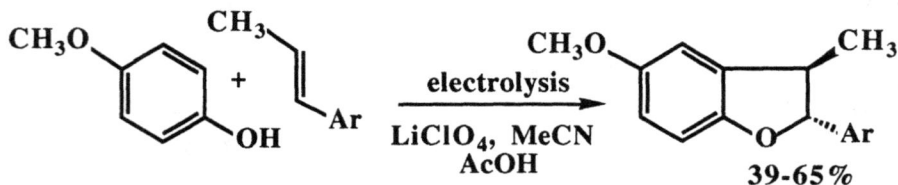

---

**IV.E-7**  F.-T. Luo et al., *J. Org. Chem.*, **57**, 2213; S. Saito, T. Moriwake et al., *Synlett*, 237; M.A. Sturgess et al., *Tetrahedron Lett.*, **33**, 7739; J.-E. Backvall and P.G. Andersson, *J. Am. Chem. Soc.*, **114**, 6374; B.M. Trost and J.A. Flygate, *ibid.*, **114**, 5476.

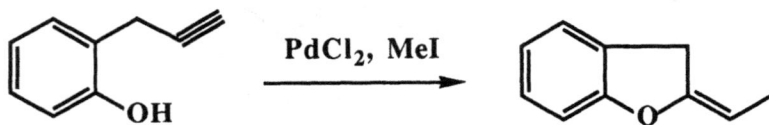

**similar  examples  of  Pd  mediated  cyclizations**          **66%**

**IV.E-8**  H. Ishii et al., *Chem. Pharm. Bull.*, **40**, 2002 and 1993 and 2003.

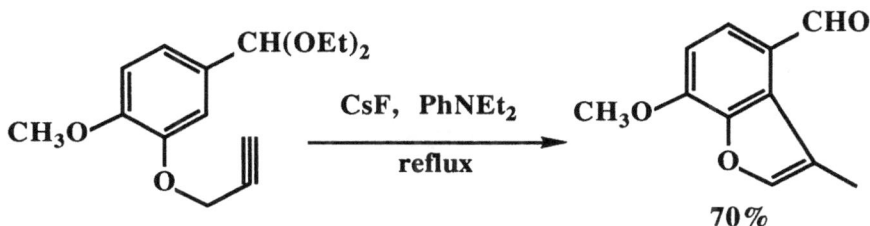

70%

---

**IV.E-9**  R.D. Walkup et al., *Tetrahedron Lett.*, **33**, 3969 and *Synth. Commun.*, **22**, 1007; I. Marek et al., *Tetrahedron Lett.*, **33**, 1747; N.E. Shore et al., *J. Am. Chem. Soc.*, **114**, 10061; B.H. Lipshutz and J.C. Barton, *ibid.*, **114**, 1084; E.D. Mihelich and G.A. Hite, *ibid.*, **114**, 7318; M. Gray et al., *Synlett*, 597; G. Pandey and B.B.V. Soma Sekhar, *J. Org. Chem.*, **57**, 4019; J.V. Comasseto and M.V.A. Grajin, *Synth. Commun.*, **22**, 949.

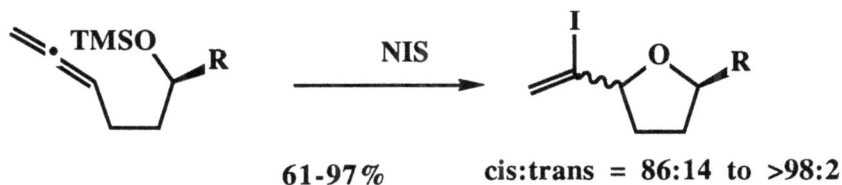

61-97%          cis:trans = 86:14 to >98:2

**similar cyclizations upon alkenes using HgX$_2$, I$_2$, ICl, N-PSP, PhSCl, PhSeSePh and ArTeCl$_3$**

---

**IV.E-10**  S. Fujiwara and A.B. Smith, III, *Tetrahedron Lett.*, **33**, 1185.

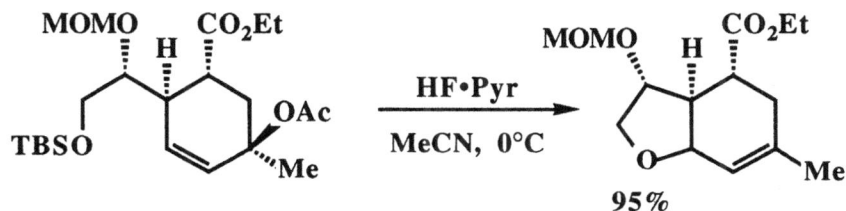

95%

**IV.E-11**   T. Lubbers and H.J. Schafer, *Synlett.*, 743.

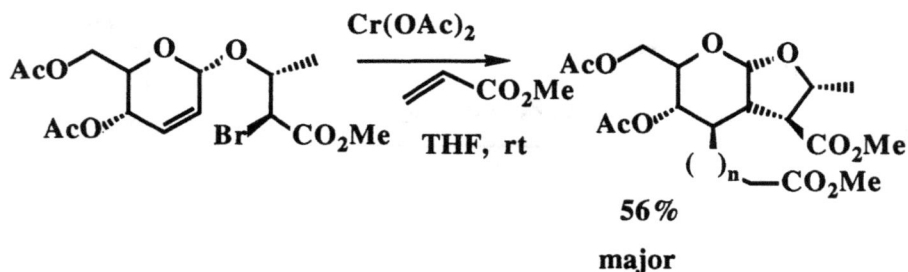

Cr(OAc)$_2$

CO$_2$Me

THF, rt

56%

major

**IV.E-12**   H.E. Zimmerman and C.W. Wright, *J. Am. Chem. Soc.*, **114**, 6603.

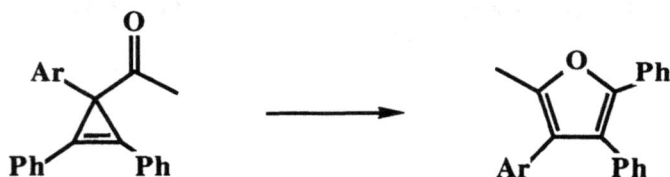

**IV.E-13**   B.M. Trost et al., *J. Am. Chem. Soc.*, **114**, 7903.

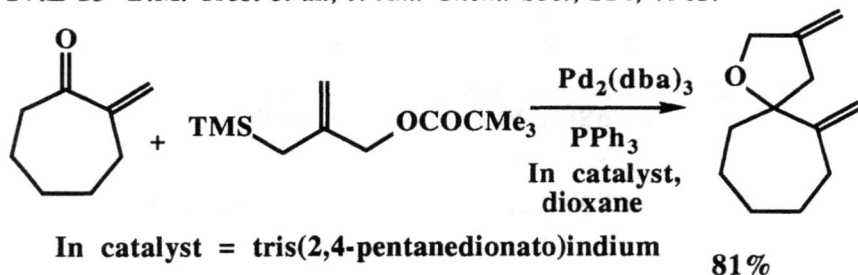

Pd$_2$(dba)$_3$

PPh$_3$

In catalyst, dioxane

In catalyst = tris(2,4-pentanedionato)indium

81%

**IV.E-14**   A. Padwa et al., *J. Chem. Soc., Perkin Trans. 1*, 2837.

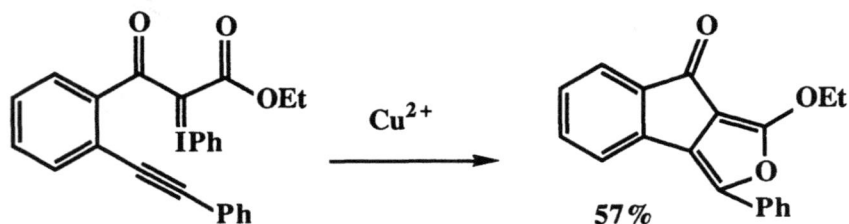

Cu$^{2+}$

57%

**IV.E-15** J.A. Marshall et al., *J. Am. Chem. Soc.*, **114**, 1450 and *J. Org. Chem.*, **57**, 3387; W. Eberbach and N. Laber, *Tetrahedron Lett.*, **33**, 57.

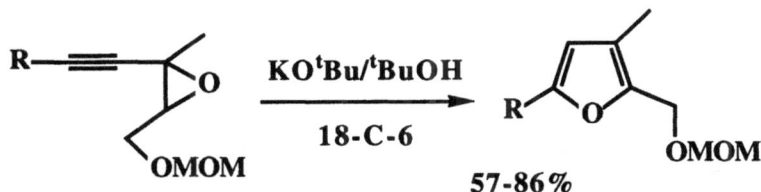

**IV.E-16** R.P. Kapoor et al., Synth. Commun., **22**, 2555; O. Prakash and S. Goyal, *Synthesis*, 629.

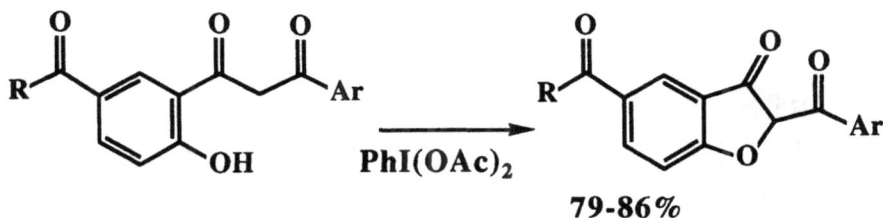

**IV.E-17** Y. Yang and H.N.C. Wong, *Chem. Commun*, 657; **see also:** K. Takai, K. Utimoto et al., *Tetrahedron*, **48**, 3495; J.D. Figueroa-Villar et al., *Heterocycles*, **34**, 891.

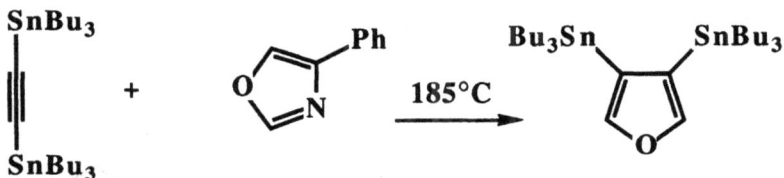

**IV.E-18** W.H. Chan, A.W.M. Lee and E.T.T. Chan, *J. Chem. Soc., Perkin Trans. 1*, 945; U.K. Nadir and B.P. Chaurasia, *Ind. J. Chem.*, **31B**, 189; K. Burger and B. Helmreich, *Chem. Commun.*, 348.

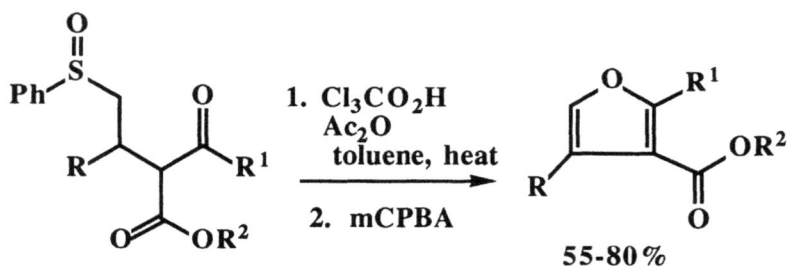

1. $Cl_3CO_2H$
   $Ac_2O$
   toluene, heat

2. mCPBA

55-80%

**via Pummerer rearrangement**

---

**IV.E-19** T.S. Balaban and M. Hiegemann, *Tetrahedron*, **48**, 9827.

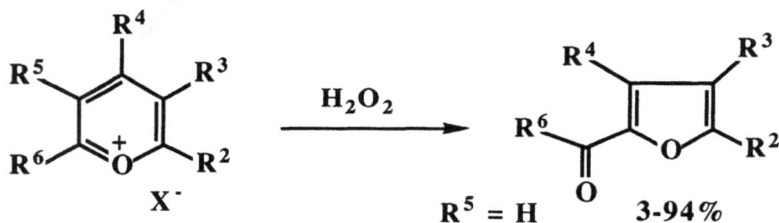

$H_2O_2$

$R^5 = H$    3-94%

---

**IV.E-20** M. Miura et al., *J. Org. Chem.*, **57**, 4754.

$Ar-X$  +

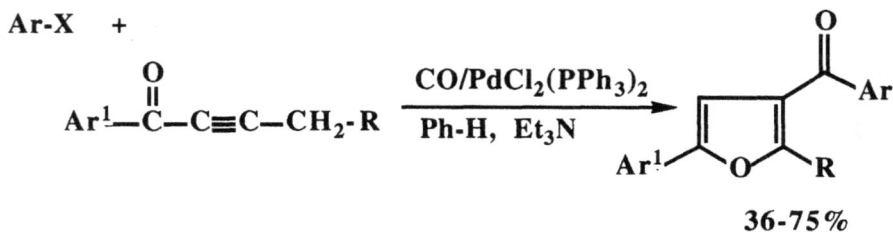

$CO/PdCl_2(PPh_3)_2$
Ph-H, $Et_3N$

36-75%

**IV.E-21** F. Barba and J.L. de la Fuente, *Tetrahedron Lett.*, **33**, 3911; J.-P. Dulcere et al., *Synlett*, 737.

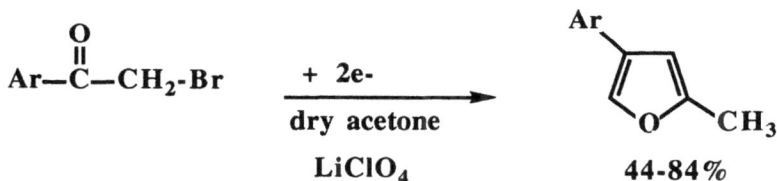

$$Ar-\overset{O}{\overset{||}{C}}-CH_2\text{-Br} \quad \xrightarrow[\substack{\text{dry acetone} \\ LiClO_4}]{+\ 2e-}$$

44-84%

**IV.E-22** J.L. Kice et al., *J. Org. Chem.*, **57**, 5270; Yu. L. Zborovskii, V.I. Staninets and L.V. Saichenko, *J. Org. Chem., USSR*, **28**, 579; T. Ogawa et al., *Chem. Lett.*, 1947.

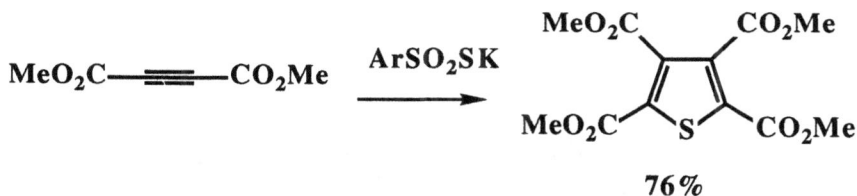

$$MeO_2C\text{———}CO_2Me \quad \xrightarrow{ArSO_2SK}$$

76%

**IV.E-23** K. Gewald and U. Hain, *Monatsh. Chem.*, **123**, 455; J.M. Quintela et al., *J. Heterocyclic Chem.*, **29**, 1693; J. Lissavetzky et al., *Synthesis*, 526; H. Ila, H. Junjappa et al., *Tetrahedron*, **48**, 10377.

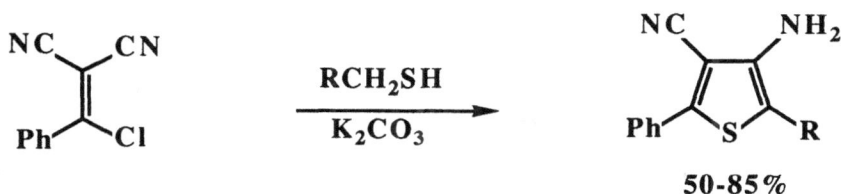

$$\xrightarrow[K_2CO_3]{RCH_2SH}$$

50-85%

**IV.E-24** F. Freeman et al., *J. Org. Chem.*, **57**, 1722; E. Waldvogel, *Helv. Chim. Acta*, **75**, 907.

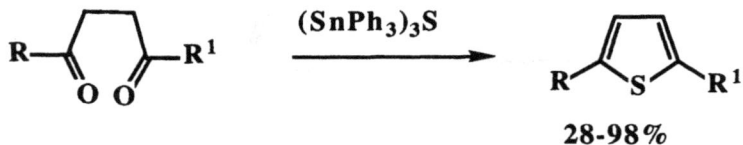

28-98%

**IV.E-25** D.H. Bremner et al., *Synthesis*, 528; A.J. Bridges et al., *Tetrahedron Lett.*, **33**, 7499 and 7495.

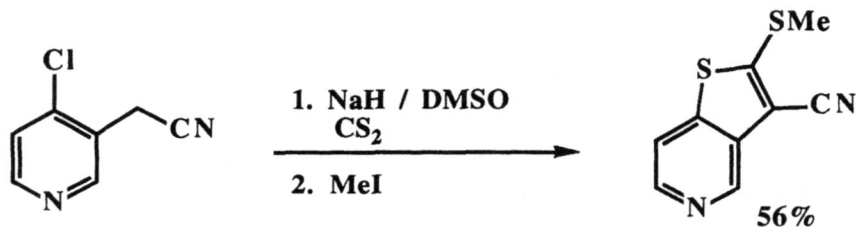

1. NaH / DMSO
   CS$_2$
2. MeI

56%

**IV.E-26** E.V. Dehmlow and R. Westerheide, *Synthesis*, 947.

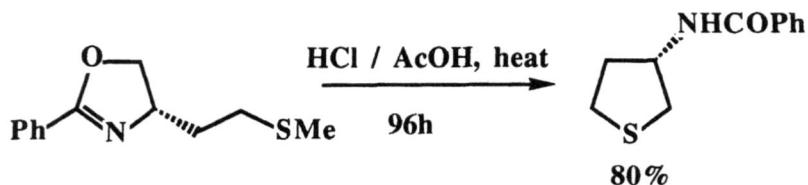

HCl / AcOH, heat
96h

80%

**IV.E-27**  C. Hsiao and T. Kolasa, *Tetrahedron Lett.*, **33**, 2629.

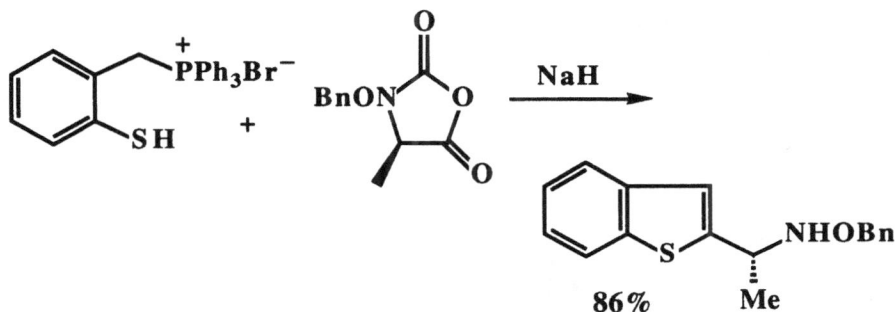

86%

**IV.E-28**  G.W.J. Fleet et al., *Tetrahedron Lett.*, **33**, 4503 and *J. Chem. Soc., Chem. Commun.* 1605; J. Inanaga et al., *Tetrahedron Lett.*, **33**, 8109.

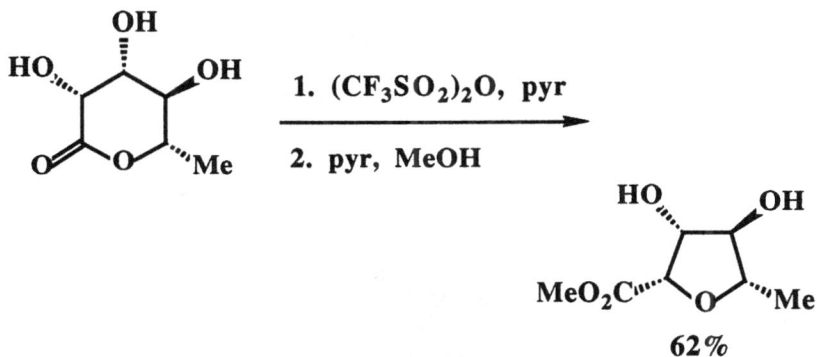

1. $(CF_3SO_2)_2O$, pyr

2. pyr, MeOH

62%

## IV.F.        Pyrroles, Indoles, etc.

**IV.F-1** R.S. Coleman and A.J. Carpenter, *J. Org. Chem.*, **57**, 5813; W.S. Murphy and P.J. O'Sullivan, *Tetrahedron Lett.*, **33**, 531 and 535.

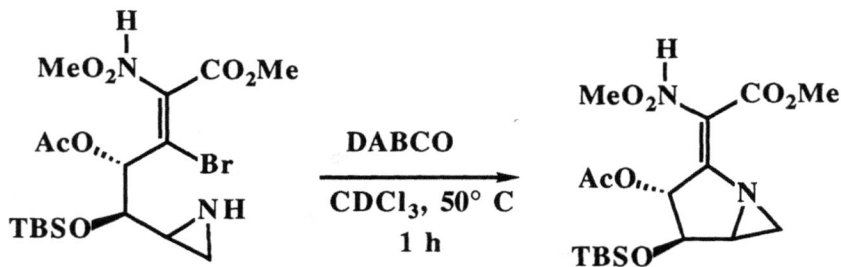

**IV.F-2** J.M. Dickinson and J.A. Murphy, *Tetrahedron*, **48**, 1317.

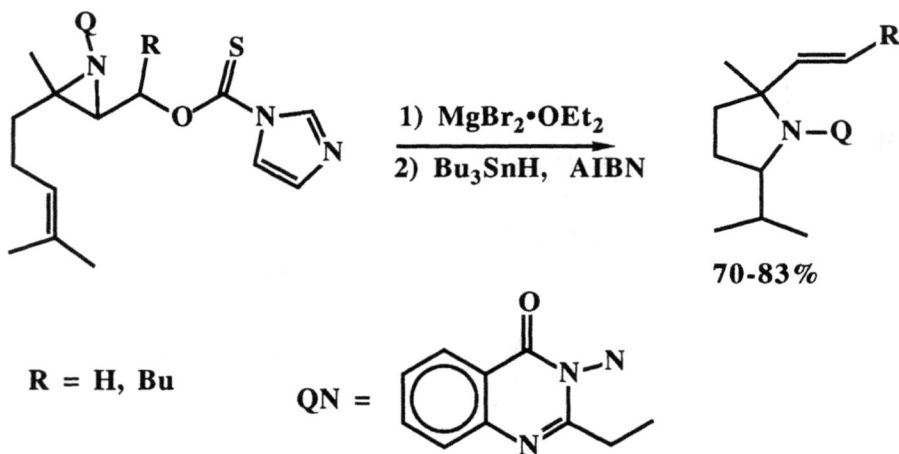

R = H, Bu

**IV.F-3** C.H. Heathcock et al., *J. Org. Chem.*, **57**, 7056; M.E. Hassan, *Gazz. Chim. Ital.*, **122**, 7; S. Takano et al., *Tetrahedron Asym.*, **3**, 681.

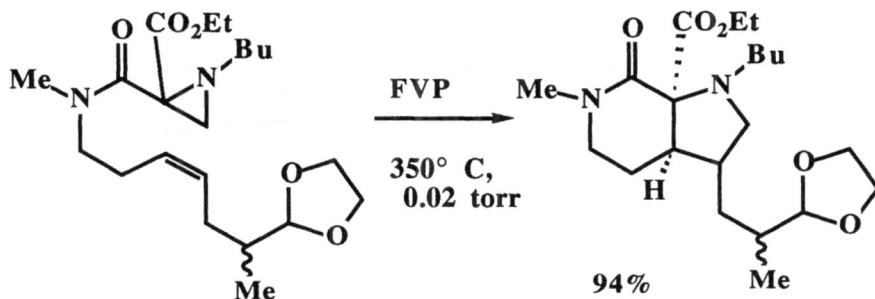

**IV.F-4** E. Laborde, *Tetrahedron Lett.*, **33**, 6607; K. Achiwa et al., *Bull. Pharm. Chem.*, **39**, 3175 (1991); R.M. Williams et al., *Tetrahedron Lett.*, **33**, 6755 and *J. Org. Chem.*, **57**, 6527; R. Grigg et al., *Tetrahedron*, **48**, 10431 and 9735.

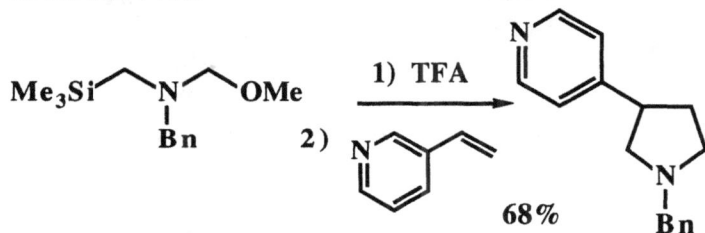

**IV.F-5** W.H. Pearson et al., *J. Org. Chem.*, **57**, 6354 and *J. Am. Chem. Soc.*, **114**, 1329.

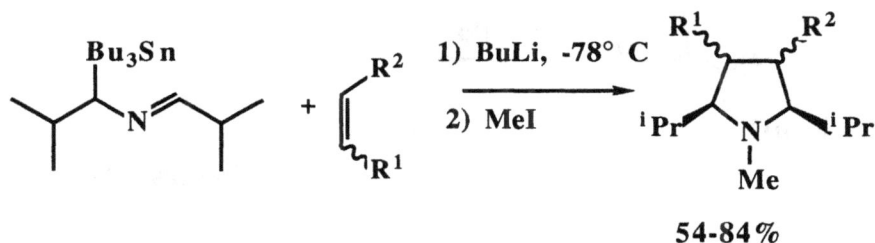

**IV.F-6**  G. Lhommet  et al., *Synthesis*, 884; H.E. Zimmerman and C.W. Wright, *J. Am. Chem. Soc.*, **114**, 363.

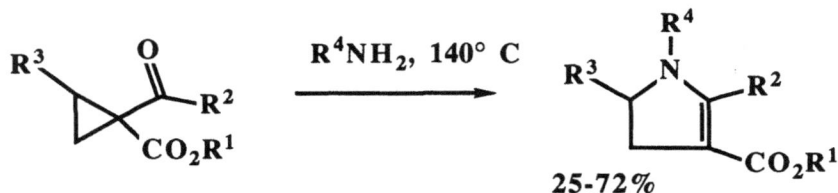

**IV.F-7**  F. Dumas and J. d'Angelo, *Tetrahedron Lett.*, **33**, 2005.

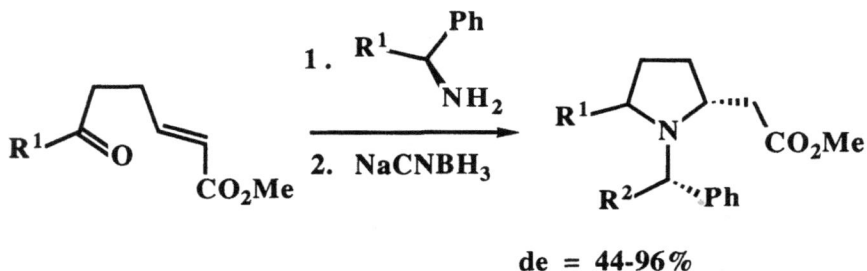

$$de = 44\text{-}96\%$$

**IV.F-8**  G.W. Fleet et al., *Tetrahedron*, **48**, 10177 and 10191; J.H. Van Boom et al., *Synth. Commun.*, **22**, 1762; M. Marzi et al., *Tetrahedron*, **48**, 10127; S.Benetti et al., *J. Org. Chem.*, **57**, 6279; W.H. Pearson et al., *ibid.*, **57**, 3977.

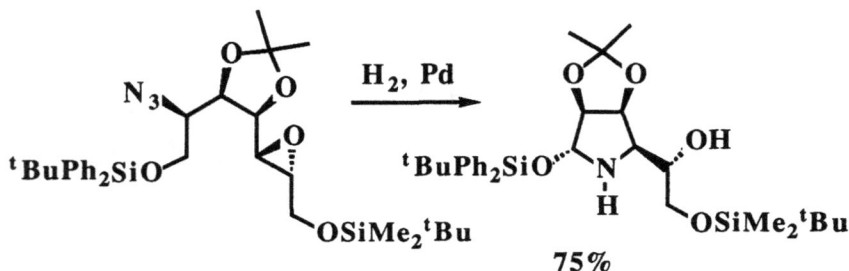

**IV.F-9**  T. Livinghouse et al., *J. Am. Chem. Soc.*, **114**, 5459 and *J. Org. Chem.*, **57**, 1323.

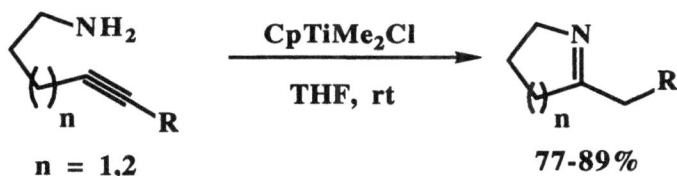

n = 1,2                                                          77-89%

**IV.F-10**  A. Toshimitsu and H. Fuji, *Chem. Lett.*, 2016; I.W. Davies et al., *J. Chem. Soc., Chem. Commun.*, 335; H. Takahata, H. Bandoh and T. Momose, *J. Org. Chem.*, **57**, 4401; T. Momose et al., *Tetrahedron Lett.*, **33**, 7893; F.-T. Luo and R.-T. Wang, *ibid.*, **33**, 6835; M. Tokuda and H. Suginome et al., *ibid.*, **33**, 6359; T.J. Marks et al., *J. Am. Chem. Soc.*, **114**, 275 and *Organomet.*, **11**, 2003.

87%

similar approaches to pyrrolidines via cyclizations upon alkenes or allenes using AgOTf, $Hg(OAc)_2$, $PdCl_2(MeCN)_2$, and lanthanide catalysts

**IV.F-11**  H. Aoyama et al., *J. Org. Chem.*, **57**, 3037.

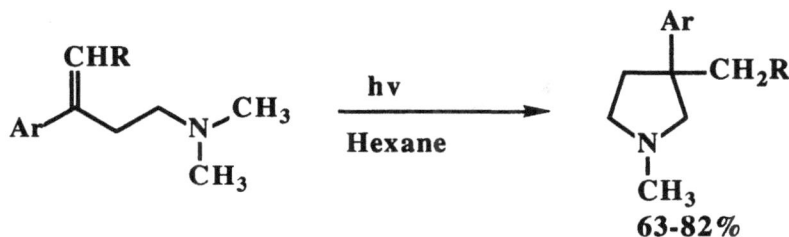

63-82%

**IV.F-12** M. Mori and S. Watanuki, *J. Chem. Soc., Chem. Commun.*, 1082; K.H. Dotz et al., *Synthesis*, 147.

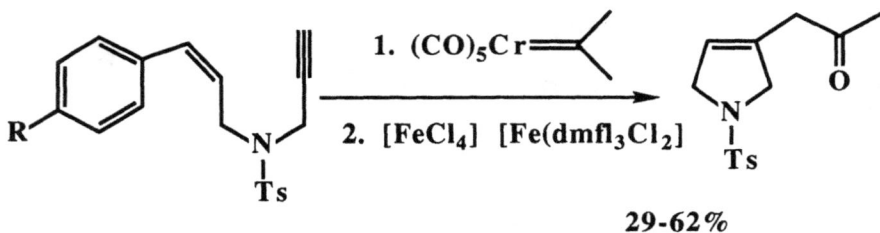

1. $(CO)_5Cr$=

2. $[FeCl_4]$  $[Fe(dmfl_3Cl_2]$

29-62%

---

**IV.F-13** R.M. Adlington and S.J. Mantell, *Tetrahedron*, **48**, 6529; D.L. Boger et al., *J. Org. Chem.*, **57**, 2873; H. Hiemstra et al., *Terahedron*, **48**, 4659.

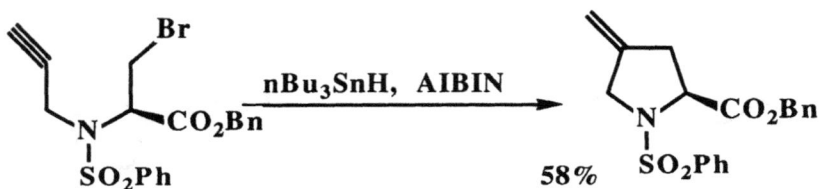

$nBu_3SnH$, AIBIN

58%

---

**IV.F-14** D.L. Flynn et al., *Tetrahedron Lett.*, **33**, 7281 and 7283.

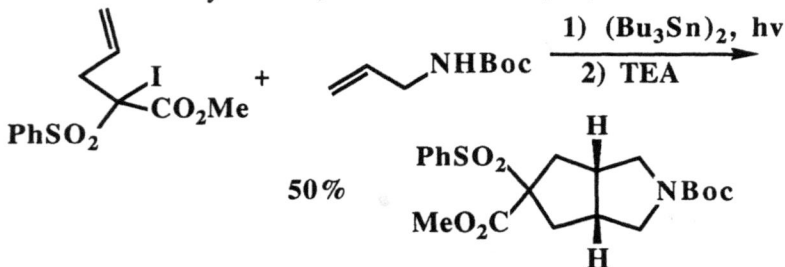

1) $(Bu_3Sn)_2$, hv

2) TEA

50%

**IV.F-15**  J.E. Baldwin et al., *Tetrahedron Lett.*, **33**, 1517.

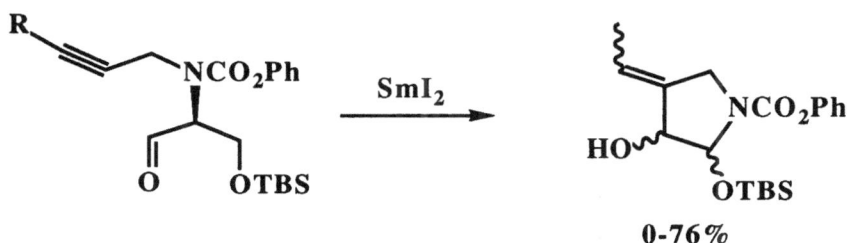

**0-76%**

**IV.F-16**  T. Momose et al., *J. Chem. Soc., Perkin Trans 1*, 509; Y. Endo and K. Shudo, *Heterocycles*, **33**, 91.

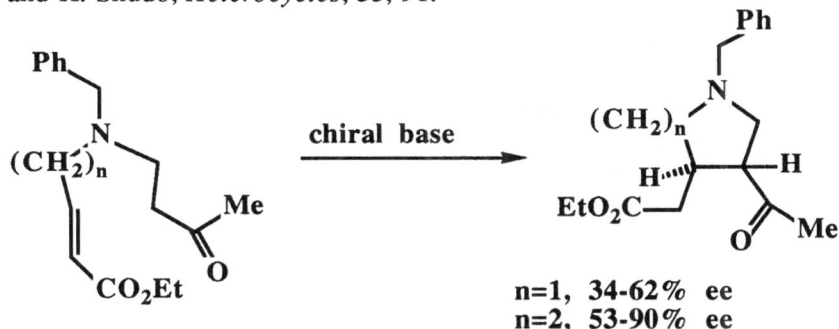

n=1, 34-62% ee
n=2, 53-90% ee

**IV.F-17**  M. Mori, N. Vesaka and M. Shibasaki, *J. Org. Chem.*, **57**, 3519; G.C. Fu and R.H. Grubbs; *J. Am. Chem. Soc.*, **114**, 7324; J.H. Tidwell and S.L. Buchwald, *J. Org. Chem.*, **57**, 6381.

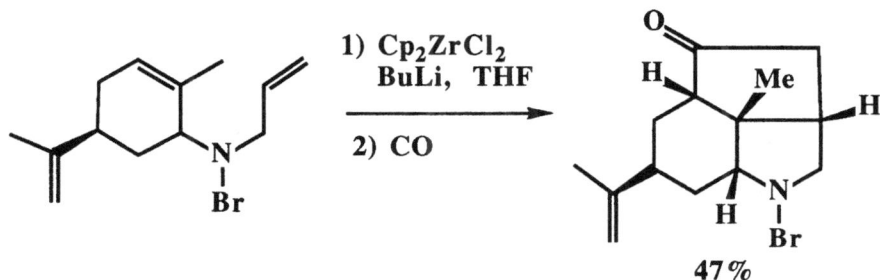

1) $Cp_2ZrCl_2$ BuLi, THF
2) CO

**47%**

**Similar cyclization using Mo catalyst**

**IV.F-18**  R. Grigg et al., *Tetrahedron*, **48**, 10399 and *J. Chem. Soc., Chem. Commun.*, 1537.

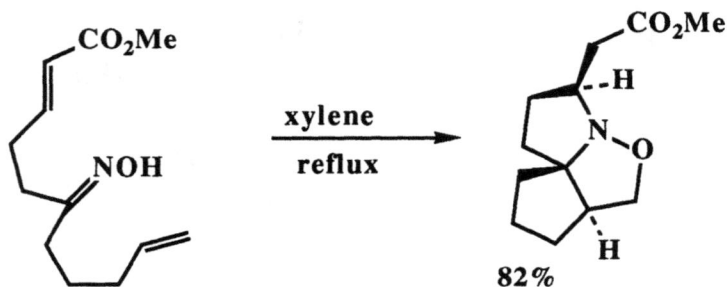

82%

---

**IV.F-19**  E.M. Campi et al., *Aust. J. Chem.*, **45**, 1167.

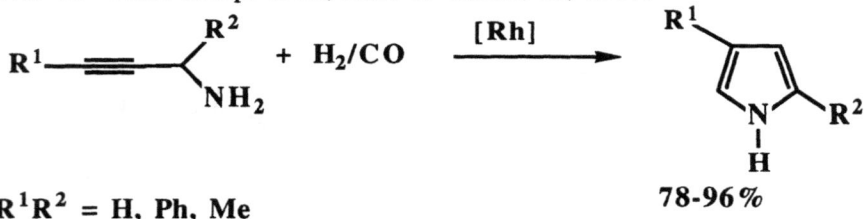

$R^1R^2$ = H, Ph, Me

78-96%

---

**IV.F-20**  R.J.P. Corriu et al., *Tetrahedron*, **48**, 6231.

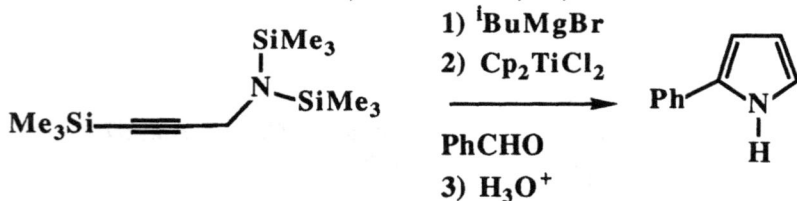

1) $^i$BuMgBr
2) Cp$_2$TiCl$_2$

PhCHO
3) H$_3$O$^+$

---

**IV.F-21**  A.M. van Leusen et al., *J. Org. Chem.*, **57**, 2245.

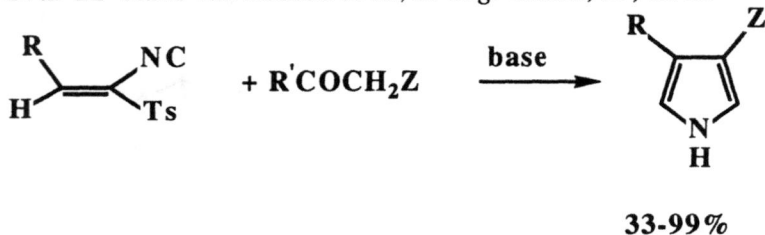

+ R'COCH$_2$Z

base

33-99%

**IV.F-22**  J.T. Gupton, J.A. Sikorski et al., *J. Org. Chem.*, **57**, 5480.

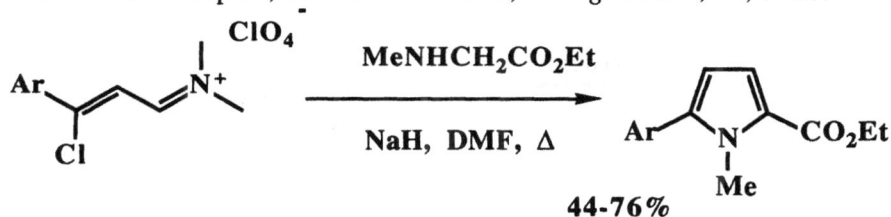

$$MeNHCH_2CO_2Et$$
$$\xrightarrow{\hspace{1cm}}$$
$$NaH, \ DMF, \ \Delta$$

44-76%

---

**IV.F-23**  M. Hojo et al., *Synthesis*, 533; R.W. Soeder et al., *Synth. Commun.*, **22**, 2737; M.T. Konieczny and M. Cushman, *Tetrahedron Lett.*, **33**, 6939; K.M. Biswas et al., *J. Chem. Soc., Perkin Trans 1*, 461.

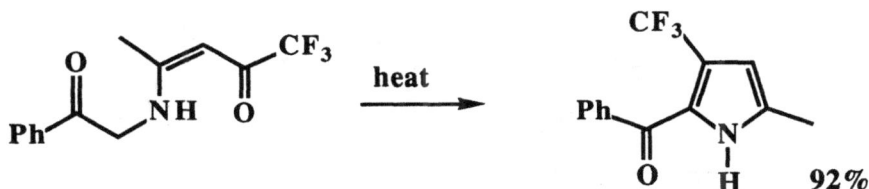

heat

92%

---

**IV.F-24**  J.M. Muchowski et al., *J. Org. Chem.*, **57**, 1653.

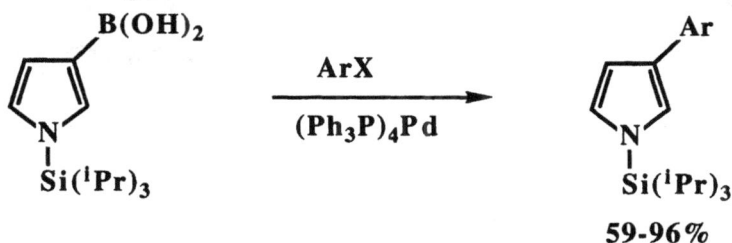

ArX
$$\xrightarrow{\hspace{1cm}}$$
$$(Ph_3P)_4Pd$$

59-96%

**IV.F-25**  C.W. Bird and L. Jiang, *Tetrahedron Lett.*, **33**, 7253.

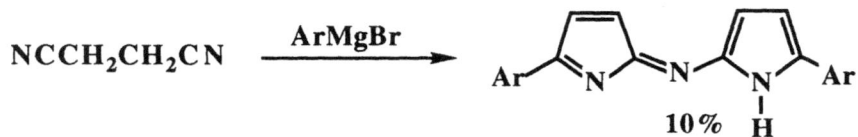

**IV.F-26**  F. Lucchesini,  *Tetrahedron*, **48**, 9951.

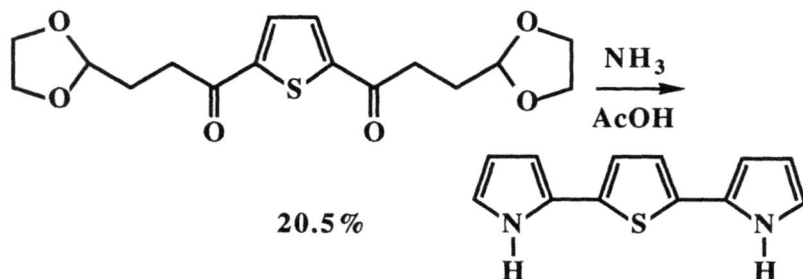

**IV.F-27**  A. Furstner and D.N. Jumbam, *Tetrahedron*, **48**, 5991.

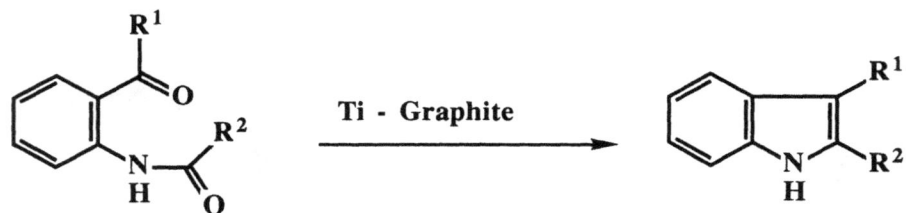

**IV.F-28**  S. Cacchi et al., *Tetrahedron Lett.*, **33**, 3915.

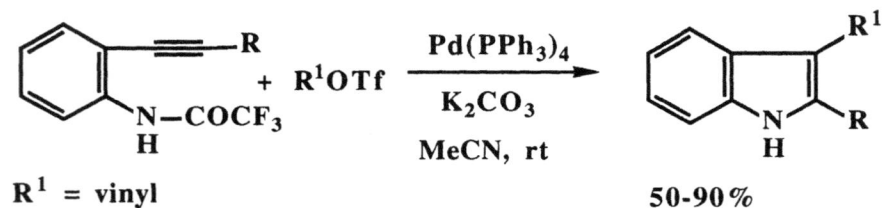

$R^1$ = vinyl

50-90%

**IV.F-29** T.M. Sielecki and A.I. Meyers, *J. Org. Chem.*, **57**, 3673.

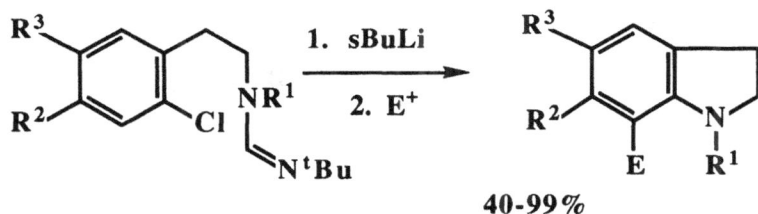

**40-99%**

---

**IV.F-30** R. Grigg et al., *Tetrahedron Lett.*, **33**, 7965; J.E. Macor et al., *ibid.*, **33**, 8011.

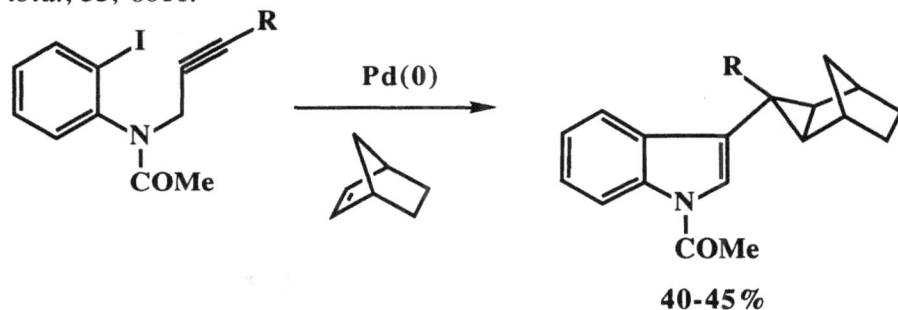

**40-45%**

---

**IV.F-31** T. Izumi et al., *J.Heterocyclic Chem.*, **29**, 899 and 1085; H.D.H. Showalter and G. Pohlmann, *Org. Prep. Proced. Int.*, **24**, 484; Y. Watanabe et al., *Chem. Lett.*, 769.

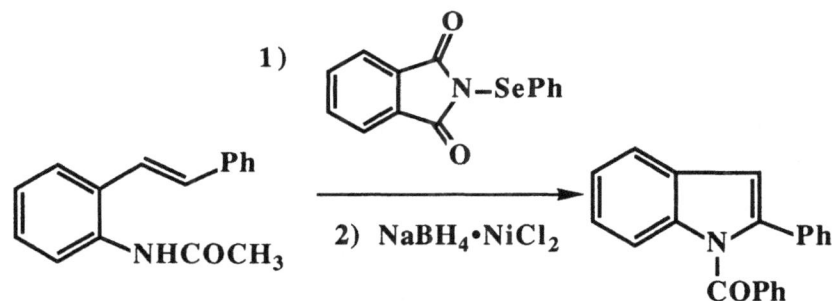

**Similarly by reduction of ortho nitro sytrenes.**

**IV.F-32**  B.J. Wakefield et al., *Tetrahedron*, **48**, 939.

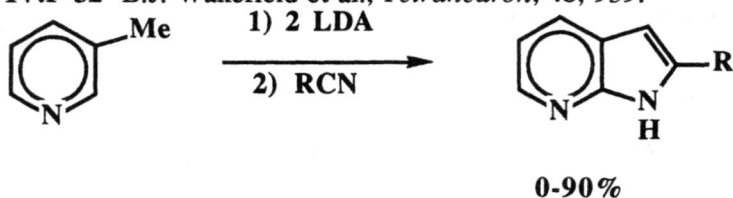

0-90%

---

**IV.F-33**  P. Molina et al., *J. Org. Chem.* , **57**, 929.

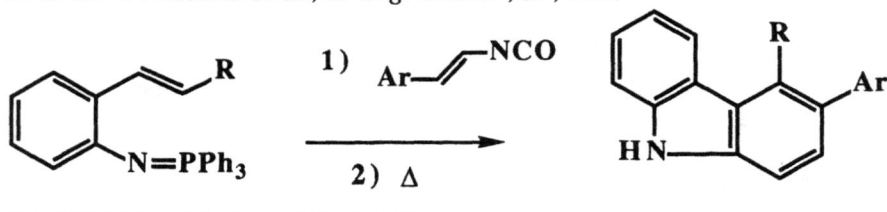

---

**IV.F-34**  D. St. C. Black et al., *Aust. J. Chem.*, **45**, 611 and 1051 and 1879.

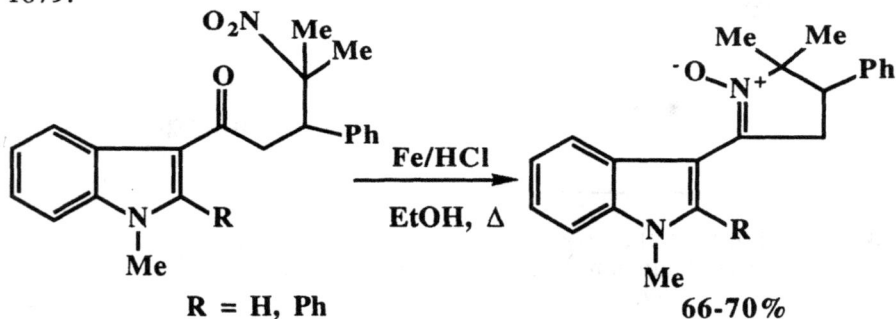

R = H, Ph                                        66-70%

**similarly for 7-substituted indoles**

---

**IV.F-35**  P. Wipf and Y. Kim, *Tetrahedron Lett.*, **33**, 5477; Y. Kita et al., *Heterocycles*, **33**, 503; J.W. Sowell et al., *J. Heterocyclic Chem.*, **29**, 51.

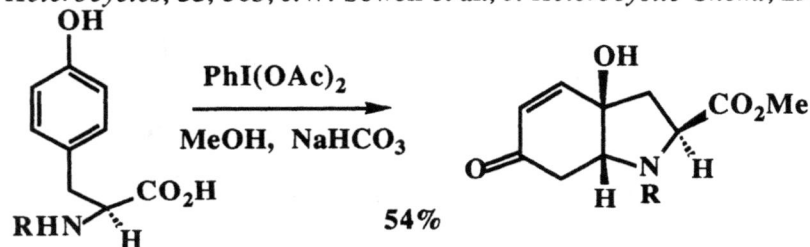

54%

## IV.G.        Pyridines, Quinolines, etc.

**IV.G-1** A.B. Holmes et al., *Tetrahedron Lett.*, **33**, 7421 and 7425.

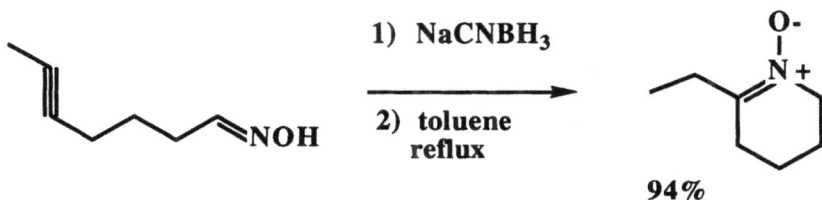

1) NaCNBH$_3$

2) toluene reflux

94%

---

**IV.G-2** G. Rassu, G. Casiraghi et al., *Tetrahedron*, **48**, 727; C.H. Heathcock et al., *J. Org. Chem.*, **57**, 2531,2544,2554,2566,2575 and 2585; M. Mori et al., *Heterocycles*, **33**, 819; H. Walter et al., *Helv. Chim. Acta*, **75**, 1274.

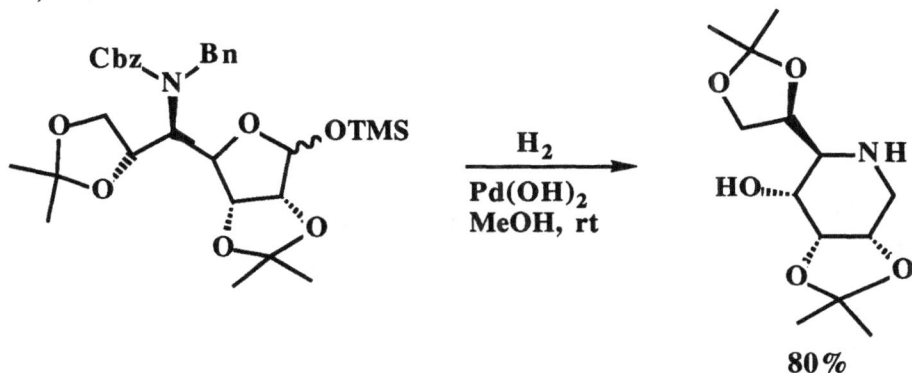

H$_2$

Pd(OH)$_2$
MeOH, rt

80%

---

**IV.G-3** A.C. Oehlschlager et al., *J. Org. Chem.*, **57**, 7226.

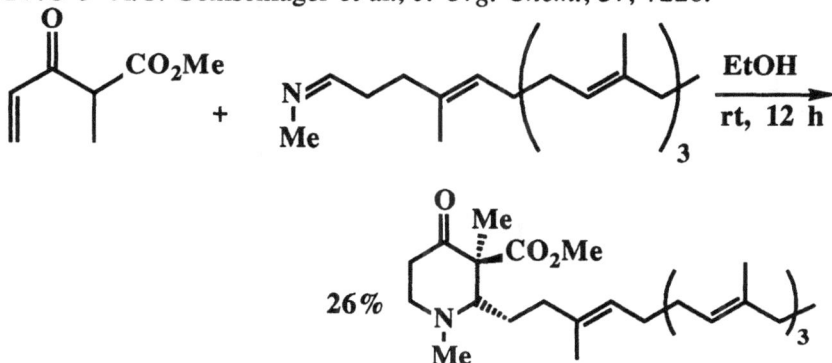

EtOH

rt, 12 h

26%

**IV.G-4**  K.-Y. Ko et al., *Tetrahedron Lett.*, **33**, 6651.

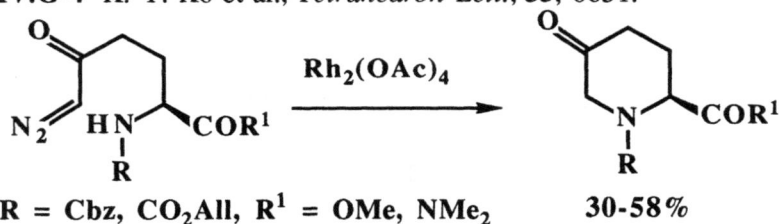

R = Cbz, CO₂All, R¹ = OMe, NMe₂        30-58%

---

**IV.G-5**  H. Takahata, T. Momose et al., *Tetrahedron Asym.*, **3**, 607 and *Heterocycles*, **34**, 435.

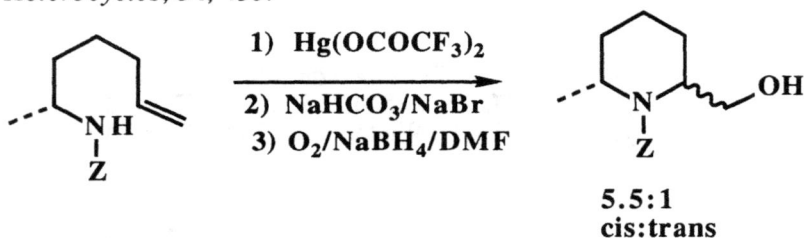

1) Hg(OCOCF₃)₂
2) NaHCO₃/NaBr
3) O₂/NaBH₄/DMF

5.5:1
cis:trans

---

**IV.G-6**  L.E. Overman and A.K. Sarkar, *Tetrahedron Lett.*, **33**, 4103.

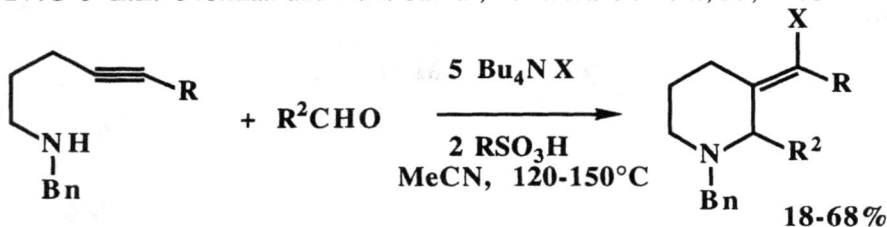

+ R²CHO

5 Bu₄N X

2 RSO₃H
MeCN,  120-150°C

18-68%

---

**IV.G-7**  L.E. Overman et al., *J. Org. Chem.*, **57**, 1179.

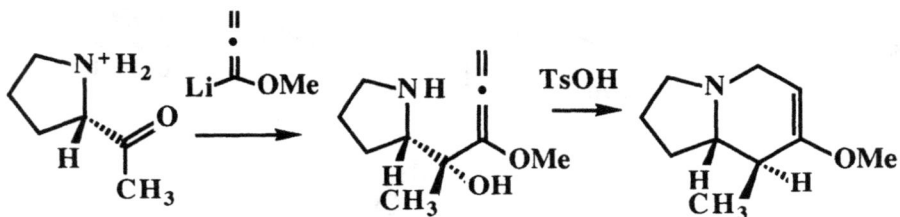

Li—C(=O)—OMe

TsOH

**IV.G-8**  S.M. Weinreb et al., *J. Org. Chem.*, **57**, 2528.

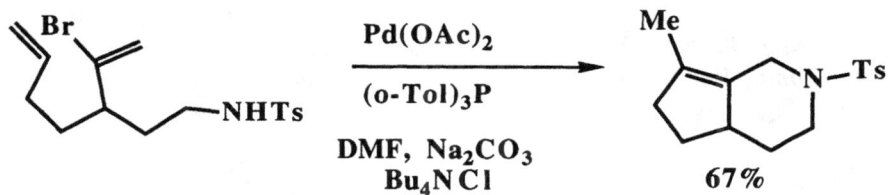

**IV.G-9**  S.R. Angle et al., *J. Org. Chem.*, **57**, 5947.

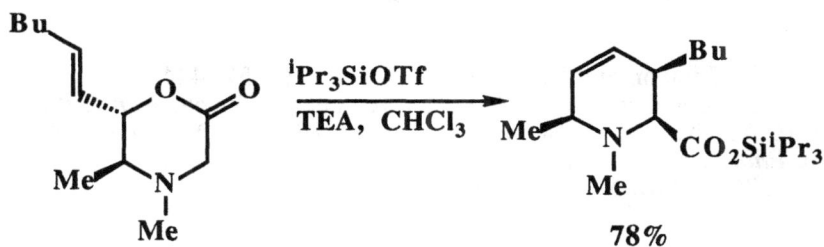

**IV.G-10**  R.P. Polniaszek and L.W. Dillard, *J. Org. Chem.*, **57**, 4103.

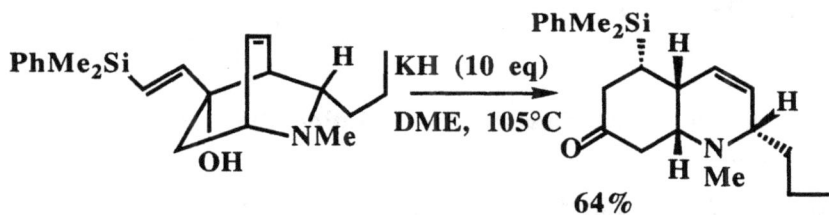

**IV.G-11**  D.L. Boger and M. Zhang, *J. Org. Chem*, **57**, 3974; A.B. Koldobskii, V.V. Lunin, S.A. Voznesevskii, *J. Org. Chem., USSR*, **28**, 620; J. Barluenga et al., *Synthesis*, 107; T.L. Gilchrist and M.A.M. Healy, *J. Chem. Soc., Perkin Trans. 1*, 749; M. Makosza and A. Tyrala, *Acta. Chem. Scand.*, **46**, 689; M. Chakrabarty et al., *Tetrahedron Lett.*, **33**, 117.

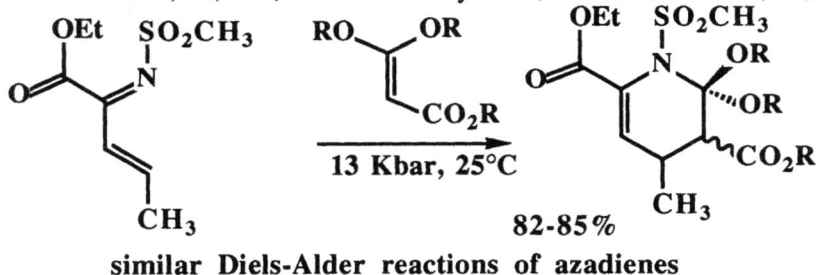

similar Diels-Alder reactions of azadienes

---

**IV.G-12**  H. Waldmann and M. Braun, *J. Org. Chem.*, **57**, 4444; A.B. Holmes et al., *Chem. Commun.*, 786; P. Herczegh et al., *Tetrahedron Lett.*, **33**, 3133; H. Abraham and L. Stella, *Tetrahedron*, **48**, 9707; N. Katagiri et al., *Chem. Pharm. Bull.*, **40**, 1737; K. Hattori and H. Yamamoto, *J. Org. Chem.*, **57**, 3264; M.M. Midland and R.N. Koops, *ibid.*, **57**, 1158.

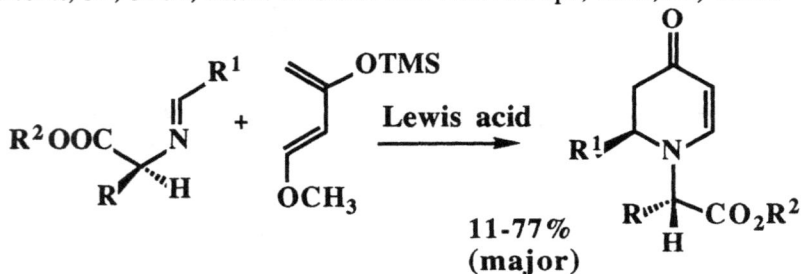

similar Diels-Alder reactions involving imino dienophiles

---

**IV.G-13**  D.L. Comins and R.S. Al-awar, *J. Org. Chem.*, **57**, 4098; M.E. Jung, Z. Longmei et al., *ibid.*, **57**, 3528.

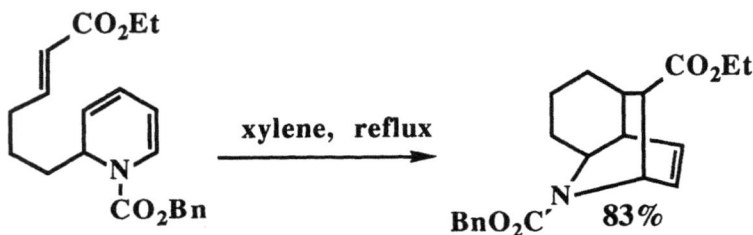

**IV.G-14** H.C. van der Plas et al., *J. Org. Chem.*, **57**, 3000 and *Tetrahedron*, **48**, 1643.

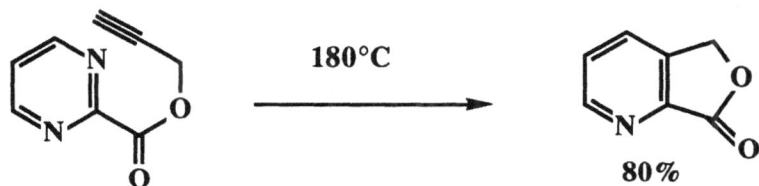

**180°C**

**80%**

---

**IV.G-15** M. Falorni et al., *Synthesis*, 972.

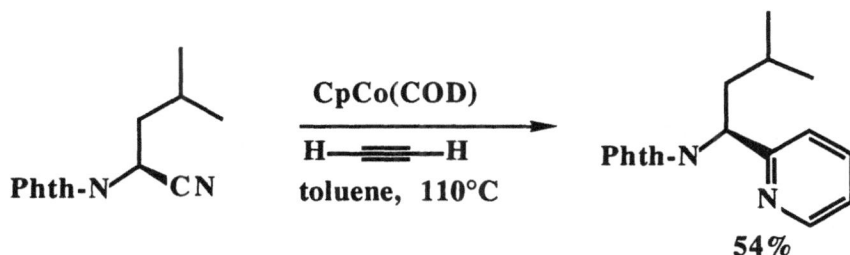

**CpCo(COD)**

**H——≡——H**

**toluene, 110°C**

Phth-N    CN

Phth-N

**54%**

---

**IV.G-16** P. Molina et al., *Synlett*, 873 and *Synthesis*, 827; M.L. Gelmi et al., *J. Chem. Soc., Perkin Trans 1*, 701; R.K. Smalley et al., *Synlett.*, 231.

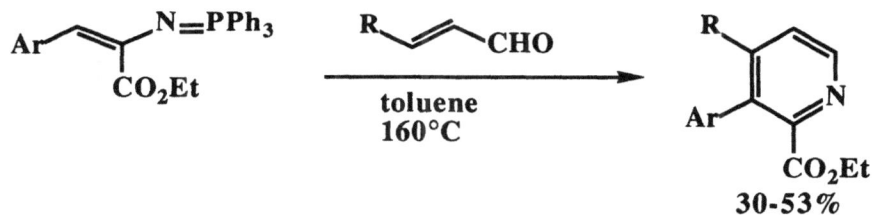

Ar    N=PPh₃

CO₂Et

R    CHO

**toluene**
**160°C**

**30-53%**

**IV.G-17** J.-J. Vanden Eynde et al., *Tetrahedron*, **48**, 1263 and *Synth. Commun.*, **22**, 3291; I.C. Ivanov et al., *Liebigs Ann. Chem.*, 203; M. Hajo, *Heterocycles*, **34**, 1927; K. Ito and S. Miyajima, *J. Heterocyclic Chem.*, **29**, 1037; A. Maccioni et al., *ibid.*, **29**, 1631; M.H. Elnagdi et al., *Gazz. Chim. Ital.*, **122**, 299; J.M. Robinson et al., *J. Org. Chem.*, **57**, 7352; C.E. Sunkel, A.G. Garcia et al., *J. Med.Chem.*, **35**, 2407; G.E.H. Elgemeie et al., *J. Chem. Soc., Perkin Trans 1*, 1073; N.A. Ismail et al., *Org. Prep. Proc. Int.*, **24**, 33.

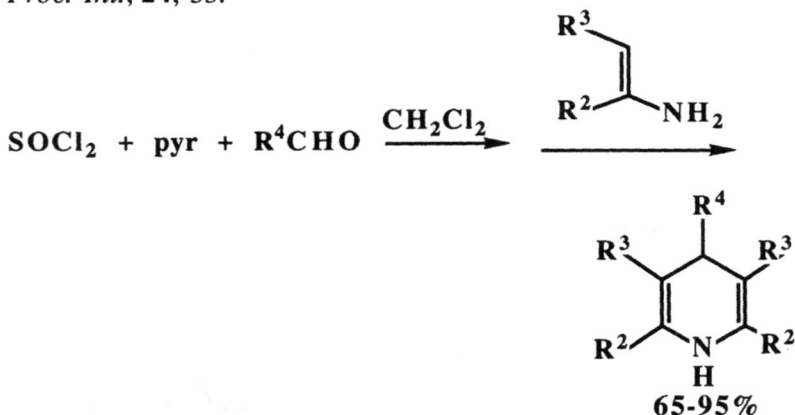

$$SOCl_2 \ + \ pyr \ + \ R^4CHO \xrightarrow{CH_2Cl_2}$$

**65-95%**

---

**IV.G-18** J.-G. Jun and H.S. Shin, *Tetrahedron Lett.*, **33**, 4593; T.S. Balaban et al., *Liebigs Ann. Chem.*, 173.

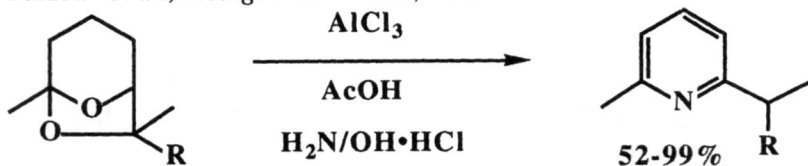

**52-99%**

---

**IV.G-19** R. Nesi, D. Giomi et al., *J. Org. Chem.*, **57**, 3713; R. Nesi et al., *Synth. Commun.*, **22**, 2349; A. Brandi et al., *Tetrahedron Lett.*, **33**, 6697.

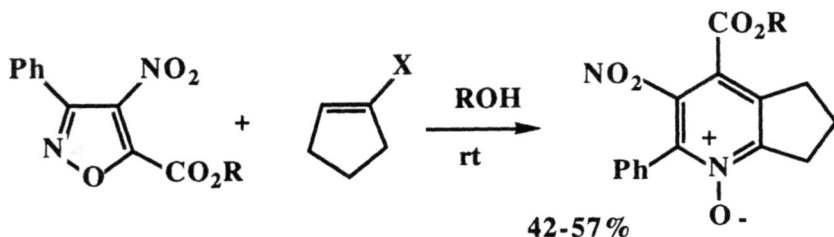

**42-57%**

**IV.G-20**  H. Inoue et al., *Can. J. Chem.*, **70**, 1.

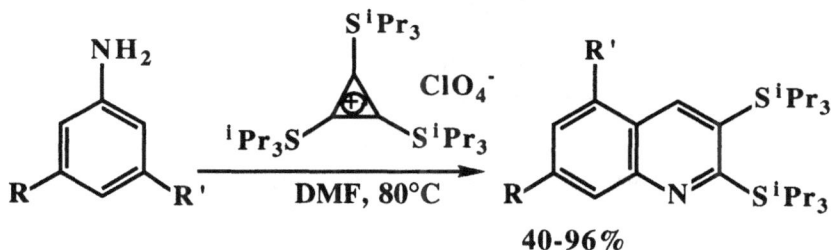

40-96%

**IV.G-21**  V.N. Kalinin et al., *Tetrahedron Lett.*, **33**, 373; S. Torii et al., *Synlett*, 513; D. Nanni et al., *J. Org. Chem.*, **57**, 1842; J.C. Jochims et al., *Synthesis*, 875.

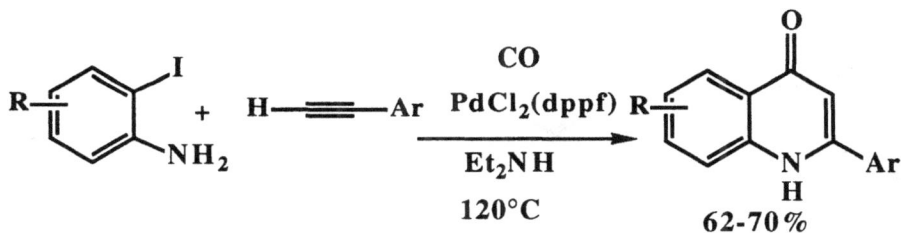

62-70%

**similar approaches utilizing other alkynes**

**IV.G-22** D.P. Curran and H. Liu, *J. Am. Chem. Soc.*, **114**, 5863.

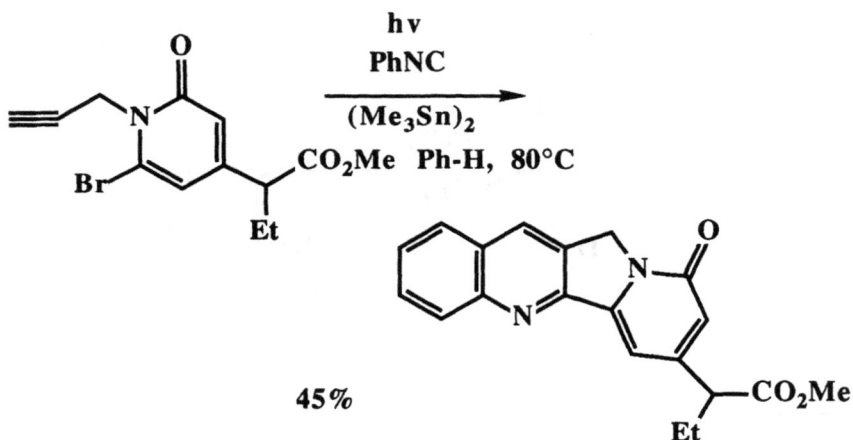

45%

**4+1 radical annulation and cyclization**

---

**IV.G-23** M. Rubiralta et al., *Synth. Commun.*, **22**, 359.

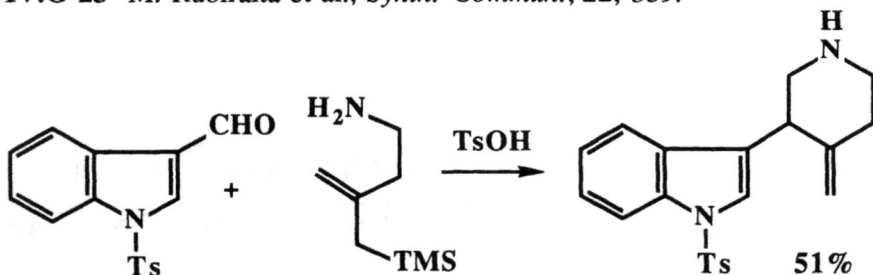

51%

---

**IV.G-24** J.M. Cook et al., *Heterocycles*, **34**, 517 and *Tetrahedron Lett.*, **33**, 4721.

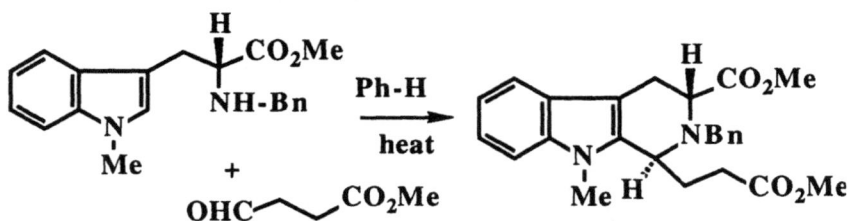

**Pictet-Spengler**

**IV.G-25** S. Hibino et al., *J. Org. Chem.*, **57**, 5917.

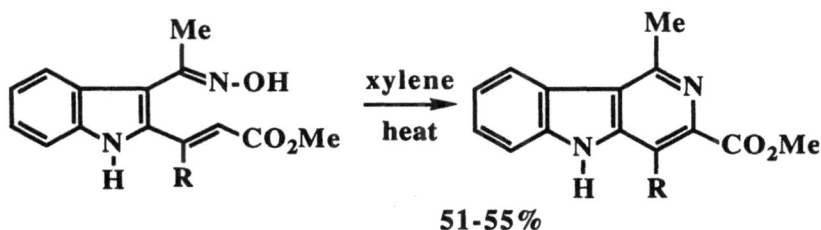

51-55%

---

**IV.G-26** A.M. van Leusen et al., *J. Chem. Soc., Chem. Commun.*, 1401.

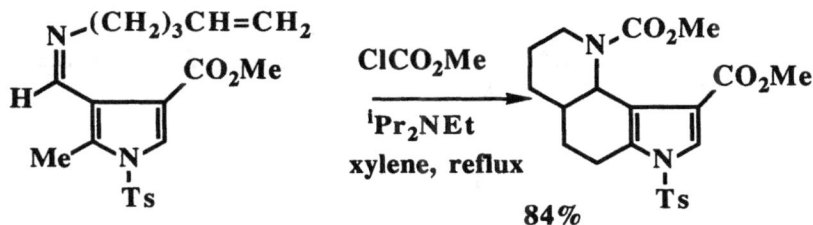

84%

**via pyrrolo quiniodimethane intermediate**

---

**IV.G-27** D. Craig et al., *Tetrahedron*, **48**, 7803.

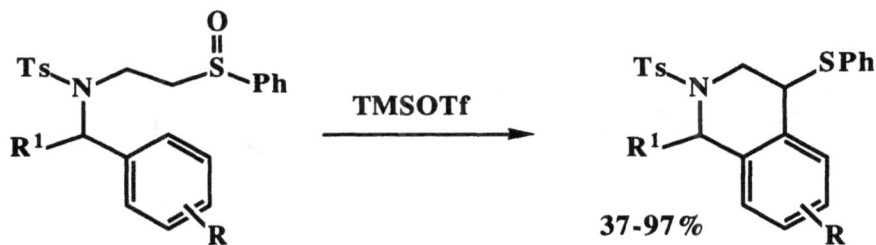

37-97%

**via Pummerer rearrangement**

## IV.H.    Pyrans, Pyrones and Sulfur Analogues

**IV.H-1**  Y.-M. Tsai et al., *J. Org. Chem.*, **57**, 7010.

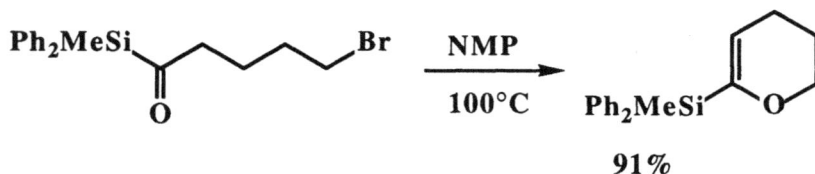

91%

**IV.H-2**  G.L. Edwards and K.A. Walker, *Tetrahedron Lett*, **33**, 1779; D.J. Hart et al., *J. Org. Chem.*, **57**, 5670.

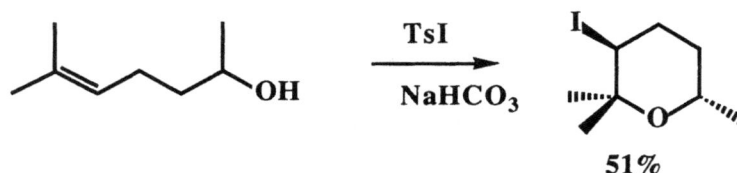

51%

**IV.H-3**  A. Krief et al., *Synlett*, 320.

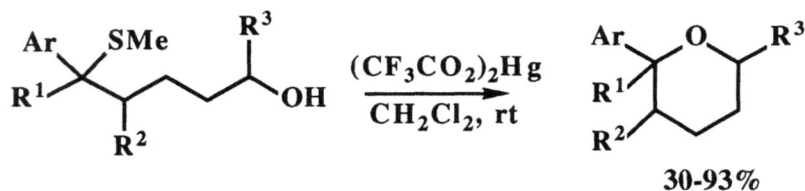

30-93%

**IV.H-4**  T. Fujisawa et al., *Bull. Chem. Soc., Jpn.*, **65**, 3487.

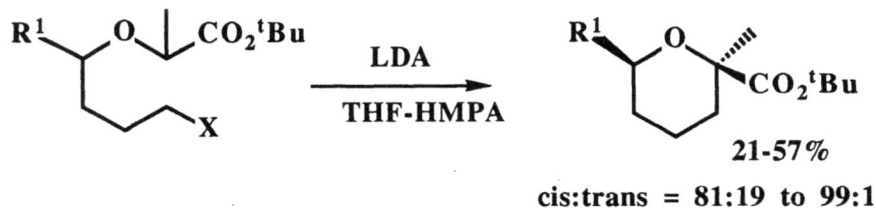

21-57%

cis:trans = 81:19 to 99:1

**IV.H-5** J. Yoshida, Y. Ishichi and S. Isoe, *J. Am Chem. Soc.*, **114**, 7594.

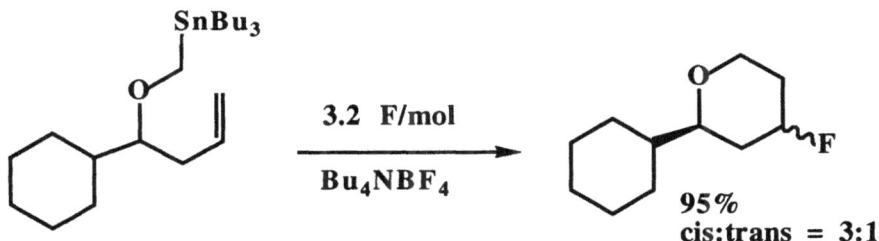

$$\text{3.2 F/mol}$$
$$\text{Bu}_4\text{NBF}_4$$

95%
cis:trans = 3:1

**IV.H-6** M.L. Gelmi and D. Pocar, *Synthesis*, 453.

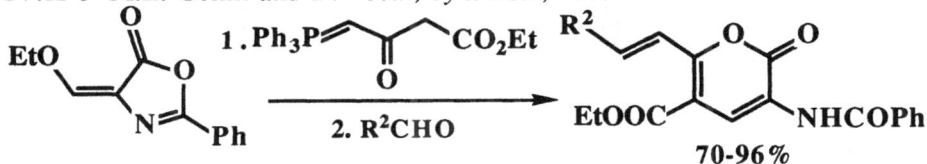

1. $\text{Ph}_3\text{P}=\!\!\!\overset{}{\underset{}{}}\!\!\!\text{CO}_2\text{Et}$

2. $\text{R}^2\text{CHO}$

70-96%

**IV.H-7** L. Gao and A. Murai, *Tetrahedron Lett.*, **33**, 4349; G.P. Moss and C.K. Oai, *J. Chem. Soc., Chem. Commun.*, 342; C. Iwata et al., *ibid.*, 516.

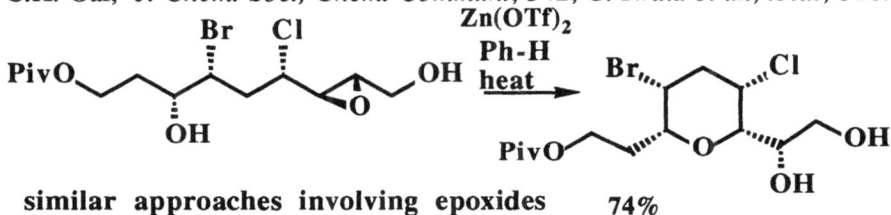

$\text{Zn(OTf)}_2$
Ph-H
heat

similar approaches involving epoxides          74%

**IV.H-8** H. Yamamoto et al., *J. Org. Chem.*, **57**, 1951; I.E. Marko and A. Mekhactia, *Tetrahedron Lett.*, **33**, 1799; M. Cinquini, F. Cozzi et al., *J. Org. Chem.*, **57**, 3605.

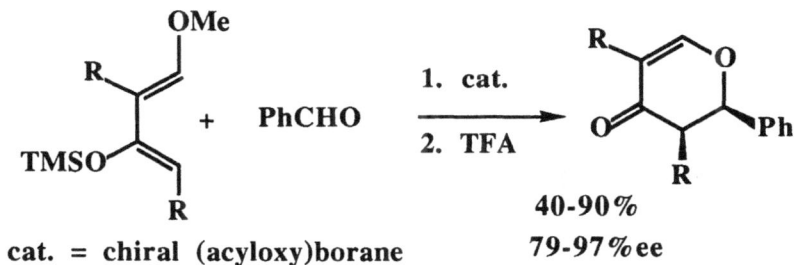

+ PhCHO

1. cat.

2. TFA

40-90%
79-97% ee

cat. = chiral (acyloxy)borane

**IV.H-9** D. Dvorak et al., *Coll. Czech. Chem. Commun.*, **57**, 2337; K. Hiroi et al., *Terahedron Lett.*, **33**, 7161; C.P. Dell, *ibid.*, **33**, 699; V. Bhat et al., *Synth. Commun.*, **22**, 97; L.F. Tietze and C. Schneider, *Synlett.*, 755.

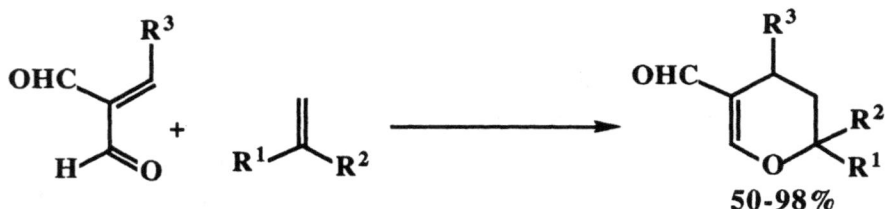

50-98%

---

**IV.H-10** C.K. Lau et al., *Can. J. Chem.*, **70**, 1717; A.R. Katritzky and X. Lan, *Synthesis*, 761; M. Weissenfels et al., *J. Prakt. Chem.*, **334**, 147; S. Inoue et al., *Chem Lett.*, 1237.

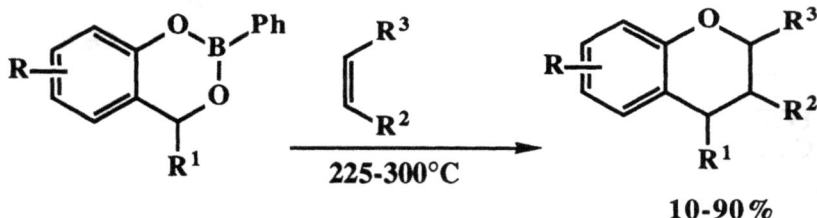

225-300°C

10-90%

similar approaches involving o-quinone methides

---

**IV.H-11** F.G. West et al., *J. Org. Chem.*, **57**, 3479; **see also:** M.A. McKervey and T. Ye, *J. Chem. Soc., Chem. Commun.*, 823; A. Padwa et al., *Tetrahedron Lett.*, **33**, 6427 and *J. Org. Chem.*, **57**, 5747.

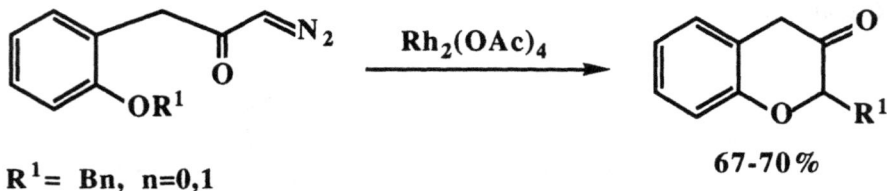

$Rh_2(OAc)_4$

67-70%

$R^1$= Bn, n=0,1

Stevens [1,2]-shift of cyclic oxonium ylides

**IV.H-12**  G.C. Fu and R.H. Grubbs, *J. Am. Chem. Soc.*, **114**, 5426.

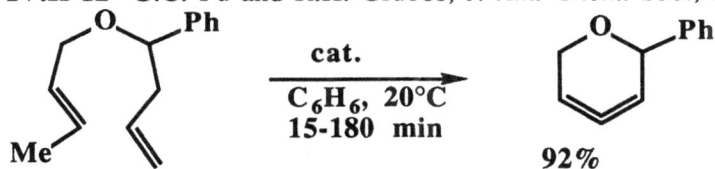

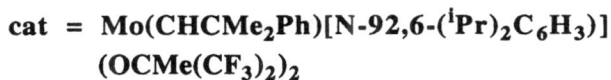

$$cat = Mo(CHCMe_2Ph)[N-92,6-(^iPr)_2C_6H_3]$$
$$(OCMe(CF_3)_2)_2$$

**similar oxygen heterocycles by ring closing metathesis**

---

**IV.H-13**  T. Balasubramanian and K.K. Balasubramanian, *J. Chem. Soc., Chem. Commun.*, 1760.

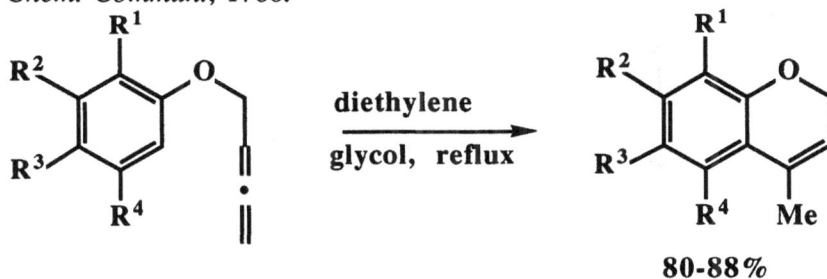

---

**IV.H-14**  D. Crich et al., *J. Am. Chem. Soc.*, **114**, 8313.

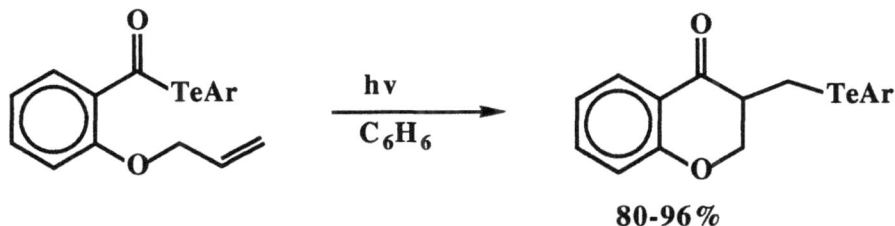

**IV.H-15**  A. Nishinaga et al., *Synthesis*, 839; M. Hesse et al., *Helv. Chim. Acta*, **75**, 457; D.M. Rao and A.V.S. Rao, *Ind. J. Chem.*, **31b**, 335; O. Prakash et al., *Synth. Commun.*, **22**, 327; N.B. Mulchandani et al., *Ind. J. Chem.*, **31b**, 338 and 341; J.D. Brion et al., *Synthesis*, 375.

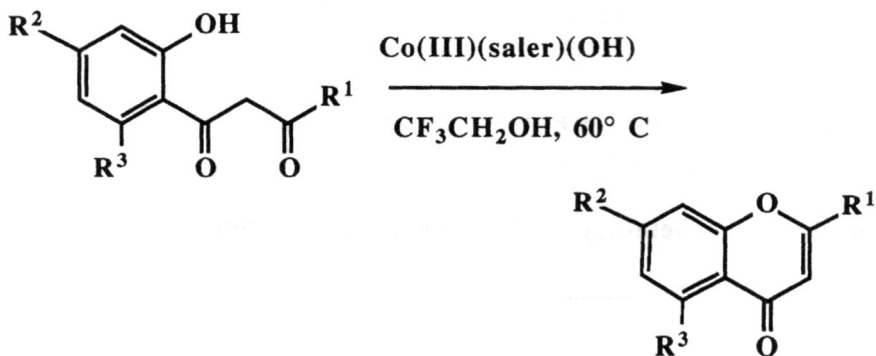

**IV.H-16**  S.A. Ahmad-Junan and D.A. Whiting, *J. Chem. Soc., Perkin Trans 1*, 675 .

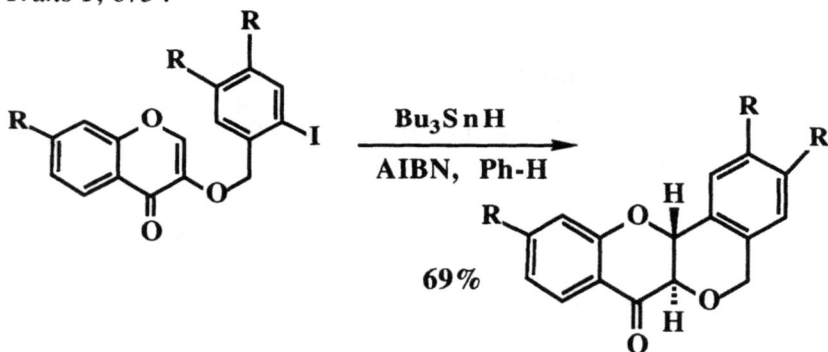

**IV.H-17**  P. Kumar et al., *J. Chem. Soc., Chem. Commun.*, 1580.

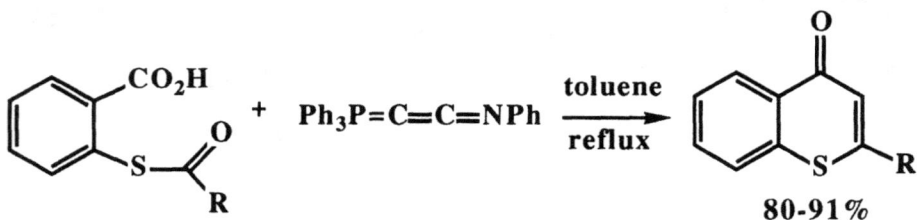

**IV.H-18** W.-D. Rudorf and J. Koditz, *Synthesis*, 667.

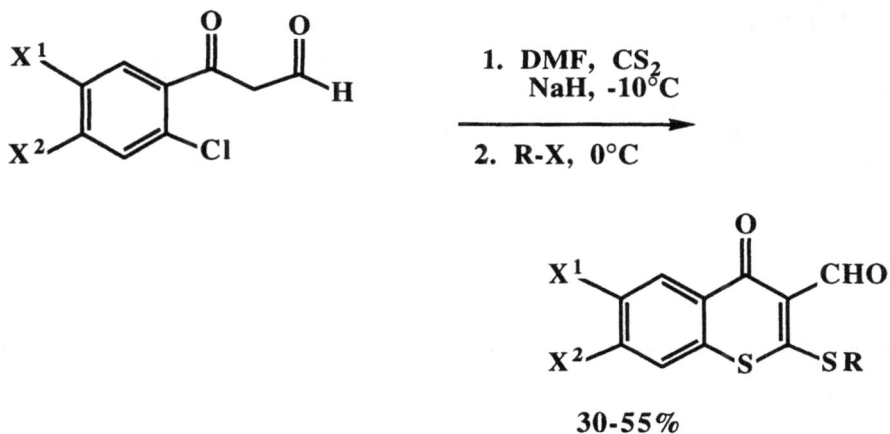

1. DMF, CS$_2$
   NaH, -10°C

2. R-X, 0°C

30-55%

---

**IV.H-19** S. Motoki, T. Saito et al., *J. Chem. Soc., Perkin Trans 1*, 2943.

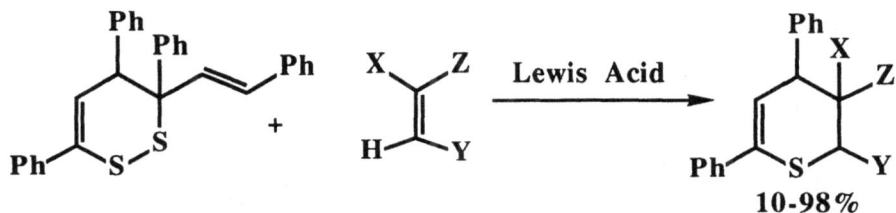

Lewis Acid

10-98%

---

**IV.H-20** I.L. Pinto et al., *Tetrahedron Lett.*, **33**, 7597; H. Shimizu et al., *J. Chem. Soc., Chem. Commun.*, 1586; G.W. Kirby et al., *J. Chem. Soc., Perkin Trans 1*, 1261.

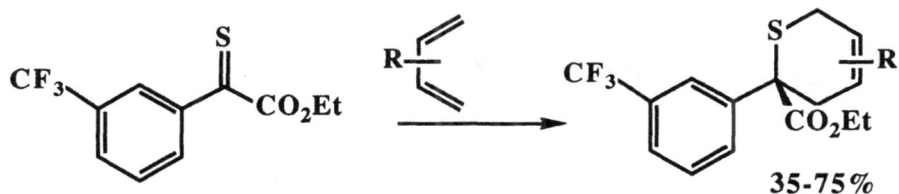

35-75%

## IV.I.  Other Heterocycles with One Heteroatom

**IV.I-1** J. Nakayama et al., *Heterocycles*, **34**, 1487.

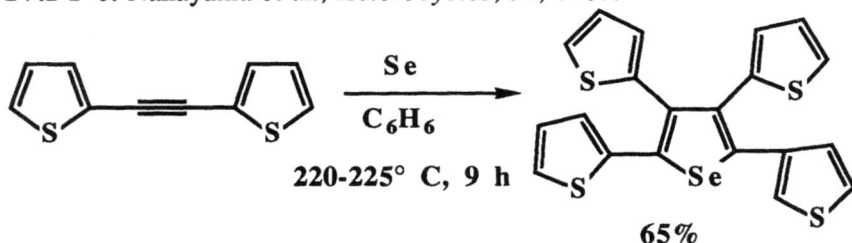

Se

C$_6$H$_6$

220-225° C, 9 h

65%

---

**IV.I-2** A.H. Moustafa et al., *Ind. J. Chem.*, **31B**, 24.

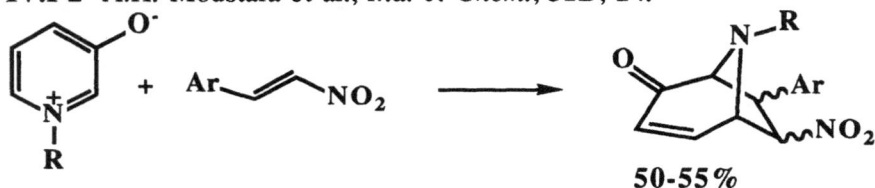

50-55%

---

**IV.I-3** J. Mann and L-C. de Almeida Barbosa, *J. Chem. Soc., Perkin Trans 1*, 787.

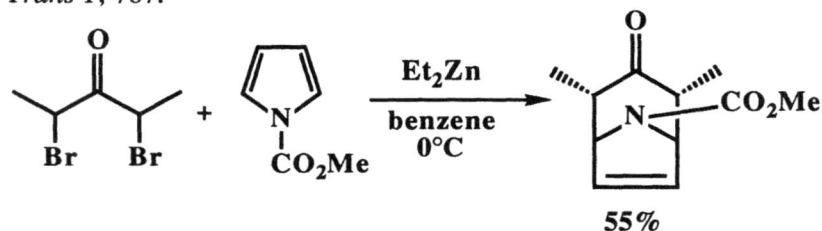

Et$_2$Zn

benzene
0°C

55%

---

**IV.I-4** H.M.L. Davies and N.J.S. Huby, *Tetrahedron Lett.*, **33**, 6935.

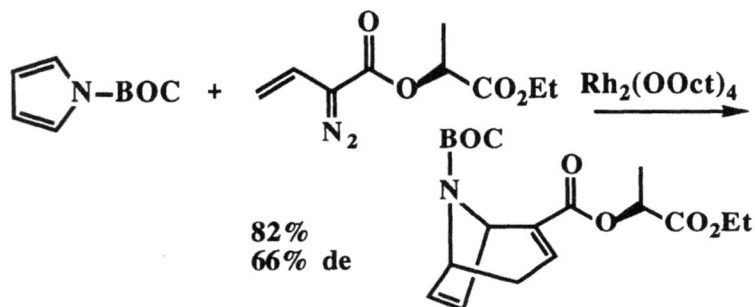

N–BOC +

Rh$_2$(OOct)$_4$

82%
66% de

**IV.I-5** F.D. Lewis and G.D. Reddy, *Tetrahedron Lett.*, **33**, 4249.

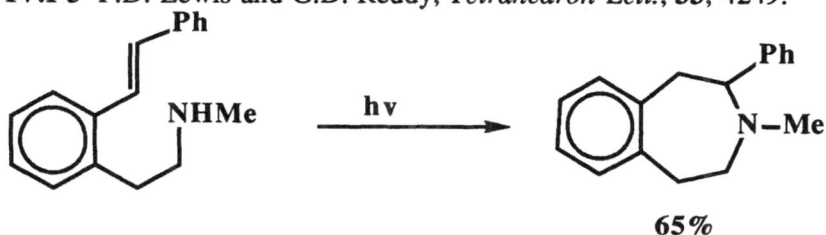

65%

---

**IV.I-6** D. Anastasiou and W.R. Jackson, *Austr. J. Chem.*, **45**, 21.

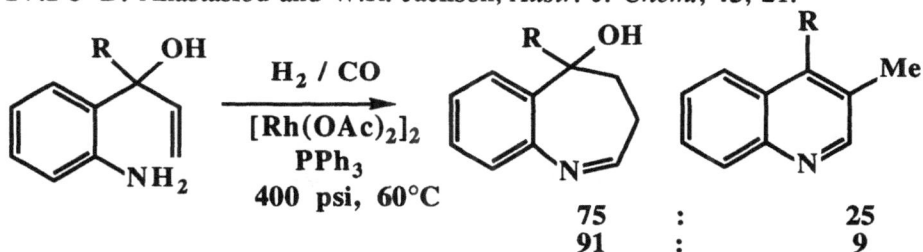

|   |   |   |
|---|---|---|
| 75 | : | 25 |
| 91 | : | 9 |

---

**IV.I-7** W.-N. Chou, J.B. White and W.B. Smith, *J. Am. Chem. Soc.*, **114**, 4658.

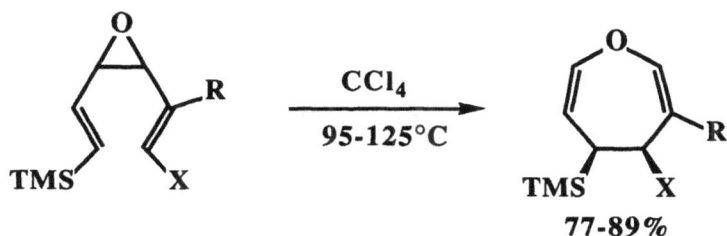

77-89%

---

**IV.I-8** B.A. Marples et al., *Synlett*, 987; D. Berger, L.E. Overman,*ibid.*, 811.

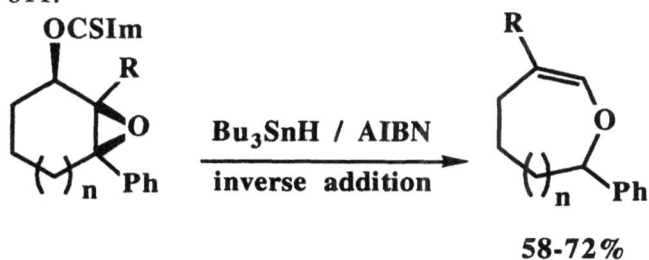

58-72%

**IV.I-9**  T. Kitamura et al., *J. Chem. Soc., Perkin Trans 1*, 1969.

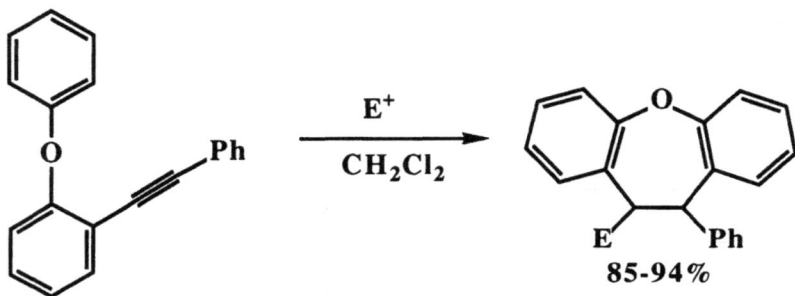

$$\xrightarrow[\text{CH}_2\text{Cl}_2]{\text{E}^+}$$

85-94%

---

**IV.I-10**  Y.Sugihara, J. Murata et al., *J. Am. Chem. Soc.*, **114**, 1479.

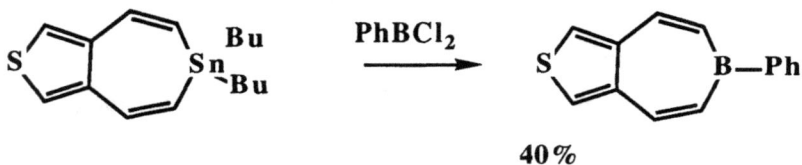

$$\xrightarrow{\text{PhBCl}_2}$$

40%

## IV.J.  Heterocycles with a Bridgehead Heteroatom

**IV.J-1**  J. Barluenga et al., *J. Chem. Soc., Chem. Commun.*, 1419.

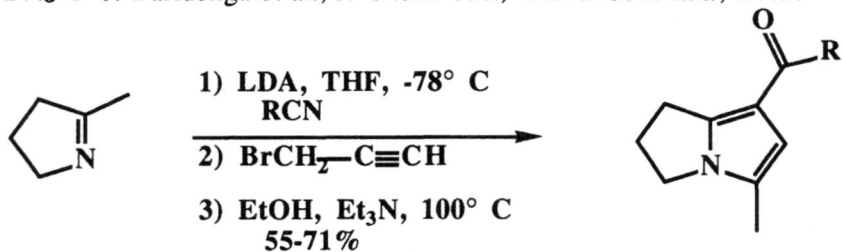

1) LDA, THF, -78° C
   RCN

2) BrCH$_2$—C≡CH

3) EtOH, Et$_3$N, 100° C
   55-71%

R = Ph, c-C$_6$H$_{11}$

**IV.J-2** S.G. Davies et al., *Tetrahedron Asym.*, **3**, 123; G. Lhommet et al., *ibid.*, **3**, 695.

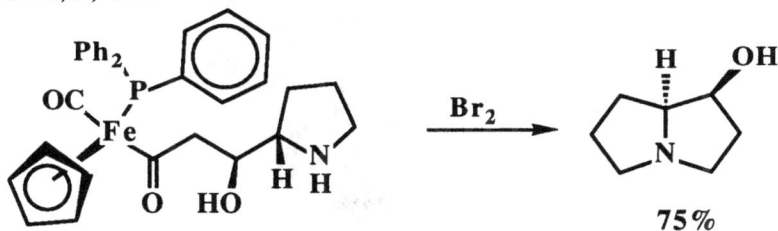

75%

---

**IV.J-3** A. Guarna et al., *J. Org. Chem.*, **57**, 4206.

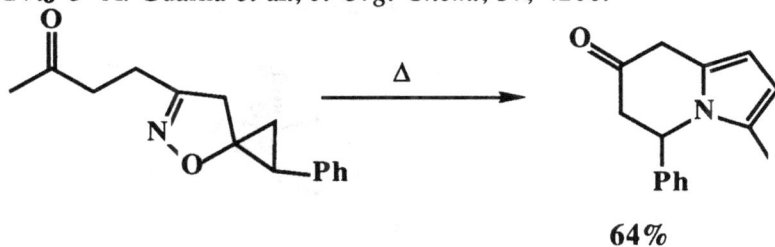

64%

---

**IV.J-4** B.R. Yerxa and H.W. Moore, *Tetrahedron Lett.*, **33**, 7811.

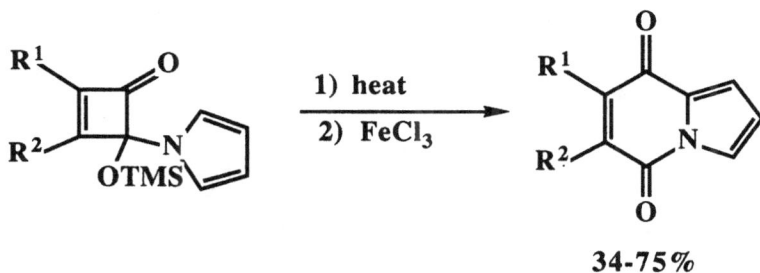

34-75%

**IV.J-5**  J.A. Seijas et al., *Tetrahedron*, **48**, 1637.

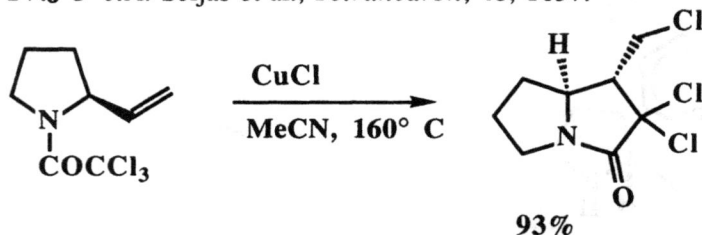

CuCl

MeCN, 160° C

93%

---

**IV.J-6**  M. Ikeda et al., *Chem. Pharm. Bull.*, **40**, 2308; P.F. Keusenkothen and M.B. Smith, *Tetrahedron*, **48**, 2977; S. Knapp and F.S. Gibson, *J. Org. Chem.*, **57**, 4802; M.D. Bachi et al., *ibid.*, **57**, 6803; C. Gennari, C. Scolastico et al., *Tetrahedron*, **48**, 3945; Y.-M. Tsai, *Tetrahderon Lett.*, **33**, 7895.

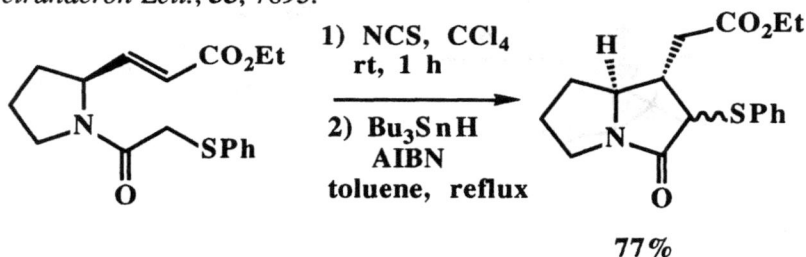

1) NCS, CCl$_4$
   rt, 1 h

2) Bu$_3$SnH
   AIBN
   toluene, reflux

77%

**Similar ring systems via radical cyclizations using Bu$_3$SnH**

---

**IV.J-7**  G. Pandey and G. Devi Reddy, *Tetrahedron Lett.*, **33**, 6533.

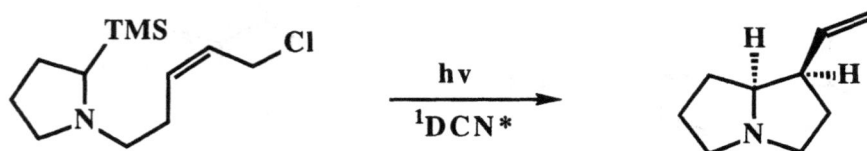

hv

$^1$DCN*

---

**IV.J-8**  B.M. Trost and C. Pedregal, *J. Am. Chem. Soc.*, **114**, 7292.

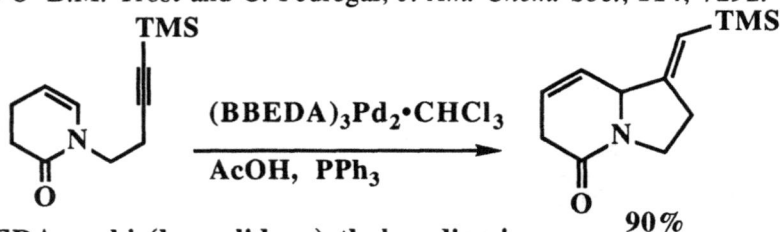

(BBEDA)$_3$Pd$_2$•CHCl$_3$

AcOH, PPh$_3$

90%

**BBEDA = bis(benzylidene)ethylenediamine**

**IV.J-9** M.P. Sibi et al., *J. Org. Chem.*, **57**, 4329; W.H. Pearson et al., *ibid.*, **57**, 3977.

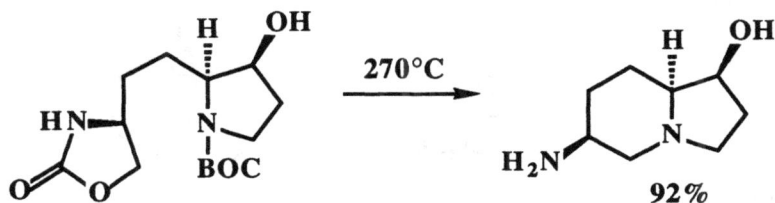

**novel thermolytic annulation of an oxazolidinine**

---

**IV.J-10** M. Pratoet al., *Tetrahedron Lett.*, **33**, 6537; **see also:** A. Padwa et al., *J. Am. Chem. Soc.*, **114**, 593 and *Tetrahedron Lett.*, **33**, 4731.

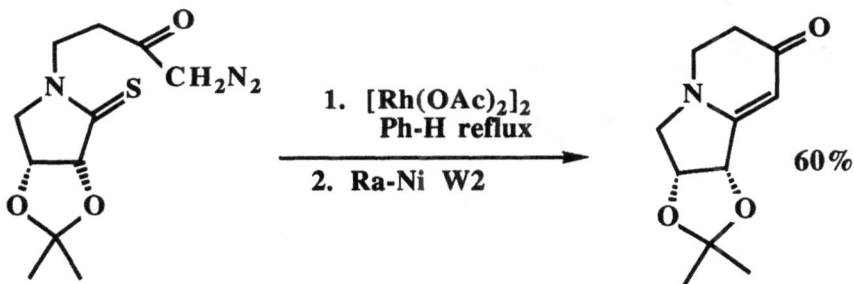

---

**IV.J-11** J.M. Muchowski et al., *Can. J. Chem.*, **70**, 1838.

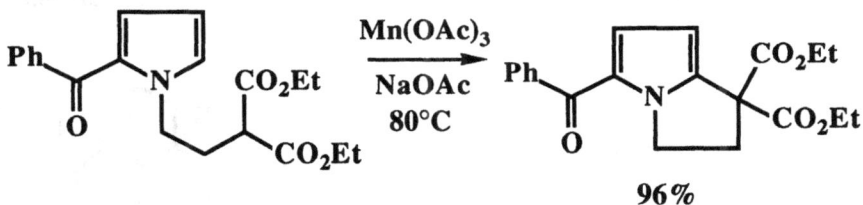

**IV.J-12**  Y. Gelas-Mialhe et al., *Tetrahedron Lett.*, **33**, 73; K.D. Moeller and S.L. Rothfus, *ibid.*, **33**, 2913; J.M. Vernon et al., *J. Chem. Soc., Perkin Trans 1*, 895.

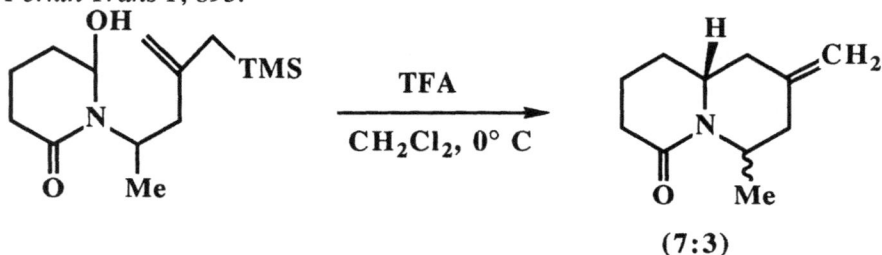

(7:3)

**similar examples of acyliminium ion cyclization**

**IV.J-13**  G.W. Gribble and B. Pelcman, *J. Org. Chem.*, **57**, 3636.

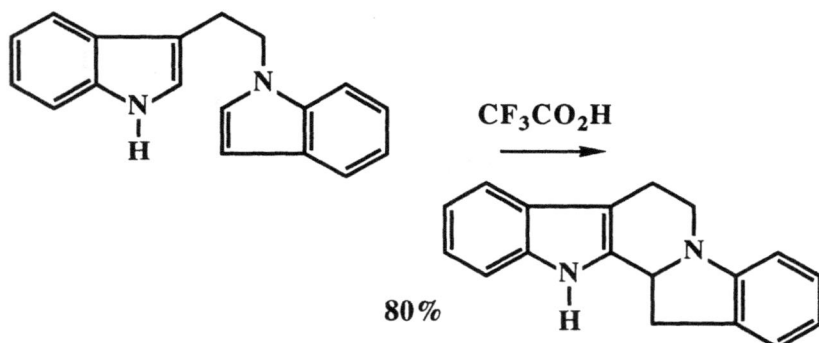

80%

**IV.J-14**  W.H. Pearson and J.M. Schkeryantz, *J. Org. Chem.*, **57**, 6783 and *Tetrahderon Lett.*, **33**, 5291; J.C. Teulade et al., *J. Heterocycles Chem.*, **29**, 691.

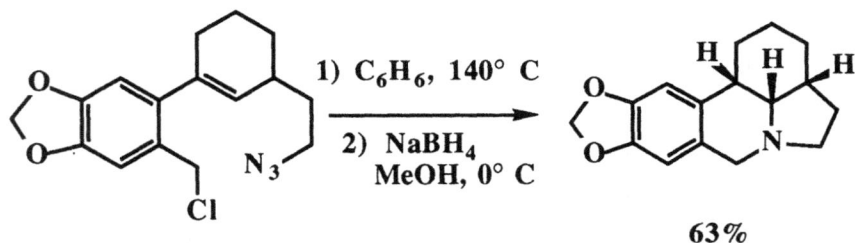

63%

**IV.J-15**  B. Abarca, G. Jones et al., *Heterocycles*, **33**, 203; Y. Hu, *Synth. Commun.*, **22**, 2103; K. Matsumoto, J.W. Lown et al., *Tetrahedron Lett.*, **33**, 7643; J. Alvarez-Builla et al., *Tetrahedron*, **48**, 8793.

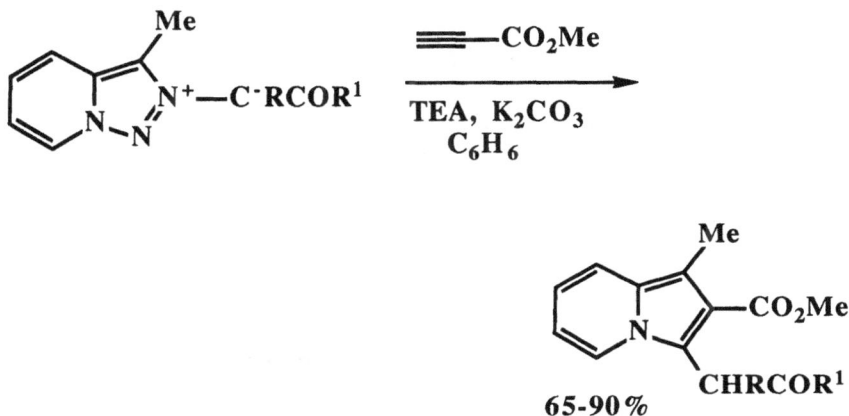

$$\equiv-CO_2Me$$

TEA, $K_2CO_3$
$C_6H_6$

65-90%

**IV.J-16**  M.D. Wang and H. Alper, *J. Am. Chem. Soc.*, **114**, 7018.

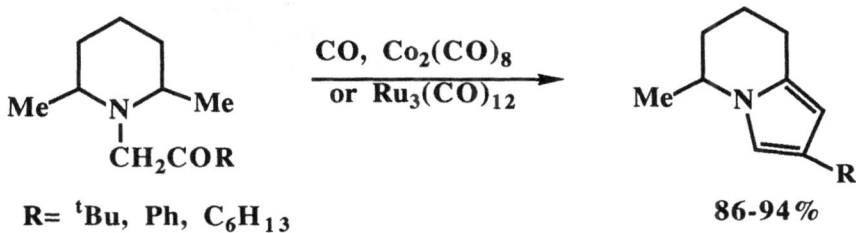

CO, $Co_2(CO)_8$
or $Ru_3(CO)_{12}$

R= $^tBu$, Ph, $C_6H_{13}$

86-94%

**IV.J-17**  D.F. Taber et al., *J. Org. Chem.*, **57**, 5990.

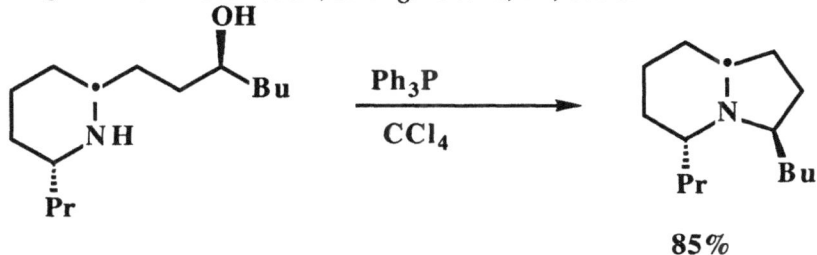

$Ph_3P$
$CCl_4$

85%

## I.V.K.    Heterocycles with Two or More Heteroatoms

## IV.K.1a.    5-Membered Heterocycles with 2 N's

**IV.K.1a-1**  R.H. Smith Jr., C.J. Michejda et al., *Tetrahedron Lett.*, **33**, 4683.

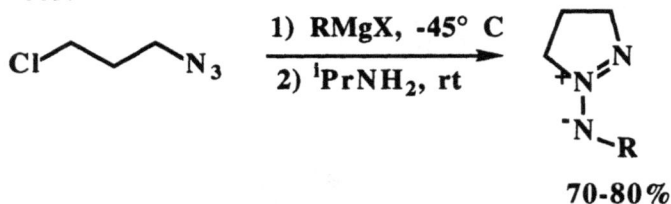

$$Cl\diagdown\diagup N_3 \xrightarrow[\text{2) }^i\text{PrNH}_2, \text{ rt}]{\text{1) RMgX, -45° C}}$$

70-80%

---

**IV.K.1a-2**  A.S. Shawli et al., *J. Chem. Res. (S)*, 360; B. Lande et al., *Can. J. Chem.*, **70**, 802; J. Vebrel et al., *Bull. Soc. Chem., Belg.*, **101**, 323.

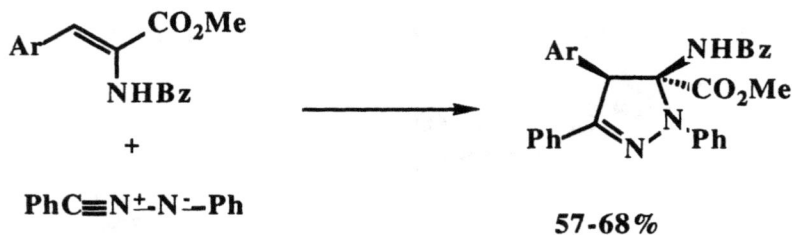

$$Ar\diagdown\diagup \overset{CO_2Me}{\underset{NHBz}{}}$$

+

PhC≡N⁺-N⁻-Ph

57-68%

---

**IV.K.1a-3**  P. Bravo et al, *J. Chem. Res. (S)*, 40; G. Adams et al., *J. Prakt. Chem.*, **334**, 227.

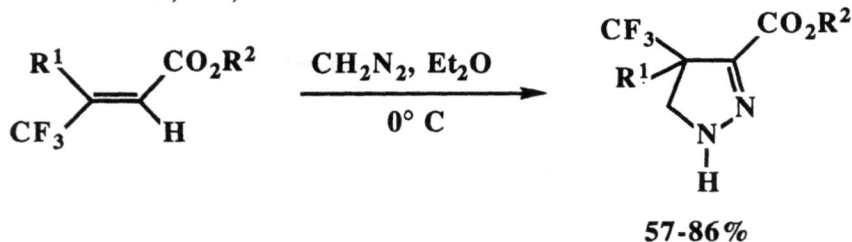

$$\underset{CF_3}{R^1}\diagup\diagdown \overset{CO_2R^2}{\underset{H}{}} \xrightarrow[\text{0° C}]{\text{CH}_2\text{N}_2, \text{Et}_2\text{O}}$$

57-86%

**IV.K.1a-4**  O.A. Attanasi et al., *Tetrahedron*, **48**, 1707; B.C. Hamper, *J. Org. Chem.*, **57**, 5680.

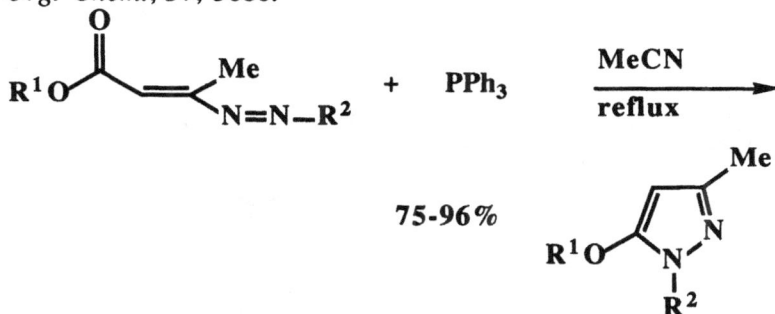

**IV.K.1a-5**  Y. Xu et al., *Tetrahedron Lett.*, **33**, 6161.

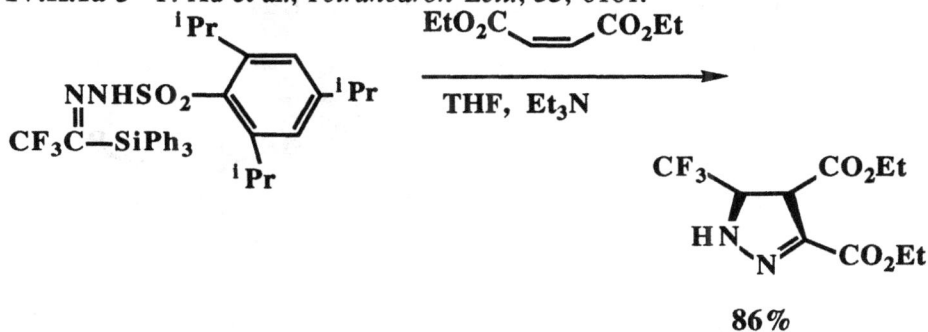

**IV.K.1a-6**  H.F. Zohdi, *J. Chem. Res (S)*, 82; M.J. Fray et al., *ibid.*, 10.

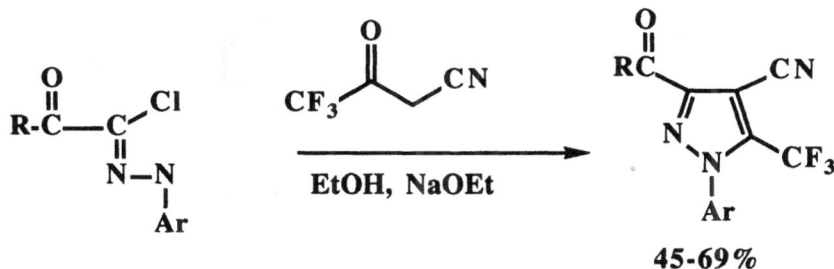

**IV.K.1a-7**  A.A. Fadda, *Ind. J. Chem.*, **30B**, 749 (1991).

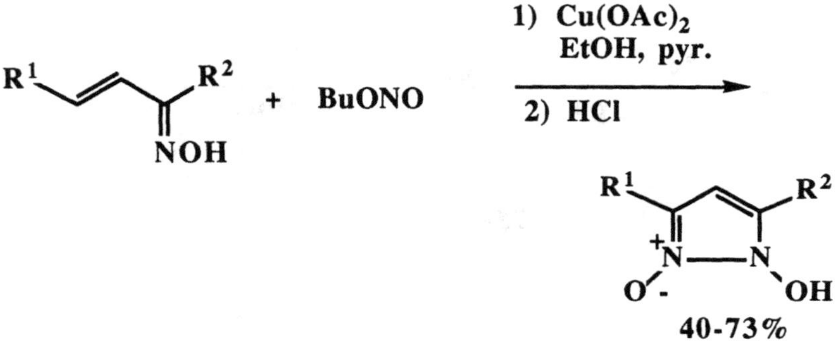

40-73%

---

**IV.K.1a-8**  G. Morel et al., *J. Org. Chem.*, **57**, 2121.

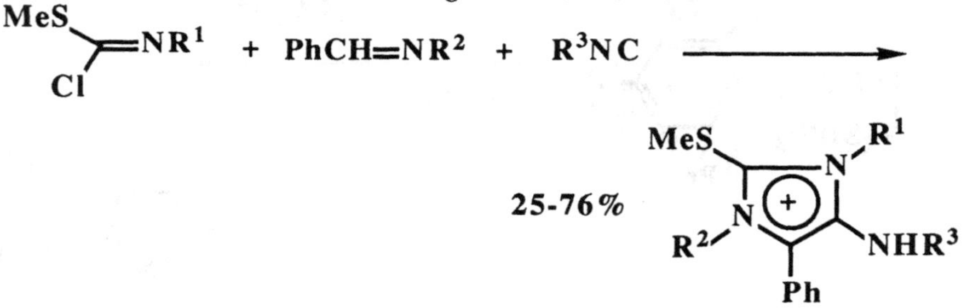

25-76%

---

**IV.K.1a-9**  K. Saito et al., *Heterocycles*, **34**, 1415; U.K. Nadir and N. Basu, *Tetrahedron Lett.*, **33**, 7949.

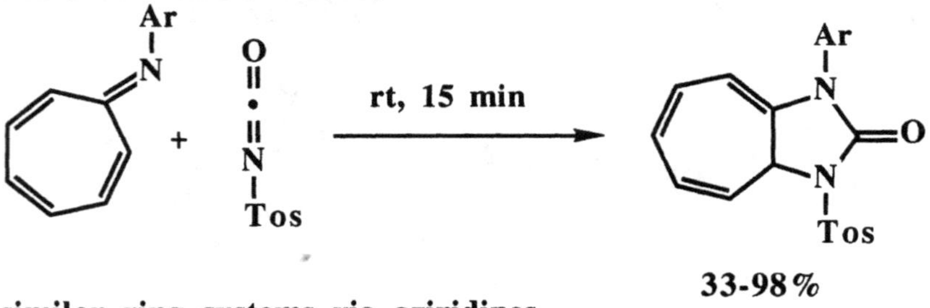

33-98%

**similar ring systems via aziridines**

**IV.K.1a-10**  Y. Watanabe et al., *J. Chem. Soc., Chem. Commun.*, 1318.

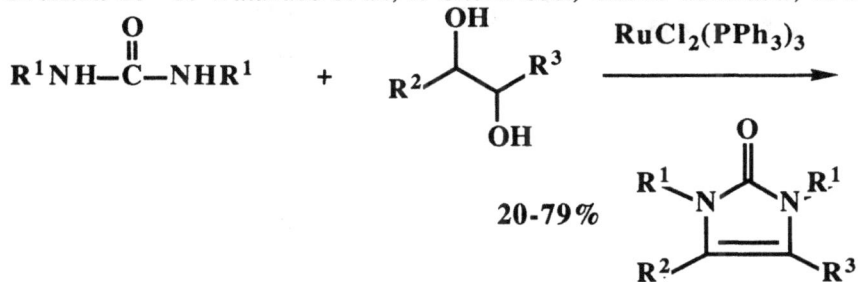

20-79%

**IV.K.1a-11**  Y. Singh and R.H. Prager, *Austr J. Chem.*, **45**, 1811.

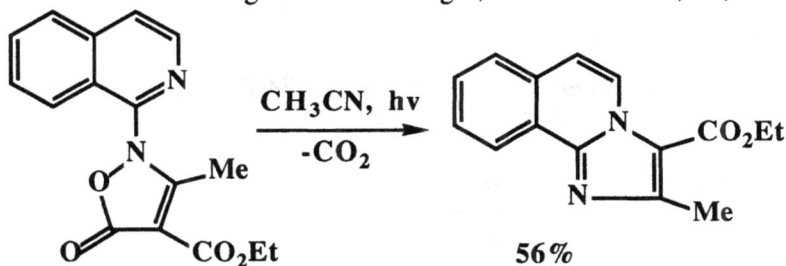

56%

**IV.K.1a-12**  H. Heimgartner et al., *Helv. Chim. Acta*, **75**, 1251.

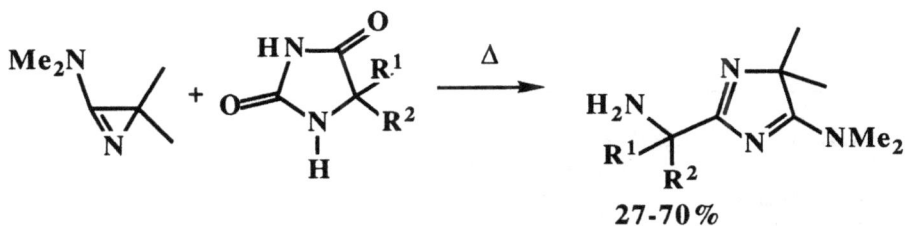

27-70%

**IV.K.1a-13**  A.T. Nielsen, A.P. Chafin et al., *J. Org. Chem.*, **57**, 6756.

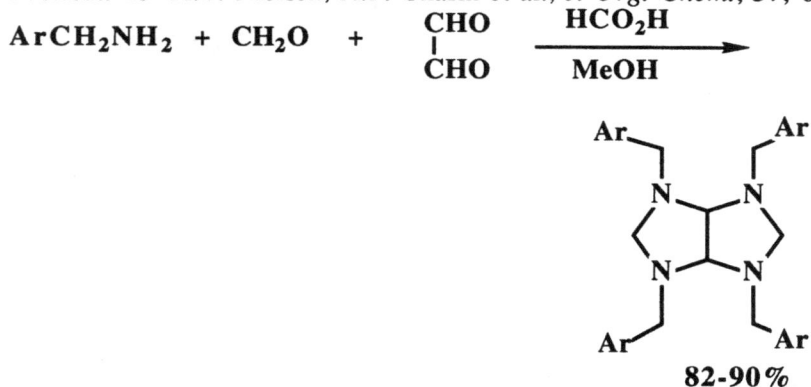

$$ArCH_2NH_2 \ + \ CH_2O \ + \ \underset{CHO}{\overset{CHO}{|}} \xrightarrow[\text{MeOH}]{HCO_2H}$$

**82-90%**

---

**IV.K.1a-14**  W.R. Jackson et al., *Tetrahedron*, **48**, 7467; T. Nagamatsu, H. Yamasaki and F. Yoneda, *Heterocycles*, **33**, 775.

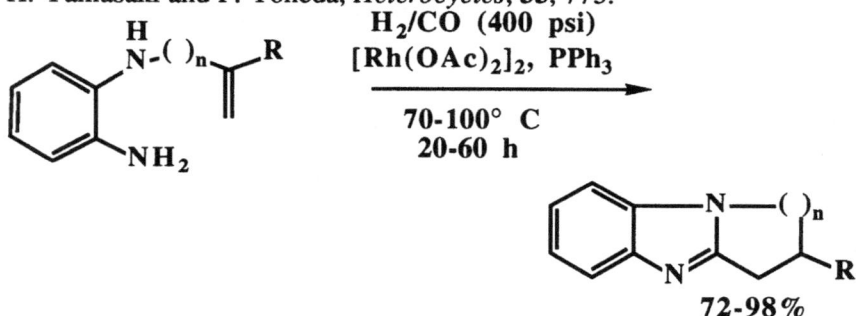

$$\xrightarrow[\begin{array}{c}\textbf{70-100° C}\\ \textbf{20-60 h}\end{array}]{\begin{array}{c}\textbf{H}_2\textbf{/CO (400 psi)}\\ \textbf{[Rh(OAc)}_2\textbf{]}_2\textbf{, PPh}_3\end{array}}$$

**72-98%**

---

**IV.K.1a-15**  M.F.J.R.P. Proenca et al., *J. Chem. Soc., Perkin Trans 1*, 913 and 2119; V. Dryanska, *Heterocycles*, **33**, 649; N.V. Nguyen and K. Baum, *Tetrahedron Lett.*, **33**, 2949; J.-J. Vanden Eynde et al., *Bull. Soc. Chem., Belg.*, **101**, 233.

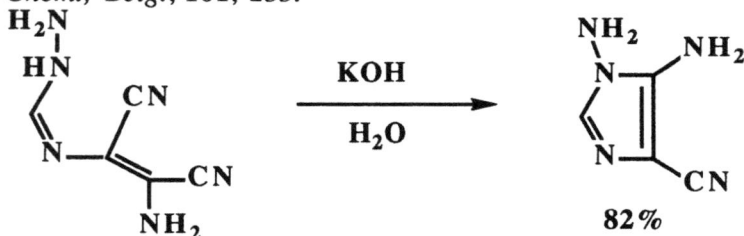

$$\xrightarrow[\text{H}_2\text{O}]{\text{KOH}}$$

**82%**

**IV.K.1a-16** I. Ojima and Y. Pei, *Tetrahderon Lett.*, **33**, 887.

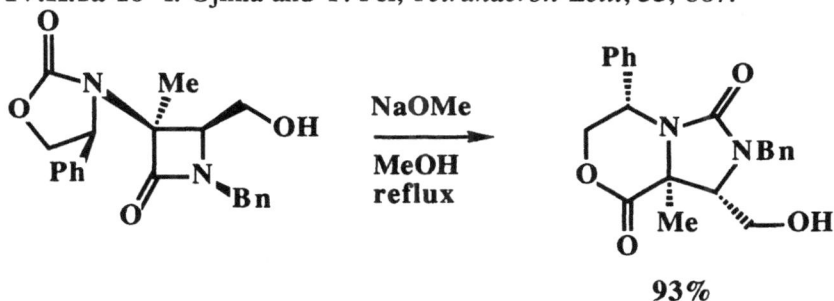

93%

## IV.K.1b. 6-Membered Heterocycles with 2 N's

**IV.K.1b-1** A.G. Martinez, M. Hanack et al., *Sytnhesis*, 1053; A.G. Martinez et al., *J. Org. Chem*, **57**, 1627.

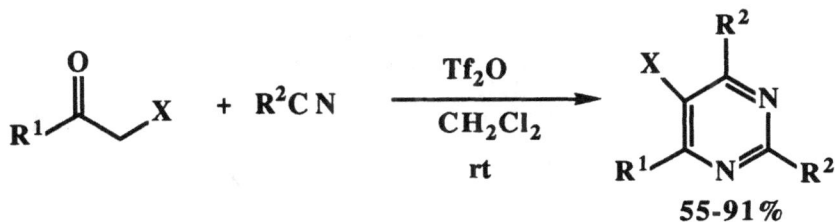

55-91%

---

**IV.K.1b-2** J.M. Muchowski et al., *Tetrahedron Lett.*, **33**, 3449.

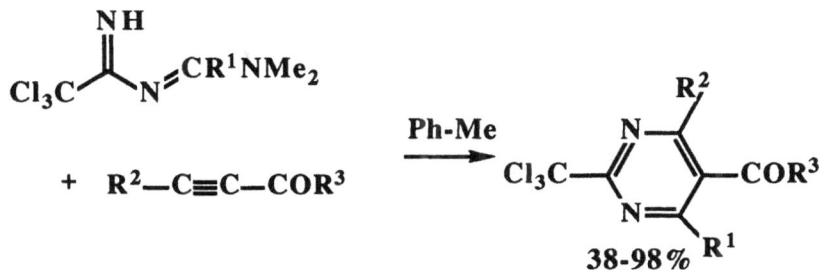

38-98%

**IV.K.1b-3** P.S. Manchand, P. Rosen et al., *J. Org. Chem.*, **57**, 3531; R. Neidlein and T. Eichinger, *Helv. Chim. Acta*, **75**, 124, 1039; H.G. McFadden and J.H. Huppatz, *Aust. J. Chem.*, **45**, 1045; A. Monge et al., *J. Heterocycles Chem.*, **29**, 1545.

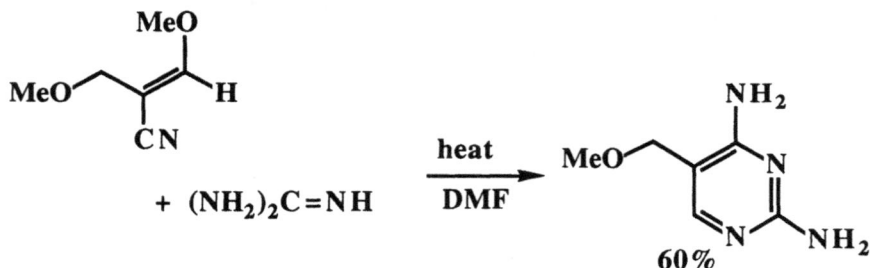

**IV.K.1b-4** M. Al-Talib, J.C. Jochims et al., *Synthesis*, 697.

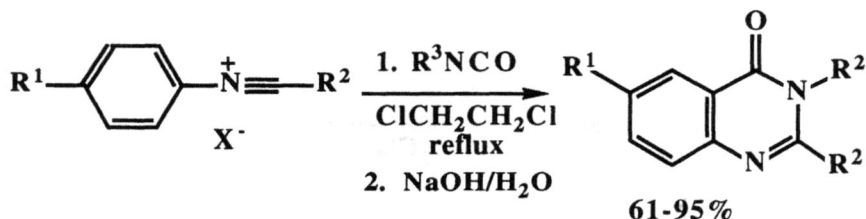

**IV.K.1b-5** D.L. Boger et al., *J. Org. Chem.*, **57**, 4333 and 1631.

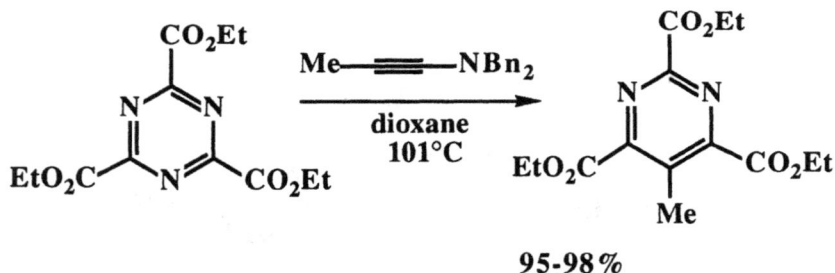

**IV.K.1b-6**  P. Molina et al., *J. Chem. Soc., Chem. Commun.*, 295.

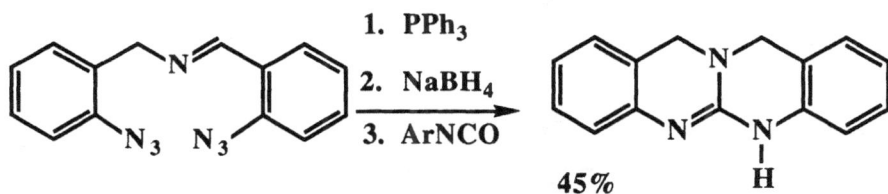

1. PPh$_3$
2. NaBH$_4$
3. ArNCO

45%

---

**IV.K.1b-7**  J. Barluenga et al., *Synlett.*, 563; J.V. Eynde et al., *Synth. Commun.*, **22**, 3141; J.-W. Chern et al., *Heterocycles*, **34**, 1133.

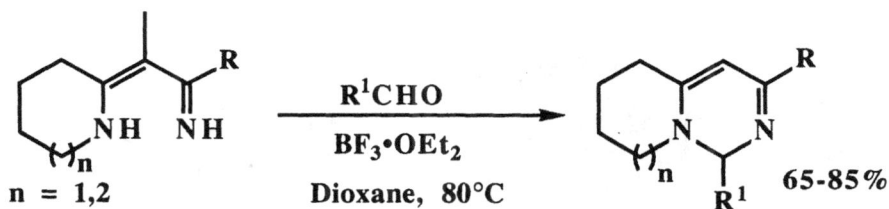

R$^1$CHO

BF$_3$•OEt$_2$

Dioxane, 80°C

n = 1,2

65-85%

---

**IV.K.1b-8**  H. Singh et al., *J. Chem. Soc., Perkin Trans 1*, 1139.

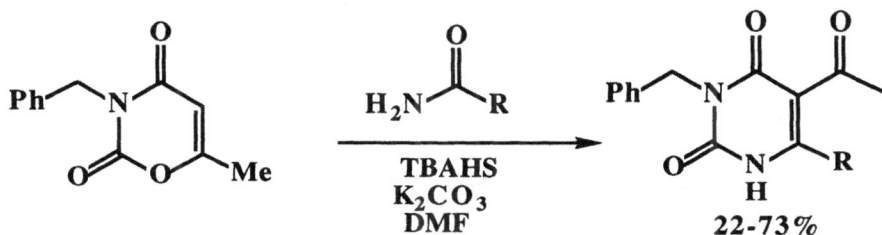

TBAHS
K$_2$CO$_3$
DMF

22-73%

**IV.K.1b-9**  S. Gronowitz et al., *J. Heterocycles Chem.*, **29**, 1049; J. Alvarez-Builla et al., *Liebigs Ann. Chem.*, 777.

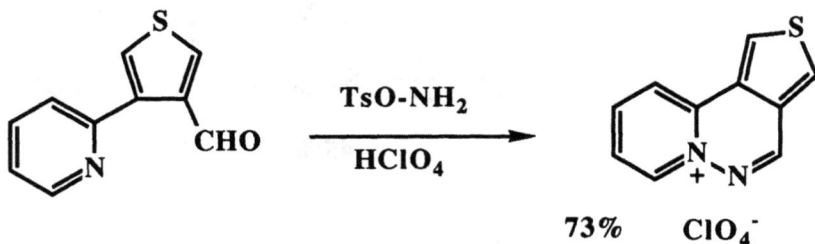

TsO-NH$_2$

HClO$_4$

73%     ClO$_4^-$

---

**IV.K.1b-10**  K.J. Hale et al., *Tetrahedron Lett.*, **33**, 7613.

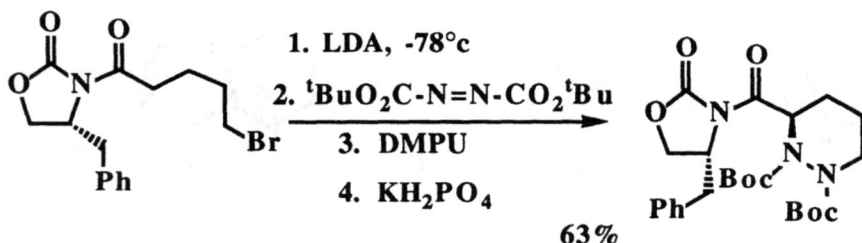

1. LDA, -78°c

2. $^t$BuO$_2$C-N=N-CO$_2$$^t$Bu

3. DMPU

4. KH$_2$PO$_4$

63%

---

**IV.K.1b-11**  K. Banert and M. Hagedorn, *Tetrahedron Lett.*, **33**, 7331.

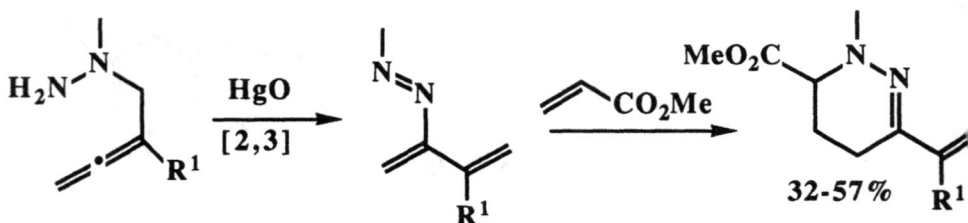

HgO

[2,3]

32-57%

**IV.K.1b-12**  A. Albini et al., *Heterocycles*, **33**, 573.

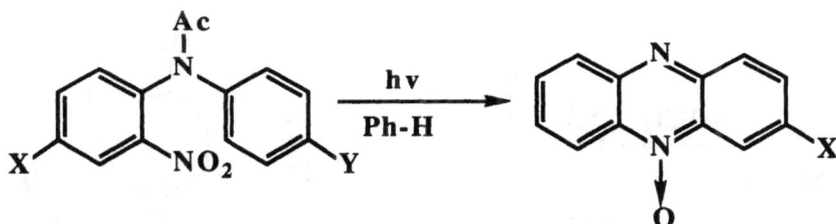

**IV.K.1b-13**  F. Compernolle et al., *J. Chem. Soc., Perkin Trans 1*, 1035; G.B. Phillips et al., *J. Med. Chem.*, **35**, 743.

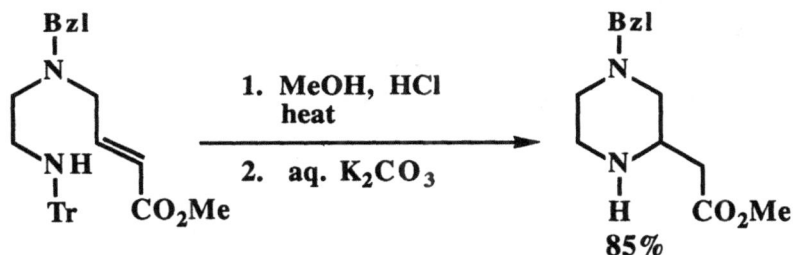

1. MeOH, HCl
   heat

2. aq. $K_2CO_3$

85%

## IV.K.1c.    7-Membered Heterocycles with 2 N's

**IV.K.1c-1**  J.A. Ortiz et al., *J. Org. Chem.*, **57**, 3535.

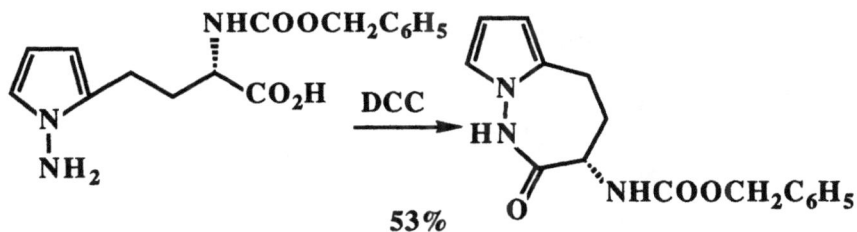

DCC

53%

**IV.K.1c-2**  J.-J. Vanden Eynde, E. Anders et al., *Bull. Soc. Chem., Belg.*, **101**, 801.

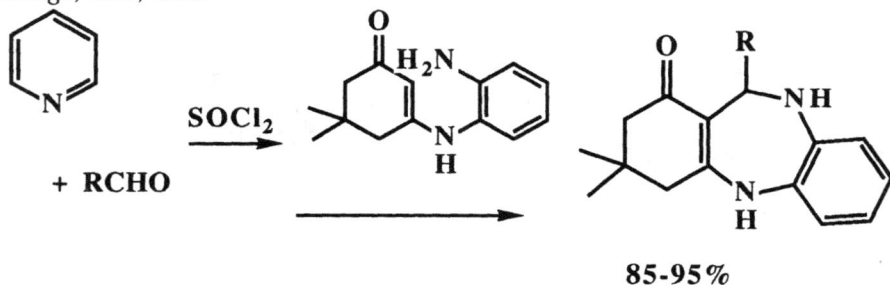

85-95%

---

**IV.K.1c-3**  S. Eguchi et al., *Synlett.*, 295.

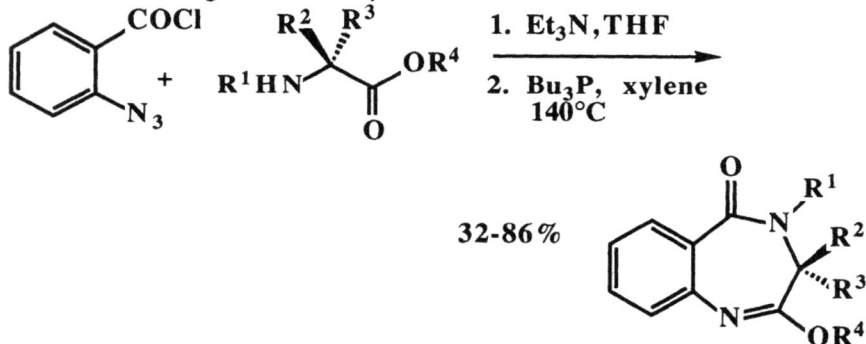

32-86%

---

**IV.K.1c-4**  M. Furukawa et al., *J. Chem. Soc., Perkin Trans 1*, 1287.

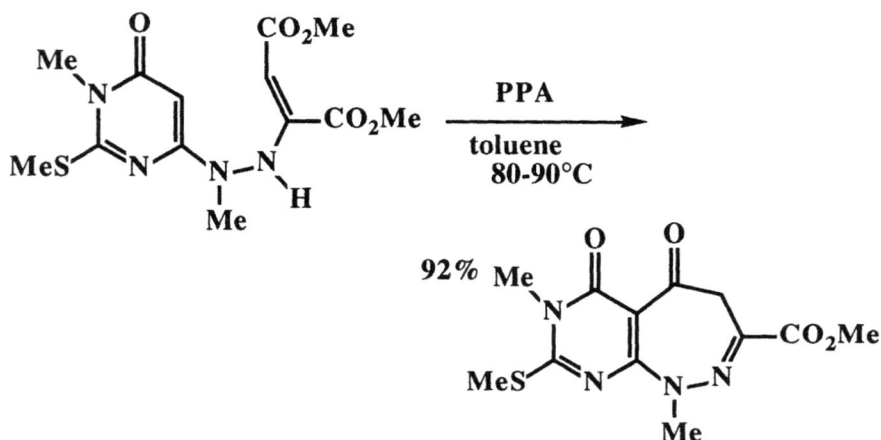

92%

**V.K.1c-5**  A. Chimirri et al., *Heterocycles*, **34**, 1191.

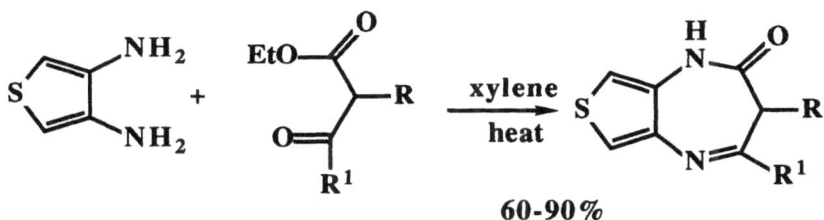

60-90%

## IV.K.2.      Heterocycles with 2 O's and 2 S's

**IV.K.2-1**  K. Kurosawa et al., *Bull. Chem. Soc., Jpn.*, **65**, 1371.

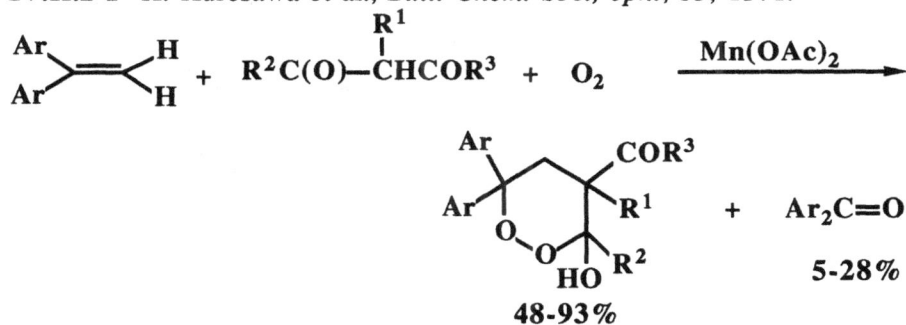

48-93%          5-28%

---

**IV.K.2-2**  W. Adam et al., *Chem. Ber.*, **125**, 1263.

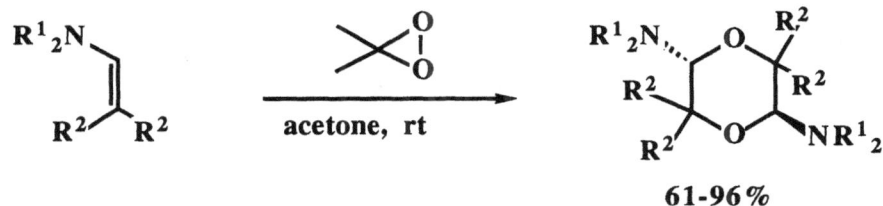

61-96%

**IV.K.2-3**  M. Casey and A.J. Culshaw, *Synlett.*, 214.

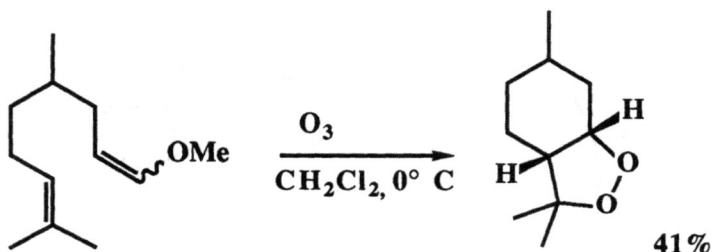

O$_3$
CH$_2$Cl$_2$, 0° C

**41%**

**IV.K.2-4**  C.W. Jefford et al., *Tetrahedron Lett.*, **33**, 7129.

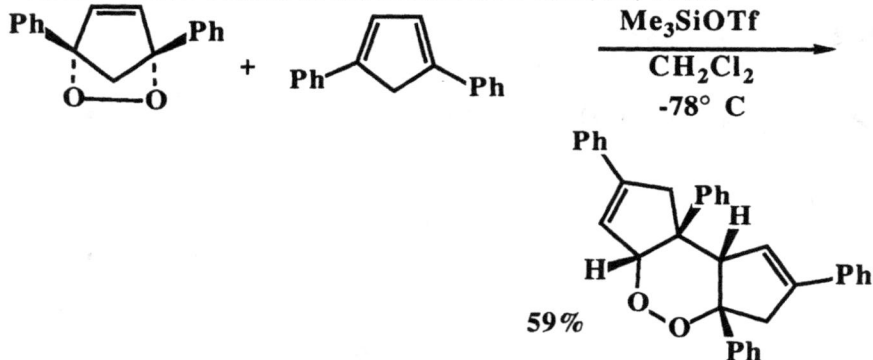

+

Me$_3$SiOTf
CH$_2$Cl$_2$
-78° C

**59%**

**IV.K.2-5**  B.B. Snider and Z. Shi, *J. Am. Chem. Soc.*, **114**, 1790.

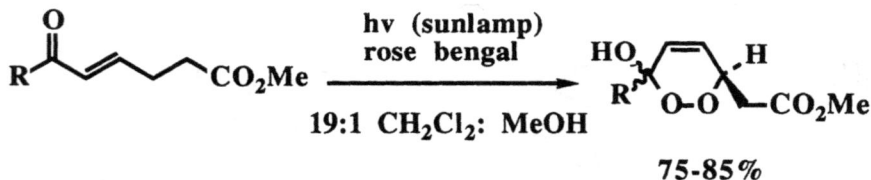

hv (sunlamp)
rose bengal

19:1 CH$_2$Cl$_2$: MeOH

**75-85%**

**IV.K.2-6**  K.T. Potts et al., *J. Org. Chem.*, **57**, 3895.

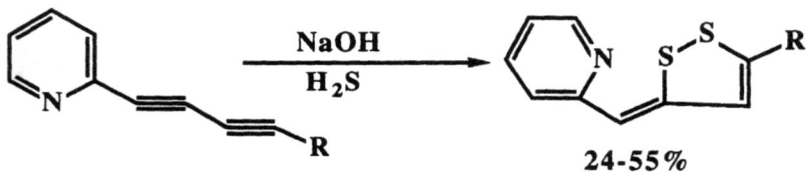

NaOH
H$_2$S

**24-55%**

**IV.K.2-7** G. Pattenden and A.J. Shuker, *J. Chem. Soc., Perkin Trans 1*, 1215.

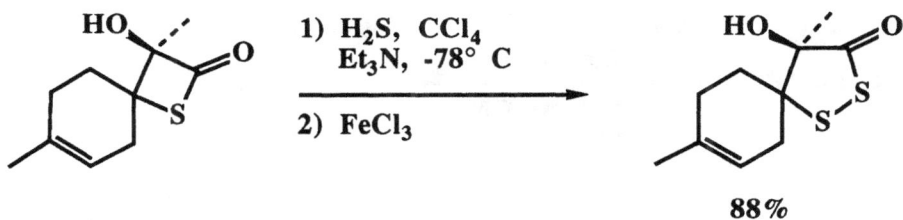

**88%**

---

**IV.K.2-8** S.L. Beaucage et al., *Org. Prep. Proced. Int.*, **24**, 488.

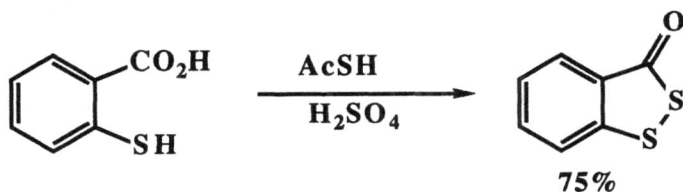

**75%**

---

**IV.K.2-9** D.S. Grierson et al., *J. Chem. Soc., Chem. Commun.*, 1531.

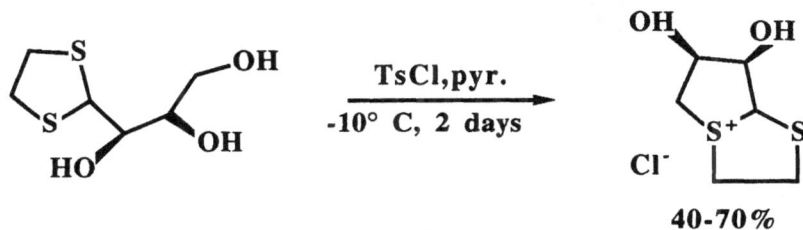

**40-70%**

---

**IV.K.2-10** L.P. Turchaninova et al., *J. Org. Chem., USSR*, **28**, 435.

$$(R^1CH_2-\underset{\underset{R^2}{|}}{C}HCH_2)_2S_n \xrightarrow{\Delta}$$

n = 3,4

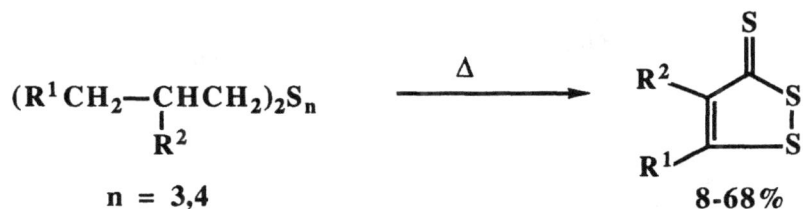

**8-68%**

## IV.K.3.    Heterocycles with 1N and 1O

**IV.K.3-1**  A.R. Hajipour and S.G. Pyne, *J. Chem. Res. (S)*, 388.

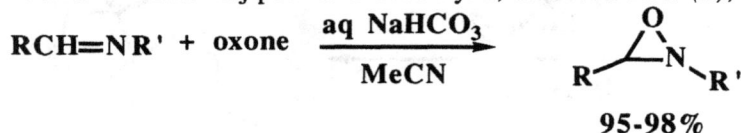

$$RCH=NR' + oxone \xrightarrow[\text{MeCN}]{\text{aq NaHCO}_3}$$

**95-98%**

---

**IV.K.3-2**  R.H. Prager et al., *Aust. J. Chem.*, **45**, 1571.

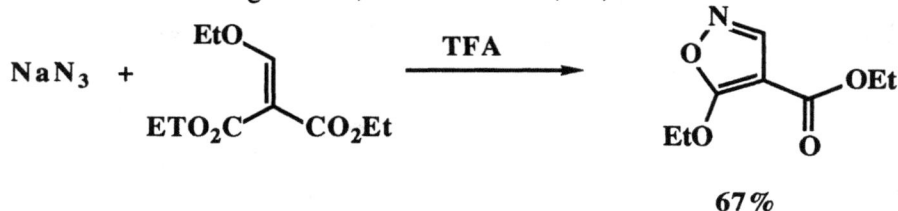

$$NaN_3 +$$

$$\xrightarrow{\text{TFA}}$$

**67%**

---

**IV.K.3-3**  R.F. Cunico and Chia P. Kuan, *J. Org. Chem.*, **57**, 3331.

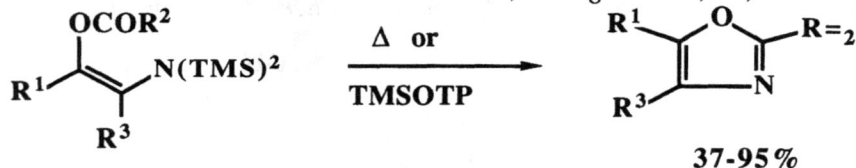

$$\xrightarrow[\text{TMSOTP}]{\Delta \ \ or}$$

**37-95%**

---

**IV.K.3-4**  K.J. Doyle and C.J. Moody, *Tetrahedron Lett.*, **33**, 7769.

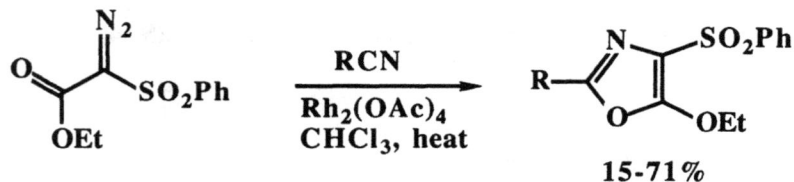

$$\xrightarrow[\substack{\text{Rh}_2(\text{OAc})_4 \\ \text{CHCl}_3, \text{heat}}]{\text{RCN}}$$

**15-71%**

**IV.K.3-5** E.L. Williams, *Tetrahderon Lett.*, **33**, 1033.

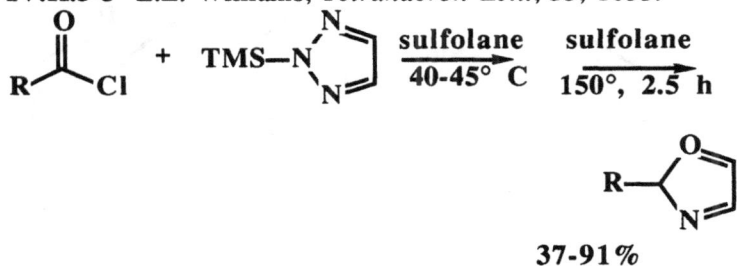

R—C(=O)—Cl + TMS—N(triazole) →[sulfolane 40-45° C] →[sulfolane 150°, 2.5 h]

R—(oxazole)

**37-91%**

---

**IV.K.3-6** H.G. Viehe et al., *Bull. Soc. Chem., Belg.*, **101**, 313.

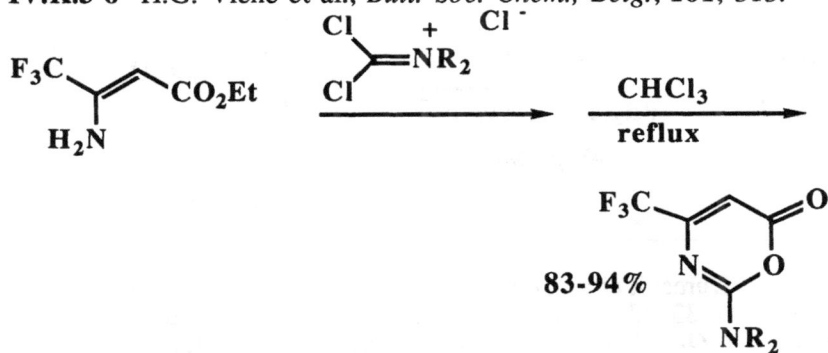

$F_3C$—CH=C($H_2N$)—$CO_2Et$  +  $Cl_2C$=$NR_2$ $Cl^-$ →  →[$CHCl_3$ reflux]

$F_3C$—(pyrimidinone)—$NR_2$=O

**83-94%**

---

**IV.K.3-7** F.Ogura et al., *J. Chem. Soc., Chem. Commun.*, 1070.

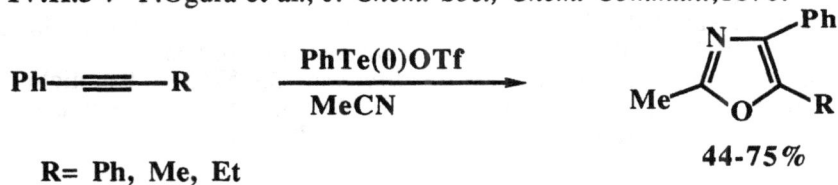

Ph—≡—R  →[PhTe(0)OTf / MeCN]  Me—(oxazole)—Ph, R

R= Ph, Me, Et

**44-75%**

---

**IV.K.3-8** H.Hiemstra et al., *J. Org. Chem.*, **57**, 6083.

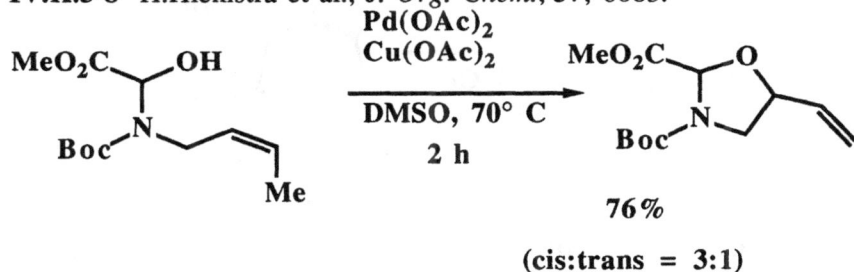

$MeO_2C$—CH(OH)—N(Boc)—CH₂—CH=CH—Me  →[Pd(OAc)₂ Cu(OAc)₂ / DMSO, 70° C 2 h]  $MeO_2C$—(oxazolidine)—CH=CH₂, N-Boc

**76%**

(cis:trans = 3:1)

**IV.K.3-9**  S.E. Denmark et al., *J. Org. Chem.*, **57**, 4912.

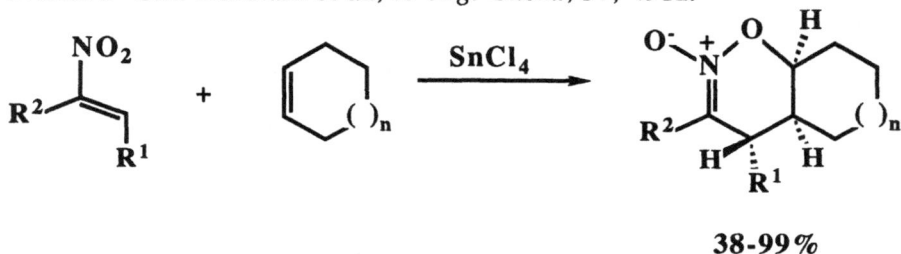

38-99%

---

**IV.K.3-10**  B.M. Trost et al., *Angew*, **31**, 228 and *J. Am. Chem. Soc.*, **114**, 8745.

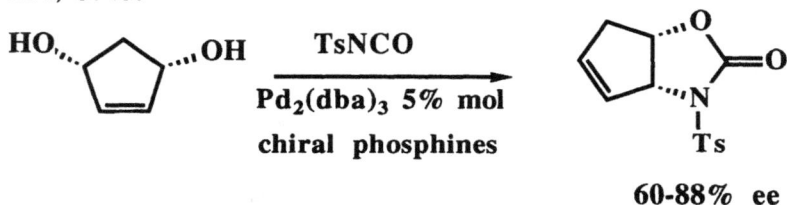

60-88% ee

---

**IV.K.3-11**  A. Dureault et al.,*Tetrahedron Lett.*, **33**, 1059; K. Shishido et al., *Heterocycles*, **33**, 73 and*Tetrahedron Lett.*, **33**, 4589; Y. Shishido and C. Kibayashi, *J. Org. Chem.*, **57**, 2876; J.N. Kim and E.K. Ryu, *Heterocyclces*, **34**, 1423; G. Resnati et al., *ibid.*, **34**, 1703; J. Svetlik et al., *Liebigs Ann. Chem.*, 591; E. Malamidou-Xenikaki and E. Coutouli-Argyropoulou, *ibid.*, 75; M. Shiozaki et al., *J. Am. Chem. Soc.*, **114**, 10065; K.M.L. Rai et al., *Org. Prep. Proceed. Int.*, **24**, 91; A. Brandt et al., *Tetrahedron*, **48**, 3323; R.M. Patton et al., *Tetrahedron*, **48**, 8053; T. Akiyama , K. Okada and S. Ozaki, *Tetrahedron Lett.*, **33**, 5763; S. Chimichi and B. Cosimelli, *Synth. Commun.*, **22**, 2909; M.P. Sibi and J.A. Gaboury, *Synlett.*, 83; J.C. Rohloff et al., *Tetrahedron Lett.*, **33**, 3113.

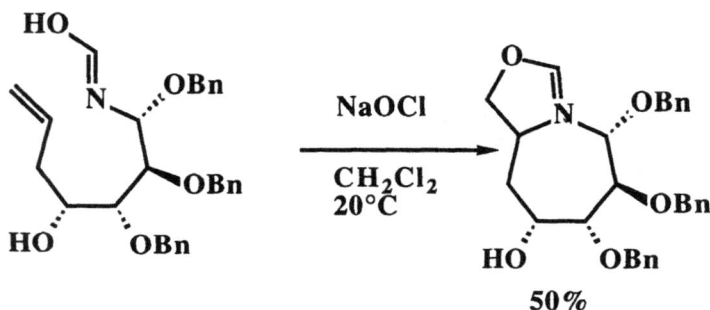

50%

similar  cyclizations  of  nitrile  oxides

**IV.K.3-12**  A.B. Holmes et al., *J. Chem. Soc., Perkin Trans 1*, 1089; M. Ikeda, *Chem Pharm. Bull.*, **40**, 2014; D.D. Dhavale and C. Trombini, *Heterocycles*, **34**, 2253; T. Hisano et al., *ibid.*, **34**, 1707; P. Bravo et al., *Tetrahedron*, **48**, 9775; A. Banerji and S. Basu, *ibid.*, **48**, 3335; H.G. Aurich et al., *ibid.*, **48**, 669; S. Kanemasa et al., *Tetrahedron Lett.*, **33**, 7889; C. Kibayashi et al., *ibid.*, **33**, 3765.

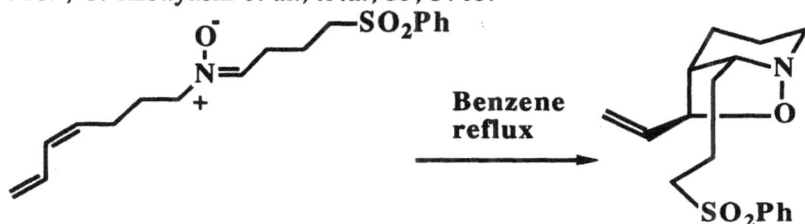

similar cyclization involving nitrones

---

**IV.K.3-13**  M. Shibasaki et al., *Heterocycles*, **33**, 161; J.H. Lee, M.J. Kurth et al., *J. Org. Chem.*, **57**, 6513.

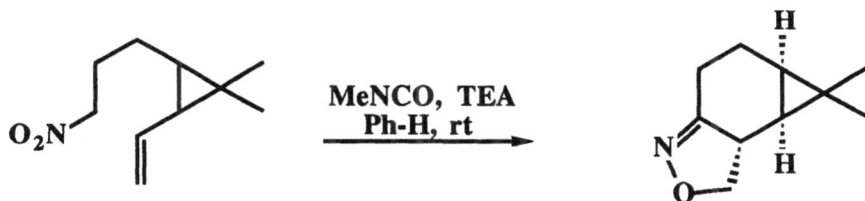

---

**IV.K.3-14**  R. Mazurkiewicz, *Synthesis*, 941; Y. Guindon, G. Jung et al., *Tetrahedron Lett.*, **33**, 4257; P. Wipf and C.P. Miller, *ibid.*, **33**, 907; J. Das et al., *ibid.*, **33**, 7835.

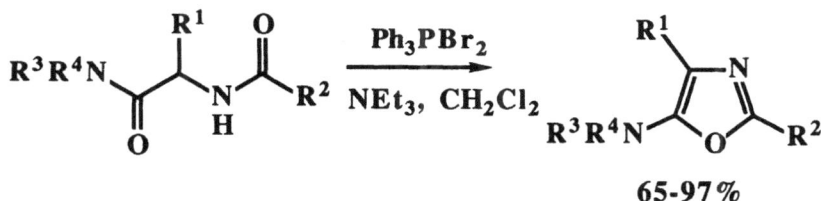

65-97%

similar cyclizations using I$_2$/AgOTf, Burgess's reagent and CuBr$_2$/DBu

**IV.K.3-15**  M. Hojo et al., *Heterocycles*, **34**, 1047.

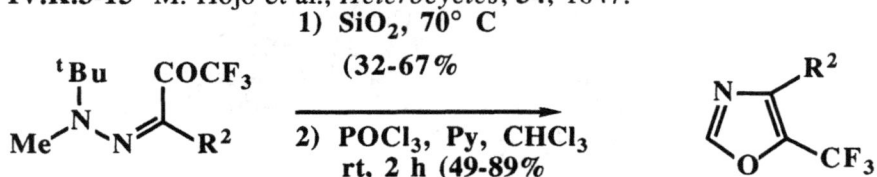

1) SiO$_2$, 70° C
(32-67%

2) POCl$_3$, Py, CHCl$_3$
rt, 2 h (49-89%

---

**IV.K.3-16**  K. Van Aken and G. Hoornaert, *J. Chem. Soc., Chem. Commun.*, 895.

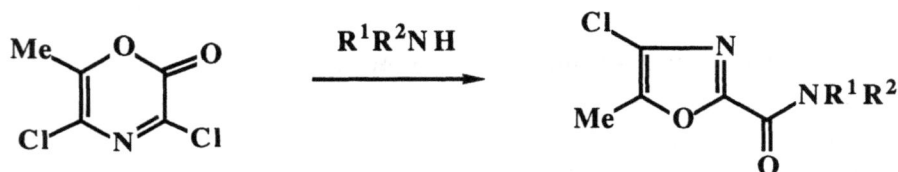

R$^1$R$^2$NH

---

**IV.K.3-17**  A.R. Katritzky et al., *J. Prakt. Chem.*, **334**, 114.

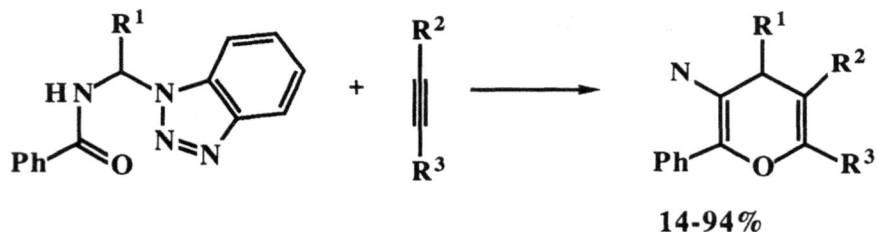

14-94%

---

**IV.K.3-18**  J. Streith et al., *Helv. Chim. Acta*, **75**, 109; J.R. Malpass and C. Smith, *Tetrahedron Lett.*, **33**, 273; J.M.J. Trouchet et al., *J. Chem. Res. (S)*, 228; S.-F. Martin et al., *Tetrahderon Lett.*, **33**, 3583.

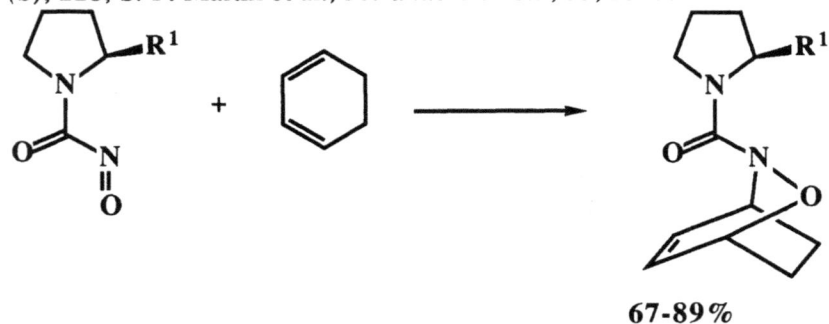

67-89%

**IV.K.3-19**   H. Reissig et al., *J. Org. Chem.*, **57**, 339 and *Angew.*, **31**, 1033.

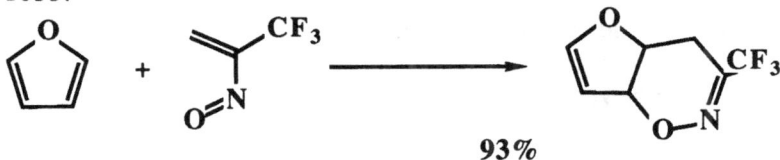

93%

---

**IV.K.3-20**   K. Harada et al., *Chem. Pharm. Bull.*, **40**, 1921.

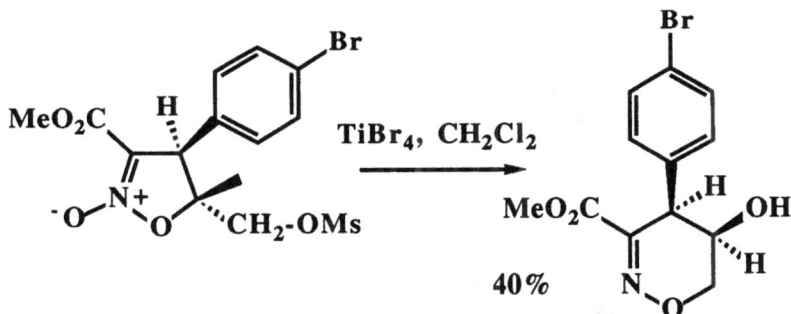

TiBr₄, CH₂Cl₂

40%

---

**IV.K.3-21**   K. Pandiarajan and J.C.N. Berry, *J. Chem. Soc., Perkin Trans 1*, 2055.

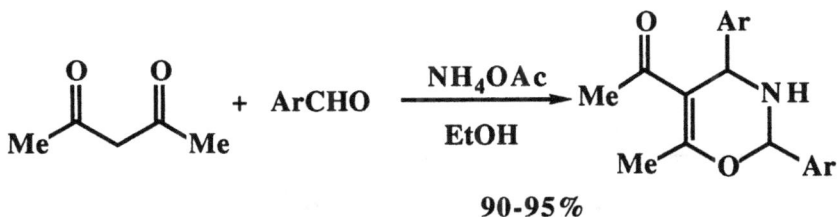

$$\text{Me} \overset{O}{\underset{}{\diagup}} \overset{O}{\underset{}{\diagdown}} \text{Me} \quad + \quad \text{ArCHO} \quad \xrightarrow[\text{EtOH}]{\text{NH}_4\text{OAc}}$$

90-95%

## IV.K.4.    Heterocycles with 1 N and 1 S

**IV.K.4-1**  M. Kidwai and R. Batra, *Ind. J. Chem.*, **30B**, 784 (1991).

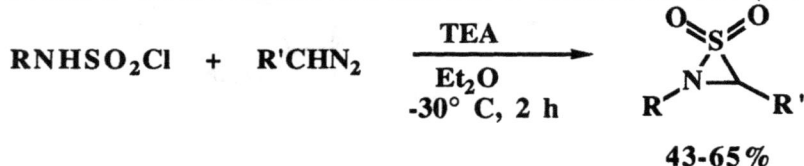

$$RNHSO_2Cl \quad + \quad R'CHN_2 \xrightarrow[\substack{Et_2O \\ -30^\circ \ C, \ 2 \ h}]{TEA}$$

43-65%

---

**IV.K.4-2**  S.W. Wright et al., *Tetrahedron Lett.*, **33**, 153.

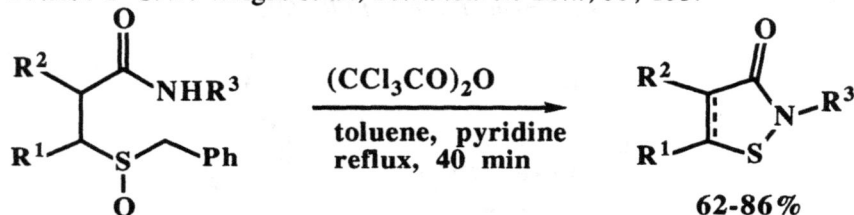

$$\xrightarrow[\substack{toluene, \ pyridine \\ reflux, \ 40 \ min}]{(CCl_3CO)_2O}$$

62-86%

---

**IV.K.4-3**  D.L. Boger and R.F. Menezes, *J. Org. Chem.*, **57**, 4331; C.W. Holzapfel et al., *Synth. Commun.*, **22**, 3029; A. Padwa et al., *Tetrahedron Lett.*, **33**, 5877.

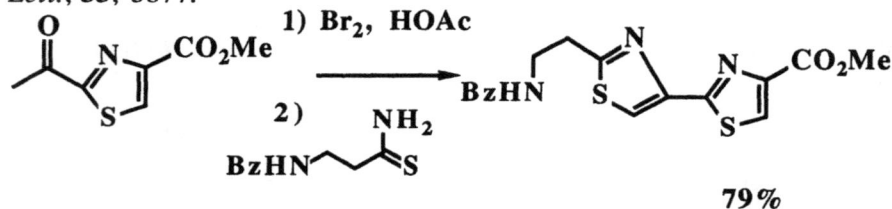

1) Br₂, HOAc

$$\xrightarrow{\quad\quad}$$

2) BzHN $\sim$ $\overset{NH_2}{\underset{S}{}}$

79%

---

**IV.K.4-4**  O. Prakash et al., *J. Chem. Soc., Perkin Trans 1*, 707; R.M. Moriarty et al., *Synthesis*, 845; D. Wobig, *Liebigs Ann. Chem.*, 415.

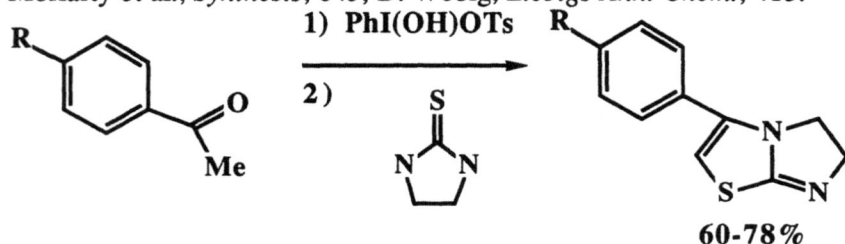

1) PhI(OH)OTs

$$\xrightarrow{\quad\quad}$$

2)

60-78%

**IV.K.4-5**  J. Barluenga, S. Fustero et al., *Tetrahedron*, **48**, 9745.

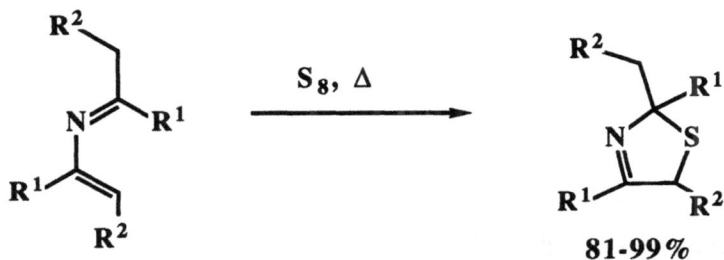

81-99%

**IV.K.4-6**  K. Banert et al., *Angew. Chem., Int. Ed. Engl.*, **31**, 90.

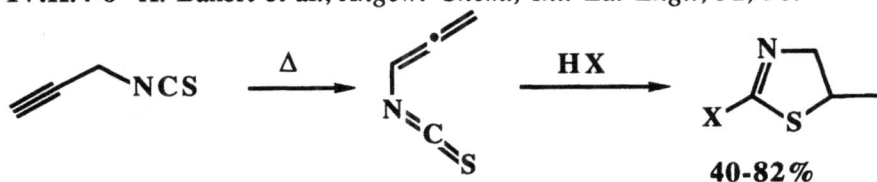

40-82%

X=OMe,  NH$_2$,  NHPh,  NPh$_2$,  SCHMe$_2$,  OPh

**IV.K.4-7**  K. Saito et al., *Heterocycles*, **34**, 1415.

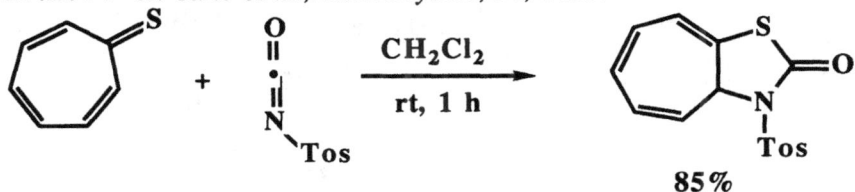

85%

**IV.K.4-8**  R. Leardini et al., *Tetrahedron*, **48**, 3961.

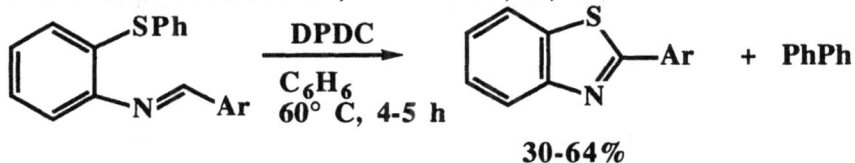

30-64%

DPDC = di-iso-propyl  peroxydicarbonate

**IV.K.4-9**  M.F. El-Zohry, *Org. Prep. Proced. Int.*, **24**, 81; A.H.H. Elghandour et al., *Tetrahedron*, **48**, 9295.

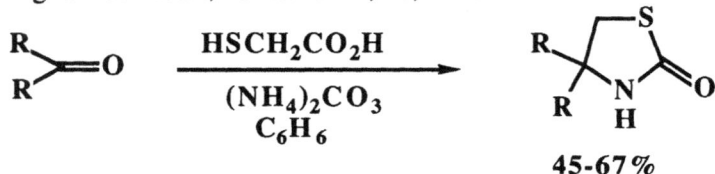

$$R_2C{=}O \xrightarrow[\substack{(NH_4)_2CO_3 \\ C_6H_6}]{HSCH_2CO_2H} \text{4,4-disubstituted thiazolidin-2-one}$$

45-67%

**IV.K.4-10**  L.D.S. Yadav and S. Sharma, *Synthesis*, 919.

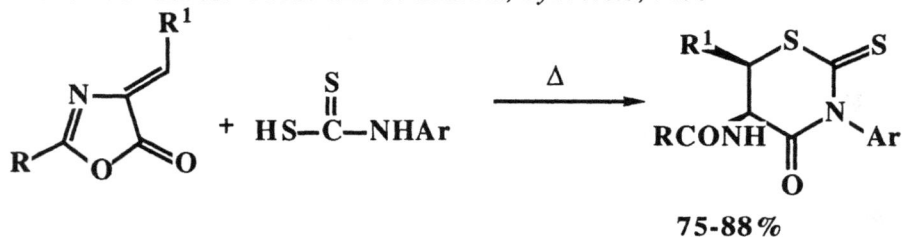

oxazolone + $HS{-}\overset{\displaystyle S}{\underset{\|}{C}}{-}NHAr \xrightarrow{\Delta}$ product

75-88%

**IV.K.4-11**  P. Hudhomme and G. Duguay, *Bull. Soc. Chim., Fr.*, **128**, 760 (1991).

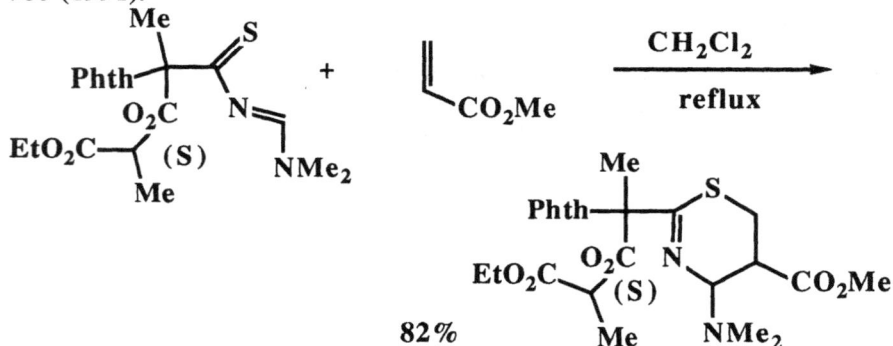

$$\text{thioamide} + \text{CH}_2{=}\text{CHCO}_2\text{Me} \xrightarrow[\text{reflux}]{CH_2Cl_2} \text{product}$$

82%

**IV.K.4-12**  A.R. Katritzky et al., *Org. Prep. Proced. Int.*, **24**, 463.

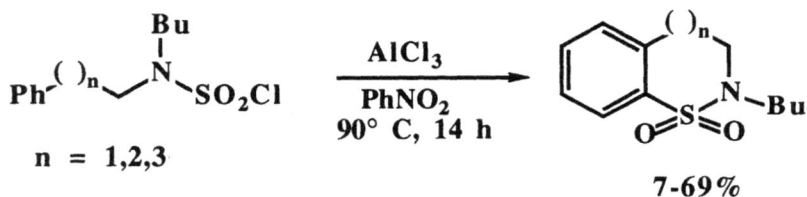

$$\text{Ph}{-}(\ )_n{-}\underset{\underset{SO_2Cl}{|}}{N}{-}Bu \xrightarrow[\substack{PhNO_2 \\ 90^\circ\ C,\ 14\ h}]{AlCl_3} \text{product}$$

n = 1,2,3

7-69%

**IV.K.4-13** H. Quast et al., *Liebigs Ann. Chem.*, 1259.

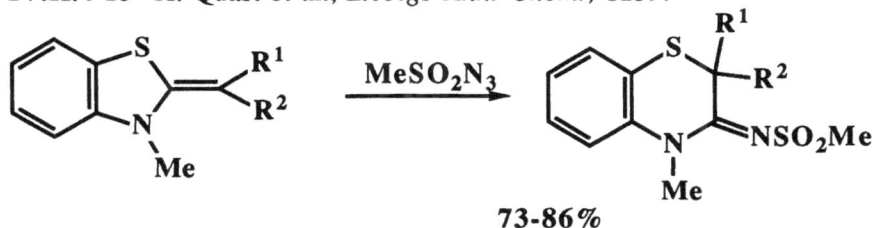

73-86%

## I.K.5    Heterocycles with 1 O and 1 S

**IV.K.5-1** T. Ibata and H. Nakano, *Bull. Chem. Soc. Jpn*, **65**, 3088.

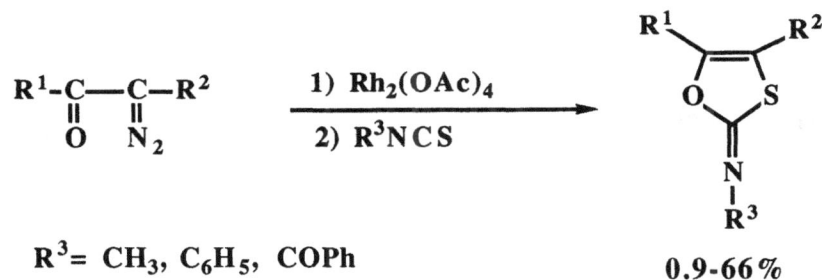

$R^3 = CH_3, C_6H_5, COPh$

0.9-66%

**IV.K.5-2** D. Villemin and A.P. Alloum, *Synth. Commun.*, **22**, 1351.

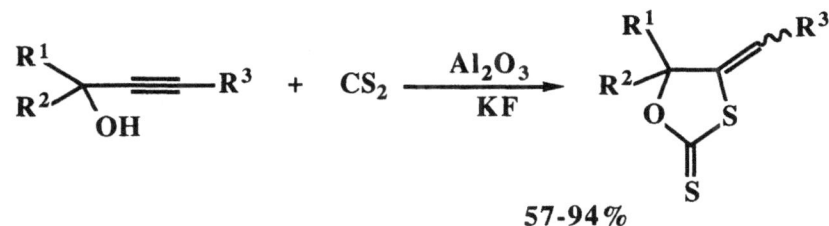

57-94%

**IV.K.5-3** W.B. Motherwell et al., *J. Chem. Soc., Chem. Commun.*, 1067.

**8-61%**

---

**IV.K.5-4** A. Hosomi et al., *Chem. Lett.*, 1073.

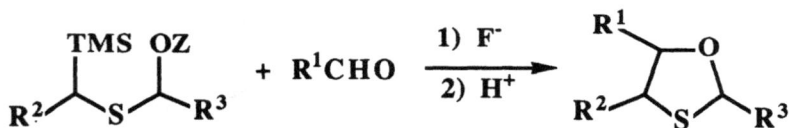

Z = alkyl or silyl

## IV.K.6.    Heterocycles with 3 or more N's

**IV.K.6-1** J.W. Benbow, G.K. Schulte and S.J. Danishefsky, *Angew. Chem., Int. Ed. Engl.*, **31**, 915; R. Carrie et al., *Bull. Soc. Chim., Fr.*, **129**, 308 and *J. Chem. Res. (S)*, 52; P.K. Kadaba, *J. Org. Chem.*, **57**, 3075; see also: H.J. Wadsworth, S.M. Jenkins et al., *J. Med. Chem.*, **35**, 1280.

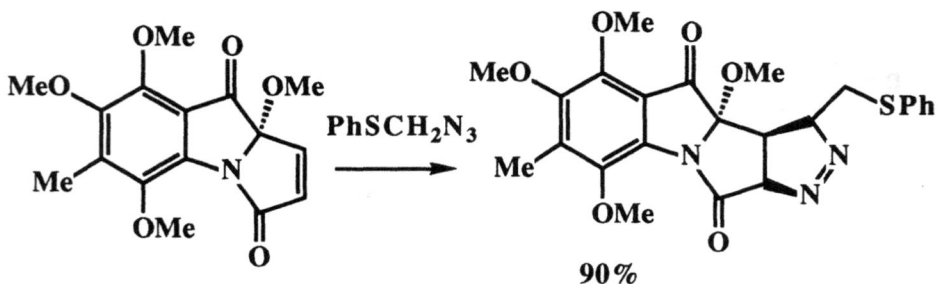

**90%**

similar reactions of azides and alkenes

**IV.K.6-2** R.K. Smalley et al., *J. Chem. Res. (S)*, 192.

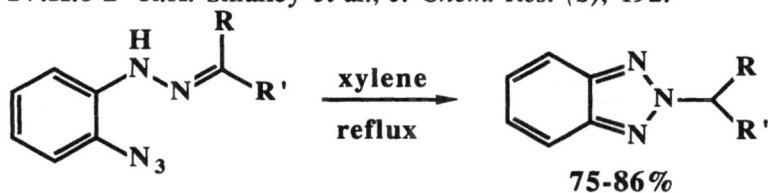

75-86%

---

**IV.K.6-3** W.S. Wilson et al., *Aust. J. Chem.*, **45**, 513 and 525; A.R. Katritzky et al., *J. Heterocyclic Chem.*, **29**, 1519.

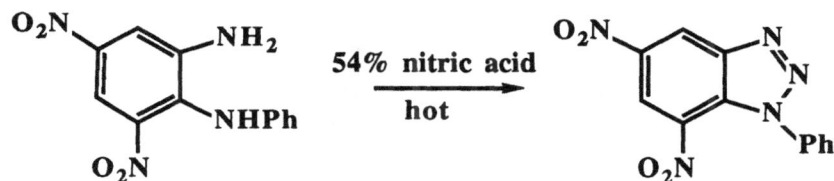

---

**IV.K.6-4** J.C. Jochims et al., *Synthesis*, 710.

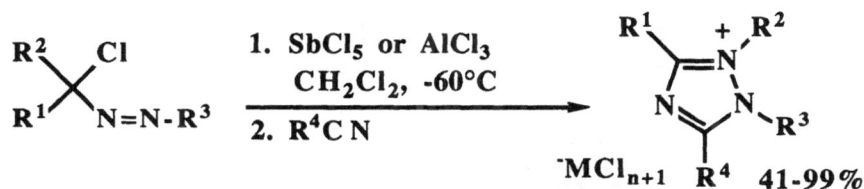

41-99%

---

**IV.K.6-5** K. Uneyama and K. Sugimoto, *J. Org. Chem.*, **57**, 6014; K.K. Reddy et al., *Ind. J. Chem.*, **31B**, 191.

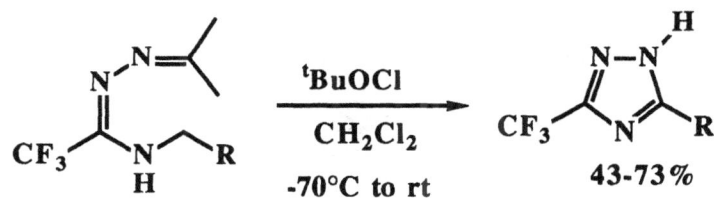

43-73%

**IV.K.6-6** M. Toselli and P. Zanirato, *J. Chem. Soc., Perkin Trans. 1*, 1101.

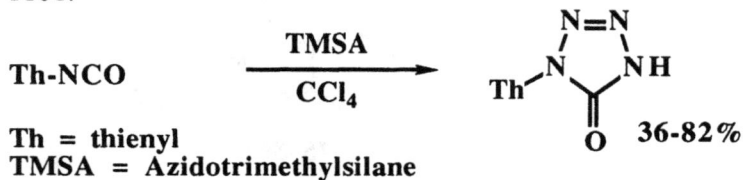

Th-NCO  →(TMSA / CCl$_4$)→ 

$$
\begin{array}{c}
\text{N=N} \\
\text{Th-N} \quad \text{NH} \\
\text{O} \quad \textbf{36-82\%}
\end{array}
$$

Th = thienyl
TMSA = Azidotrimethylsilane

---

**IV.K.6-7** H. Quast and T. Hergenrother, *Liebigs Ann. Chem.*, 581.

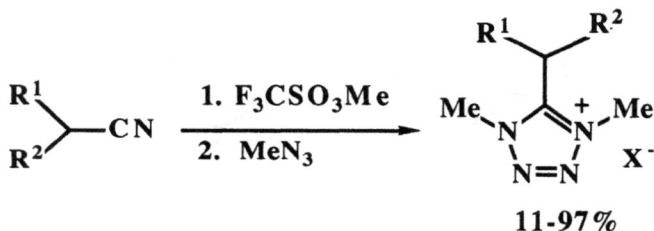

$$
\begin{array}{c}
\text{R}^1 \\
\quad \diagdown \\
\quad \text{CN} \\
\diagup \\
\text{R}^2
\end{array}
\xrightarrow[\text{2. MeN}_3]{\text{1. F}_3\text{CSO}_3\text{Me}}
$$

$$
\begin{array}{c}
\text{R}^1 \quad \text{R}^2 \\
\text{Me-N} \overset{+}{\diagup} \text{N-Me} \quad \text{X}^- \\
\text{N=N}
\end{array}
$$

**11-97%**

---

**IV.K.6-8** H. Neunhoeffer et al., *Synthesis*, 637 and *Heterocycles*, **33**, 893 and *Tetrahedron*, **48**, 5227.

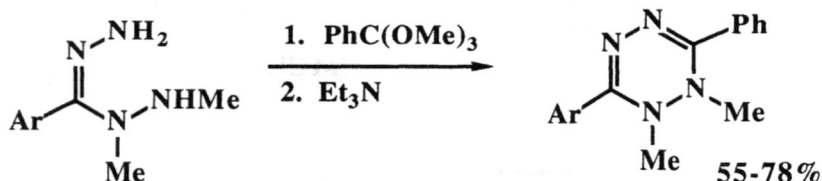

$$
\begin{array}{c}
\text{NH}_2 \\
\text{N} \\
\text{Ar} \quad \text{N-NHMe} \\
\text{Me}
\end{array}
\xrightarrow[\text{2. Et}_3\text{N}]{\text{1. PhC(OMe)}_3}
$$

$$
\begin{array}{c}
\text{N-N} \quad \text{Ph} \\
\text{N} \\
\text{Ar} \quad \text{N-N-Me} \\
\text{Me}
\end{array}
$$

**55-78%**

---

**IV.K.6-9** B.E. Love et al., *Synth. Commun.*, **22**, 1597.

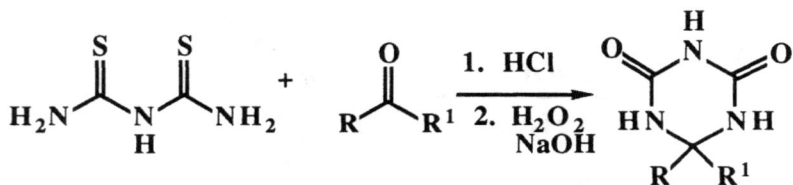

$$
\begin{array}{c}
\text{S} \quad\quad \text{S} \\
\text{H}_2\text{N} \quad \text{N} \quad \text{NH}_2 \\
\text{H}
\end{array}
+
\begin{array}{c}
\text{O} \\
\text{R} \quad \text{R}^1
\end{array}
\xrightarrow[\text{2. H}_2\text{O}_2 \; \text{NaOH}]{\text{1. HCl}}
$$

$$
\begin{array}{c}
\text{H} \\
\text{O} \quad \text{N} \quad \text{O} \\
\text{HN} \quad \text{NH} \\
\text{R} \quad \text{R}^1
\end{array}
$$

## IV.K.7.    Heterocycles with 2 N's and 1 O

**IV.K.7-1** E.N. Beal and K. Turnbull, *Synth. Commun.*, **22**, 673 and 1515.

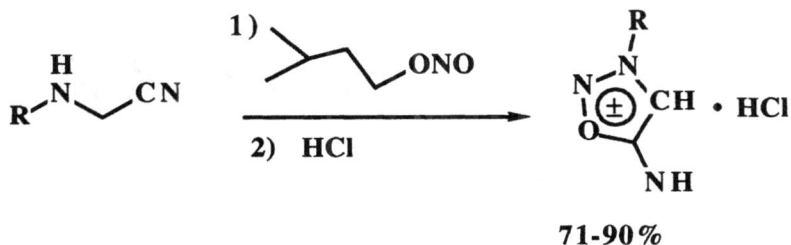

71-90%

---

**IV.K.7-2** T. Chiba and M. Okimoto, *J. Org. Chem.*, **57**, 1375.

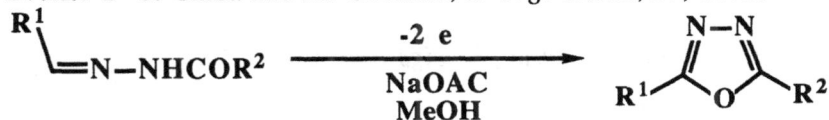

---

**IV.K.7-3** J.M. Kane and M.A. Staeger, *Synth. Commun.*, **22**, 1.

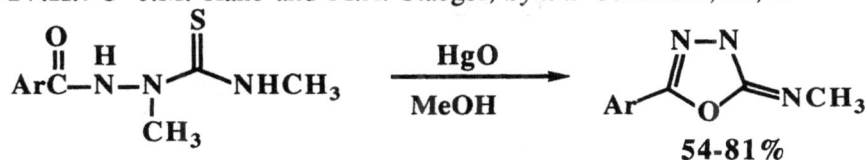

54-81%

---

**IV.K.7-4** L.K. Dyall et al., *Aust. J. Chem.*, **45**, 371.

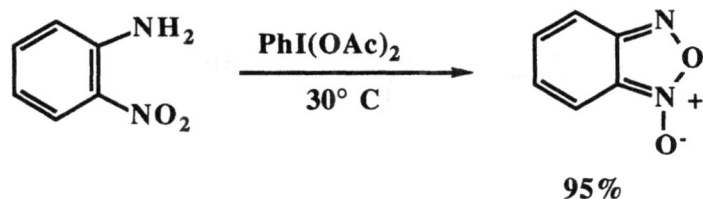

95%

**IV.K.7-5**  C. LaRosa et al., *J. Chem. Res (S)*, 32.

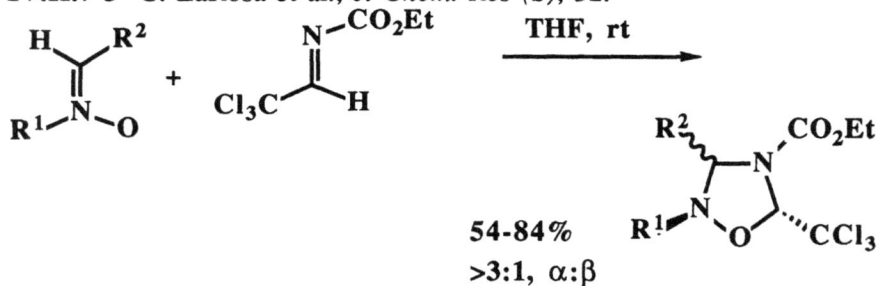

54-84%
>3:1, α:β

## IV.K.8.    Heterocycles with 2 N's and 1S or 1 Se

**IV.K.8-1**  H. Sonnenschein et al., *Liebigs Ann. Chem.*, 287.

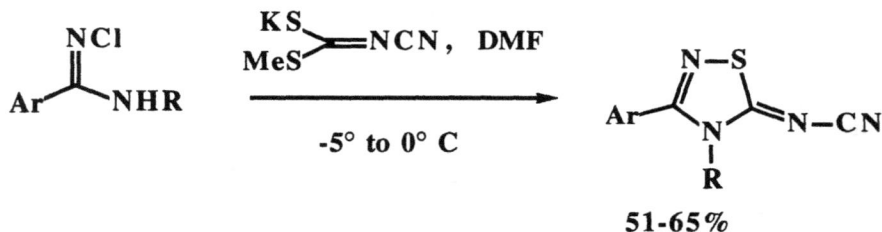

51-65%

**IV.K.8-2**  G. L' abbe and S. Leurs, *J. Chem. Soc., Perkin Trans 1*, 181 and *J. Heterocyclic Chem.*, **29**, 17.

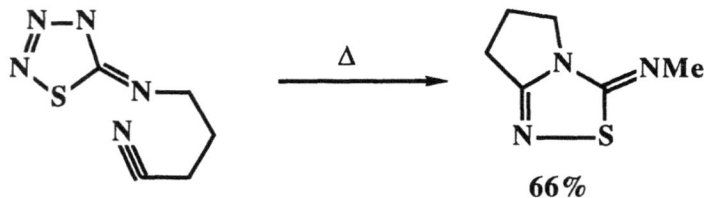

66%

**IV.K.8-3** Y. Sanemitsu et al., *J. Org. Chem.*, **57**, 1053.

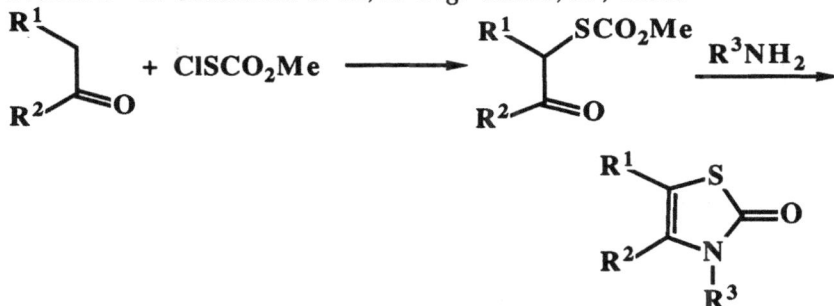

**IV.K.8-4** K. Nagarajan et al., *Ind. J. Chem.*, **31B**, 310.

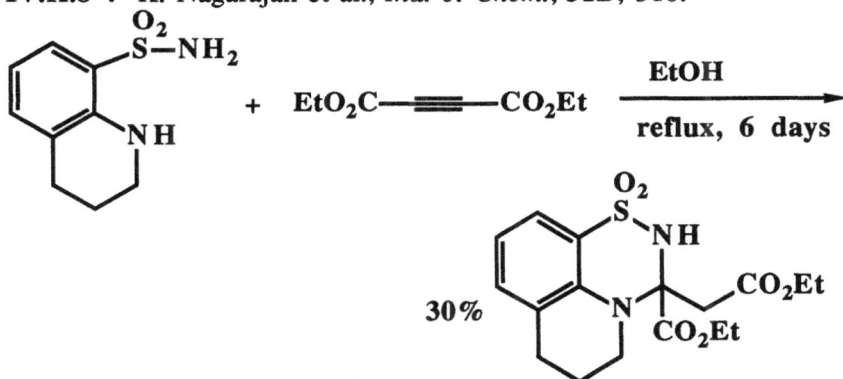

## IV.L. Other Heterocycles

**IV.L-1** M. Regitz et al., *Tetrahedron Lett.*, **33**, 1049; R.K. Bansal et al., *Ind. J. Chem.*, **31B**, 254.

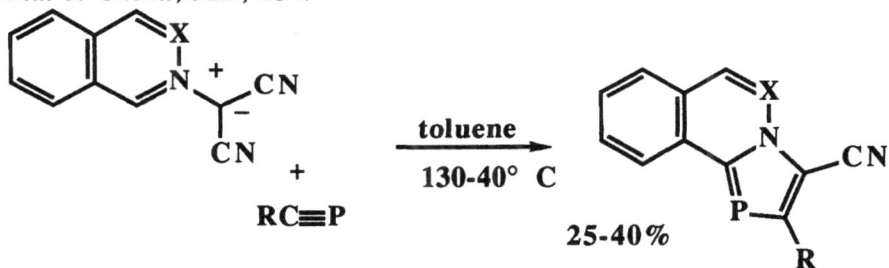

X = CH,N

**IV.L-2** S. Shatzmiller and S. Bercovici, *Liebigs Ann. Chem.*, 997.

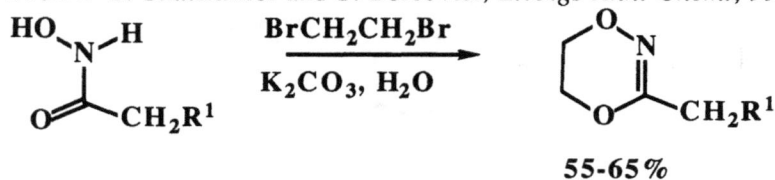

55-65%

---

**IV.L-3** M.R. Bryce et al., *J. Chem. Soc., Perkin Trans. 1*, 2295.

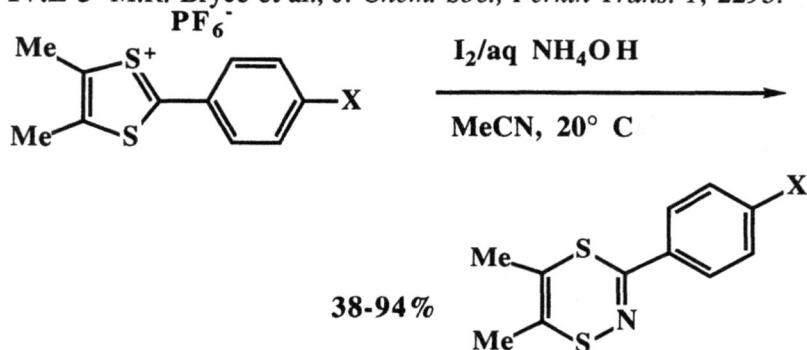

38-94%

---

**IV.L-4** R. Sato et al., *Tetrahedron Lett.*, **33**, 947; E.I. Troyansky, G.I. Nikishin et al., *Synlett*, 233; R.M. Pagni, G.W. Kabalka et al., *Tetrahedron Lett.*, **33**, 7709.

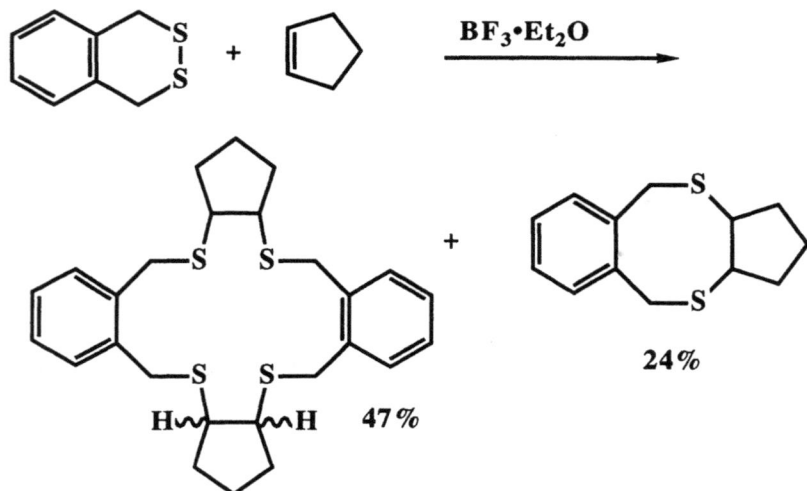

47%

24%

**IV.L-5**  M. Tanaka et al., *Chem. Lett.*, 45.

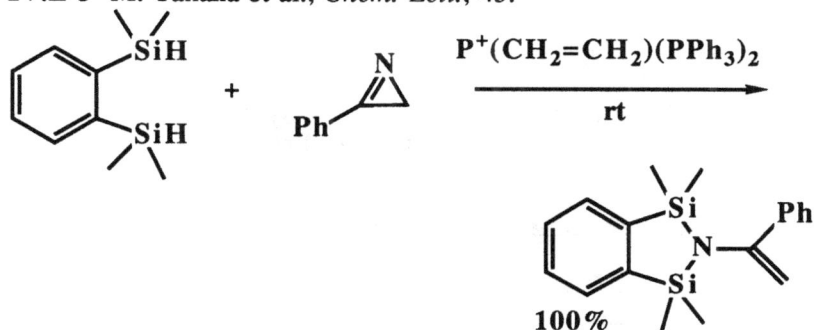

100%

---

**IV.L-6**  W. Adams and R. Albert, *Tetrahedron Lett.*, **33**, 8015.

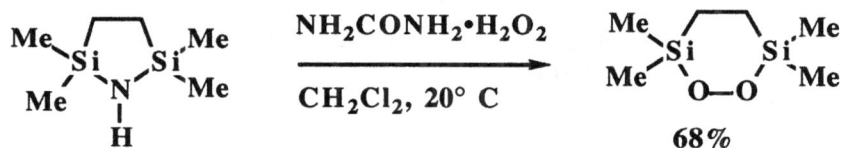

68%

---

**IV.L-7**  H. Fujihara et al., *J. Chem Soc., Perkin Trans.1*, 2583.

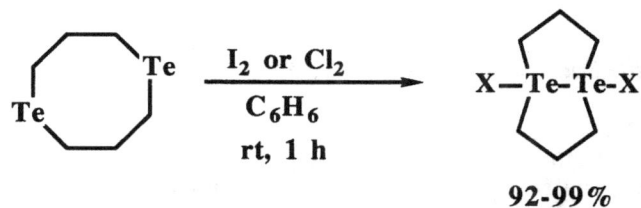

92-99%

## IV.M.    Reviews

**IV.M-1**  X. Creary, *Acc. Chem. Res.*, **25**, 31.

Review:  "Reactions of Halodiazirines by Sn2 and Electron Transfer Initiated Processes."

**IV.M-2**  C.H. Heathcock, *Angew. Chem. Int. Ed. Engl.*, **31**, 665.

Review:   "The Enchanting Alkaloids of Yusuriha."

**IV.M-3**  G. Hoornaert, *Bull. Soc. Chem., Belg.*, **101**,53.

Lecture:    "2(1H-Pyrazinones and 2H-1,4-Oxazin-2 ones as Synthetic Tools for other Heterocycles."

**IV.M-4**  E. Schmitz et al., *Bull. Soc. Chem., Belg.*, **101**, 61.

Lecture:   "Ring Transformations of Pyrazolones via Azo-Olefins."

**IV.M-5**  E.V. Babaev and N.S. Zefirov, *Bull. Soc. Chem., Belg.*, **101**, 67.

Review:   "Ring-Transformation Graphs in Heterocylic Chemistry."

**IV.M-6**  K. Akiba, *Bull. Soc. Chem., Belg.*, **101**, 339.

Lecture:   "Heteroaromatic Cations as Key Intermediates for the Synthesis of Functionalized Dihydro-Heteroaromatic Compounds."

**IV.M-7**  A.R. Katritzky, *Bull. Soc. Chem., Belg.*, **101**, 409.

Lecture:   "N-Substituted Benzotriazoles: Properties, Reactivties and Synthetic Utility."

**IV.M-8** J. Chuche et al., *Bull. Soc. Chem., Belg.*, **101**, 415.

Lecture: **"New Syntheses and Reactions of Chiral Acetylenic Oxiranes."**

**IV.M-9** A. Dondoni, *Bull. Soc. Chem., Belg.*, **101**, 433.

Lecture: **"Acyclic Diastereoselective Synthesis Using Functionalized Thiazoles. Routes to Carbohydrates and Related Natural Products."**

**IV.M-10** N. DeKimpe and C. Stevens, *Bull. Soc. Chem., Belg.*, **101**, 569.

Lecture: **"Synthesis of Azaheterocycles from Functionalized Imines."**

**IV.M-11** J. Sauer, *Bull. Soc. Chem., Belg.*, **101**, 521.

Lecture: **"From Heterocycles to New Heterocycles and Carbocycles by Pericyclic Reactions."**

**IV.M-12** G. Heinisch, *Bull. Soc. Chem., Belg.*, **101**, 579.

Lecture: **"Recent Advances in Pyridazine Chemistry."**

**IV.M-13** H.T. Teunissen and F. Bickelhaupt, *Bull. Soc. Chem., Belg.*, **101**, 609.

Lecture: **"New Low Coordination Phosphorus Heterocycles."**

**IV.M-14** G. Hajos et al., *Bull. Soc. Chem., Belg.*, **101**, 597.

Lecture: **"New Fused Triazines with Bridgehead Nitrogen: Synthesis and Reactivity."**

**IV.M-15** J. Bergman, G. Lidgren and A. Gogoll, *Bull. Soc. Chem., Belg.*, **101**, 643.

Lecture: "Synthesis and Reactions of Oxazolones from L-Tryptophan and α-Haloacetic Anhydrides."

**IV.M-16** S.V. Ley, *Bull. Soc. Chem., Belg.*, **101**, 641.

Lecture: "New Routes to Biologically Active Heterocyclic Natural Products."

**IV.M-17** L.F. Tietze, M. Buback et al., *Chem. Ber.*, **125**, 2249.

Review: "Effects of High Pressure on the Stereoselectivity of Intermolecular Hetero Diels-Alder Reactions."

**IV.M-18** J. Roncali, *Chem. Rev.*, 711.

Review: "Conjugated Poly(thiophenes): Synthesis Functionalzation and Applications."

**IV.M-19** A. Guggisberg and M. Hesse, *Helv. Chim. Acta*, **75**, 647.

Review: "Meilensteine der Alkaloids Forschung in der Helvetica Chimica Acta, 1918-1991."

**IV.M-20** D. Seebach et al., *Helv. Chim. Acta*, **75**, 913.
Review: "Structure and Reactivity of Five- and Six Ring N,N-, N,O- and O,O- Acetals: A Lesson in Allylic 1,3-Strain ."

**IV.M-21** K. Afarinkia, *Heterocycles*, **34**, 369.

Review: "Four-Memebered Heterocycles containing one Phosphorus and One Other Heteroatom."

**IV.M-22**  T. Nomura et al., *Heterocycles*, **33**, 405.

Review:    "Mass Spectrometry of Prenylated
           Flavonoids."

---

**IV.M-23**  K. Achiwa et al., *Heterocycles*, **33**, 435.

Review:    "Development of Modified Chiral Dioxolane
           Bishosphine Ligands and Their Use in
           Efficient Asymmetric Synthesis of Naturally
           Occurring Lignans."

---

**IV.M-24**  T. Okuda et al., *Heterocycles*, **33**, 463.

Review:    "Oligomeric Hydrolyzable Tannins- Their [1]H
           NMR Spectra and Partial Degradation."

---

**IV.M-25**  O. Chupakhin et al., *Heterocycles*, **33**, 931

Review:    "Recent Advances in the Chemistry of as -
           Triazinium salts."

---

**IV.M-26**  W. Boczon, *Heterocycles*, **33**, 1101.

Review:    "Some Stereochemical Aspects of
           Bisquinolizidine Alkaloids Sparteine Type."

---

**IV.M-27**  J.C. Palacios et al., *Heterocycles*, **33**, 973.

Review:    "Haloalkyl IsoThiocyanates, Useful and
           Versatile Reagents in Heterocyclic
           Chemistry."

---

**IV.M-28**  A.R. Katritzky and J.N. Lam, *Heterocyles*, **33**, 1011.

Review:    "Heterocyclic N-Oxides and N-Imides."

**IV.M-29**  T. Hosokawa and S.-I. Murahashi, *Heterocycles*, **33**, 1079.

> **Review:**  **"Synthesis of Oxygen-Containing Heterocycles using Palladium (II) Catalysts."**

**IV.M-30**  M. Begtrup, *Heterocycles*, **33**, 1129.

> **Review:**  **"A General Description of the Reactivity of Heteroaromatic Compouds Based on the Donor Acceptor Concept."**

**IV.M-31**  A. Albini, *Heterocycles*, **34**, 1973.

> **Review:**  **"Alkaliod N-oxides"**

**IV.M-32**  R.E. Niziurski-Mann and M.P. Cava, *Heterocycles*, **34**, 2003.
> **Review:**  **"Synthesis of Mixed Thiophene-Pyrrole Heterocycles."**

**IV.M-33**  S. Radl and D. Bouzard, *Heterocycles*, **34**, 2143.

> **Review:**  **"Recent Advances in the Synthesis of Antibacterial Quinolones."**

**IV.M-34**  A.R. Katritzky and W.-Q. Fan, *Heterocycles*, **34**, 2179.

> **Review:**  **"Mechanisms and Rates of the Electrophilic Substitution Reactions of Heterocycles."**

**IV.M-35**  M. Alvarez and J.A. Joule, *Heterocycles*, **34**, 2385.

> **Review:**  **"Synthesis of Pyridoacridines."**

**IV.M-36**  G. Pattenden, *J. Heterocylic Chem.*, **29**, 607.

> **Review:**  **"Synthesic Studies with Natural Oxazoles and Thiazoles."**

**IV.M-37**  C.W. Reese, *J. Heterocylic Chem.*, **29**, 639.

   **Review:**  **"Polysulfur-Nitrogen Heterocyclic Chemistry."**

---

**IV.M-38**  W.N. Speckamp, *J. Heterocylic Chem.*, **29**, 653.

   **Review:**  **" New Aspects of Old Reaction.  Cations and Radicals in Heterocyclic Synthesis."**

---

**IV.M-39**  P. Catsoulacos and D. Catoulacos, *J. Heterocylic Chem.*, **29**, 675.

   **Review:**  **"On the Synthesis of Pyrido[3,2,1kl]phenothiazine Quino[γ,1-bc][1,4]benzothiazepine and Their Derivatives."**

---

**IV.M-40**  M.P. Georgiadis et al., *Org. Prep. Proced. Int.*, **24**, 95.

   **Review:**  **"Oxidative Rearrangement of Furylcarbinols to 6-Hydroxy-2H-Pyran-3(6H)-ones, a Useful Synthon for the Preparation of a Variety of Heterocyclic Compounds."**

---

**IV.M-41**  S. Eguchi et al., *Org. Prep. Proced. Int.*, **24**, 209.

   **Review:**  **"The Aza-Wittig Reaction in Heterocyclic Synthesis."**

---

**IV.M-42**  C.-L.J. Wang and M.A. Wuonola, *Org. Prep. Proced. Int.*, **24**, 583.

   **Review:**  **" Recent Progress in the Synthesis and Reactions of Substituted Piperides."**

---

**IV.M-43**  R.D. Katsarava and D.P. Kharadze, *Russ. Chem. Rev.*, **61**, 87.

   **Review:**  **"Heterocyclic Bifunctional Monomers in the Synthesis of Polymers."**

**IV.M-44**  F.A. Lakhvich et al., *Russ. Chem. Rev.*, **61**, 243.

   Review:  "Heteroprostanoids: Synthesis and Biological Activity."

**IV.M-45**  M.G. Voronkov and V.I. Knutov, *Russ. Chem. Rev.*, **60**, 1293 (1991).

   Review:  "Nitrogen and Sulphur Containing Macroheterocycles and their Complexes. New Low Molecular Models of Enzymes."

**IV.M-46**  M.V. Ul'ev et al., *Russ. Chem. Rev.*, **60**, 1309 (1991).

   Review:  "Thiophenethiols."

**IV.M-47**  R.A. Kuroyan, *Russ. Chem. Rev.*, **60**, 1368 (1991).

   Review:  "Synthesis of Spiro[piperidine-4,c'n hetero(carbo)cycles]."

**IV.M-48**  V.N. Kalinin, *Synthesis*, 413.

   Review:  "Carbon-Carbon Bond Formation in Heterocycles using Ni- and Pd- Catalyzed Reactions."

**IV.M-49**  K.M. Doxsee et al., *Synlett*, 13.

   Review:  "Titanium Metallacycles as Intermediates in the Synthesis of Acyclic and Heterocyclic Compounds."

**IV.M-50**  H.-J. Knolker, *Synlett*, 371.

   Review:  "Iron-Mediated Synthesis of Heterocyclic Ring Systems and Applications in Alkaloid Chemistry."

**IV.M-51**  C.M. Lukehart, *Synlett*, 681.

    **Review:**    **"Transformation of Dialkynylelementa Species to Exoalkylidene-1-elementacyclobut-2-enyl ligands: A Novel Route to Unsaturated Four Membered Heterocycles."**

---

**IV.M-52**  G.H. Posner et al., *Tetrahedron*, **48**, 9111.

    **Review:**    **"Diels-Alder Cycloaddtions of 2-Pyrones and Pyridones."**

# V

# PROTECTING GROUPS

## V.A. Hydroxyl Protecting Groups

**V.A-1** T. Akiyama, H. Shima and S. Ozaki, *Synlett*, 415; T. Ohsawa et al., *Tetrahedron Lett.*, **33**, 5555; R.B. Lennox, G. Just et al., *J. Org. Chem.*, **57**, 1777.

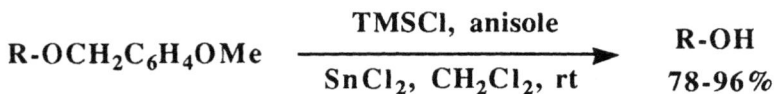

$$\text{R-OCH}_2\text{C}_6\text{H}_4\text{OMe} \xrightarrow[\text{SnCl}_2,\ \text{CH}_2\text{Cl}_2,\ \text{rt}]{\text{TMSCl, anisole}} \begin{array}{c} \text{R-OH} \\ 78\text{-}96\% \end{array}$$

---

**V.A-2** S.V. Ley et al., *Tetrahedron Lett.*, **33**, 4767; M. Cai et al., *Synth. Commun.*, **22**, 2653; J. Gras and D. Poucet, *ibid.*, **22**, 405.

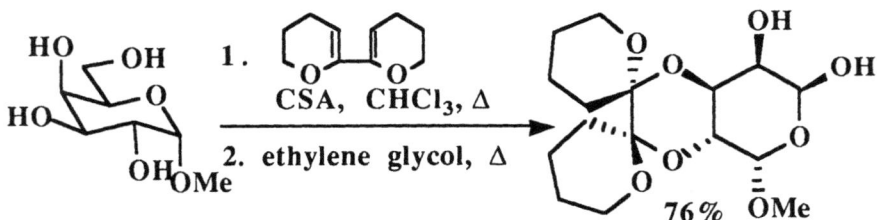

---

**V.A-3** T. Ziegler et al., *Synthesis*, 1013; J.S. Davies et al., *J.Chem. Soc., Perkin Trans. 1*, 3043.

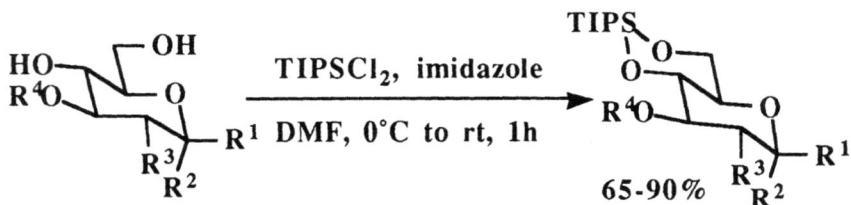

**V.A-4** S. Colin-Messager et al., *Tetrahedron Lett.*, **33**, 2689.

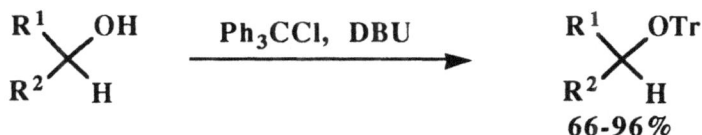

$$\underset{R^2}{\overset{R^1}{\diagdown}}\!\!\!\!\overset{OH}{\underset{H}{\diagup}} \quad \xrightarrow{Ph_3CCl, \ DBU} \quad \underset{R^2}{\overset{R^1}{\diagdown}}\!\!\!\!\overset{OTr}{\underset{H}{\diagup}}$$

66-96%

---

**V.A-5** P. Klein and W.L. Nelson, *J. Med. Chem.*, **35**, 4589.

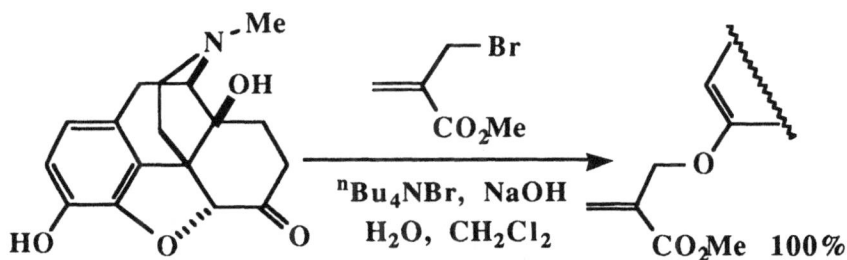

With reagents: $CH_2{=}C(CH_2Br)CO_2Me$; $^nBu_4NBr$, NaOH, $H_2O$, $CH_2Cl_2$ → 100%

---

**V.A-6** T. Kusama et al., *Chem. Pharm. Bull.*, **40**, 1718; A. Banerji et al., *Tetrahedron Lett.*, 33, 5129; E. Dunach et al., *ibid.*, **33**, 2485; B.A. Marples et al., *Synlett.*, 646.

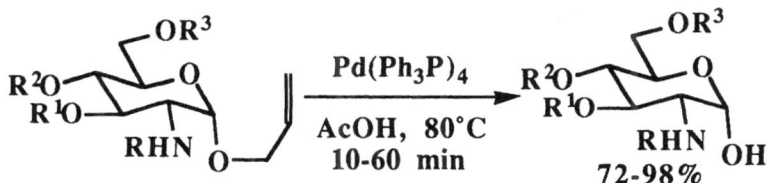

Reagents: $Pd(Ph_3P)_4$, AcOH, 80°C, 10-60 min → 72-98%

---

**V.A-7** T. Ziegler et al., *Tetrahedron Lett.*, **33**, 4413.

**Diastereoselective Formation of Pyruvylated Glycosides from Partly Protected Sugars and Methyl Pyruvate.**

---

**V.A.8** M.R. Hale and A.H. Hoveyda, *J. Org. Chem.*, **57**, 1643.

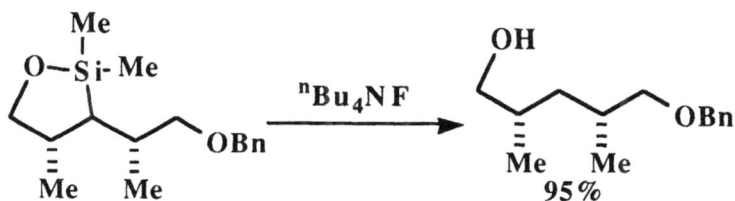

Reagents: $^nBu_4NF$ → 95%

**V.A-9**  D.H.R. Barton and J. Zhu, *Tetrahedron*, **48**, 8337.

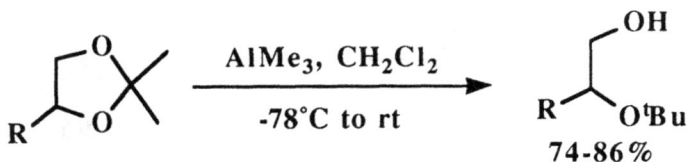

74-86%

**V.A-10**  P.K. Chowdhury, *J. Chem. Res. (S)*, 68.

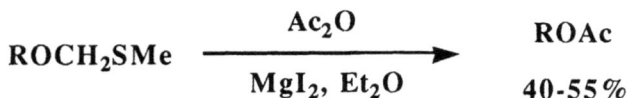

$$ROCH_2SMe \xrightarrow[\text{MgI}_2,\ \text{Et}_2O]{\text{Ac}_2O} ROAc$$

40-55%

**V.A-11**  K. Tanemura et al., *Bull. Chem. Soc. Jpn.*, **65**, 304; T. Nishiguchi et al., *J. Chem. Soc., Perkin Trans. 1*, 153; F. Chavey and R. Godinez, *Synth. Commun.*, **22**, 159; V. Bhuma and M.L. Kantam, *ibid.*, **22**, 2941; R.K. Guy and R.A. DiPietro, *ibid.*, **22**, 687; J. Brussee et al., *ibid.*, **22**, 2781.

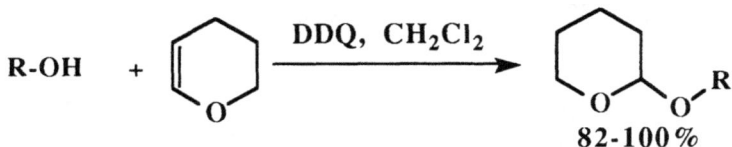

82-100%

**V.A-12**  E.J. Corey and G.B. Jones, *J. Org. Chem*, **57**, 1028.

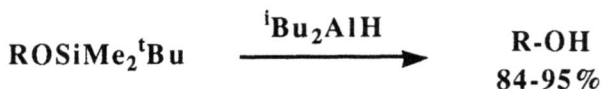

$$ROSiMe_2{}^tBu \xrightarrow{{}^iBu_2AlH} R\text{-}OH$$

84-95%

**V.A-13**  Y. Watanabe et al., *J. Chem. Soc., Chem. Commun.*, 681.

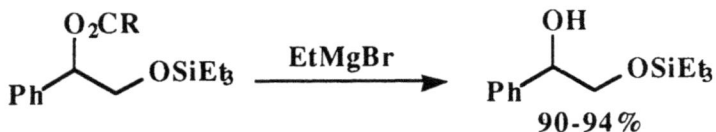

90-94%

**V.A-14**  S. Ray et al., *Synth. Commun.*, **22**, 2703.

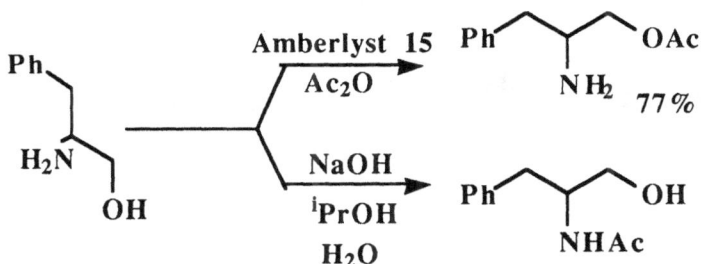

**V.A-15**  H. Seto and L.N. Mander, *Synth. Commun.*, **22**, 2823.

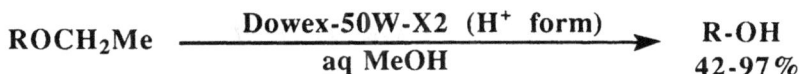

$$\text{ROCH}_2\text{Me} \xrightarrow[\text{aq MeOH}]{\text{Dowex-50W-X2 (H}^+\text{ form)}} \begin{array}{c}\text{R-OH} \\ \text{42-97\%}\end{array}$$

**V.A-16**  H.K. Patney, *Synlett.*, 567;  D. Dawkins and P.R. Jenkins, *Tetrahedron Asym.*, **3**, 833.

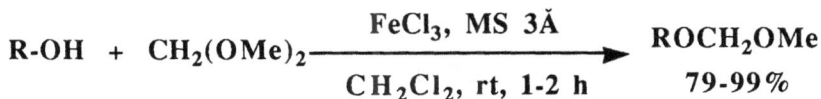

$$\text{R-OH} + \text{CH}_2(\text{OMe})_2 \xrightarrow[\text{CH}_2\text{Cl}_2,\ \text{rt, 1-2 h}]{\text{FeCl}_3,\ \text{MS 3Å}} \begin{array}{c}\text{ROCH}_2\text{OMe} \\ \text{79-99\%}\end{array}$$

**V.A-17**  P. Braun, H. Waldmann and H. Kunz, *Synlett*, 39;  B. Wirz and W. Walther, *Tetrahedron Asym.*, **3**, 1049;  G. Pedrocchi-Fentoni and S. Servi, *J. Chem. Soc., Perkin Trans. 1*, 1029.

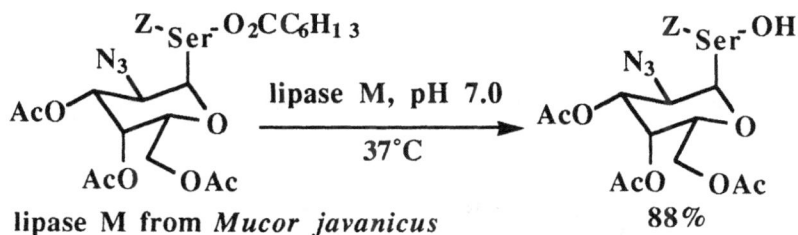

lipase M from *Mucor javanicus*     88%

**V.A-18**  K. Tanemura et al., *J. Chem. Soc., Perkin Trans. 1*, 2997.

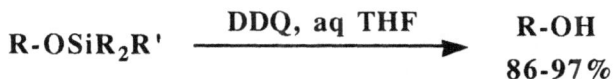

$$\text{R-OSiR}_2\text{R}' \xrightarrow{\text{DDQ, aq THF}} \begin{array}{c}\text{R-OH} \\ \text{86-97\%}\end{array}$$

**V.A-19**  P. Deshong et al., *J. Org. Chem.*, **57**, 2492;  W. Zhang and M.J. Robins, *Tetrahedron Lett.*, **33**, 1177;  E.J. Corey and K.Y. Yi, *ibid.*, **33**, 2289.

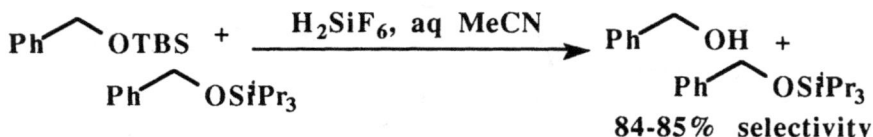

$$\text{Ph}\diagup\text{OTBS} + \text{Ph}\diagup\text{OSiPr}_3 \xrightarrow{\text{H}_2\text{SiF}_6,\ \text{aq MeCN}} \text{Ph}\diagup\text{OH} + \text{Ph}\diagup\text{OSiPr}_3$$

**84-85% selectivity**

## V.B.    Amine Protecting Groups

**V.B-1**  J.D. Prugh et al., *Synth. Commun.*, **22**, 2357.

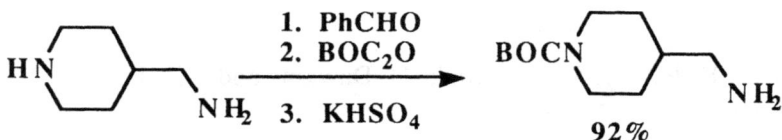

$$\xrightarrow[\text{3. KHSO}_4]{\substack{\text{1. PhCHO} \\ \text{2. BOC}_2\text{O}}}$$

**92%**

---

**V.B-2**  A.P. Davies and T.J. Egan, *Tetrahedron Lett.*, **33**, 8125.

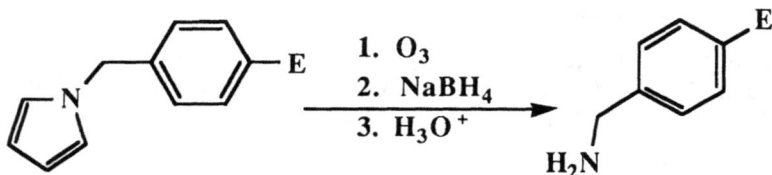

$$\xrightarrow[\text{3. H}_3\text{O}^+]{\substack{\text{1. O}_3 \\ \text{2. NaBH}_4}}$$

---

**V.B-3**  S. Knapp, K.Y. Chen, et al., *J. Org. Chem.*, **57**, 6239.

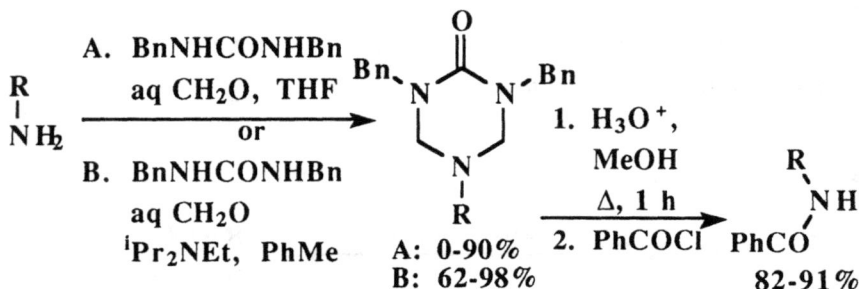

A. BnNHCONHBn
   aq CH$_2$O, THF
   **or**
B. BnNHCONHBn
   aq CH$_2$O
   $^i$Pr$_2$NEt, PhMe

A: 0-90%
B: 62-98%

1. H$_3$O$^+$,
   MeOH
   $\Delta$, 1 h
2. PhCOCl

**82-91%**

**V.B-4**  S.-I. Murahashi et al., *Tetrahedron Lett.*, **33**, 6991.

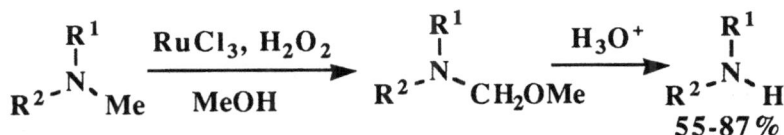

$$R^2\text{-}N(R^1)\text{-}Me \xrightarrow[\text{MeOH}]{RuCl_3,\ H_2O_2} R^2\text{-}N(R^1)\text{-}CH_2OMe \xrightarrow{H_3O^+} R^2\text{-}N(R^1)\text{-}H$$

**55-87%**

---

**V.B-5**  A. Merzouk and F. Guibé, *Tetrahedron Lett.*, **33**, 477.

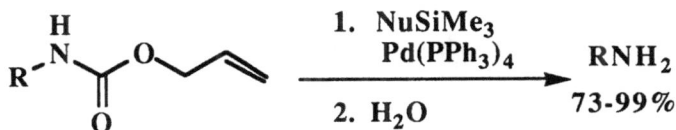

$$R\text{-}NH\text{-}C(=O)\text{-}O\text{-allyl} \xrightarrow[\text{2. } H_2O]{\substack{\text{1. NuSiMe}_3 \\ \text{Pd(PPh}_3)_4}} RNH_2$$

**73-99%**

---

**V.B-6**  F.M. Moracci et al., *J. Chem. Soc., Perkin Trans. 1*, 2001; E. Ravina et al., *Heterocycles*, **34**, 1303.

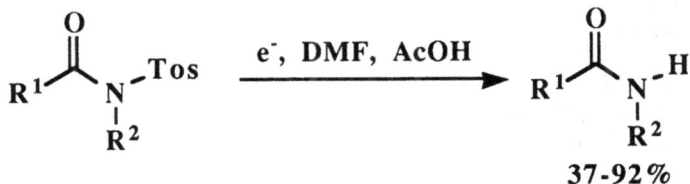

$$R^1\text{-}C(=O)\text{-}N(Tos)(R^2) \xrightarrow{e^-,\ DMF,\ AcOH} R^1\text{-}C(=O)\text{-}N(H)(R^2)$$

**37-92%**

---

**V.B-7**  K. Fukumoto et al., *Heterocycles*, **33**, 851.

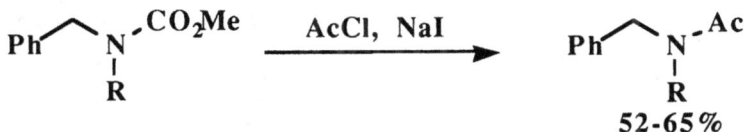

$$Ph\text{-}CH_2\text{-}N(CO_2Me)(R) \xrightarrow{AcCl,\ NaI} Ph\text{-}CH_2\text{-}N(Ac)(R)$$

**52-65%**

---

**V.B-8**  T. Pietgonka and D. Seebach, *Angew. Chem., Int. Ed. Engl.*, **31**, 1481; U.R. Kalkote et al., *Org. Prep. Proc. Int.*, **24**, 83; C. Safak et al, *J. Med. Chem.*, **35**, 1297.

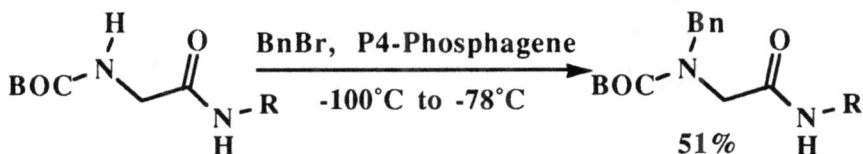

$$BOC\text{-}NH\text{-}CH_2\text{-}C(=O)\text{-}NH\text{-}R \xrightarrow[\text{-100°C to -78°C}]{BnBr,\ P4\text{-}Phosphagene} BOC\text{-}N(Bn)\text{-}CH_2\text{-}C(=O)\text{-}NH\text{-}R$$

**51%**

**V.B-9**  F. Johnson et al., *J. Am. Chem. Soc.*, **114**, 4923.

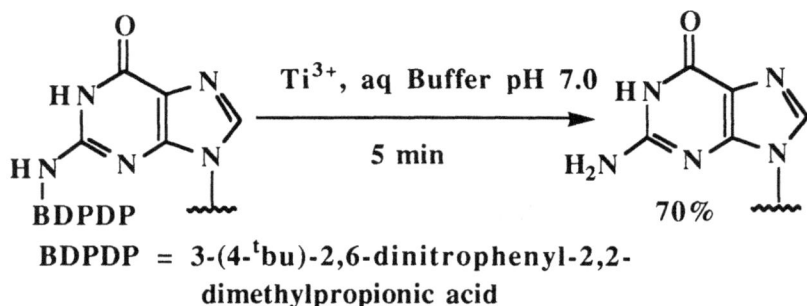

BDPDP = 3-(4-$^t$bu)-2,6-dinitrophenyl-2,2-
             dimethylpropionic acid

## V.C.    Carboxyl Protecting Groups

**V.C-1**  F.J. Urban and B.S. Moore, *J. Heterocyclic Chem.* **29**, 431;  R.
Chenevert and R. Martin, *Tetrahedron Asym.*, **3**, 199;  M. Mekrami and S.
Sicsic, *ibid.*, **3**, 431;  C. Fuganti et al, *ibid.*, **3**, 383;  G. Jaouen, J.A.S.
Howell et al., *ibid.*, **3**, 375;  T. Anthonsen et al, *ibid.*, **3**, 65;  B. Wirz et al.,
*ibid.*, **3**, 137;  A.E. Walts et al., *J. Org. Chem.*, **57**, 3525;  K. Achiwa et al.,
*Chem. Pharm. Bull.*, **40**, 1083.

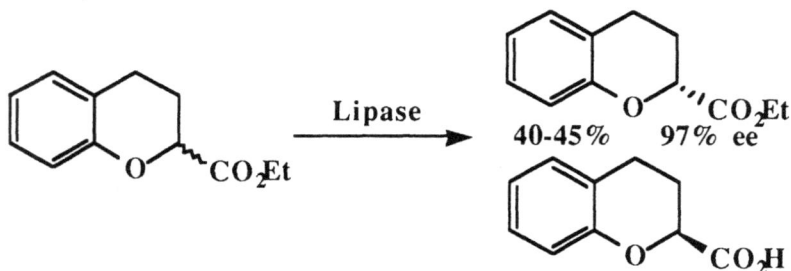

**V.C-2**  D.P. Phillion and J.K. Pratt, *Synth. Commun.*, **22**, 13

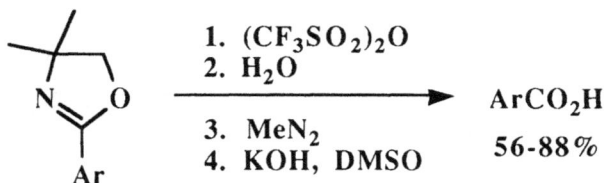

**V.C-3**  H.M. Hugel et al., *Synth. Commun.*, **22**, 693.

$$RCO_2H \underset{Na_2S}{\overset{ClCH_2CN, \ Et_3N}{\rightleftharpoons}} RCO_2CH_2CN$$

74-90%                                           78-96%

---

**V.C-4**  C. Deb and B. Basu, *Ind. J. Chem.*, **31B**, 131;  G.W. Kabalka et al., *Synth. Commun.*, **22**, 1792.

$$RCO_2R' \xrightarrow[\text{2. } H_3O^+, \ rt]{\substack{\text{1. } (Bu_3Sn)_2O \\ CCl_4, \ \Delta}} RCO_2H$$

65-95%

---

**V.C-5**  C.R. Schmid, *Tetrahedron Lett.*, **33**, 757.

$$\xrightarrow[\Delta, \ 3 \ h]{90\% \ HCO_2H} RCO_2H$$

80-95%

## V.D.     Protecting Groups for Aldehydes and Ketones

**V.D-1**  B. Kumar et al., *Ind. J. Chem.*, **30B**, 869 (91);  Y.H. Kim et al., *Synth. Commun.*, **22**, 1585.

$$\xrightarrow[MeCN]{Fe(ClO_4)_3}$$

70-89%

---

**V.D-2**  D.L. Comins and A. Dehghani, *Tetrahedron Lett.*, **33**, 6299;  S.C. Conway and G.W. Gribble, *Synth. Commun.*, **22**, 2987.

$$\xrightarrow[LDA, \ -78°C, \ THF]{}$$

73-92%

**V.D-3**  Y. Kume and H. Ohta, *Tetrahedron Lett.*, **33**, 6367.

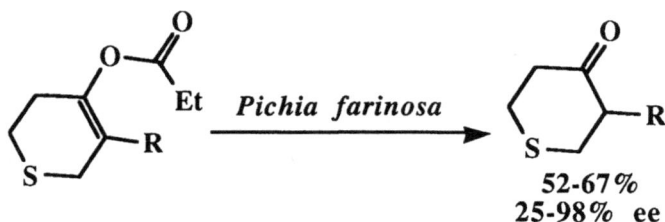

52-67%
25-98%  ee

---

**V.D-4**  A. Tandon and S. Ray, *Ind. J. Chem.*, **31B**, 58;  M.C. Pirrung and D.S. Nunn, *Tetrahedron Lett.*, **33**, 6591;  K.S. Bruzik and M.-D. Tsai, *J. Am. Chem. Soc.*, **114**, 6361;  P. Hudson and P.J. Parsons, *Synlett.*, 867; T. Harada, A. Oku et al., *J. Org. Chem.*, **57**, 1637 and *Tetrahedron*, **48**, 8621;  J. Otera et al., *ibid.*, **48**, 1449.

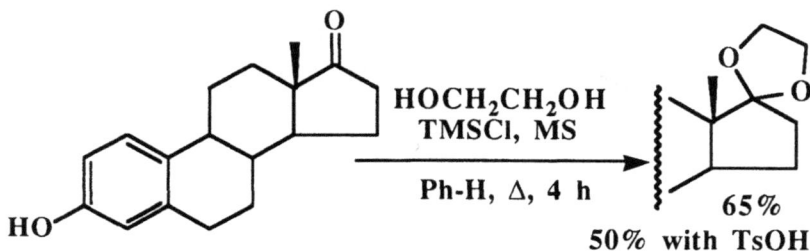

HOCH$_2$CH$_2$OH
TMSCl, MS

Ph-H, Δ, 4 h

65%
50% with TsOH

---

**V.D-5**  P. Kumar et al., *Tetrahedron Lett.*, **33**, 825;  D. Villemin et al., *J. Chem. Soc., Chem. Commun.*, 1192;  T. Gallagher et al., *J. Chem. Soc., Perkin Tran. 1*, 1901.

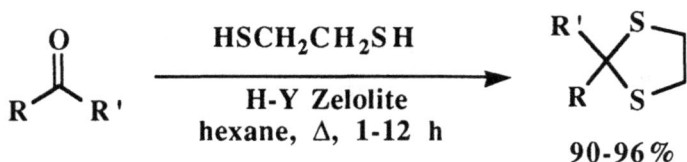

HSCH$_2$CH$_2$SH

H-Y Zelolite
hexane, Δ, 1-12 h

90-96%

---

**V.D-6**  R. Curci et al., *Tetrahedron Lett.*, **33**, 4225;  K. Tanemura et al., *J. Chem. Soc., Chem. Commun.*, 979.

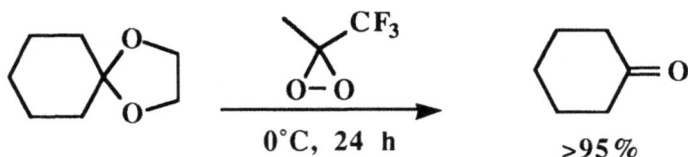

CF$_3$

O–O

0°C, 24 h

>95%

**V.D-7** G.A. Epling and Q. Wang, *Synlett.*, 335 and *Tetrahedron Lett.*, **33**, 5909;  M. Kamata et al., *ibid.*, **33**, 5085;  H. Fla, H. Junjappa et al., *ibid.*, **33**, 8163;  J.A. Soderquist and E. L. Miranda, *J. Am. Chem. Soc.*, **114**, 10078.

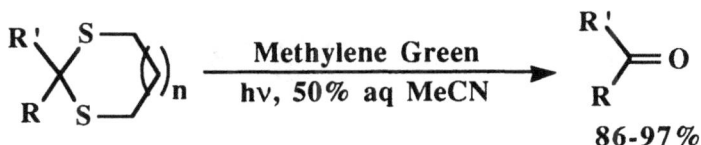

86-97%

**V.D-8** S.S. Elmorsy et al., *Tetrahedron Lett.*, **33**, 1657;  K.L. Ford and E.J. Roskamp, *ibid.*, **33**, 1135;  P. Cotelle and J.-P. Catteau, *ibid.*, **33**, 3855;  S. Yue et al., *Synth. Commun.*, **22**, 1217;  N.N. Rao et al., *ibid.*, **22**, 1299.

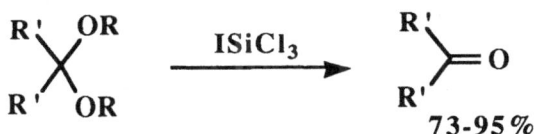

73-95%

## V.E.    Amino Acid Protection

**V.E-1** J.S. Bajwa et al., *Tetrahedron Lett.*, **33**, 2299 and 2955;  A.J. Pallenberg *ibid.*, **33**, 7693;  J.-P. Mazaleyrat et al., *ibid.*, **33**, 4301.

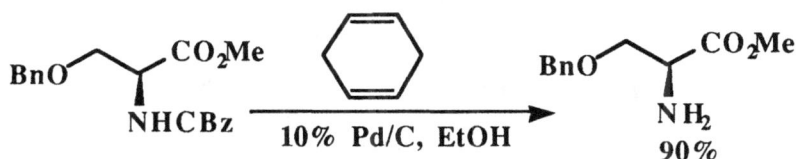

**V.E-2** W. Oppolzer et al., *Helv. Chim. Acta*, **75**, 2572 and 1965.

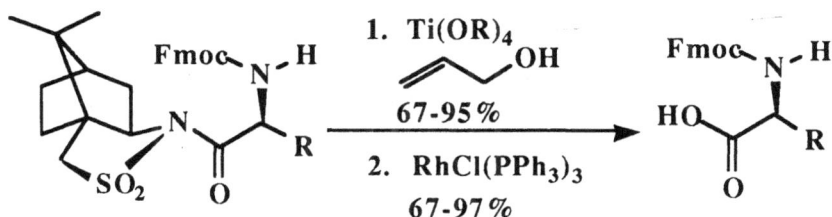

**V.E.-3**  M. Pugniere et al., *Tetrahedron Asym.*, **3**, 1015.

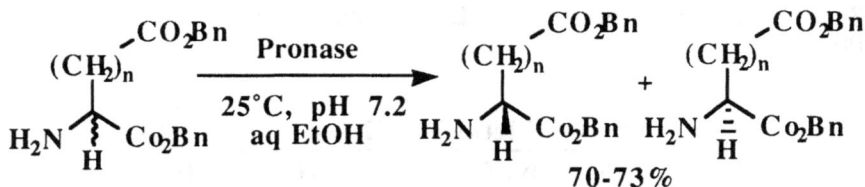

**V.E-4**  R. Ramage and G. Raphy, *Tetrahedron Lett.*, **33**, 385.

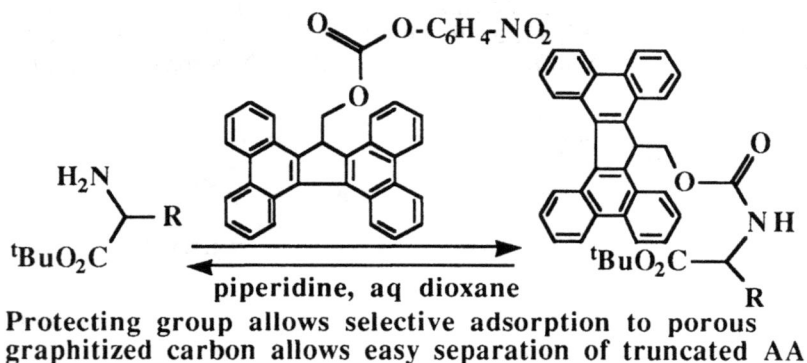

Protecting group allows selective adsorption to porous
graphitized carbon allows easy separation of truncated AA

**V.E-5**  M.J. Crossley and C.W. Tansey, *Aust. J. Chem.*, **45**, 479.

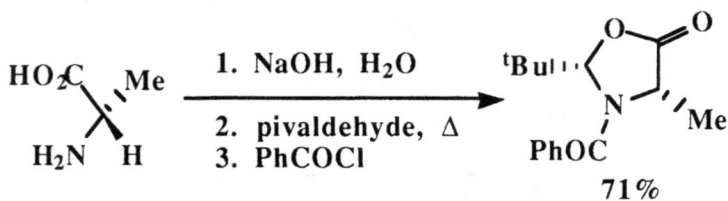

**V.E-6**  U. Schöllkopf et al., *Liebigs Ann. Chem.*, 523.

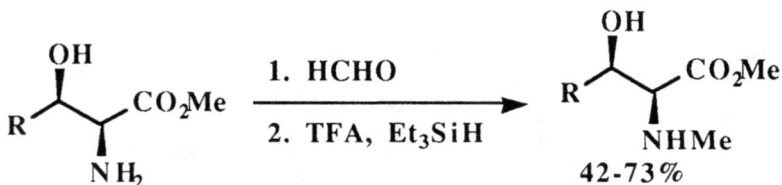

**V.E-7** A.A. Mazurov et al., *Coll. Czech. Chem. Commun.*, **57**, 1495;
A.M. Castano and A.M. Echavarren, *Tetrahedron*, **48**, 3377.

> **p-Azobenzene Carboxamidomethyl Esters - New Colored**
> **Hydrophobic Carboxyl Protecting Groups in Peptide**
> **Synthesis**

**V.E-8** M. Patek and M. Lebl, *Coll. Czech. Chem. Commun.*, **57**, 508.

> **"Safety-Catch" Protecting Groups in Peptide Synthesis**

**V.E-9** J. McNulty and I.W.J. Still, *Synth. Commun.*, **22**, 979.

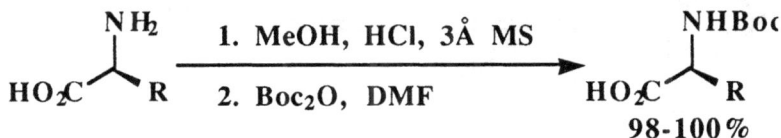

$$\text{HO}_2\text{C} \overset{\text{NH}_2}{\diagup} \text{R} \quad \xrightarrow[\text{2. Boc}_2\text{O, DMF}]{\text{1. MeOH, HCl, 3Å MS}} \quad \text{HO}_2\text{C} \overset{\text{NHBoc}}{\diagup} \text{R}$$
**98-100%**

**V.E-10** G. Barany et al., *J. Org. Chem.*, **57**, 3013.

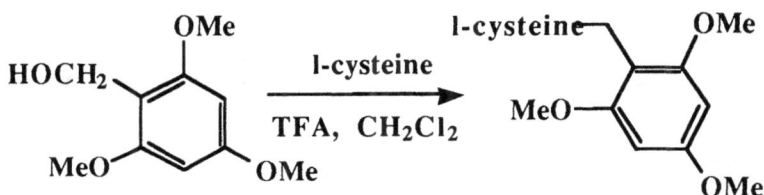

## V.F.    Other Protecting Groups

**V.F-1** W. Holzer et al., *Heterocycles*, **34**, 303.

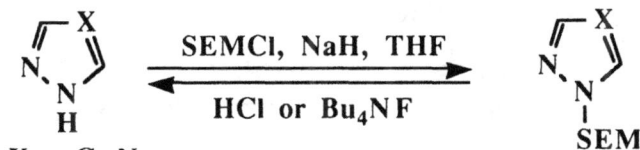

X = C, N
**Suitable N' protecting group in lithiation**
**reactions of pyrazoles and 1,2,4-triazoles**

**V.F-2**  P. Pellon, *Tetrahedron Lett.*, **33**, 4451;  A.G. Mitchell et al., *J. Chem. Soc., Perkin Trans. 1*, 2345;  S. Lagar and G. Guillaumet, *Synth. Commun.*, **22**, 923.

> **Phosphine - Boranes in Synthesis.  Borane as an Efficient Protecting Group in the Preparation of Functionalized Phosphines**

---

**V.F-3**  M.B. Smith et al., *Chem. Lett.*, 247 and *Synth. Commun.*, **22**, 2865.

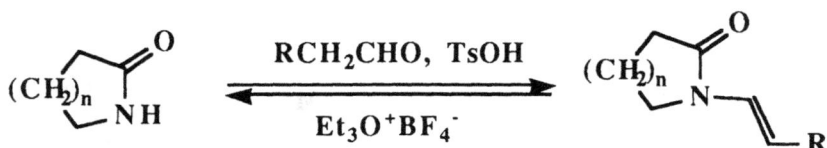

---

**V.F-4**  T.R. Hoye and N.E. Witowski, *J. Am. Chem. Soc.*, **114**, 7291.

**A Protecting Group Strategy for Desymmetrization**

---

**V.F-5**  N. Vijayashree and A.G. Samuelson, *Tetrahedron Lett.*, **33**, 559.

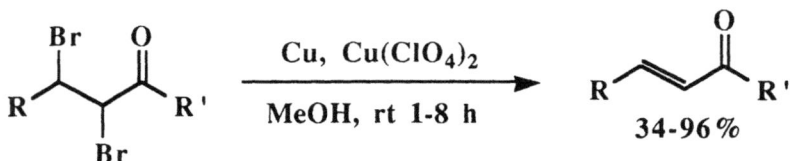

---

**V.F-6**  R.K. Russell et al., *Synth. Commun.*, **22**, 3221;  K. Misra et al., *Ind. J. Chem.*, **31B**, 326.

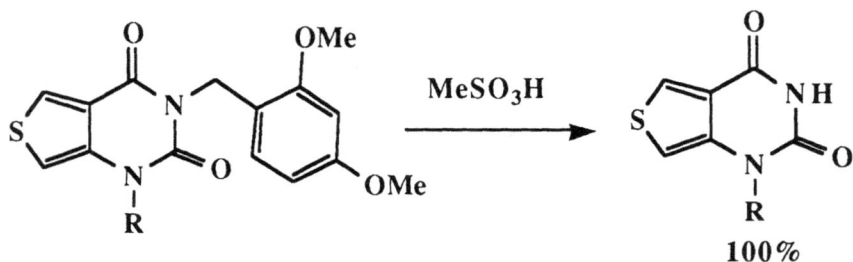

# VI

# USEFUL SYNTHETIC PREPARATIONS

## VI.A.  Functional Group Preparations

## VI.A.I.  Acetals and Ketals

**VI.A.1-1**  K. Higashi et al., *Chem. Pharm. Bull.*, **40**, 2019 and 1042;  T. Mukaiyama et al., *Chem. Lett.*, 2105, 1041 and 1401;  S. Ikegami et al., *Tetrahedron Lett.*, **33**, 3523;  K. Nakanishi et al., *ibid.*, **33**, 4295;  H. Kung et al., *ibid.*, **33**, 1969;  P. Sinay et al., *J. Am. Chem. Soc.*, **114**, 6354; R.R. Schmidt et al., *Liebigs Ann. Chem.*, 371;  K. Briner and A. Vasella, *Helv. Chim. Acta*, **75**, 621.

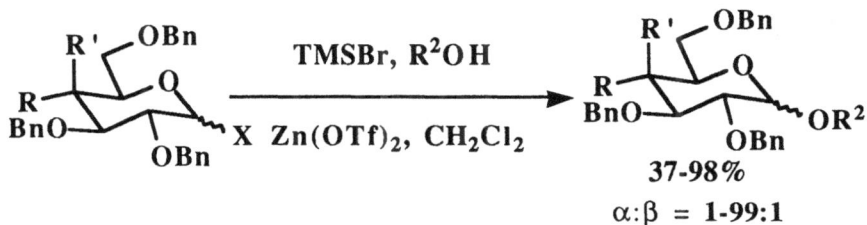

$\alpha{:}\beta = 1\text{-}99{:}1$

---

**VI.A.1-2**  E. Suarez et al., *Angew. Chem., Int. Ed. Engl.*, **31**, 772;  M.A. Brimble and G.M. Williams, *J. Org. Chem.*, **57**, 5818.

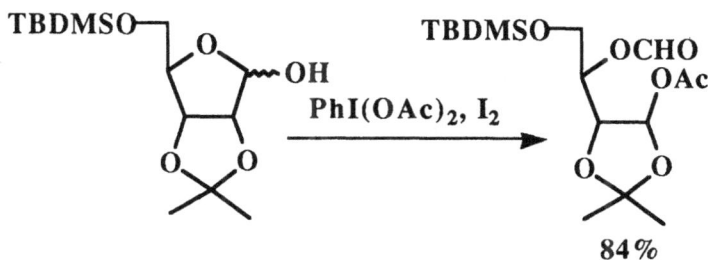

**VI.A.1-3**  J. Yoshida, S. Isoe et al., *J. Org. Chem.*, **57**, 1321.

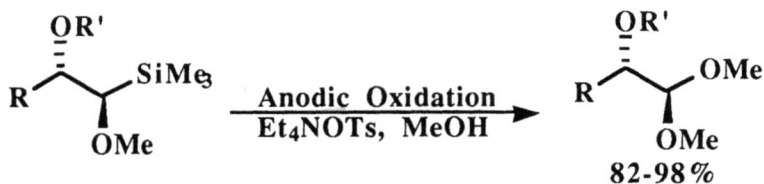

$$\underset{\text{Et}_4\text{NOTs, MeOH}}{\xrightarrow{\text{Anodic Oxidation}}}$$

82-98%

---

**VI.A.1-4**  S. Kusumoto et al., *Tetrahedron Lett.*, **33**, 7165;  E.J. Thomas et al., *J. Chem. Soc., Perkin Trans. 1*, 2223;  G. Stork and G. Kim, *J. Am. Chem. Soc.*, **114**, 1087.

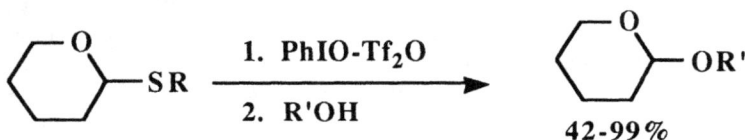

1. PhIO-Tf$_2$O

2. R'OH

42-99%

---

**VI.A.1-5**  B.M. Trost et al., *Angew. Chem., Int. Ed. Engl.*, **31**, 1335.

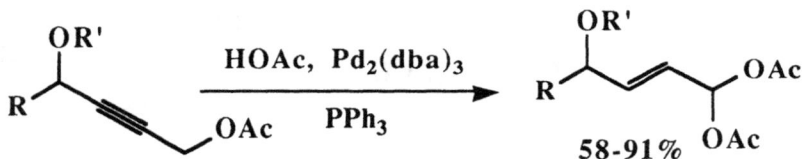

$$\underset{\text{PPh}_3}{\xrightarrow{\text{HOAc, Pd}_2(\text{dba})_3}}$$

58-91%

---

**VI.A.1-6**  B.C. Ranu et al., *J. Org. Chem.*, **57**, 7349.

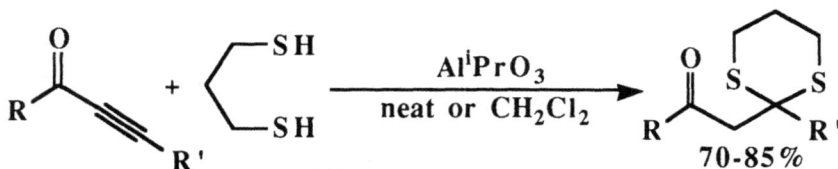

$$\underset{\text{neat or CH}_2\text{Cl}_2}{\xrightarrow{\text{Al}^i\text{PrO}_3}}$$

70-85%

---

**VI.A.1-7**  T. Igumi et al., *J. Heterocyclic Chem.*, **29**, 1625.

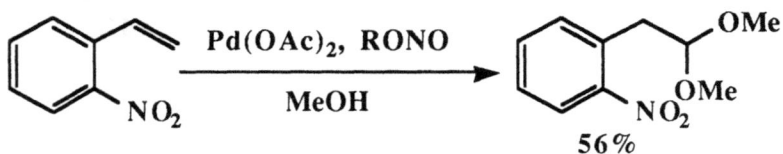

$$\underset{\text{MeOH}}{\xrightarrow{\text{Pd(OAc)}_2,\text{ RONO}}}$$

56%

**VI.A.1-8**  N. Taubken and J. Thiem, *Synthesis*, 517.

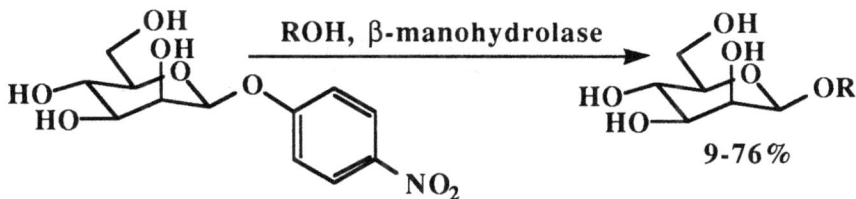

ROH, β-manohydrolase

9-76%

---

## VI.A.2.    Acids and Anhydrides

(see also I.G.2)

**VI.A.2-1**  R.S. Varma and M.E. Hagan, *Tetrahedron Lett.*, **33**, 7719; B.Bhatia and J. Iqbal, *ibid.*, **33**, 7961;  A.K. Singh et al., *ibid.*, **33**, 2307; see also H.C. Brown et al., *ibid.*, **33**, 1037;  R. Hernandez et al., *Synthesis*, 653.

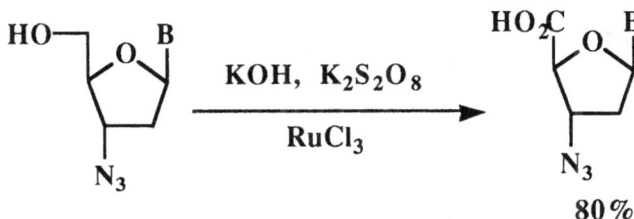

KOH,  $K_2S_2O_8$

$RuCl_3$

80%

---

**VI.A.2-2**  H. Griengl et al., *Synthesis*, 365;  F. Le Gottic et al., *Synth. Commun.*, **22**, 1149 and 1155;  A.L. Gutman et al., *J. Org. Chem.*, **57**, 1063.

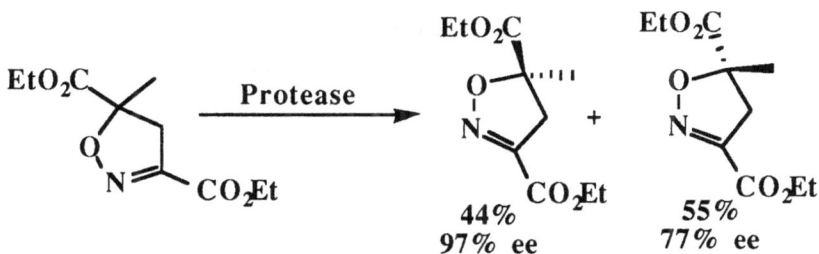

Protease

44%
97% ee

+

55%
77% ee

**VI.A.2-3**  M.J. Dabdoub and L.H. Viana, *Synth. Commun.*, **22**, 1619.

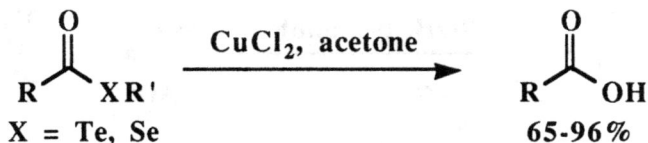

X = Te, Se

65-96%

---

**VI.A.2-4**  E.F. Llama et al., *Org. Prep. Proc. Int.*, **24**, 165;  Y. Goldberg and H. Alper, *J. Org. Chem.*, **57**, 3731.

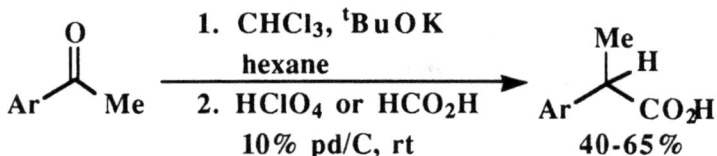

1. CHCl₃, ᵗBuOK
   hexane
2. HClO₄ or HCO₂H
   10% pd/C, rt

40-65%

---

**VI.A.2-5**  K. Jones et al., *Synth. Commun.*, **22**, 3089;  E. Dunach et al., *Tetrahedron*, **48**, 5235.

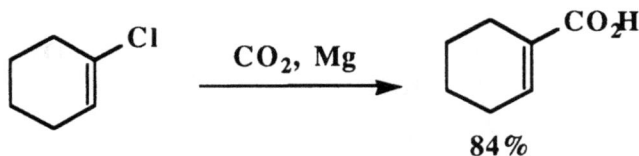

CO₂, Mg

84%

---

**VI.A.2-6**  J.-F. Biellmann et al., *Tetrahedron Lett.*, **33**, 4911;  R.J. Linderman and K. Chen; *ibid.*, **33** 6767.

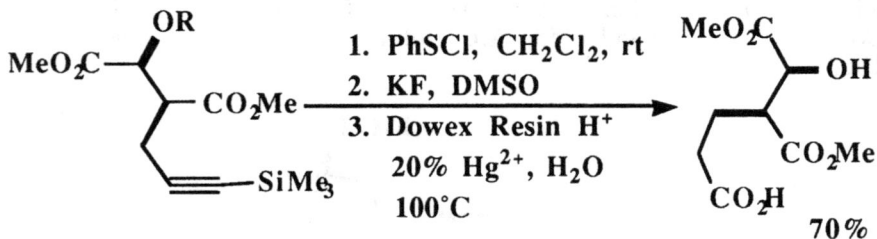

1. PhSCl, CH₂Cl₂, rt
2. KF, DMSO
3. Dowex Resin H⁺
   20% Hg²⁺, H₂O
   100°C

70%

**VI.A.2-7** D.H.R. Barton et al., *Tetrahedron Lett.*, **33**, 5017 and 5013; J.Cs. Jaszberenyi et al.,*ibid.*, **33**, 7299.

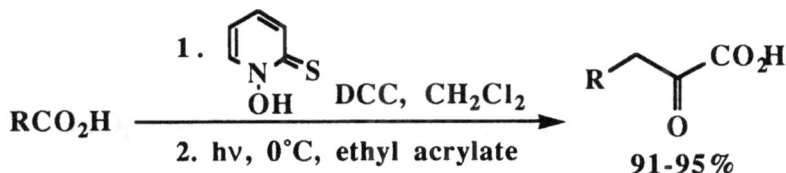

RCO$_2$H $\xrightarrow{\begin{array}{c}1.\ \text{[pyridine-N-OH=S]}\ \text{DCC, CH}_2\text{Cl}_2 \\ \hline 2.\ \text{hv, 0°C, ethyl acrylate}\end{array}}$ R—CH$_2$—C(=O)—CO$_2$H

91-95%

**VI.A.2-8** R.R. Srivastava and G.W. Kabalka, *Tetrahedron Lett.*, **33**, 593.

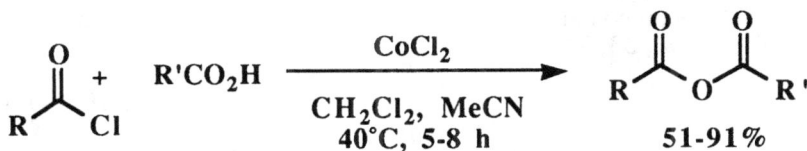

R—C(=O)—Cl + R'CO$_2$H $\xrightarrow{\begin{array}{c}\text{CoCl}_2 \\ \hline \text{CH}_2\text{Cl}_2,\ \text{MeCN} \\ 40°C,\ \text{5-8 h}\end{array}}$ R—C(=O)—O—C(=O)—R'

51-91%

**VI.A.2-9** G.B. Gill et al., *J. Chem. Soc., Perkin Trans. 1*, 2367; G.I. Nikishin et al., *Synthesis*, 917.

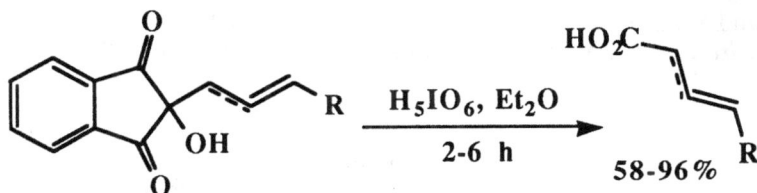

[indandione-OH with diene-R] $\xrightarrow{\begin{array}{c}\text{H}_5\text{IO}_6,\ \text{Et}_2\text{O} \\ \hline \text{2-6 h}\end{array}}$ HO$_2$C—diene—R

58-96%

## VI.A.3.   Alcohols and Related Species

(see also II.B.1, III.A)

**VI.A.3-1** D.J. Ramon and M. Yus, *Tetrahedron*, **48**, 3585;  T.Akiyama, H. Hirofuji, S. Ozaki, *Bull. Chem. Soc. Jpn.*, **65**, 1932.

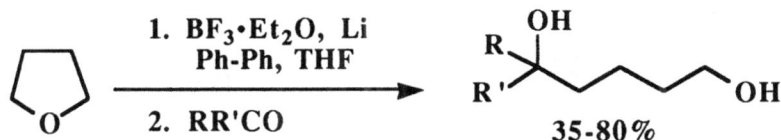

[THF] $\xrightarrow{\begin{array}{c}1.\ \text{BF}_3\cdot\text{Et}_2\text{O, Li} \\ \text{Ph-Ph, THF} \\ \hline 2.\ \text{RR'CO}\end{array}}$ RR'C(OH)—CH$_2$CH$_2$CH$_2$—OH

35-80%

**VI.A.3-2**  M. Bessodes et al., *Tetrahedron Lett.*, **33**, 4317;  T. Tanase, T. Takei, M. Hidai and S. Yano, *J. Chem. Res. (S)*, 252.

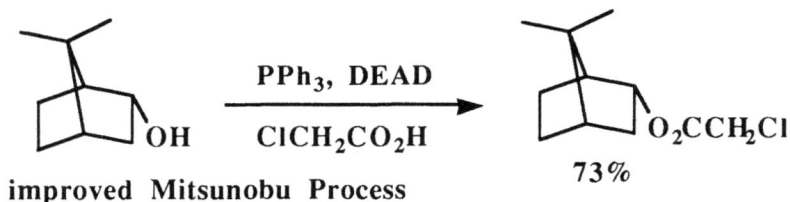

**improved Mitsunobu Process**

---

**VI.A.3-3**  P. Mohr, *Tetrahedron Lett.*, **33**, 2455;  T. Hayashi et al., *ibid.*, **33**, 7185;  M. Koreeda and D.S. Visger, *ibid.*, **33**, 6603;  F.S. Guziec, Jr. and D. Wei, *ibid.*, **33**, 7465.

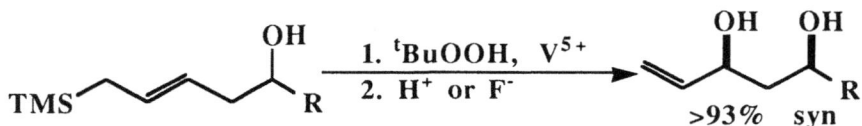

---

**VI.A.3-4**  G.A. Molander and K.L. Bobbitt, *J. Org. Chem.*, **57**, 5031;  H.-J. Liu and W. Luo, *Can. J. Chem.*, **70**, 128;  P. Nussbaumer and A. Stutz, *Tetrahedron Lett.*, **33**, 7507;  A. Mordini et al., *Synlett.*, 803;  S. Maiorana et al., *ibid.*, 315.

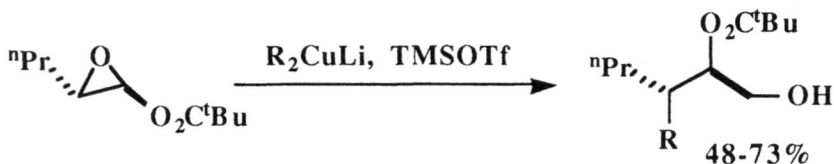

---

**VI.A.3-5**  M. Kato et al., *J. Chem. Soc., Chem. Commun.*, 697.

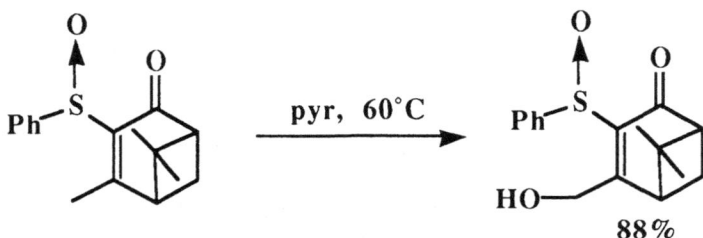

**VI.A.3-6** M. Hirobe et al., *Tetrahedron Lett.*, **33**, 4949; R.B. Rookhuigen et al., *ibid.*, **33**, 1633.

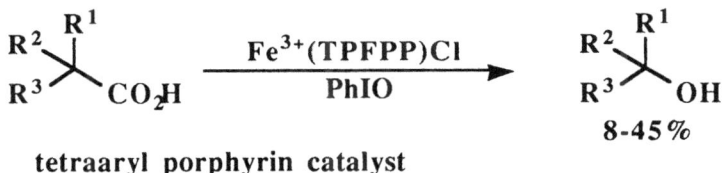

$$R^2\!\!-\!\!\overset{R^1}{\underset{R^3}{C}}\!\!-\!\!CO_2H \quad \xrightarrow[\text{PhIO}]{Fe^{3+}(TPFPP)Cl} \quad R^2\!\!-\!\!\overset{R^1}{\underset{R^3}{C}}\!\!-\!\!OH$$

8-45%

**tetraaryl porphyrin catalyst**

---

**VI.A.3-7** G.M. Sosnouskii et al., *J. Org. Chem. (USSR)*, **28**, 671; C. Bonini et al., *Tetrahedron Lett.*, **33**, 7429; C.M. Rayner and A.D. Westwell, *ibid.*, **33**, 2409; C. Moberg et al., *ibid.*, **33**, 2191; F. Sato et al., *Tetrahedron*, **48**, 5639; C. Bonini and G. Righi, *ibid.*, **48**, 1531; T. Mukaiyama et al., *Chem. Lett.*, 231;

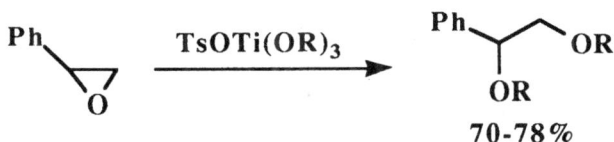

$$\text{Ph}\!-\!\overset{\triangle}{\underset{O}{}} \quad \xrightarrow{TsOTi(OR)_3} \quad \text{Ph}\!-\!\underset{OR}{CH}\!-\!CH_2\!-\!OR$$

70-78%

---

**VI.A.3-8** H.-J. Gain et al., *Synthesis*, 169; T. Fujisawa et al., *Chem. Lett.*, 107; G. Guanti et al., *J. Org. Chem.*, **57**, 1540; H.E. Schink and J.E. Backvall, *ibid.*, **57**, 1588; C. De Micheli, G. Carrea et al., *ibid.*, **57**, 2825; U. Ader and M.P. Schneider, *Tetrahedron Asym.*, **3**, 201 and 205; K. Sakai et al., *ibid.*, **3**, 297; J.-L. Brevet and K. Mori, *Synthesis*, 1007; D. Basavaiah and P.R. Krishna, *Pure Appl. Chem.*, **64**, 1067; L.T. Kanerva et al., *J. Chem. Soc., Perkin Trans. 1*, 1759; D.M. LeGrand and S.M. Roberts, *ibid.*, 1751; D. O'Hagan and N.A. Zaidi, *ibid.*, 947; H. Tsukube et al., *J. Chem. Soc., Chem. Commun.*, 1751; D. Bianchi et al., *Tetrahedron Lett.*, **33**, 3231; S.W. Schneller et al., *ibid.*, **33**, 2249.

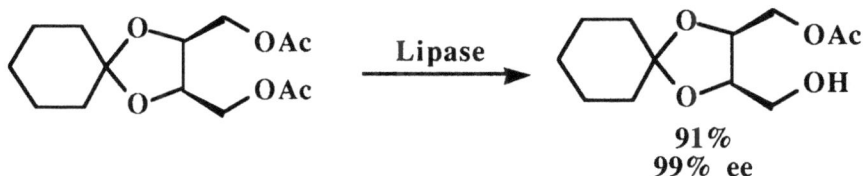

91%
99% ee

**VI.A.3-9** J.O. Metzger and U. Biermann *Synthesis*, 463; C. Fréchou et al., *Tetrahedron Lett.*, **33**, 5067.

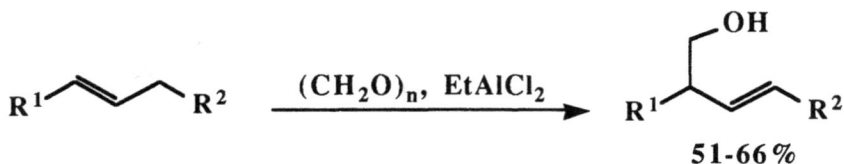

$$R^1 \diagup\diagdown R^2 \quad \xrightarrow{\text{(CH}_2\text{O)}_n,\ \text{EtAlCl}_2} \quad R^1\text{...CH}_2\text{OH...}R^2$$

**51-66%**

## VI.A.4.    Aldehydes and Ketones

(see also I.A.1, II.A.1, V.E.)

**VI.A.4-1** R. Menicagli et al., *Tetrahedron Lett.*, **33**, 2867; A.R. Katritzky et al., *Can. J. Chem.*, **70**, 2040; H. Takayama et al., *J. Org. Chem.*, **57**, 2173; J.-P. Beque et al., *J. Org. Chem.*, **57**, 3807.

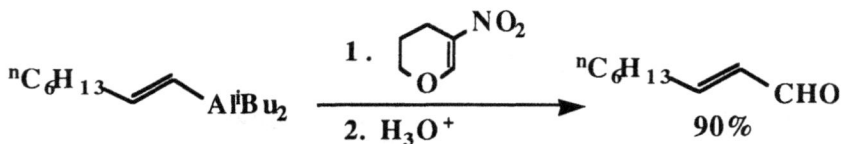

$$^n\text{C}_6\text{H}_{13}\diagup\diagdown\text{Al}^i\text{Bu}_2 \quad \xrightarrow[\text{2. H}_3\text{O}^+]{\text{1. (NO}_2 \text{ dihydropyran)}} \quad ^n\text{C}_6\text{H}_{13}\diagup\diagdown\text{CHO}$$

**90%**

---

**VI.A.4-2** S.Z. Zard et al., *Tetrahedron Lett.*, **33**, 1285.

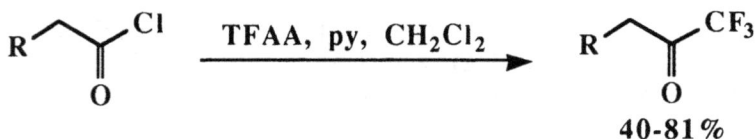

$$R\diagup\underset{O}{\overset{}{C}}\diagdown\text{Cl} \quad \xrightarrow{\text{TFAA, py, CH}_2\text{Cl}_2} \quad R\diagup\underset{O}{\overset{}{C}}\diagdown\text{CF}_3$$

**40-81%**

---

**VI.A.4-3** J. Sehwortz et al., *Tetrahedron Lett.*, **33**, 6787, 6783 and 6791; E.K. Ryu, *ibid.*, **33**, 3141.

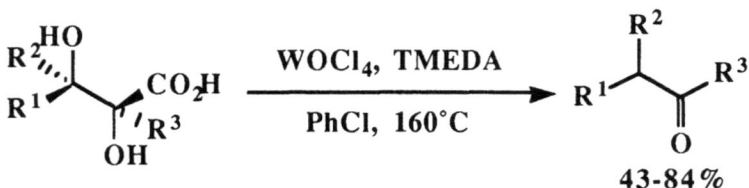

$$R^2\text{...HO...}R^1\text{...CO}_2\text{H...}R^3\text{...OH} \quad \xrightarrow[\text{PhCl, 160°C}]{\text{WOCl}_4,\ \text{TMEDA}} \quad R^1\text{...}R^2\text{...}\underset{O}{\overset{}{C}}\text{...}R^3$$

**43-84%**

**VI.A.4-4**  F. Effenberger et al., *Tetrahedron Lett.*, **33**, 5157.

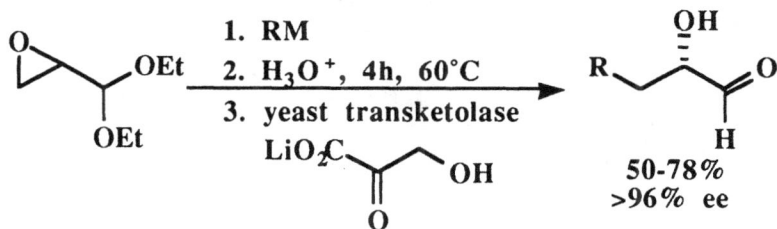

**VI.A.4-5**  G.W. Kabalka et al., *Synth. Commun.*, **22**, 2587;  B.C. Ranu et al., *ibid.*, **22**, 1523;  D.P. Curren and M. Yu, *Synthesis*, 123;  N. Iranpoor and E. Mottaghinejad, *J. Organomet. Chem.*, **423**, 399;  V. Rukachaisirikul and R.W. Hoffmann, *Tetrahedron*, **48**, 10563;  O. Piva, *Tetrahedron Lett.*, **33**, 2459;  W. Shaozu et al., *Gazz. Chim. Ital.*, **121**, 519 (91).

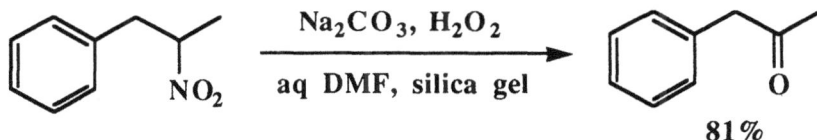

**VI.A.4-6**  D. Enders et al., *Synlett.*, 897;  T. Ye and M.A. Mckervey, *Tetrahedron*, **48**, 8007;  see also R.D. Rieke and W.R. Klein, *Synth. Commun.*, **22**, 2635;  H.H. Wasserman et al., *Tetrahedron Lett.*, **33**, 6003 and *Tetrahedron*, **48**, 7071.

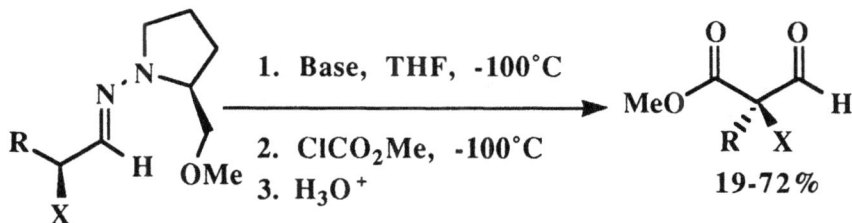

**VI.A.4-7**  Y.H. Kim et al., *Chem. Ind.*, 31;  J.M. Moreto et al., *Tetrahedron Lett.*, **33**, 3021;  M.J. Miller et al., *J. Org. Chem.*, **57**, 3546; J.S. Cha et al., *Org. Prep. Proc. Int.*, **24**, 331.

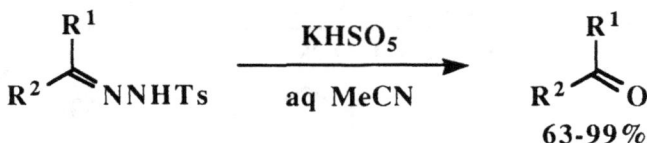

$$R^2 \underset{\displaystyle NNHTs}{\overset{\displaystyle R^1}{=}} \quad \xrightarrow[\text{aq MeCN}]{\text{KHSO}_5} \quad R^2 \underset{\displaystyle O}{\overset{\displaystyle R^1}{=}}$$

**63-99%**

---

**VI.A.4-8**  L.S. Hegedus et al., *J. Org. Chem.*, **57**, 1461;  C. Wedler and H. Schick, *Synthesis*, 543;  C.R. Schmid and D.A. Bradley, *ibid.*, 587; Y.N. Polivia et al., *J. Org. Chem. (USSR)*, **28**, 672.

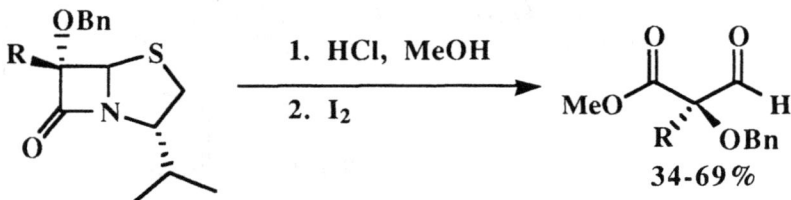

1. HCl, MeOH
2. $I_2$

$$\text{MeO} \overset{O}{-} \underset{\overset{|}{R} \ OBn}{C} \overset{O}{-} H$$

**34-69%**

## VI.A.5.    Amides

**VI.A.5-1**  H. Kotsuki et al., *Tetrahedron Lett.*, **33**, 4945.

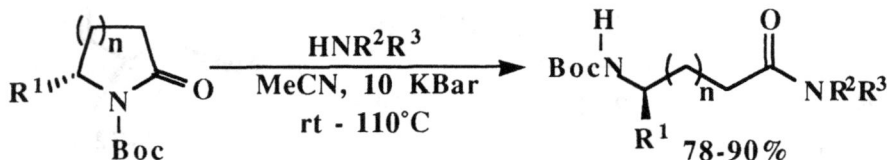

$$\xrightarrow[\substack{\text{MeCN, 10 KBar} \\ \text{rt - 110°C}}]{\text{HNR}^2\text{R}^3}$$

BocN—...—NR$^2$R$^3$    **78-90%**

---

**VI.A.5-2**  H.S.P Rao et al., *Synth. Commun.*, **22**, 1339.

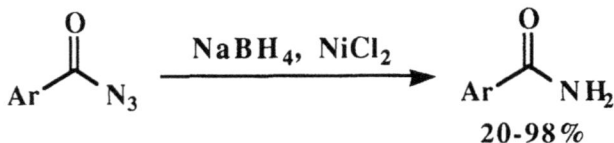

$$\text{Ar} \overset{O}{-} N_3 \quad \xrightarrow{\text{NaBH}_4, \ \text{NiCl}_2} \quad \text{Ar} \overset{O}{-} NH_2$$

**20-98%**

**VI.A.5-3** K.A. Hackl and H. Falk, *Monatsh. Chem.*, **123**, 599; J.P. Gesson et al., *Bull. Soc. Chim. Fr.*, **129**, 227; E. Diez-Barra et al., *Synth. Commun.*, **22**, 1661.

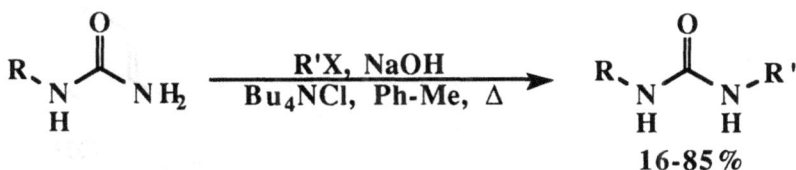

**VI.A.5-4** M. Larcheveque et al., *Tetrahedron Lett.*, **33**, 4313; S. Knapp et al., *ibid.*, **33**, 1025; T.W. Hart and B. Vacher, *ibid.*, **33**, 3009; S.V. Kessar et al., *J. Chem. Soc., Chem. Commun.*, 840; W. Wang and V. Snieckus, *J. Org. Chem.*, **57**, 424; A.R. Katritzky et al., *J. Chem. Soc., Perkin Trans. 1*, 3055.

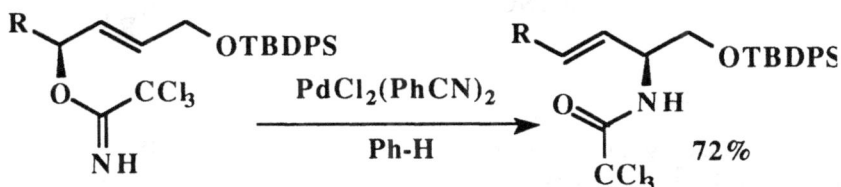

**VI.A.5-5** T. Hosokawa, S.-I. Murahashi et al., *Tetrahedron Lett.*, **33**, 6643; A.R. Katritzky et al., *J. Org. Chem.*, **57**, 547.

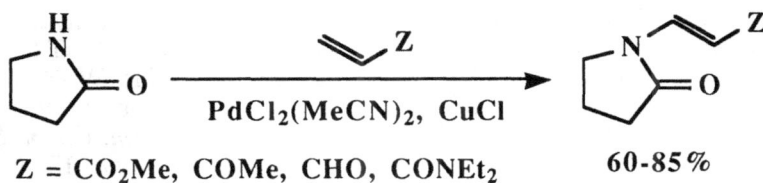

Z = CO₂Me, COMe, CHO, CONEt₂

**VI.A.5-6** T.W. Leung and B.D. Dombek, *J. Chem. Soc., Chem. Commun.*, 205.

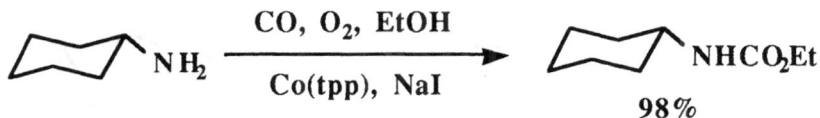

**VI.A.5-7**  A.I. Meyers et al., *Tetrahedron Lett.*, **33**, 1181.

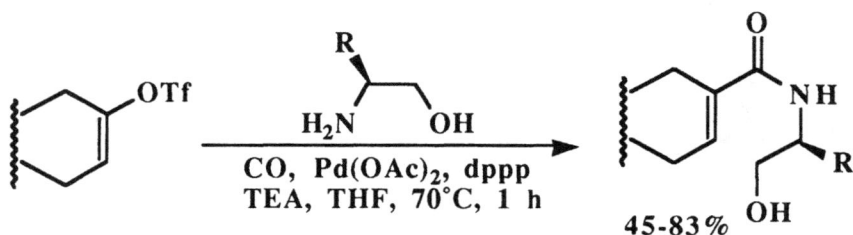

CO, Pd(OAc)$_2$, dppp
TEA, THF, 70°C, 1 h

45-83%

**VI.A.5-8**  A. Battaglia et al., *J. Org. Chem.*, **57**, 5128.

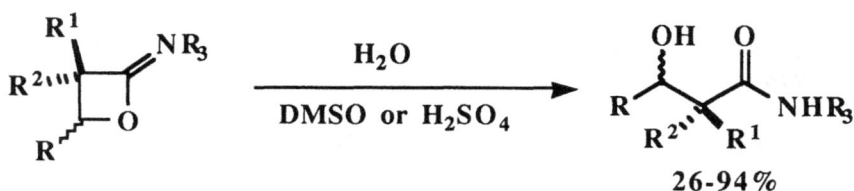

H$_2$O

DMSO or H$_2$SO$_4$

26-94%

**VI.A.5-9**  E. Deutsch et al., *J. Med. Chem.*, **35**, 274;  G.A. Showell et al., *ibid.*, **35**, 911;  G.M.F. Bisset et al., *ibid.*, **35**, 859;  T. Kunieda et al., *Heterocycles*, **33**, 131;  M. Riviere-Baudet et al., *Tetrahedron Lett.*, **33**, 6453;  A.W. Schwabacher and R.A. Bychowski, *ibid.*, **33**, 21;  S. Chen and J. Xu, *ibid.*, **33**, 647;  C. Mioskowski et al, *ibid.*, **33**, 5055;  J.J. Landi, Jr. and H.R. Brinkman, *Synthesis*, 1093;  C. Thom and P. Kocienski, *ibid.*, 582;  S.E. Kelly and T.G. LaCour, *Synth. Commun.*, **22**, 859;  M. Veda and H. Mori, *Bull. Chem. Soc. Jpn.*, **65**, 1636;  V. Gotor et al., *J. Chem. Soc., Perkin Trans. 1*, 2885 and 2891 and *Tetrahedron Asym.*, **3**, 1519;  T.F. Braish and D.E. Fox, *Synlett.*, 979;  M.R. Shipton, *ibid.*, 491;  W.-B. Wang and E.J. Roskamp, *J. Org. Chem.*, **57**, 6101;  A.O. Stewart and D.W. Brooks, *ibid.*, **57**, 5020;  A.S. Goldman, J. Kohn et al., *J. Am. Chem. Soc.*, **114**, 6649;  A.K. Ghosh et al., *J. Chem. Soc., Chem. Commun.*, 1308 and *Tetrahedron Lett.*, **33**, 2781.

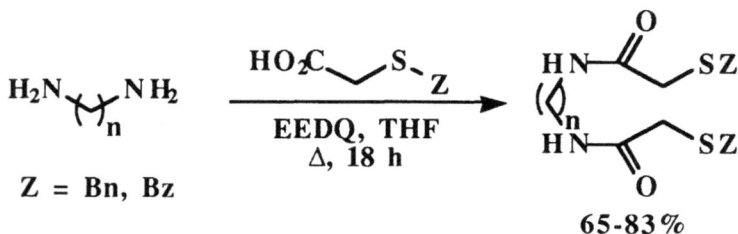

EEDQ, THF
Δ, 18 h

Z = Bn, Bz

65-83%

**VI.A.5-10**  C. Baldoli, P. Del Buttero and S. Maiorana, *Tetrahedron Lett.*, **33**, 4049.

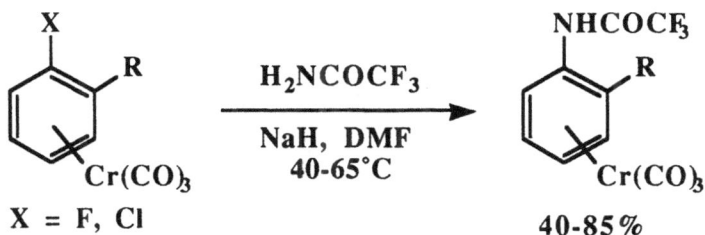

X = F, Cl

40-85%

---

**VI.A.5-11**  R.V. Hoffman and J.M. Salvador, *J. Org. Chem.*, **57**, 4487; M.B. Berry et al., *Synlett.*, 659;  Y. Kikugawa et al., *Synthesis*, 1058.

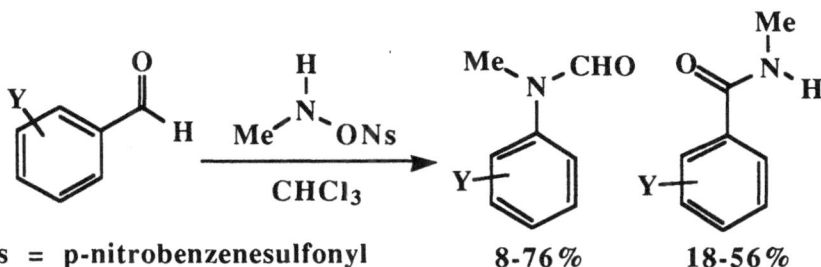

Ns = p-nitrobenzenesulfonyl

8-76%          18-56%

---

**VI.A.5-12**  S. Murahashi et al., *J. Org. Chem.*, **57**, 2521;  H.C. Griengl et al., *J. Chem. Soc., Perkin Trans. 1*, 137.

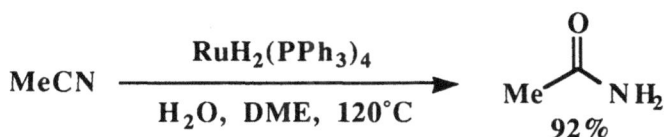

92%

---

**VI.A.5-13**  K. Takai, K. Utimoto et al., *Bull. Chem. Soc. Jpn.*, **65**, 1543; Y. Liu et al., *Chem. Lett.*, 1143.

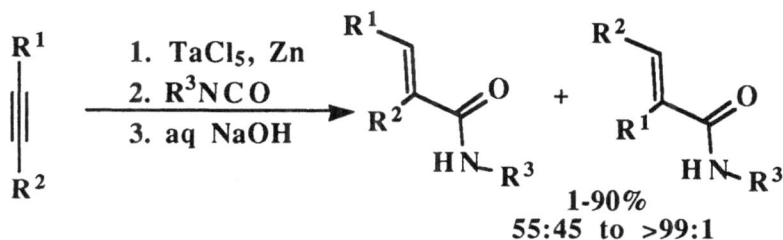

1-90%
55:45 to >99:1

**VI.A.5-14** K. Yamakawa et al., *Tetrahedron Lett.*, **33**, 1455; T. Koch and M. Hesse, *Synthesis*, 931; J. Aube et al., *J. Org. Chem.*, **57**, 1635

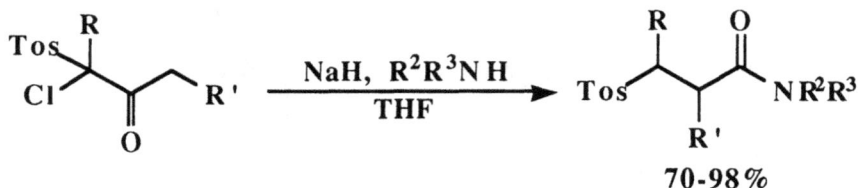

$$\text{Tos}\overset{R}{\underset{O}{\overset{|}{C}}}\text{Cl} \quad \xrightarrow[\text{THF}]{\text{NaH, } R^2R^3NH} \quad \text{Tos}\overset{R}{\smile}\overset{O}{\underset{R'}{\smile}}NR^2R^3$$

70-98%

## VI.A.6.    Amine and Carbamates

**VI.A.6-1** C.-G. Cho and G.H. Posner, *Tetrahedron Lett.*, **33**, 3599.

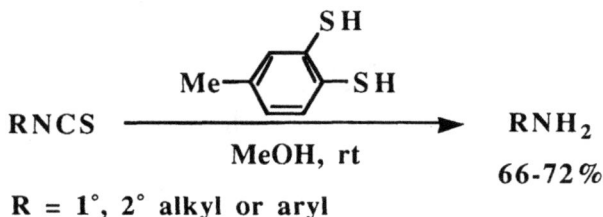

$$\text{RNCS} \quad \xrightarrow[\text{MeOH, rt}]{\text{Me}-\overset{SH}{\underset{SH}{\bigcirc}}} \quad RNH_2$$

66-72%

R = 1°, 2° alkyl or aryl

**VI.A.6-2** ; M. Yasuda et al., *Heterocycles*, **34**, 965; D. Sinou et al., *Tetrahedron Lett.*, **33**, 2481; P. Kalck et al., *J. Chem. Soc., Chem. Commun.*, 1552; J.M. Hawkins and T.A. Lewis, *J. Org. Chem.*, **57**, 2114; R.G. Beryman et al., *J. Am. Chem. Soc.*, **114**, 1708.

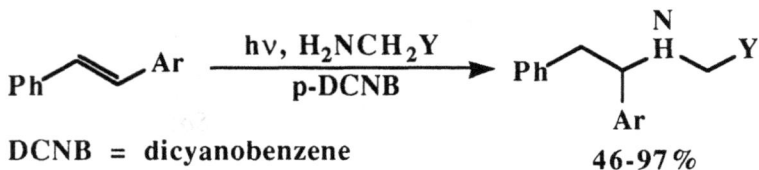

$$\text{Ph}\diagup\diagdown\text{Ar} \quad \xrightarrow[\text{p-DCNB}]{hv, \ H_2NCH_2Y} \quad \text{Ph}\diagdown\overset{\underset{\textstyle Ar}{}}{\diagup}\overset{N}{\underset{H}{}}\diagup Y$$

DCNB = dicyanobenzene                    46-97%

**VI.A.6-3** E. Ciganek, *J. Org. Chem.*, **57**, 4521.

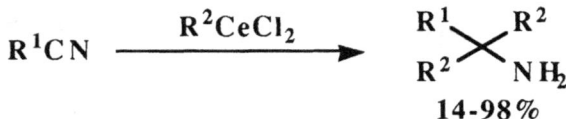

$$R^1CN \quad \xrightarrow{R^2CeCl_2} \quad \overset{R^1}{\underset{R^2}{>}}\overset{R^2}{\underset{NH_2}{<}}$$

14-98%

**VI.A.6-4** N. Katagiri, C. Kaneko and A. Toyota, *Chem. Pharm. Bull.*, **40**, 1039; P. Goya et al., *J. Chem. Res. (S)*, 216; M.A. Poss and J.A. Reid, *Tetrahedron Lett.*, **33**, 7291; I.A. O'Neil and K.M. Hamilton, *Synlett.*, 791; A.R. Katritzky et al., *Rec. Trav. Chim.*, **110**, 369 (91).

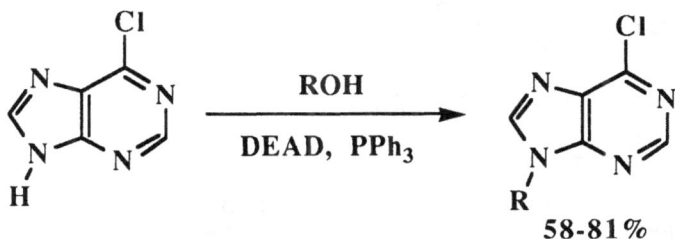

58-81%

---

**VI.A.6-5** M. Makosza and M. Bialecki, *J. Org. Chem.*, **57**, 4784; K. Matsumoto et al., *Synth. Commun.*, **22**, 785; T.F. Braish and D.E. Fox, *ibid.*, **22**, 3067; P. Lopez-Alvarado et al., *Tetrahedron Lett.*, **33**, 6875; Y. Endo et al., *J. Am. Chem. Soc.*, **114**, 9795; M.K. Stern et al., *ibid.*, **114**, 9237.

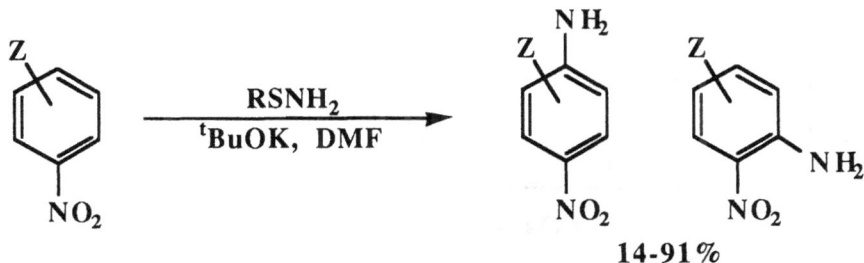

14-91%

---

**VI.A.6-6** P. Gmeiner and B. Bollinger, *Liebigs Ann. Chem.*, 273; S. Nagarajan et al., *Synth. Commun.*, **22** 1191; A. Alexakis et al., *J. Org. Chem.*, **57**, 4563; S. Itsuno et al., *Tetrahedron Lett.*, **33**, 627; see also M.A. Schwartz et al., *ibid.*, **33**, 1689.

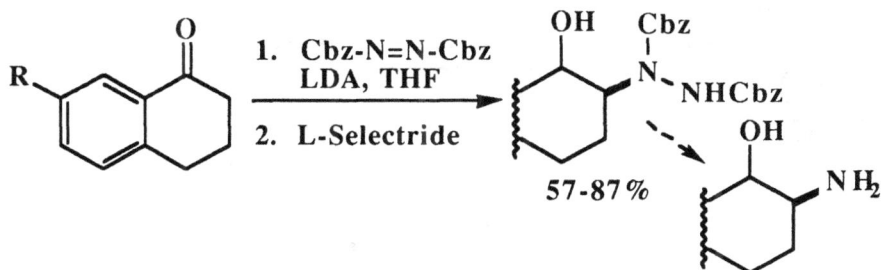

57-87%

**VI.A.6-7** K.M. Nicholas et al., *J. Chem. Soc., Chem. Commun.*, 853; R.E. Dolle et al., *J. Org. Chem.*, **57**, 128.

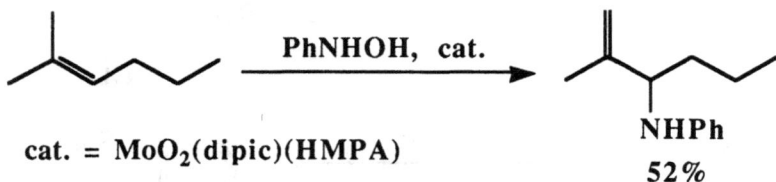

PhNHOH, cat.

cat. = MoO$_2$(dipic)(HMPA)

NHPh

52%

---

**VI.A.6-8** E.W. Thomas et al., *J. Med. Chem.*, **35**, 1233; C.F. Nutaitis, *Synth. Commun.*, **22**, 1081; J.L. Garcia Ruano et al., *Tetrahedron Lett.*, **33**, 5637; H.J. Geise et al., *Bull. Soc. Chim. Belg.*, **101**, 503; M.H. Nantz et al., *J. Org. Chem.*, **57**, 6653.

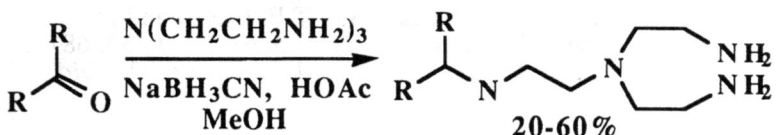

R
R    O

N(CH$_2$CH$_2$NH$_2$)$_3$
NaBH$_3$CN, HOAc
MeOH

R
R    N

N    NH$_2$
NH$_2$

20-60%

---

**VI.A.6-9** D. Tanner et al., *Tetrahedron*, **48**, 6069.

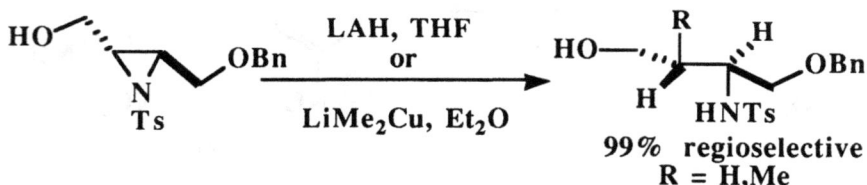

HO                OBn
       N
       Ts

LAH, THF
or
LiMe$_2$Cu, Et$_2$O

R    H
HO              OBn
H    HNTs

99% regioselective
R = H,Me

---

**VI.A.6-10** R.K. Dieter et al., *J. Org. Chem.*, **57**, 1663; J.C. Arnould et al., *Tetrahedron Lett.*, **33**, 7133; D.J. Hlasta et al., *Heterocycles*, **34**, 1897.

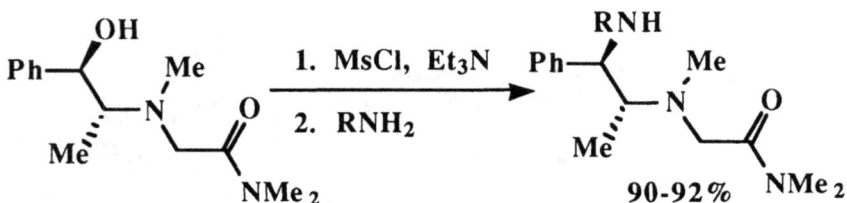

OH
Ph        Me
      N        O
Me
         NMe$_2$

1. MsCl, Et$_3$N
2. RNH$_2$

RNH
Ph        Me
      N        O
Me
         NMe$_2$

90-92%

**VI.A.6-11**  J. Genet et al., *Tetrahedron Lett.*, **33**, 2677;  R.F. Smith et al., *Synth. Commun.*, **22**, 381.

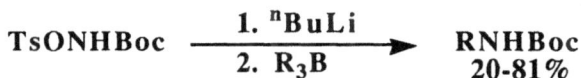

$$\text{TsONHBoc} \xrightarrow[\text{2. }R_3B]{\text{1. }^nBuLi} \begin{array}{c} \text{RNHBoc} \\ \textbf{20-81\%} \end{array}$$

**VI.A.6-12**  A. Kamal et al., *Tetrahedron Lett.*, **33**, 4077;  A. Malik et al., *J. Chem. Res. (S)*, 124.

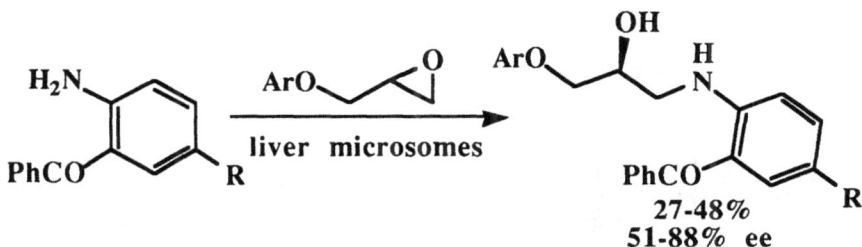

**VI.A.6-13**  M. Vaultier et al., *Synth. Commun.*, **22**, 665;  A. Zwierjak et al., *ibid.*, **22**, 1929;  D. Misiti et al., *ibid.*, **22**, 883;  A. Koziara and A. Zwierzak, *Synthesis*, 1063.

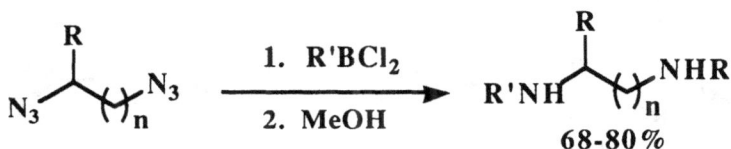

**VI.A.6-14**  J. Beque et al., *Tetrahedron Lett.*, **33**, 1879.

## VI.A.7.    Amino Acid Derivatives

**VI.A.7-1**  U. Ragnarsson et al., *J. Chem. Soc., Perkin Trans. 1*, 245.

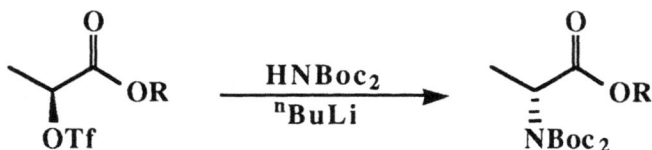

**VI.A.7-2**  J. Brussee et al., *Tetrahedron Asym.*, **3**, 769.

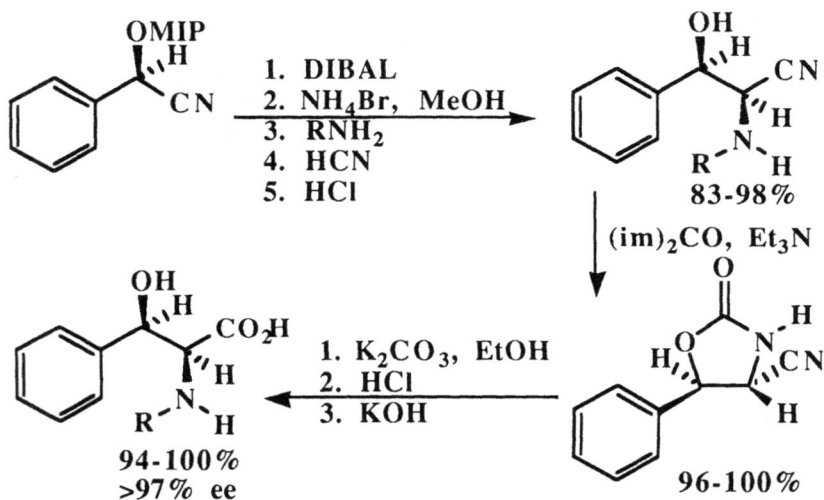

**VI.A.7-3**  F. Couty et al., *Synlett.*, 847;  L.M. Harwood et al., *Tetrahedron Asym.*, **3**, 1127.

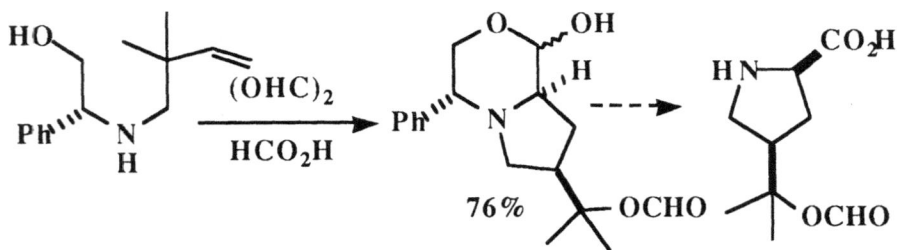

**VI.A.7-4**  C. Palomo et al., *Tetrahedron Lett.*, **33**, 4823 and 4819;  I. Ojima et al., *ibid.*, **33**, 5737.

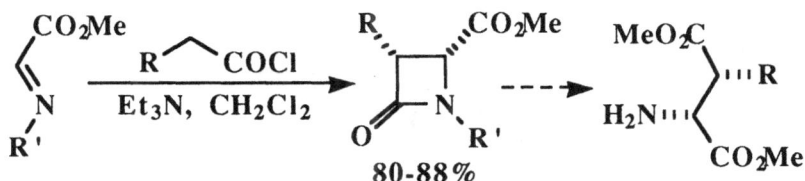

**VI.A.7-5**  C. Cativiela et al., *Synlett.*, 579 and *Synth. Commun.*, **22**, 2955.

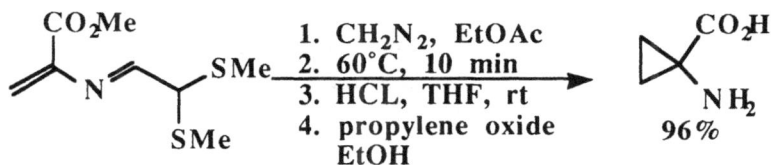

**VI.A.7-6**  W.R. Baker et al., *Tetrahedron Lett.*, **33**, 1573;  D.P.G. Hamon et al., *Tetrahedron*, **48**, 5163;  D.A. Evans et al., *J. Am. Chem. Soc.*, **114**, 9434;  A. de Meijere et al., *Chem. Ber.*, **125**, 867;  M.J. O'Donnell and S. Wu, *Tetrahedron Asym.*, **3**, 591.

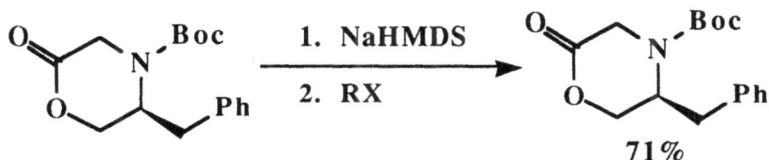

**VI.A.7-7**  E.J. Corey and J.O. Link, *J. Am. Chem. Soc.*, **114**, 1906.

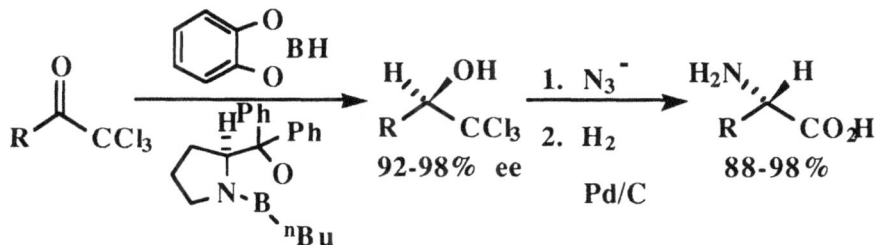

**VI.A.7-8**  R. Pedrosa et al., *Synlett.*, 45 and *Tetrahedron Lett.*, **33**, 2895; C.J. Sih et al., *ibid.*, **33**, 1953.

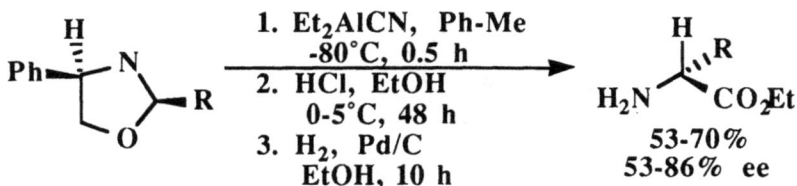

1. Et$_2$AlCN,  Ph-Me
   -80°C, 0.5 h
2. HCl, EtOH
   0-5°C, 48 h
3. H$_2$, Pd/C
   EtOH, 10 h

53-70%
53-86%  ee

**VI.A.7-9**  S. Kotha and A. Kuki, *Tetrahedron Lett.*, **33**, 1565.

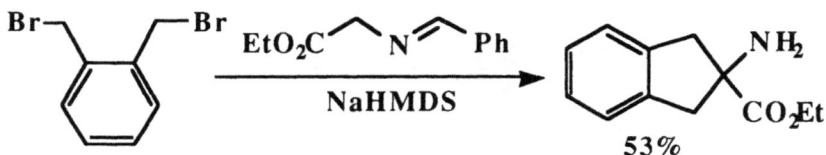

EtO$_2$C N Ph

NaHMDS

53%

**VI.A.7-10**  J. Ezquerra et al., *Tetrahedron Lett.*, **33**, 5589.

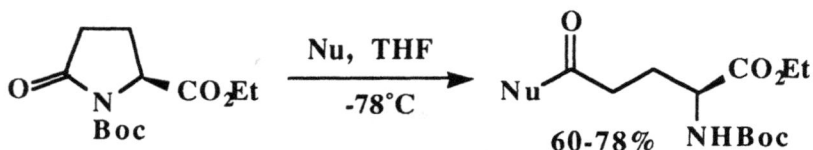

Nu, THF

-78°C

60-78%   NHBoc

**VI.A.7-11**  A. Trzeciak and W. Bannwarth, *Tetrahedron Lett.*, **33**, 4557; R. Ramage et al., *Tetrahedron*, **48**, 499; W. Voelter et al., *Coll. Czech. Chem. Commun.*, **57**, 1707; D.T. Elmore et al., *J. Chem. Soc., Chem. Commun.*, 1033; S.B. Katti et al., *ibid.*, 843.

**Synthesis of "Head-to-Tail" Cyclized Peptides on Solid Support by FMOC Chemistry**

**VI.A.7-12**  V. Soloshonok et al., *Synlett.*, 657.

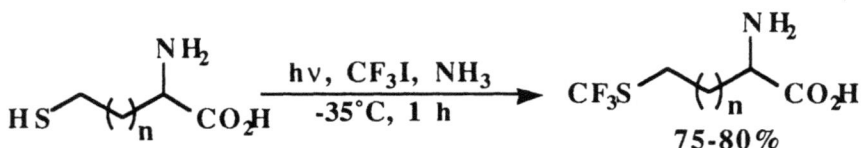

hν, CF$_3$I, NH$_3$

-35°C, 1 h

75-80%

**VI.A.7-13** C. Cativiela et al., *Tetrahedron Asym.*, **3**, 567.

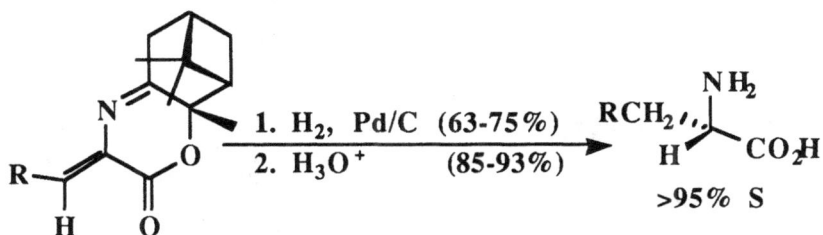

**VI.A.7-14** A. Abiko and S. Masamune, *Tetrahedron Lett.*, **33**, 5517.

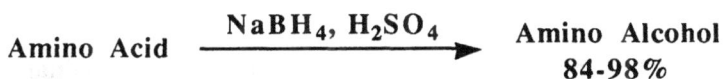

$$\text{Amino Acid} \xrightarrow{\text{NaBH}_4, \text{H}_2\text{SO}_4} \underset{\textbf{84-98\%}}{\text{Amino Alcohol}}$$

**VI.A.7-15** J.S. Nowick et al., *J. Org. Chem.*, **57**, 7364.

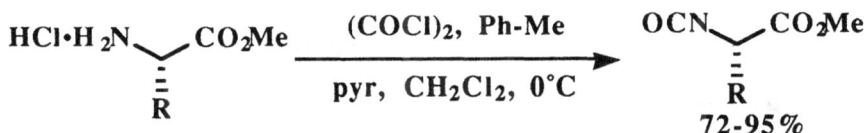

**VI.A.7-16** J. Marquet et al., *Tetrahedron*, **48**, 1333.

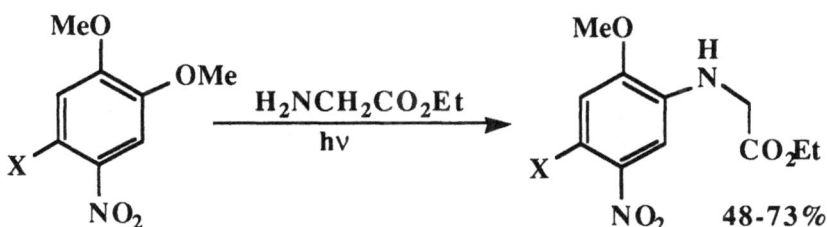

**VI.A.7-17** G.T. Crisp and P.T. Glink, *Tetrahedron*, **48**, 3541.

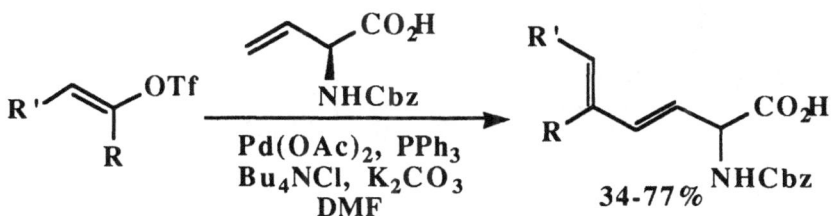

**VI.A.7-18**  L.S. Hegedus et al., *J. Am. Chem. Soc.*, **114**, 5602.

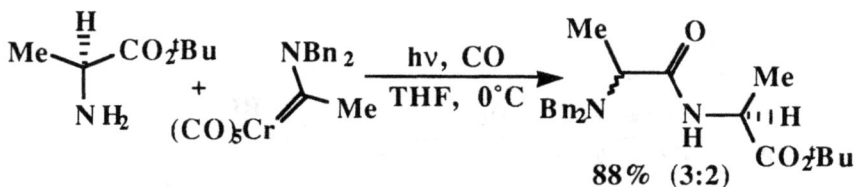

88% (3:2)

## VI.A.8.    Azides

**VI.A.8-1**  P.J. Kropp et al., *J. Org. Chem.*, **57**, 6646;  E. Napolitano and R. Fiaschi, *Gazz. Chim. Ital.*, **122**, 233.

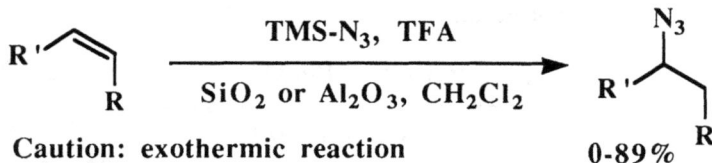

$$\text{TMS-N}_3, \text{ TFA}$$
$$\text{SiO}_2 \text{ or Al}_2\text{O}_3, \text{ CH}_2\text{Cl}_2$$

Caution: exothermic reaction

0-89%

**VI.A.8-2**  W.A. Nugent, *J. Am. Chem. Soc.*, **114**, 2768;  J. Plumet et al., *Tetrahedron Lett.*, **33**, 7417;  A. Guy et al., *Synthesis*, 821 and *Tetrahedron Asym.*, **3**, 247.

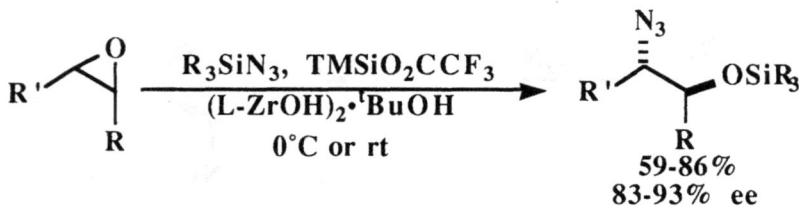

$$\text{R}_3\text{SiN}_3, \text{ TMSiO}_2\text{CCF}_3$$
$$(\text{L-ZrOH})_2 \cdot {}^t\text{BuOH}$$
$$0°\text{C or rt}$$

59-86%
83-93% ee

**VI.A.8-3**  R.M. Giuliano and F. Duarte, *Synlett.*, 419.

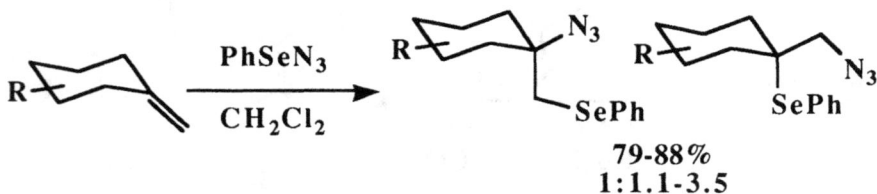

$$\text{PhSeN}_3$$
$$\text{CH}_2\text{Cl}_2$$

79-88%
1:1.1-3.5

**VI.A.8-4**  S. Zhu, *Tetrahedron Lett.*, **33**, 6503.

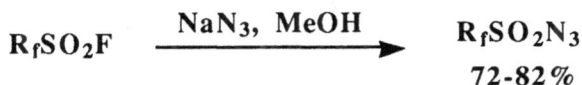

$$R_fSO_2F \xrightarrow{\text{NaN}_3, \text{ MeOH}} \begin{array}{c} R_fSO_2N_3 \\ \textbf{72-82\%} \end{array}$$

---

**VI.A.8-5**  P. Magnus and J. Lacour, *J. Am. Chem. Soc.*, **114**, 3993 and 767.

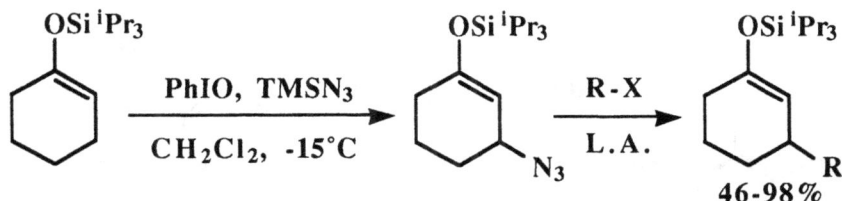

---

**VI.A.8-6**  D.A. Evans et al., *Tetrahedron Lett.*, **33**, 1189;  O.P. Goel et al., *ibid.*, **33**, 3293;  A.I. Meyers, A. de Meijere et al., *Synlett.*, 962.

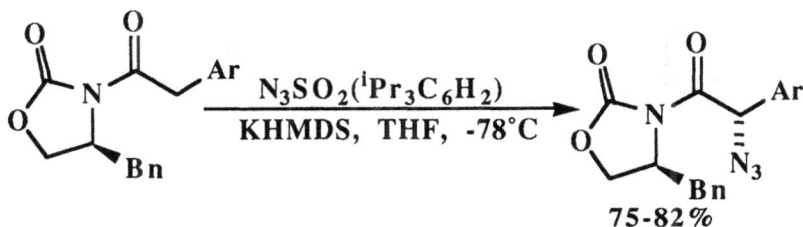

---

**VI.A.8-7**  T. Gajda and M. Matusiak, *Synthesis*, 367.

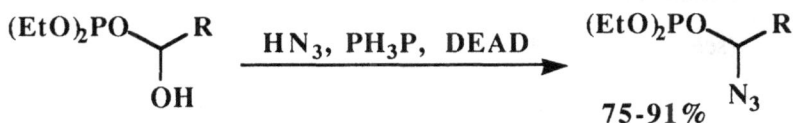

---

**VI.A.8-8**  R. Roy et al., *Synthesis*, 618.

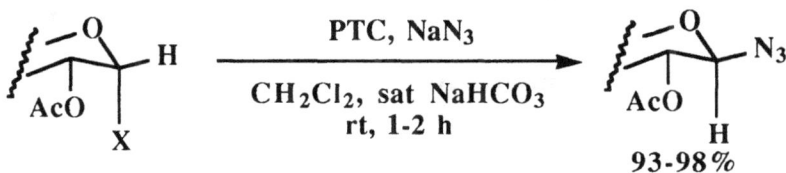

**VI.A.8-9** J.G. Lee and K.H. Kwak, *Tetrahedron Lett.*, **33**, 3165.

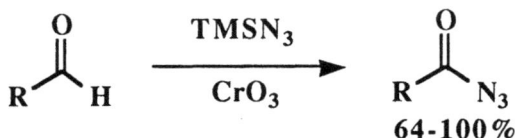

$$\underset{R}{\overset{O}{\|}}\underset{H}{\overset{}{}} \quad \xrightarrow[\text{CrO}_3]{\text{TMSN}_3} \quad \underset{R}{\overset{O}{\|}}\underset{N_3}{\overset{}{}}$$
**64-100%**

---

**VI.A.8-10** M. Weißenfels et al., *Liebigs Ann. Chem.*, 23.

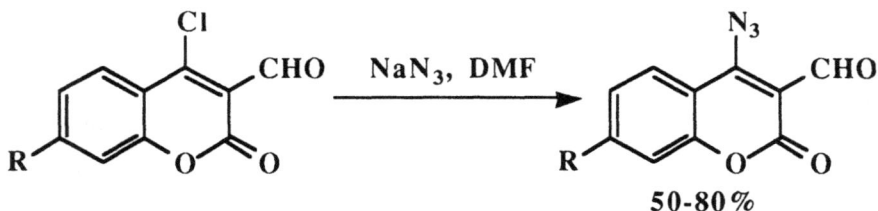

**50-80%**

---

**VI.A.8-11** P. Magnus and L. Barth, *Tetrahedron Lett.*, **33**, 2777.

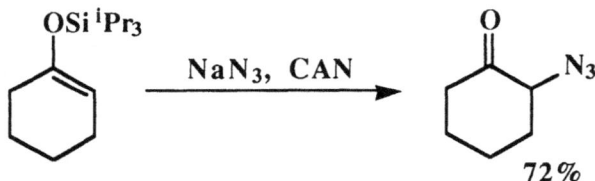

**72%**

## VI.A.9.   Esters

(see also: I.G.2, IV.D, V.C.)

**VI.A.9-1** C.J. Kowalski and R.E. Reddy, *J. Org. Chem.*, **57**, 7194.

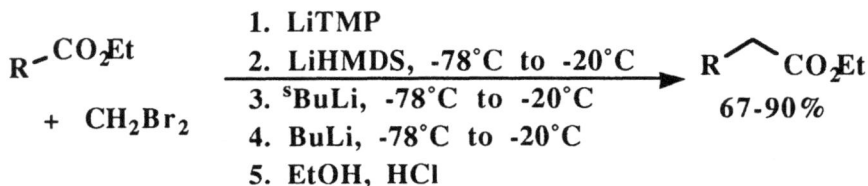

$$R^{-CO_2Et} + CH_2Br_2 \quad \xrightarrow{\begin{array}{l} \text{1. LiTMP} \\ \text{2. LiHMDS, -78°C to -20°C} \\ \text{3. }^s\text{BuLi, -78°C to -20°C} \\ \text{4. BuLi, -78°C to -20°C} \\ \text{5. EtOH, HCl} \end{array}} \quad R\text{—CO}_2Et$$
**67-90%**

**VI.A.9-2**  I. Shibata et al., *Tetrahedron Lett.*, **33**, 7149.

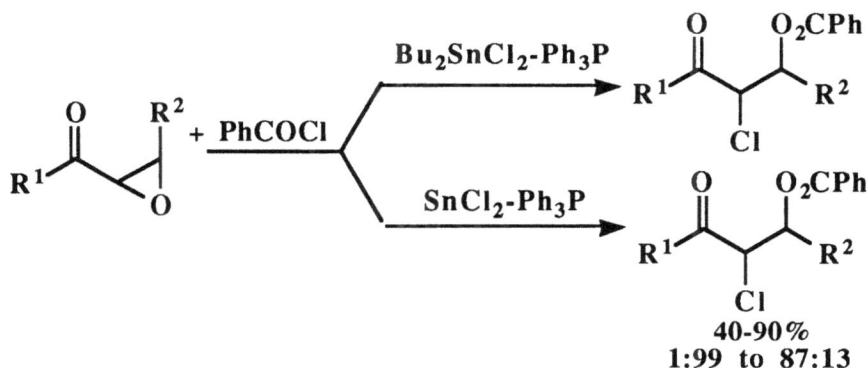

40-90%
1:99 to 87:13

---

**VI.A.9-3**  J.R. Falck et al., *Tetrahedron Lett.*, **33**, 2091;  D. Camp and I.D. Jenkins, *Aust. J. Chem.*, **45**, 47.

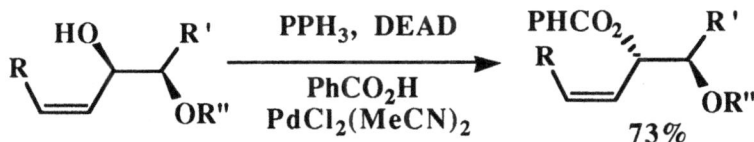

73%

---

**VI.A.9-4**  J.-i. Yoshida, S. Isoe et al., *J. Org. Chem.*, **57**, 4877;  A. Delgado and J. Clardy, *Tetrahedron Lett.*, **33**, 2789;  B.C. Soderberg and B.A. Bowden, *Organometallics*, **11**, 2220;  M. Boeykens and N. De Kimpe, *Synth. Commun.*, **22**, 3285;  S. Yamada and M. Matsumoto, *Chemistry Lett.*, 2273.

69-92%

---

**VI.A.9-5**  S. Yamada et al., *J. Org. Chem.*, **57**, 1591 and *Tetrahedron Lett.*, **33**, 2171;  J. Iqbal and R.R. Srivastava, *J. Org. Chem.*, **57**, 2001.

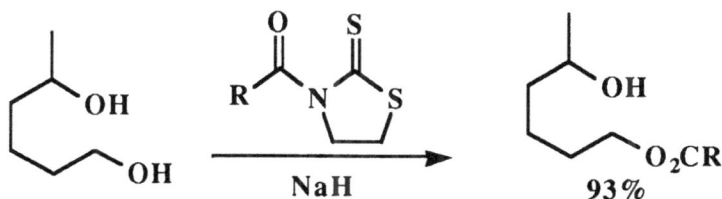

93%

**VI.A.9-6**  C. Mioskowski, J.R. Falck et al., *Tetrahedron Lett.*, **33**, 5205.

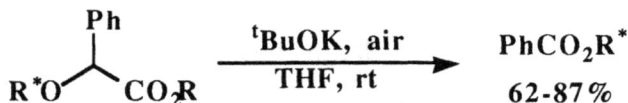

**VI.A.9-7**  Y. Tsuda et al., *Chem. Pharm. Bull.*, **40**, 1033.

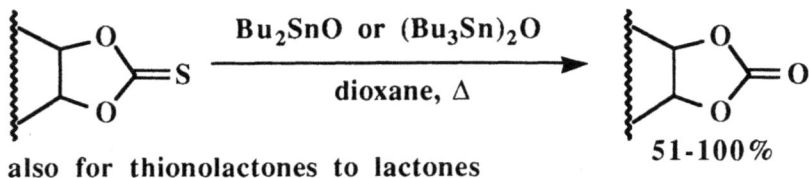

also for thionolactones to lactones

**VI.A.9-8**  J.A. Marshall et al., *Synlett.*, 643.

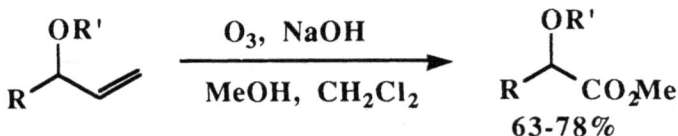

**VI.A.9-9**  T. Durst and K. Koh, *Tetrahedron Lett.*, **33**, 6799.

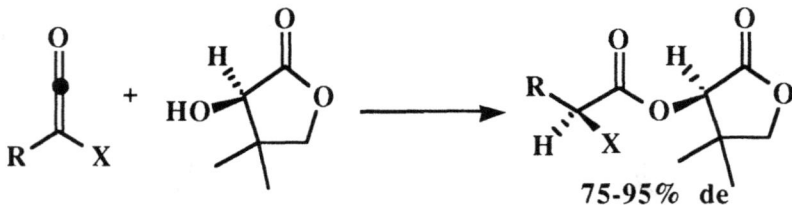

**VI.A.9-10**  H. Mestdagh, C. Rolando et al., *J. Chem. Soc., Chem. Commun.*, 1678.

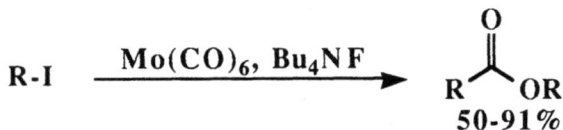

**VI.A.9-11** M.J. Taschner and L. Peddada, *J. Chem. Soc., Chem. Commun.*, 1384; P. Vogel et al., *Tetrahedron*, **48**, 10621; S.-I. Murahashi et al., *Tetrahedron Lett.*, **33**, 7557; B.W. Horrom and H. Mazdiyasni, *Org. Prep. Proc. Int.*, **24**, 696.

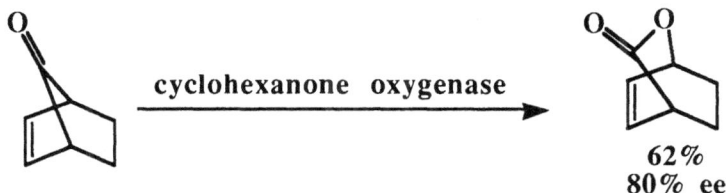

cyclohexanone oxygenase

62%
80% ee

---

**VI.A.9-12** B. Kumar et al., *Synth. Commun.*, **22**, 1087; Y. Yuan and Y. Jiang, *ibid.*, **22**, 3109; P. Klan and P. Benovsky, *Monatsh. Chem.*, **123**, 469; A. Bohac and P. Hrnciar, *Coll. Czech. Chem. Commun.*, **56**, 2879 (91); I. Shiina and T. Mukaiyama, *Chem. Lett.*, 2319; J. Otera et al, *J. Org. Chem.*, **57**, 2166; see also V. Gotor et al., *Tetrahedron*, **48**, 6477; A.L. Gutman et al., *ibid.*, **48**, 8775.

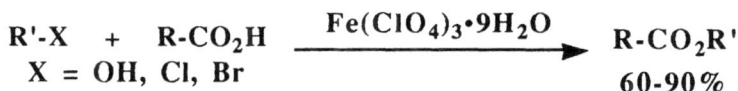

$$R'\text{-}X \; + \; R\text{-}CO_2H \xrightarrow{\text{Fe(ClO}_4)_3\cdot 9H_2O} R\text{-}CO_2R'$$

X = OH, Cl, Br        60-90%

---

**VI.A.9-13** A.S. Gopalan et al., *Tetrahedron*, **48**, 8891; F. Moris and V. Gotor, *ibid.*, **48**, 9869; B. Herradon *Tetrahedron Asym.*, **3**, 209; G. Carrea et al., *ibid.*, **3**, 267; M. Sato, C. Kaneko et al., *ibid.*,, **3**, 313; E. Santaniello et al., *J. Chem. Soc., Perkin Trans. 1*, 1159; G. Iacazio et al., *ibid.*, 661; S.M. Roberts et al., *ibid.*, 589; D. Bianchi et al., *Pure Appl. Chem.*, **64**, 1073; I. Vesely et al., *Coll. Czech. Chem. Commun.*, **57**, 357; C.R. Johnson and S.J. Bis, *Tetrahedron Lett.*, **33**, 7287; F. Theil et al., *ibid.*, **33**, 3457; H. Takahata, Y. Uchida and T. Momose, *ibid.*, **33**, 3331; M.D. Ennis and D.W. Old, *ibid.*, **33**, 6283; M.-J. Kim and H. Cho, *J. Chem. Soc., Chem. Commun.*, 1411; S. Ikegami et al., *Synth. Commun.*, **22**, 2717; V. Gotor and F. Moris, *Synthesis*, 626 and *J. Org. Chem.*, **57**, 2490; J.C. Carretero and E. Dominguez, *ibid.*, **57**, 3867.

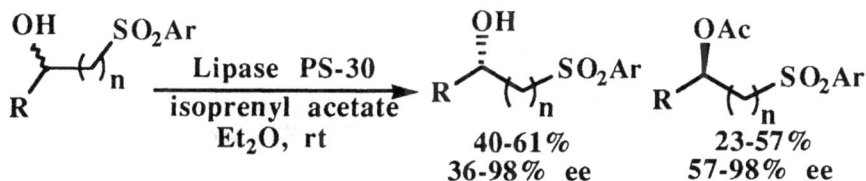

Lipase PS-30
isoprenyl acetate
Et$_2$O, rt

40-61%
36-98% ee

23-57%
57-98% ee

**VI.A.9-14**  T. Fuchigami et al., *J. Org. Chem.*, **57**, 2946;  K.C. Nicolaou et al., *J. Am. Chem. Soc.*, **114**, 8891.

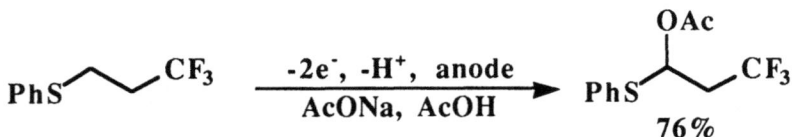

$$
\text{PhS}\diagdown\diagup\text{CF}_3 \quad \xrightarrow[\text{AcONa, AcOH}]{-2e^-, -H^+, \text{ anode}} \quad \text{PhS}\diagdown\overset{\overset{\displaystyle OAc}{|}}{\phantom{x}}\diagup\text{CF}_3
$$

76%

**VI.A.9-15**  S. Yamada et al., *Tetrahedron Lett.*, **33**, 4329;  S. Chandrasekaran et al., *J. Org. Chem.*, **57**, 5013;  J. Mlochowski et al., *Synth. Commun.*, **22**, 1851.

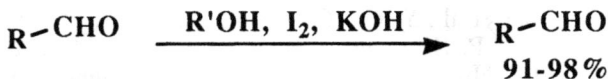

$$
\text{R}^-\text{CHO} \quad \xrightarrow{\text{R'OH, I}_2,\ \text{KOH}} \quad \text{R}^-\text{CHO}
$$

91-98%

## VI.A.10.    Ethers

**VI.A.10-1**  Y. Kita et al., *Chem. Pharm. Bull.*, **40**, 1044.

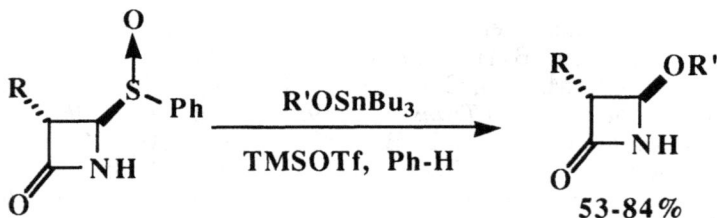

53-84%

**VI.A.10-2**  K. Smith and D. Jones, *J. Chem. Soc., Perkin Trans. 1*, 407; D.L. Boger et al., *J. Org. Chem.*, **57**, 1319;  J.-J. Brunet et al., *Tetrahedron Lett.*, **33**, 4435.

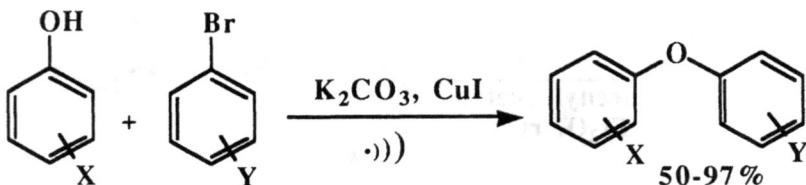

50-97%

**VI.A.10-3** T. Hiyama et al., *Tetrahedron Lett.*, **33**, 4173 and 4177.

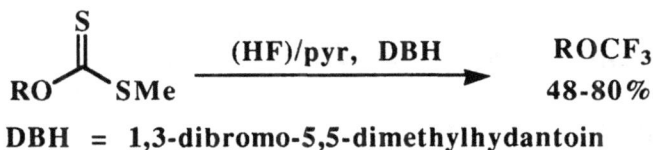

$$
\underset{\underset{RO}{\phantom{a}}\overset{S}{\overset{\|}{\phantom{a}}}\,SMe}{} \xrightarrow[\phantom{xxxxx}]{(HF)/pyr,\ DBH} \underset{48\text{-}80\%}{ROCF_3}
$$

**DBH = 1,3-dibromo-5,5-dimethylhydantoin**

---

**VI.A.10-4** J.F. McGarrity et al., *Synthesis*, 391.

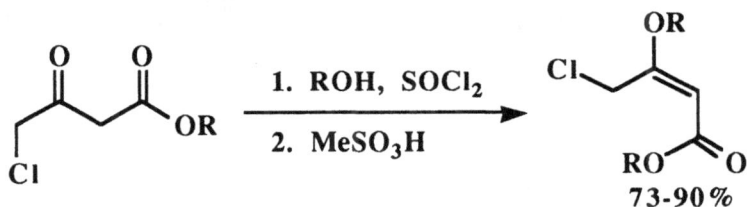

1. ROH, SOCl$_2$
2. MeSO$_3$H

73-90%

---

**VI.A.10-5** S. Mataka et al., *Heterocycles*, **33**, 791; W. Adam et al., *J. Org. Chem.*, **57**, 2680; S. Rozen et al., *J. Am. Chem. Soc.*, **114**, 7643.

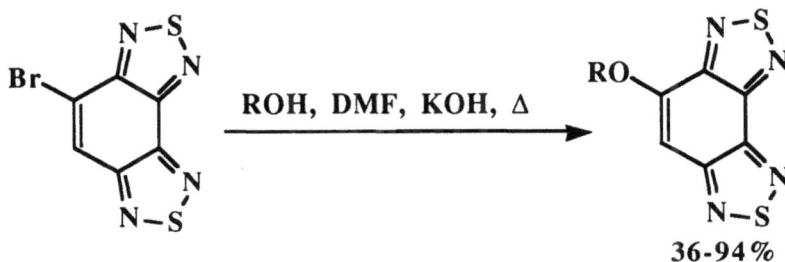

ROH, DMF, KOH, Δ

36-94%

---

**VI.A.10-6** C.J. Moody et al., *Synlett.*, 975; J. Hlavacek and V. Kral, *Coll. Czech. Chem. Commun.*, **57**, 525.

$$
\underset{X\phantom{aa}Y}{\overset{N_2}{\overset{\|}{\phantom{a}}}} \xrightarrow[\text{$^{i}$PrOH}]{Rh_2(NHCOCF_3)_4} \underset{\underset{74\text{-}85\%}{Y\phantom{aa}O^{i}Pr}}{\overset{X\phantom{aa}H}{\diagup\!\diagdown}}
$$

**VI.A.10-7** L. Brandsma et al., *Tetrahedron*, **48**, 3633; A.J. Pecrson et al., *J. Org. Chem.*, **57**, 3583.

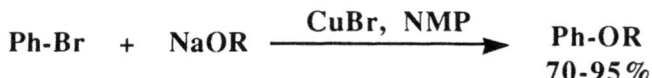

$$\text{Ph-Br} \ + \ \text{NaOR} \ \xrightarrow{\text{CuBr, NMP}} \ \underset{\textbf{70-95\%}}{\text{Ph-OR}}$$

**VI.A.10-8** N. Hartz, G.K.S. Prakash and G.A. Olah, *Synlett.*, 569; T. Harada and T. Mukaiyama, *Chem. Lett.*, 1901.

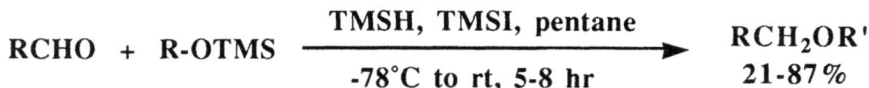

$$\text{RCHO} \ + \ \text{R-OTMS} \ \xrightarrow[\text{-78°C to rt, 5-8 hr}]{\text{TMSH, TMSI, pentane}} \ \underset{\textbf{21-87\%}}{\text{RCH}_2\text{OR}'}$$

## VI.A.11.    Halides

**VI.A.11-1** M.A. Walters et al., *Synth. Commun.*, **22**, 2829; D.L. Comins et al., *Tetrahedron Lett.*, **33**, 7635.

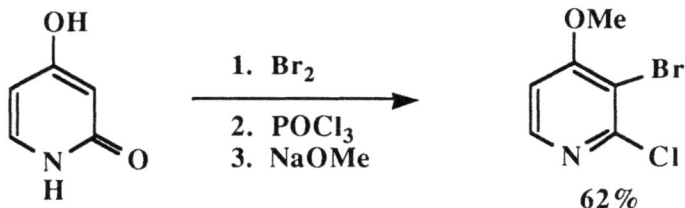

**VI.A.11-2** T. Fujisawa et al., *Synlett.*, 204; K.-B. Chai and P. Sampson, *Tetrahedron Lett.*, **33**, 585; V.K. Singh et al., *ibid.*, **33**, 6021; M.W. Hager and D.C. Liotta, *ibid.*, **33**, 7083; C. Bonini et al., *Synth. Commun.*, **22**, 1863; J. Muzart and A. Riahi, *J. Organomet. Chem.*, **433**, 323; A. Robert et al., *Tetrahedron*, **48**, 1585.

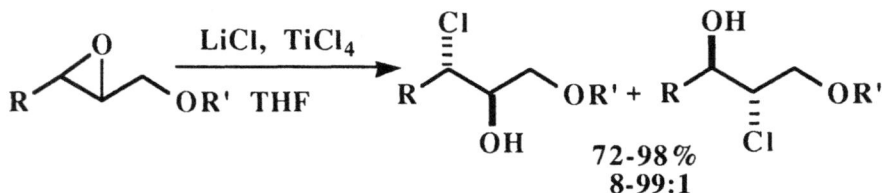

**VI.A.11-3** Y. Shen and M. Qi, *J. Chem. Soc., Perkin Trans. 1*, 3; S. Tomoda et al., *J. Chem. Soc., Chem. Commun.*, 1148.

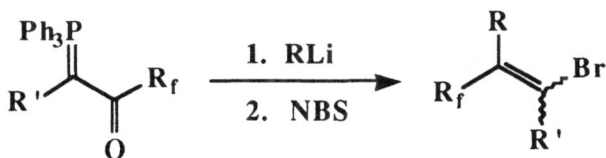

**VI.A.11-4** M. Matsui et al., *Chem. Ber.*, **125**, 467.

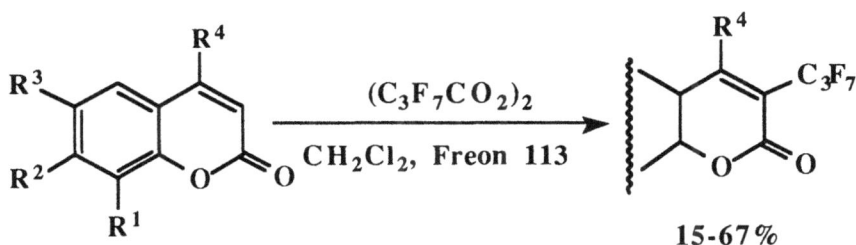

15-67%

**VI.A.11-5** T. Tsukamoto and T. Kitazume, *Synlett.*, 977; K. Iseki et al., *Chem. Pharm. Bull.*, **40**, 1346.

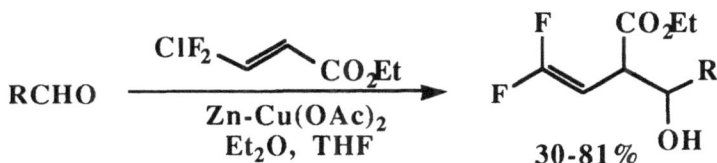

30-81%

**VI.A.11-6** T. Fuchigami et al., *Tetrahedron Lett.*, **33**, 7017; F.W. Litchtenthaler et al., *Synthesis*, 179; M. Gautschi and D. Seebach, *Angew. Chem., Int. Ed. Engl.*, **31**, 1083; Z. Mouloungui et al., *Synth. Commun.*, **22**, 1923; R.E. Banks et al, *J. Chem. Soc., Chem. Commun.*, 595; K. Fukumoto et al., *J. Chem. Soc., Perkin Trans. 1*, 221; J. Ichihara et al., *Chem. Lett.*, 1161; S. Kajigaeshi et al., *Bull. Chem. Soc. Jpn.*, **65**, 1731.

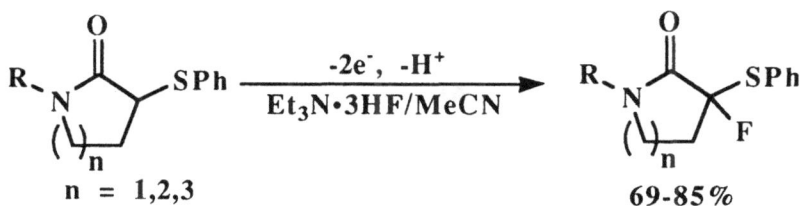

n = 1,2,3

69-85%

**VI.A.11-7**  Z.-Y. Yang and D.J. Burton, *J. Org. Chem.*, **57**, 5144 and *J. Org. Chem.*, **57**, 4676.

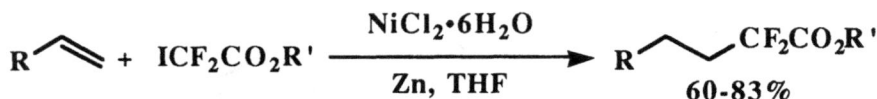

$$R\diagdown\diagup + ICF_2CO_2R' \xrightarrow[\text{Zn, THF}]{\text{NiCl}_2\cdot 6H_2O} R\diagup\diagdown\diagup CF_2CO_2R'$$

60-83%

**VI.A.11-8**  M.A. Tius and J.K. Kawakami, *Synth. Commun.*, **22**, 1461; D.A. Widdowson et al., *Synlett.*, 831;  A.G. Martinez and S.M. Gongales, *Tetrahedron Lett.*, **33**, 2043.

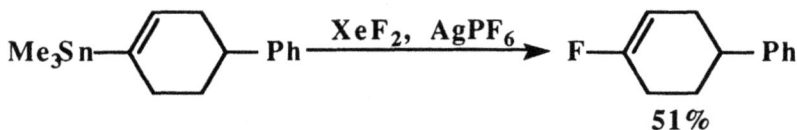

$$Me_3Sn-\text{⬡}-Ph \xrightarrow{XeF_2,\ AgPF_6} F-\text{⬡}-Ph$$

51%

**VI.A.11-9**  S. Stauber et al., *J. Org. Chem.*, **57**, 5334;  for acid bromide see  R.R. Srivastava and G.W. Kabalka, *Tetrahedron Lett.*, **33**, 593.

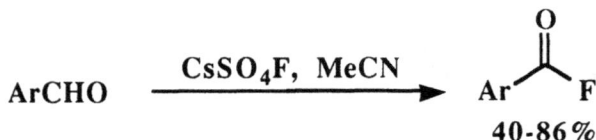

$$ArCHO \xrightarrow{CsSO_4F,\ MeCN} Ar\overset{O}{\underset{}{\diagup\!\!\diagdown}}F$$

40-86%

**VI.A.11-10**  Y. Langlois et al., *Synth. Commun.*, **22**, 2543;  W.-W. Weng and T.-Y. Luh, *J. Org. Chem.*, **57**, 2760;  S. Futamura and Z.-M. Zong, *Bull. Chem. Soc. Jpn.*, **65**, 345.

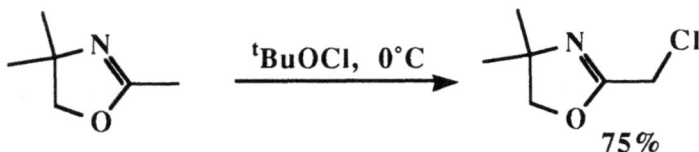

$$\text{⬠}\xrightarrow{^tBuOCl,\ 0°C}\text{⬠}Cl$$

75%

**VI.A.11-11**  J. Leroy, *Synth. Commun.*, **22**, 567;  Q.Y. Chen et al., *J. Chem. Soc., Chem. Commun.*, 807;  R.J. Lagow et al., *ibid.*, 811.

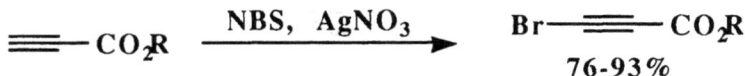

$$\equiv\!\!-CO_2R \xrightarrow{NBS,\ AgNO_3} Br-\!\!\equiv\!\!-CO_2R$$

76-93%

**VI.A.11-12** B. Schmidt and H.M.R. Hoffmann, *Chem. Ber.*, **125**, 1501; J. Barluenga et al., *J. Chem. Soc., Chem. Commun.*, 1016; For bromides see K. Fukui and T. Nonaka, *Bull. Chem. Soc. Jpn.*, **65**, 943; H.A. Muathen, *J. Org. Chem.*, **57**, 2740; K. Smith et al., *J. Chem. Soc., Perkin Trans. 1*, 1877 and *Tetrahedron*, **48**, 7479; Y. Murakami et al., *Heterocycles*, **34**, 2349.

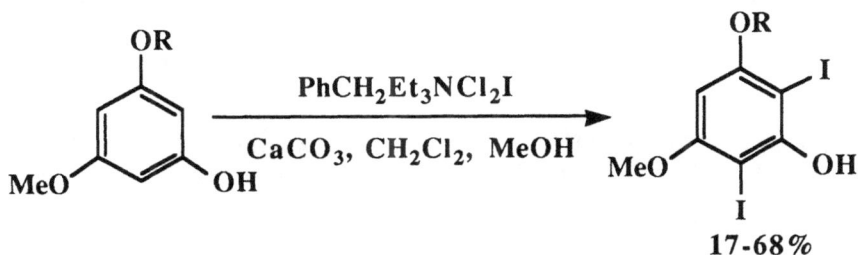

**VI.A.11-13** D.J. Milner, *Synth. Commun.*, **22**, 73; Y. Kikugawa et al., *J. Chem. Soc., Chem. Commun.*, 921; H. Yoshioka et al., *Synlett.*, 345; For chlorides see J.G. Lee and H.T. Cha, *Tetrahedron Lett.*, **33**, 3167; K.K. Park and H. Rapoport, *J. Heterocyclic Chem.*, **29**, 1031; R.P. Polniasjek and C.F. Lichti, *Synth. Commun.*, **22**, 171; M. Yoshida et al., *Chem. Lett.*, 227.

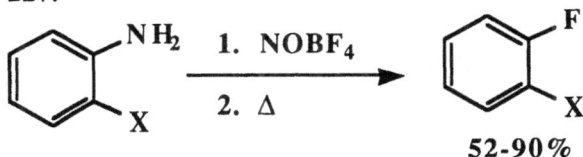

**VI.A.11-14** T. Yoshiyama and T. Fuchigami, *Chem. Lett.*, 1995; M. Kuroboshi and T. Hiyama, *ibid.*, 827; A.G. Martinez et al., *Tetrahedron Lett.*, **33**, 7787.

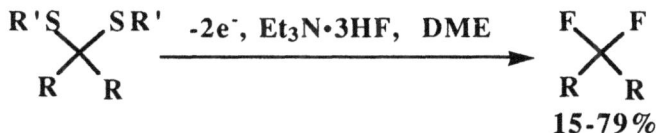

**VI.A.11-15** D.H.R. Barton et al., *Tetrahedron Lett.*, **33**, 3413; L. De Buyck, *Bull. Soc. Chim. Belg.*, **101**, 303; E.W. Della and N.J. Head, *J. Org. Chem.*, **57**, 2850; T. Wakasugi et al., *Chem. Lett.*, 171.

**Studies on the Bromination of Saturated Hydrocarbons under GoAgg[III] Conditions.**

**VI.A.11-16**  W. Sy, *Synth. Commun.*, **22**, 3215;  B.C. Ranu et al., *ibid.*, **22**, 1095;  M. Kodomari et al., *Bull. Chem. Soc. Jpn.*, **65**, 306.

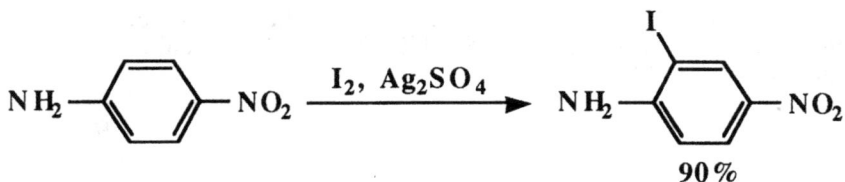

$$NH_2 \underbrace{\qquad}_{} NO_2 \xrightarrow{\ I_2,\ Ag_2SO_4\ } NH_2 \underbrace{\qquad}_{} NO_2$$

**90%**

---

**VI.A.11-17**  J. Mas and P. Metivier, *Synth. Commun.*, **22**, 2187;  J. Sandri and J. Viala, *ibid.*, **22**, 2945;  N.S. Mani et al., *ibid.*, **22**, 2175;  D.Y. Chi et al., *ibid.*, **22**, 2815;  M. Benazza et al., *Tetrahedron Lett.*, **33**, 3129 and 4901;  Ch.K. Reddy and M. Periasamy, *Tetrahedron*, **48**, 8329;  Y. Ukaji and K. Inomata, *Chem. Lett.*, 2353;  T. Hata et al., *Chem. Lett.*, 1505.

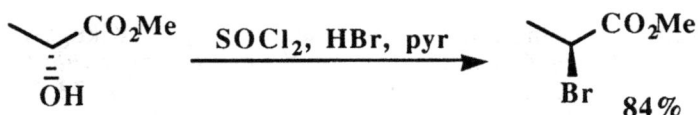

$$\underset{\overset{|}{OH}}{\diagdown}CO_2Me \xrightarrow{\ SOCl_2,\ HBr,\ pyr\ } \underset{\overset{|}{Br}}{\diagdown}CO_2Me \quad 84\%$$

---

**VI.A.11-18**  F. Tellier and R. Sauvetre, *Tetrahedron Lett.*, **33**, 3643;  J. Srogl and P. Kocovsky, *ibid.*, **33**, 5991;  M. Yamazaki et al., *Liebigs Ann. Chem.*, 1109;  B. Loubinoux and R. Schneider, *Synth. Commun.*, **22**, 2343.

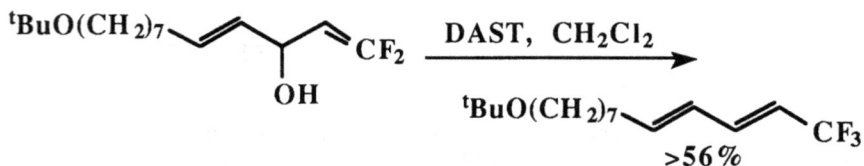

$${}^tBuO(CH_2)_7 \diagup\diagdown\diagup CF_2 \xrightarrow{\ DAST,\ CH_2Cl_2\ } {}^tBuO(CH_2)_7 \diagup\diagdown\diagup CF_3$$

with OH group

**>56%**

## VI.A.12.    Nitriles and Imines

**VI.A.12-1** M. Soufiaoui et al., *Tetrahedron*, **48**, 8935.

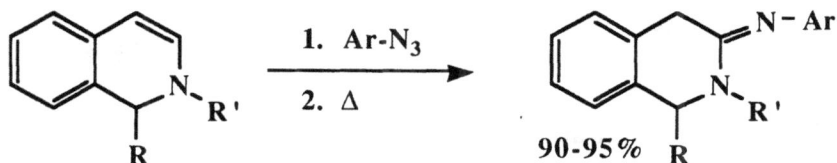

**VI.A.12-2** J. Oda et al., *J. Org. Chem.*, **57**, 5643.

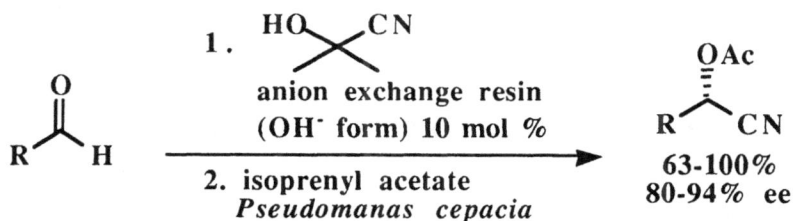

**VI.A.12-3** M.A. Poss et al., *Tetrahedron Lett.*, **33**, 5933.

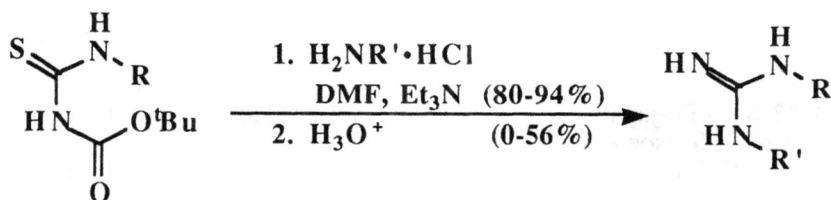

**VI.A.12-4** D.H.R. Barton et al., *J. Am. Chem. Soc.*, **114**, 5904.

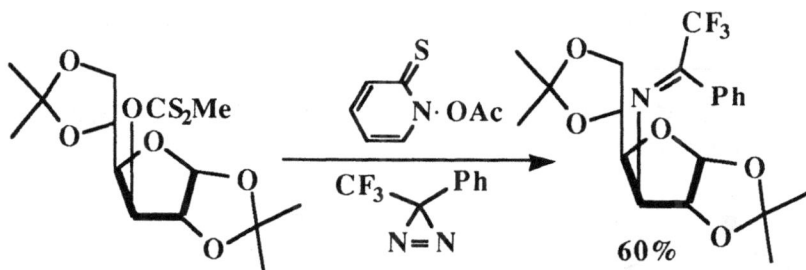

**VI.A.12-5**  K. Kim et al., *Tetrahedron Lett.*, **33**, 4963.

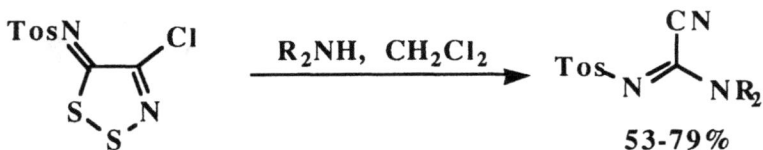

53-79%

---

**VI.A.12-6**  D.L. Boger and W.L. Corbett, *J. Org. Chem.*, **57**, 4777.

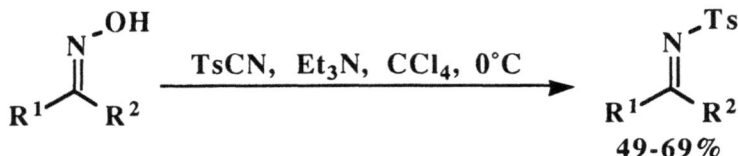

49-69%

---

**VI.A.12-7**  D.D. Bankston and M.R. Almond, *J. Heterocyclic Chem.*, **29**, 1405.

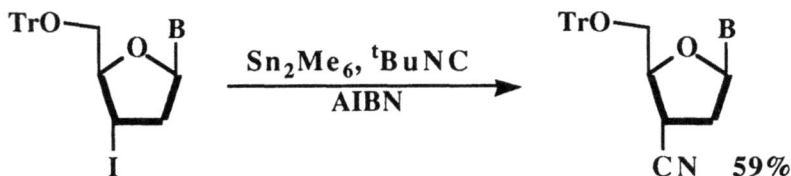

CN  59%

---

**VI.A.12-8**  D.Prajapati and J.S. Sandhu, *Ind. J. Chem.*, **30B**, 1065 (91); J.-i. Yoshida, S. Isoe et al., *J. Org. Chem.*, **57**, 4877;  E. Yoshii et al., *ibid.*, **57**, 2888;  N. Suzuki and C. Nakaya, *Synthesis*, 641, H.M. Meshram, *ibid.*, 943.

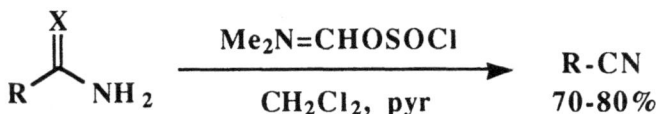

R-CN
70-80%

---

**VI.A.12-9**  D. Mitchell and T.M. Koenig, *Tetrahedron Lett.*, **33**, 3281;  J. Sapi et al., *Synthesis*, 383.

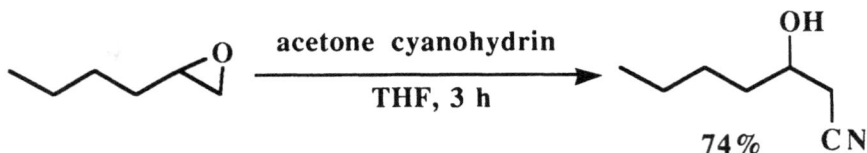

74%  CN

**VI.A.12-10**  R. Carlson et al., *Acta Chem. Scand.*, **46**, 1211.

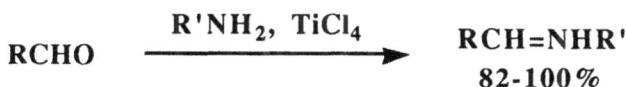

$$\text{RCHO} \xrightarrow{\text{R'NH}_2, \text{ TiCl}_4} \begin{array}{c} \text{RCH=NHR'} \\ \text{82-100\%} \end{array}$$

---

**VI.A.12-11**  L. Eberson and F. Radner, *Acta Chem. Scand.*, **46**, 312;  J.E.
Macor et al., *J. Heterocyclic Chem.*, **29**, 1465;  A. Guy et al., *Synlett.*, 821.

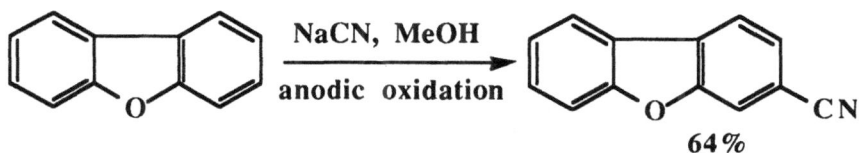

64%

## VI.A.13.    Other N-Containing Functional Groups

**VI.A.13-1**  P.R. Dave et al., *Tetrahedron*, **48**, 5839.

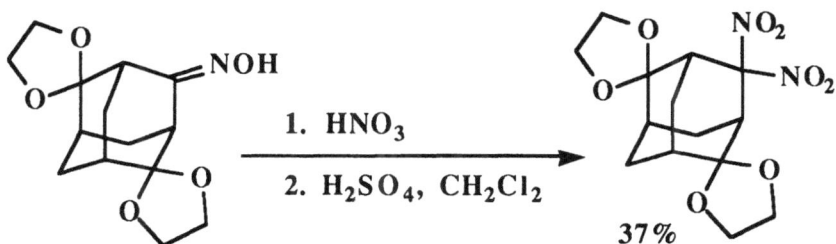

1. $HNO_3$

2. $H_2SO_4$, $CH_2Cl_2$

37%

---

**VI.A.13-2**  R.W. Stephens et al., *Tetrahedron Lett.*, **33**, 733.

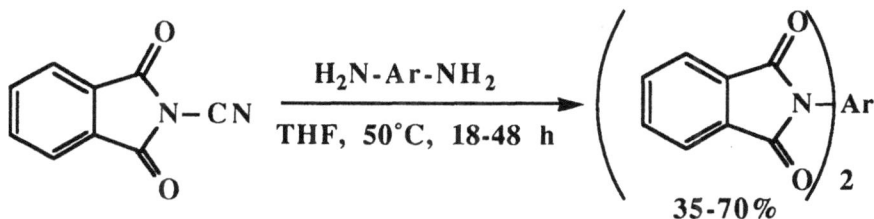

$H_2N$-$Ar$-$NH_2$

THF, 50°C, 18-48 h

35-70%

**VI.A.13-3**  H. Mayr and K. Grimm, *J. Org. Chem.*, **57**, 1057.

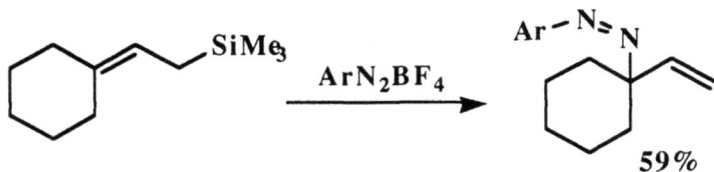

**VI.A.13-4**  A. Ogawa et al., *J. Am. Chem. Soc.*, **114**, 8729.

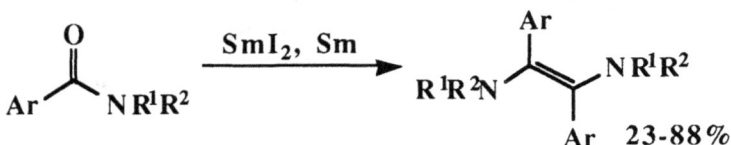

**VI.A.13-5**  H. Kisch et al., *Angew. Chem., Int. Ed. Engl.*, **31**, 1039.

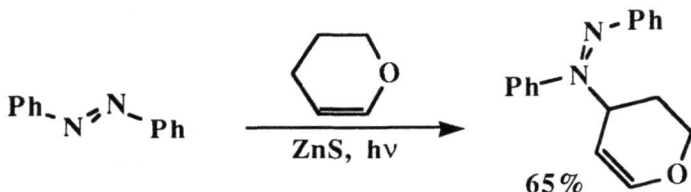

**VI.A.13-6**  B.J. Wakefield et al., *Tetrahedron*, **48**, 7619.

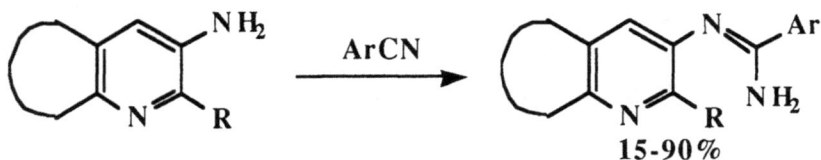

**VI.A.13-7**  A. Guirado et al., *Tetrahedron Lett.*, **33**, 4779.

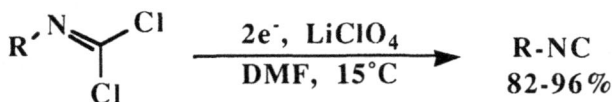

**VI.A.13-8**  G. Tojo et al., *Synth. Commun.*, **22**, 677.

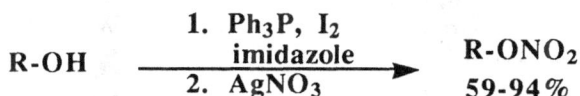

$$R\text{-}OH \xrightarrow[\substack{\textbf{2. AgNO}_3}]{\substack{\textbf{1. Ph}_3\textbf{P, I}_2 \\ \textbf{imidazole}}} \begin{array}{c} R\text{-}ONO_2 \\ \textbf{59-94\%} \end{array}$$

**VI.A.13-9**  H. Yamazaki et al., *Chem. Pharm. Bull.*, **40**, 1025.

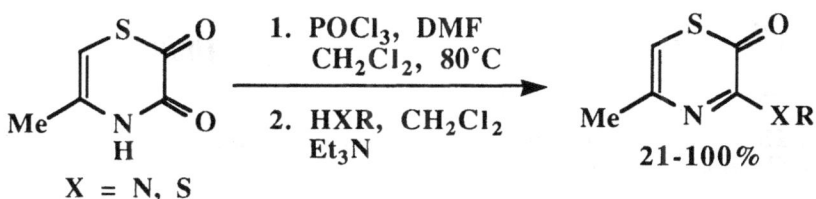

**VI.A.13-10**  G.A. Olah et al., *Synthesis*, 1085 and 1087.

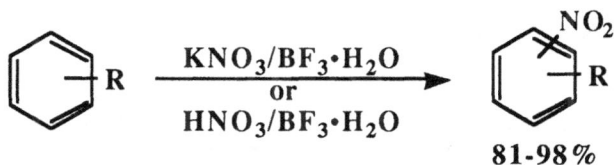

**VI.A.13-11**  M. Puciova and S. Toma, *Coll. Czech. Chem. Commun.*, **57**, 2407;  see also M.-C. Senechal-Tocquer et al., *J. Organomet. Chem.*, **433**, 261.

### Synthesis of Oximes in the Microwave Oven

**VI.A.13-12**  M.A. Aramendia et al., *Synth. Commun.*, **22**, 3263;  M.S. Holden and K.A. Cole, *ibid.*, **22**, 2579;  E.K. Ryu et al., *ibid.*, **22**, 1427.

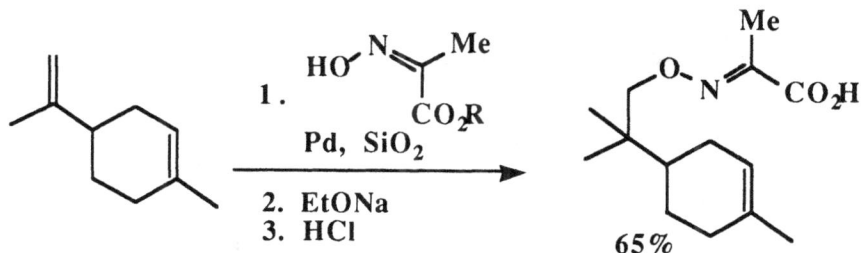

**VI.A.13-13**  U. Chiacchio et al., *Tetrahedron*, **48**, 9473;  see also  S. Prabhakar et al., *ibid.*, **48**, 6335;  W. Oppolzer et al., *J. Am. Chem. Soc.*, **114**, 5901;  L. Greci, *Synth. Commun.*, **22**, 201;  W.W. Wood and J.A. Wilkin, *ibid.*, **22**, 1683.

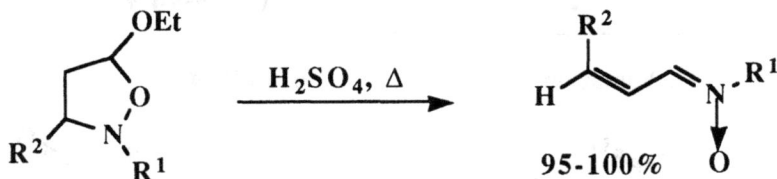

**VI.A.13-14**  B.E. Love and L. Tsai, *Synth. Commun*, **22**, 3101 and 165; R. Neidlein and Z. Sui, *.ibid.*, **22**, 229;  .

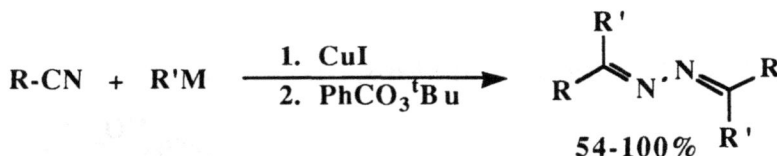

**VI.A.13-15**  J. Nowakowski, *J. Prakt. Chem.*, **334**, 187;  M.V. Vork, *J. Org. Chem. (USSR)*, **28**, 494;  S.-I. Fujiwora and T. Shin-Ike, *Tetrahedron Lett.*, **33**, 7021;  T. Yamamoto et al., *Org. Prep. Proc. Int.*, **24**, 346.

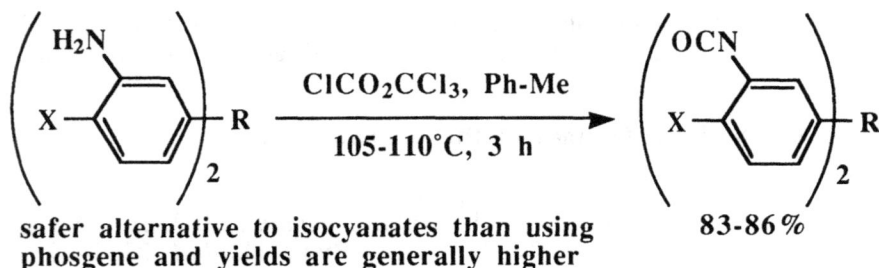

safer alternative to isocyanates than using phosgene and yields are generally higher

**VI.A.13-16**  C. Burnell-Curty and E.J. Roskamp, *J. Org. Chem.*, **57**, 5063.

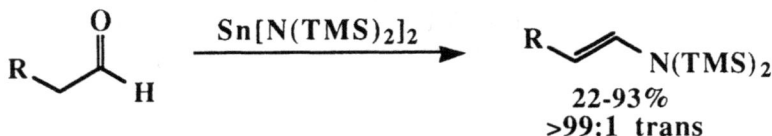

## VI.B.    Additions to Alkenes and Alkynes

**VI.B-1** L.E. Overman and I.M. Rodriguez-Campos, *Synlett.*, 995;  A. Goosen et al., *J. Chem. Soc., Perkin Trans. 1*, 627.

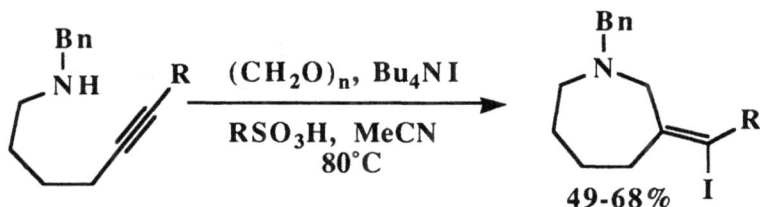

**VI.B-2** E.K. Ryu et al., *Synth. Commun.*, **22**, 2521;  F. Ghelfi et al., *ibid.*, **22**, 1101;  D. J. Burton et al., *Tetrahedron Lett.*, **33**, 2137;  F. Oberdorfer, *ibid.*, **33**, 2435;  H. Blancou, *ibid.*, **33**, 2489;  J.L. Kice et al., *ibid.*, **33**, 1949;  A.B. Smith III et al., *ibid.*, **33**, 6439;  F. Oberdorfer et al., *ibid.*, **33**, 2435;  J.-P. Dulcère and J. Rodriguez, *Synlett.*, 347;  see also M. Kotora et al, *Coll. Czech. Chem. Commun.*, **57**, 2622 and 393;  S.P. Elvey and D.R. Mootos, *J. Am. Chem. Soc.*, **114**, 9685;  R.W. Franck et al., *J. Org. Chem.*, **57**, 2084.

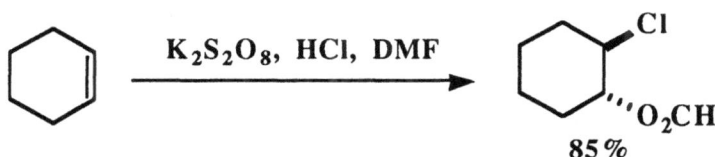

**VI.B-3** P.A. Grieco et al., *J. Am. Chem. Soc.*, **114**, 2764;  H. Meier et al., *Angew. Chem., Int. Ed. Engl.*, **31**, 791;  S.T. Kabanyane and D.I. Magee, *Can. J. Chem.*, **70**, 2758;  L. Brandsma et al., *Synth. Commun.*, **22**, 1563;  Q.B. Broxterman, B. Kaptein et al., *J. Org. Chem.*, **57**, 6286;  W. Schroth et al., *J. Prakt. Chem.*, **334**, 141.

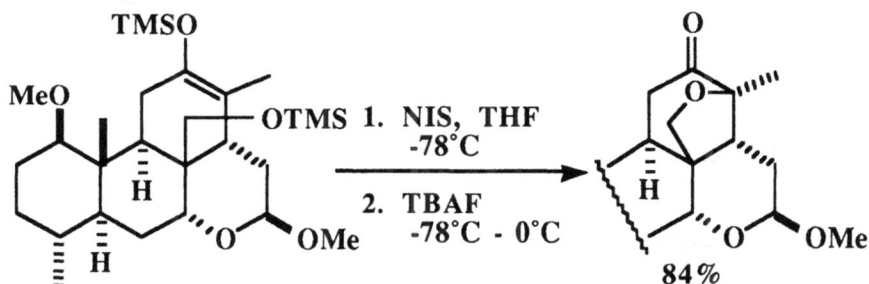

**VI.B-4** L. Lejeune and J.Y. Lallemand, *Tetrahedron Lett.*, **33**, 2977; X. Lu et al., *J. Org. Chem.*, **57**, 709.

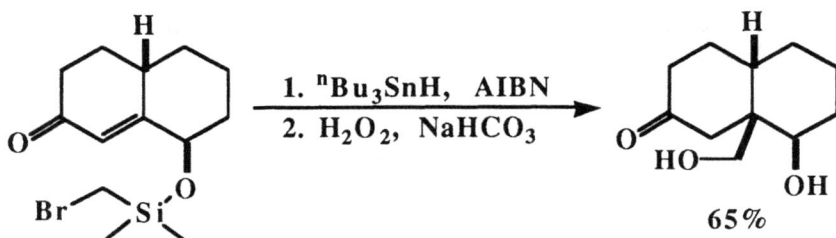

$$\xrightarrow[\text{2. } H_2O_2, \text{ NaHCO}_3]{\text{1. } {}^nBu_3SnH, \text{ AIBN}}$$

65%

**VI.B-5** R.T. Baker et al., *J. Am. Chem. Soc.*, **114**, 8863; K. Burgen et al., *ibid.*, **114**, 9350.

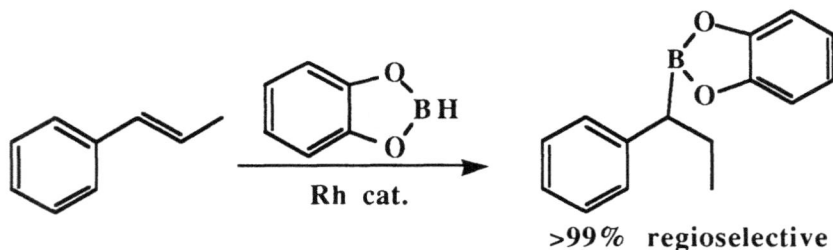

$$\xrightarrow{\text{Rh cat.}}$$

>99% regioselective

**VI.B-6** E.J. Enholm and J.D. Burrott, *Tetrahedron Lett.*, **33**, 1835; Y. Chen and W. Lin, *ibid.*, **33**, 1749; see also T. Taguchi et al., *ibid.*, **33**, 2107; J. Cossy and A. Bouzide, *ibid.*, **33**, 2505; B.M. Trost and L. Zhi, *ibid.*, **33**, 1831.

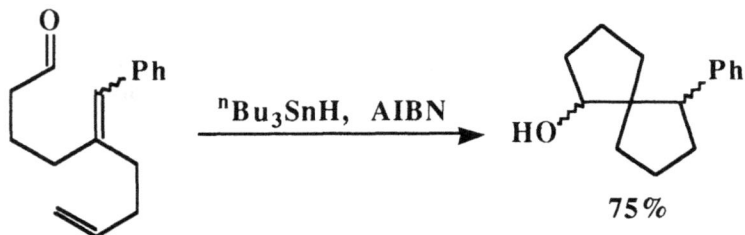

$$\xrightarrow{{}^nBu_3SnH, \text{ AIBN}}$$

75%

**VI.B-7** K. Inomata et al., *Bull. Chem. Soc. Jpn.*, **65**, 1379.

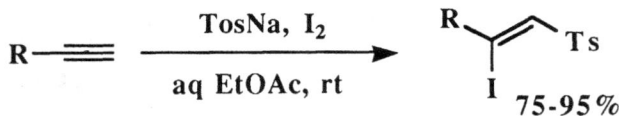

$$R\!-\!\!\!\equiv\ \xrightarrow[\text{aq EtOAc, rt}]{\text{TosNa, } I_2}\ $$

75-95%

**VI.B-8** B.M. Trost and M.C. Matelich, *Synthesis*, 151.

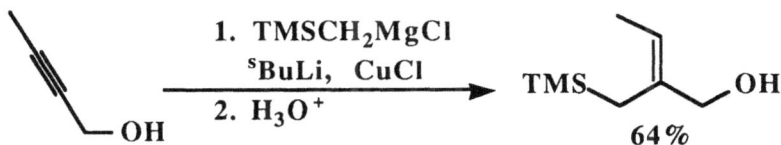

64%

---

**VI.B-9** P. Bovonsombat and E. McNelis, *Tetrahedron Lett.*, **33**, 4123 and *Synth. Commun.*, **22**, 2361.

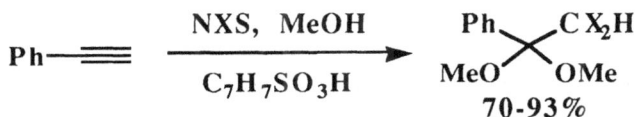

70-93%

## VI.C. Nucleotides, Etc.

**VI.C-1** K. Hirota et al., *J. Org. Chem.*, **57**, 5268; T.T. Nikiforou and B.D. Connolly, *Tetrahedron Lett.*, **33**, 2379.

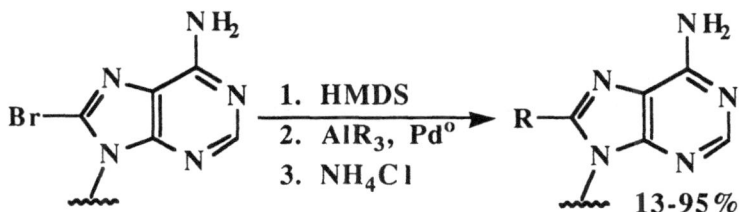

1. HMDS
2. $AlR_3$, $Pd^o$
3. $NH_4Cl$

13-95%

---

**VI.C-2** T. Benneche et al., *Tetrahedron Lett.*, **33**, 1085; T. Miyasaka et al., *ibid.*, **33**, 2841; R.M. Riggs and P.A. Crooks, *Chem. Ind.*, 220.

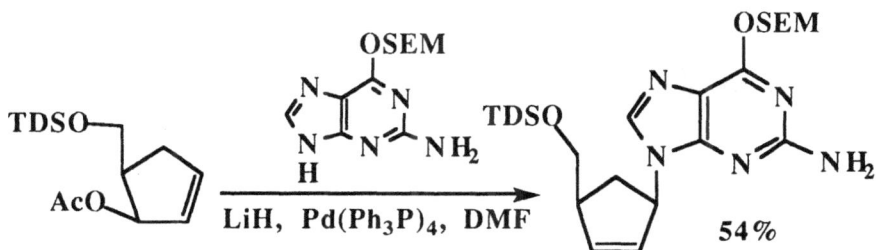

LiH, $Pd(Ph_3P)_4$, DMF

54%

**VI.C-3**  W. Samstag and J.W. Engels, *Angew. Chem., Int. Ed. Engl.*, **31**, 1386.

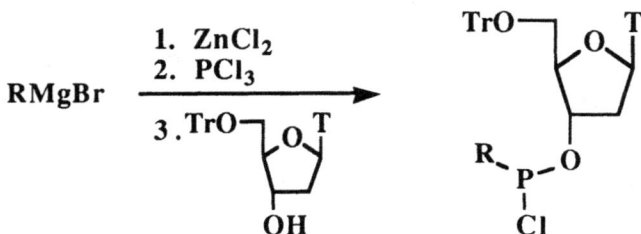

RMgBr  $\xrightarrow{\begin{array}{l}\text{1. ZnCl}_2 \\ \text{2. PCl}_3 \\ \text{3. TrO—\llap{O} T} \\ \quad\quad\text{OH}\end{array}}$

---

**VI.C-4**  E. Dorland and P. Seratinowski, *Synthesis*, 477;  V. Bhat et al., *Synth. Commun.*, 1481.

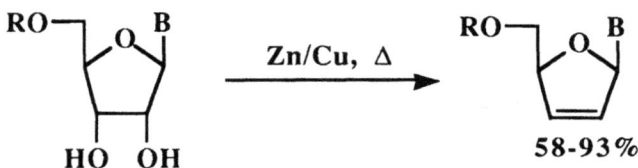

$\xrightarrow{\text{Zn/Cu, }\Delta}$

58-93%

---

**VI.C-5**  S. Castillion et al., *Tetrahedron Lett.*, **33**, 1093;  R.A. Cadenas et al., *J. Heterocyclic Chem.*, **29**, 401;  A.K. Jhingan and T. Meehan, *Synth. Commun.*, **22**, 3129;  H. Togo et al., *Chem. Lett.*, 1673;  K. Suzuki et al., *J. Am. Chem. Soc.*, **114**, 3568.

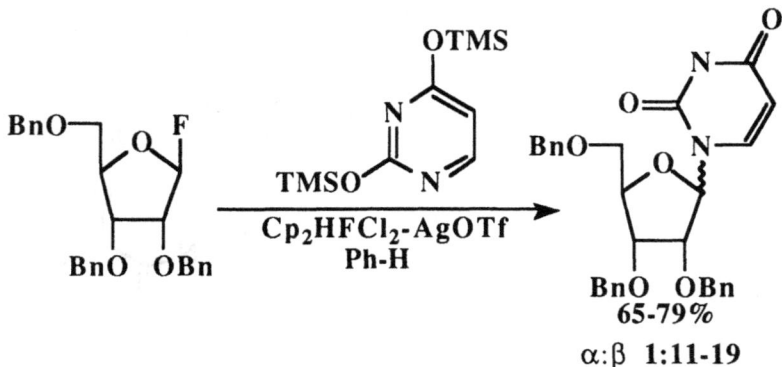

$\xrightarrow{\text{Cp}_2\text{HFCl}_2\text{-AgOTf} \atop \text{Ph-H}}$

65-79%

α:β 1:11-19

## VI.D.    Phosphorus, Selenium and Tellurium Compounds

**VI.D-1**  E. Abushanab et al., *Tetrahedron Lett.*, **33**, 5491;  S.F. Martin et al., *ibid.*, **33**, 1839;  F. Dónotrio et al., *Synth. Commun.*, **22**, 699.

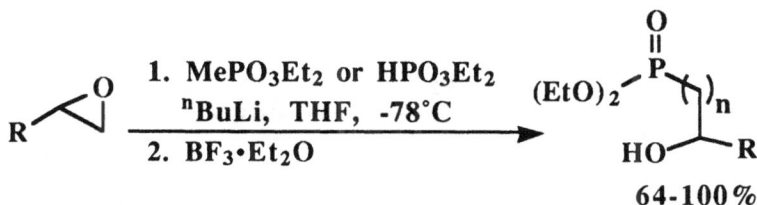

64-100%

**VI.D-2**  A.M. Modro and T.A. Modro, *Org. Prep. Proc. Int.*, **24**, 57;  R.A. McClelland and T.A. Modro, *ibid.*, **24**, 197;  O. Hindsgaul et al., *J. Am. Chem. Soc.*, **114**, 5891;  P.A. Bartlett et al., *ibid.*, **114**, 3535;  B.L. Feringa et al., *Angew. Chem., Int. Ed. Engl.*;  **31**, 1092;  C. McGuigan and B. Swords, *J. Chem. Soc., Perkin Trans. 1*, 51;  M. Vaultier et al., *Bull. Soc. Chim. Fr.*, **129**, 71.

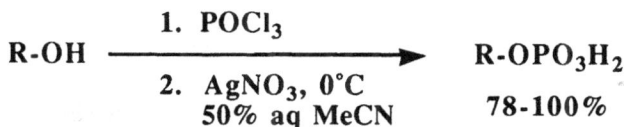

$$R\text{-OH} \xrightarrow[\substack{\text{2. } AgNO_3, \text{ } 0°C \\ \text{50\% aq MeCN}}]{\text{1. } POCl_3} R\text{-}OPO_3H_2$$

78-100%

**VI.D-3**  J.-Z. You and Z.-C. Chen, *Synth. Commun.*, **22**, 1441 and *Synthesis*, 633;  A. Krief et al., *ibid.*, 933

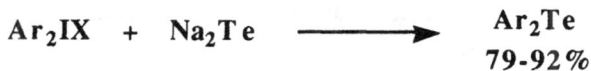

$$Ar_2IX \text{ + } Na_2Te \longrightarrow Ar_2Te$$

79-92%

**VI.D-4**  M.L. Pedersen and D.B. Berkowitz, *Tetrahedron Lett.*, **33**, 7315;  Y. Vallée et al., *ibid.*, **33**, 6131.

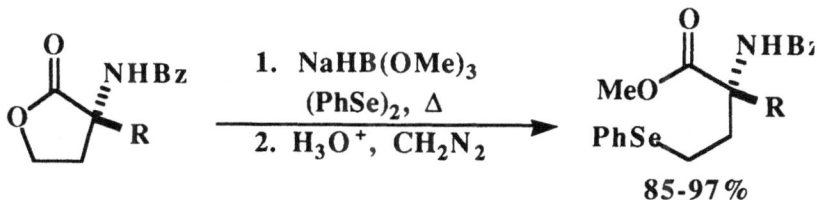

85-97%

**VI.D-5**  K. Suzuki et al., *Synlett.*, 340;  K. Afarinkia et al., *Synlett.*, 123 and 124;  D.Y. Kim et al., *J. Chem. Soc., Perkin Trans. 1*, 2451;  P.P. Gaspar et al., *J. Am. Chem. Soc.*, **114**, 8526;  L.K. Lukarov and A.P. Venkov *Synthesis*, 263;  M. Moreno-Manas and R. Pleixats, *Synth. Commun.*, **22**, 2219;  A. Couture et al., *ibid.*, **22**, 2381;  S.D. Pastor et al., *Tetrahedron*, **48**, 2911.

73-80%

---

**VI.D-6**  N. Collignon et al., *J. Organomet. Chem.*, **440**, 297;  R.A. Bartsch et al., *J. Heterocyclic Chem.*, **29**, 867.

84-99%

---

**VI.D-7**  A. Ogawa, N. Sonoda et al., *Tetrahedron Lett.*, **33**, 1329 and 5525 and *Chem. Lett.*, 2241 (1991);  for tellurium see M.J. Dabdoub et al., *Tetrahedron Lett.*, **33**, 2261;  N. Kambe, N. Sonoda et al., *J. Am. Chem. Soc.*, **114**, 7591;  R.W. Gedridge, Jr., L. Brandsma et al., *Organometallics*, **11**, 418.

44-62%

---

**VI.D-8**  Y. Ueno et al., *Synlett.*, 965;  S. Czernecki et al., *ibid.*, 967.

62-97%

**VI.D-9** M.A. Cooper and A.D. Ward, *Tetrahedron Lett.*, **33**, 5999; L. Hevesi et al., *ibid.*, **33**, 4629.

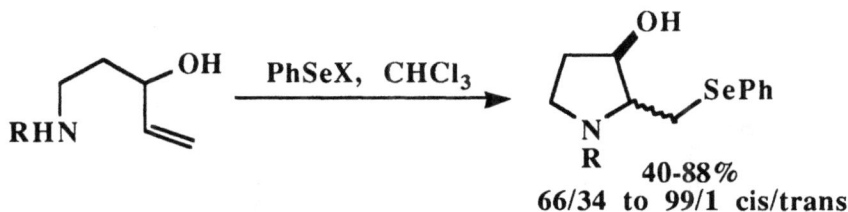

40-88%
66/34 to 99/1 cis/trans

## VI.E. Silicon Compounds

**VI.E-1** J.-B. Verlhac et al., *J. Organomet. Chem.*, **437**, C13; I. Fleming et al., *J. Chem. Soc., Perkin Trans. 1*, 3331; T.Ohno, I. Nishiguchi et al., *Tetrahedron Lett.*, **33**, 5515; Y. Tsuji et al., *Organometallics*, **11**, 2353.

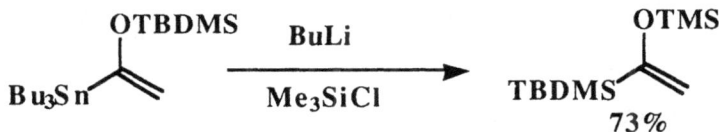

73%

**VI.E-2** N. Furukawa et al., *Heterocycles*, **34**, 1085; M. Bordeau, C. Biran, J. Donogués et al., *J. Org. Chem.*, **57**, 4705.

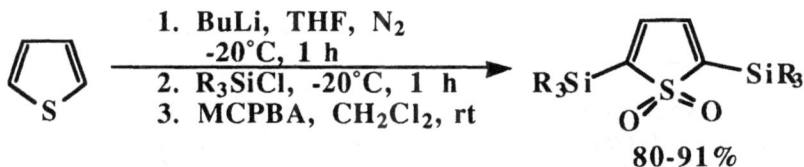

1. BuLi, THF, $N_2$
   -20°C, 1 h
2. $R_3SiCl$, -20°C, 1 h
3. MCPBA, $CH_2Cl_2$, rt

80-91%

**VI.E-3** M. Ishikawa et al., *Organometallics*, **11**, 2708.

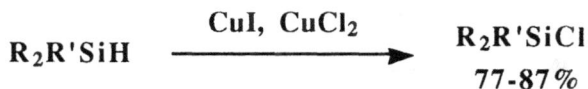

$R_2R'SiH$ → (CuI, CuCl$_2$) → $R_2R'SiCl$

77-87%

**VI.E-4** S. Yamazaki et al., *J. Org. Chem.*, **57**, 5610.

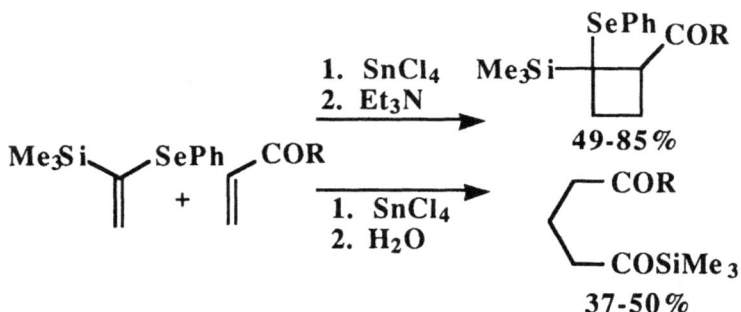

**VI.E-5** I. Matsuda, H. Nagashima et al., *Tetrahedron Lett.*, **33**, 5799.

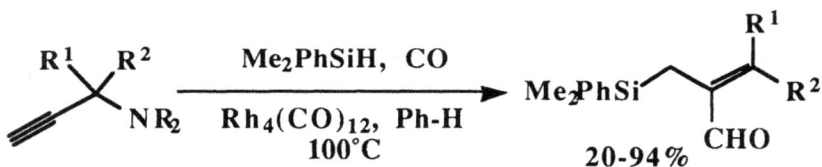

**VI.E-6** T. Nakai et al., *Synlett.*, 189.

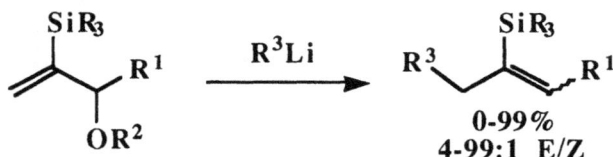

**VI.E-7** T. Hiyama et al., *Bull. Chem. Soc. Jpn.*, **65**, 1280; M.G. Voronkov et al., *J. Organomet. Chem.*, **427**, 289; L.N. Lewis et al., *ibid.*, **427**, 165; K. Yamamoto and T. Tabei, *ibid.*, **428**, *Can. J. Chem.*; V. Gevorgyan et al., *ibid.*, **424**, 15; M.R. Kesti and R.M. Waymouth, *Organometallics*, **11**, 1095; M.P. Doyle et al., *ibid.*, **11**, 549.

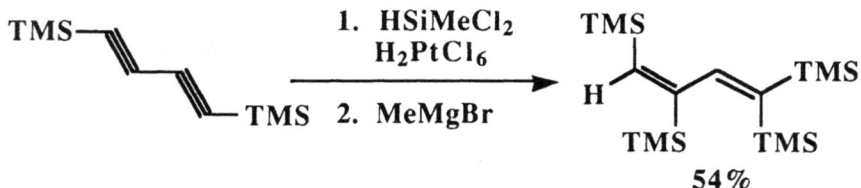

**VI.E-8** Y.Xu et al., *Tetrahedron Lett.*, **33**, 1221; B. Bosnich et al., *J. Am. Chem. Soc.*, **114**, 2121 and 2129.

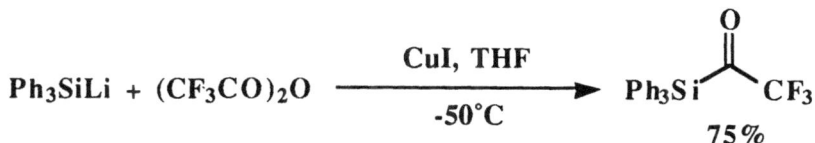

$$\text{Ph}_3\text{SiLi} + (\text{CF}_3\text{CO})_2\text{O} \xrightarrow[\text{-50°C}]{\text{CuI, THF}} \underset{\underset{75\%}{}}{\text{Ph}_3\text{Si}}\overset{\overset{O}{\|}}{\text{C}}\text{CF}_3$$

## VI.F.    Sulfur Compounds

**VI.F-1** J.M. Khurana et al., *Synth. Commun.*, **22**, 1691; A.R. Ramesha and S. Chandrasekaren, *ibid.*, **22**, 3277; N. Jayasuriya and S.L. Regen, *Tetrahedron Lett.*, **33**, 451; C.J. Li and D.N. Harpp, *ibid.*, **33**, 7293.

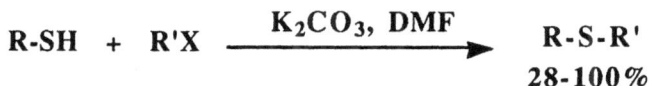

$$\text{R-SH} + \text{R'X} \xrightarrow{\text{K}_2\text{CO}_3, \text{DMF}} \underset{28\text{-}100\%}{\text{R-S-R'}}$$

**VI.F-2** M. Sprecher and E. Nov, *Synth. Commun.*, **22**, 2949; S. Rajappa et al., *Tetrahedron Lett.*, **33**, 2857; J. You and Z. Chen, *Synthesis*, 521; T. Mukaiyama et al, *Chem. Lett.*, 1747.

$$\text{R-SH} + \text{HCO}_2\text{H} \xrightarrow{\text{DCC}} \underset{30\text{-}60\%}{\text{H}\overset{\overset{O}{\|}}{\text{C}}\text{SR}}$$

**VI.F-3** P.L. Folkins and D.N. Harpp, *J. Org. Chem.*, **57**, 2013; S. Chandrasekaran et al., *ibid.*, **57**, 1699; R. Polt et al., *Tetrahedron Lett.*, **33**, 2961; T. Saito et al., *J. Chem. Soc., Perkin Trans. 1*, 600.

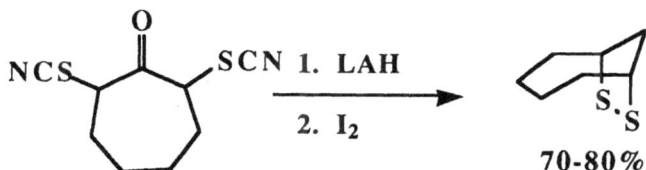

**VI.F-4**  G.A. Olah et al., *Synthesis*, 405.

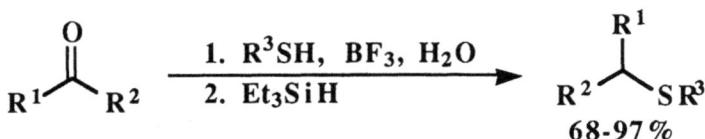

$$\begin{array}{c}\text{O}\\\parallel\\R^1\diagdown\!\!\!\diagup R^2\end{array}\xrightarrow[\text{2. Et}_3\text{SiH}]{\text{1. R}^3\text{SH, BF}_3\text{, H}_2\text{O}}\begin{array}{c}R^1\\\mid\\R^2\diagdown\!\!\!\diagup SR^3\end{array}$$

**68-97%**

---

**VI.F-5**  J. Huang and T.S. Widlanski, *Tetrahedron Lett.*, **33**, 2657;  A.L. Schwan and R. Dufault, *ibid.*, **33**, 3973;  T. Ando et al., *Chem. Lett.*, 891;  Y.H. Kim et al., *ibid.*, 1483.

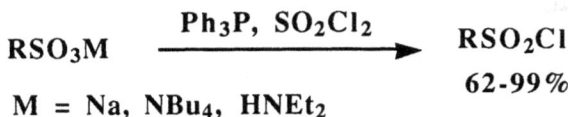

$$\text{RSO}_3\text{M}\xrightarrow{\text{Ph}_3\text{P, SO}_2\text{Cl}_2}\text{RSO}_2\text{Cl}$$

$$M = Na,\ NBu_4,\ HNEt_2$$

**62-99%**

---

**VI.F-6**  P. Beslin and P. Marion, *Tetrahedron Lett.*, **33**, 935;  W.-D. Rudorf et al., *Liebigs Ann. Chem.*, 387 and 395.

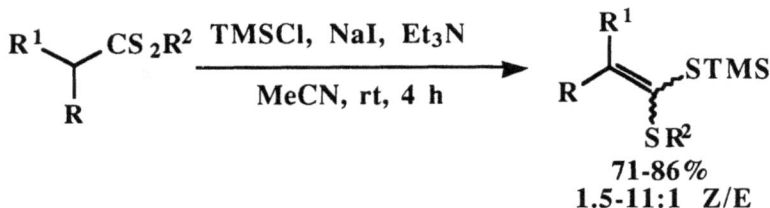

$$\begin{array}{c}R^1\\\diagdown\!\!\!\diagup CS_2R^2\\R\end{array}\xrightarrow[\text{MeCN, rt, 4 h}]{\text{TMSCl, NaI, Et}_3\text{N}}\begin{array}{c}R^1\\\diagdown\qquad\diagup STMS\\R\diagup\qquad\diagdown\\SR^2\end{array}$$

**71-86%**
**1.5-11:1  Z/E**

---

**VI.F-7**  L.J. Wilson and D.C. Liotta, *J. Org. Chem.*, **57**, 1948;  D. Villemin et al., *Synth. Commun.*, **22**, 1359;  J. Alvarez-Builla et al., *Tetrahedron Lett.*, **33**, 3677;  F. Naso et al., *ibid.*, **33**, 5121;  G. Capozzi et al., *Tetrahedron*, **48**, 9023;  S. Menichetti et al., *Synthesis*, 643;  J.P. Marino et al., *J. Am. Chem. Soc.*, **114**, 5566;  D.A. Evans et al., *ibid.*, **114**, 5977;  J.-B. Baudin et al., *Synlett.*, 911;  K. Takagi et al., *Chem. Lett.*, 509.

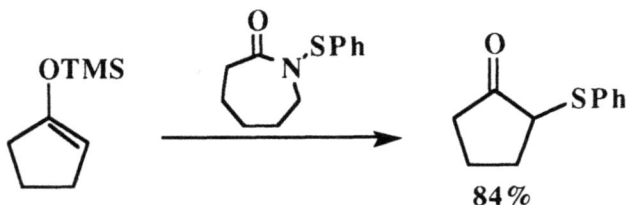

**84%**

**VI.F-8** R. Breitschuh and D. Seebach, *Synthesis*, 83; P. Lopez-Alvarado et al., *Synth. Commun.*, **22**, 2329.

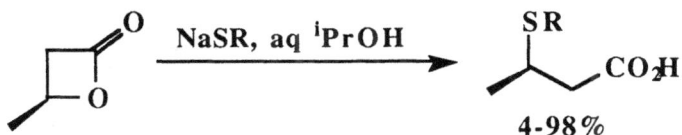

$$\text{4-98\%}$$

**VI.F-9** A.-M. Le Nocher and P. Metzner, *Tetrahedron Lett.*, **33**, 6151; K.-F. Wai and M.P. Sammes, *J. Med. Chem.*, **35**, 2065; D. Weiß et al., *Synthesis*, 75; R. Köster and R. Kucznierz, *Liebigs Ann. Chem.*, 1081.

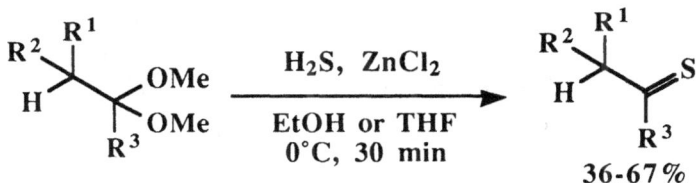

$$\text{36-67\%}$$

**VI.F-10** P. Hamel et al., *J. Org. Chem.*, **57**, 2694; H.M. Zuurmond et al., *Tetrahedron Lett.*, **33**, 2063.

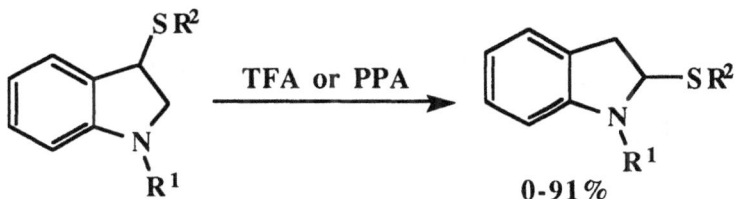

$$\text{0-91\%}$$

**VI.F-11** A. Degl'Innocenti et al., *Synlett.*, 880; E. Cortes et al., *J. Heterocyclic Chem.*, **29**, 1617; A. Bulpin and S. Masson, *J. Org. Chem.*, **57**, 4507.

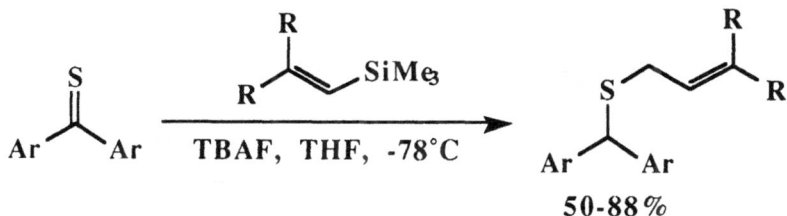

$$\text{50-88\%}$$

**VI.F-12**  P. Crotti et al., *Synlett.*, 303.

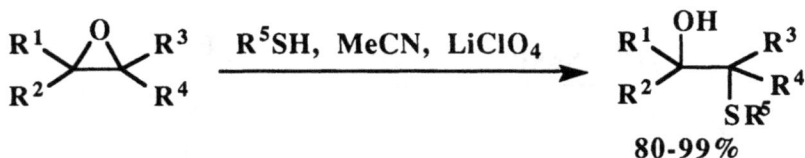

80-99%

---

**VI.F-13**  L.M. Yagupolskii et al., *Synthesis*, 749.

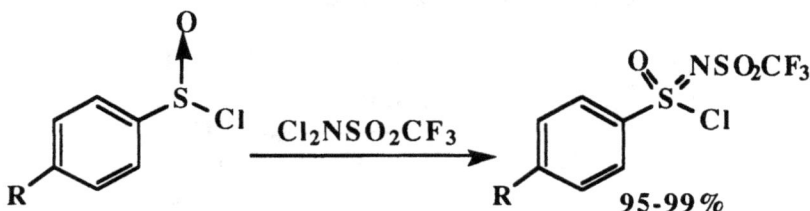

95-99%

---

**VI.F-14**  A. Kohrt and K. Hartke, *Liebigs Ann. Chem.*, 595;  M.D. Pujol and G. Guilaumet, *Synth. Commun.*, **22**, 1231;  D. Brillon, *ibid.*, **22**, 1397.

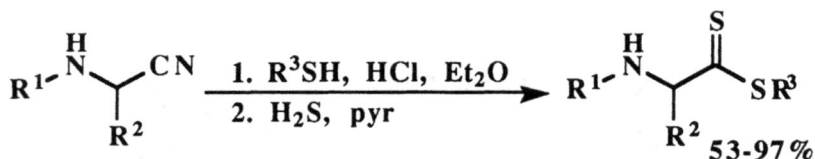

53-97%

---

**VI.F-15**  S. Torii et al., *Tetrahedron Lett.*, **33**, 7029;  G.C. Rovnyak et al., *J. Med. Chem.*, **35**, 3254;  A. Rodriguez, A. Padwa et al., *ibid.*, **33**, 5917.

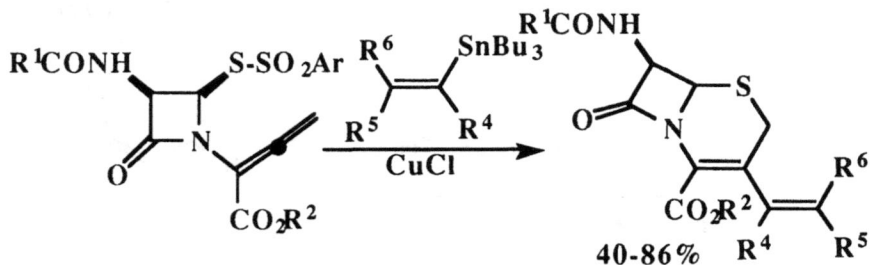

40-86%

**VI.F-16** D. Sinou et al., *Tetrahedron Lett.*, **33**, 8099; V. Fiandanese and L.Mazzone, *ibid.*, **33**, 7067; A. Degl'Innocenti et al., *Synlett.*, 499.

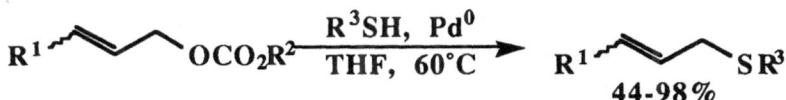

$$R^1 \diagup\!\!\diagdown\!\!\diagup OCO_2R^2 \xrightarrow[\text{THF, 60°C}]{R^3SH, \ Pd^0} R^1 \diagup\!\!\diagdown\!\!\diagup SR^3$$

**44-98%**

**VI.F-17** J.F. O'Connell and H. Rapoport, *J. Org. Chem.*, **57**, 4775.

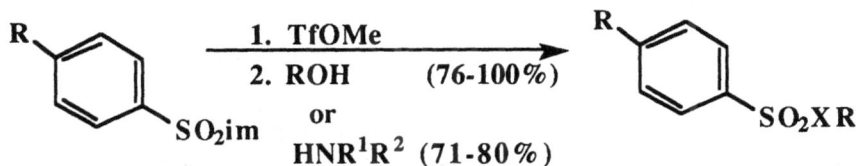

$$R \diagdown C_6H_4 SO_2im \xrightarrow[\substack{2. \ ROH \\ \text{or} \\ HNR^1R^2 \ (71\text{-}80\%)}]{1. \ TfOMe \quad (76\text{-}100\%)} R \diagdown C_6H_4 SO_2XR$$

**VI.F-18** K.V. Srinivasan, S.M. Kumar and N.R. Ayyangar, *Synthesis*, 825; D.M. Giolando and K. Kirschbaum, *ibid.*, 451; Z. Chen et al., *Synth. Commun.*, **22**, 2903; J. Chen and G.T. Crisp, *ibid.*, 683; H. Takeuchi et al., *J. Chem. Soc., Chem. Commun.*, 916; J. Dunogues et al., *Bull. Soc. Chim. Fr.*, **129**, 25; M. Tanaka and Y. Souma, *J. Org. Chem.*, **57**, 3738.

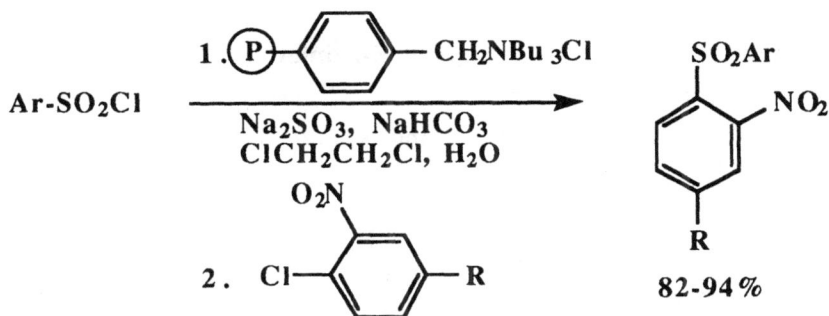

$$Ar\text{-}SO_2Cl \xrightarrow[\substack{Na_2SO_3, \ NaHCO_3 \\ ClCH_2CH_2Cl, \ H_2O}]{\substack{1. \ (P)\text{-}C_6H_4\text{-}CH_2NBu_3Cl \\ \quad \\ 2. \ O_2N\text{-}C_6H_3(Cl)\text{-}R}} \text{product}$$

**82-94%**

**VI.F-19** Z. Liu and Z. Chen, *Synth. Commun.*, **22**, 1997; P.J. Stang et al., *Synlett.*, 535.

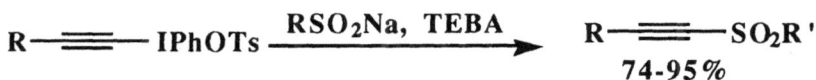

$$R\!-\!\!\equiv\!\!-IPhOTs \xrightarrow{RSO_2Na, \ TEBA} R\!-\!\!\equiv\!\!-SO_2R'$$

**74-95%**

## VI.G.    Tin Compounds

**VI.G-1** P. Quayle et al., *Tetrahedron Lett.*, **33**, 405;  J.A. Marshall and G.S. Welmaker, *J. Org. Chem.*, **57**, 7158.

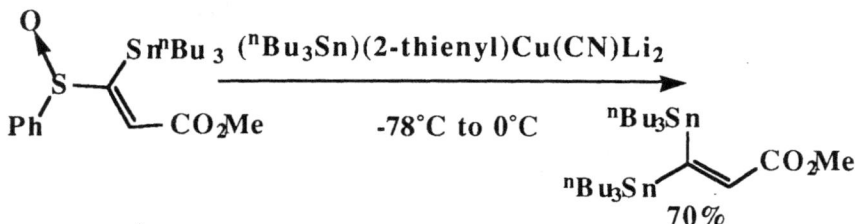

$$\underset{Ph}{\overset{O}{\|}}S-\underset{CO_2Me}{\overset{Sn^nBu_3}{\diagup}} \quad \xrightarrow[-78°C \text{ to } 0°C]{(^nBu_3Sn)(2\text{-thienyl})Cu(CN)Li_2}$$

$$\underset{^nBu_3Sn}{\overset{^nBu_3Sn}{\diagdown}}\diagup\diagdown CO_2Me$$

70%

**VI.G-2** E. Piers et al., *Can. J. Chem.*, **70**, 2058;  B.H. Lipshutz et al., *Tetrahedron Lett.*, **33**, 5861;  J.C. Podesta et al., *J. Organomet. Chem.*, **434**, 269.

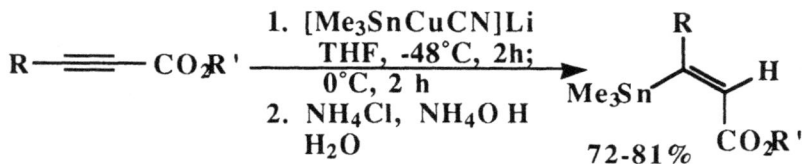

$$R \equiv\!\!\!=\!\!\!= CO_2R' \quad \xrightarrow[\substack{2.\ NH_4Cl,\ NH_4OH \\ H_2O}]{\substack{1.\ [Me_3SnCuCN]Li \\ THF,\ -48°C,\ 2h; \\ 0°C,\ 2\ h}} \quad \underset{Me_3Sn}{\overset{R}{\diagdown}}\!\!=\!\!\underset{CO_2R'}{\overset{H}{\diagup}}$$

72-81%

**Use of alcohol in step 1 gives E-isomer**

**VI.G-3** B.L. Williamson and P.J. Stang, *Synlett.*, 199;  K. Müllen, *Angew. Chem., Int. Ed. Engl.*, **31**, 1588.

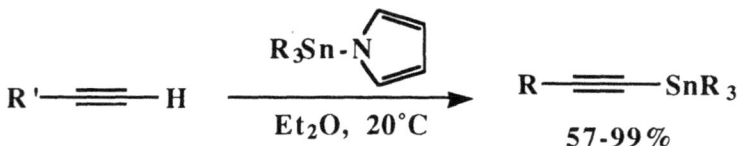

$$R' \equiv\!\!\!=\!\!\!= H \quad \xrightarrow[Et_2O,\ 20°C]{R_3Sn-N\text{(pyrrole)}} \quad R \equiv\!\!\!=\!\!\!= SnR_3$$

57-99%

**VI.G-4** H. Miyake and K. Yamamura, *Chem. Lett.*, 507 and 1099.

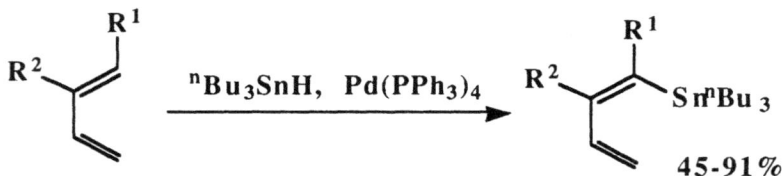

$$\underset{R^2}{\overset{R^1}{\diagdown}}\!\!\diagup\!\!\diagdown\!\! \diagup \quad \xrightarrow{^nBu_3SnH,\ Pd(PPh_3)_4} \quad \underset{R^2}{\overset{R^1}{\diagdown}}\!\!\diagup\!\! Sn^nBu_3$$

45-91%

**VI.G-5** D.M. Hodgson, *Tetrahedron Lett.*, **33**, 5603;  T. Takeda et al.,
*ibid.*, 819.

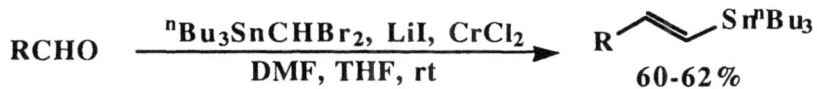

$$\text{RCHO} \xrightarrow[\text{DMF, THF, rt}]{^{n}Bu_3SnCHBr_2, \ LiI, \ CrCl_2} R\diagup\!\!\diagdown\!\!\diagup S n^{n}Bu_3$$

60-62%

# VII
# REVIEWS

## VII.A    Techniques

**VII.A-1** S.E. Denmark, ed., *Tetrahedron*, **48**, 1959-2222.

**Symposia-in-print      "New Synthetic Methods"
Number 46:**

---

**VII.A-2** H. Vorbruggen, *Bull. Soc. Chim. Belg.*, **101**, 407.

**Lecture:    "The Removal of Water in Organic Reactions -
Trimethylsilanol as a Leaving Group."**

---

**VII.A-3** T.A. Mastryukova and M.I. Kabachnik, *Russ. Chem. Rev.*, **60**, 1115 (1991).

**Review:    "Enolisation of the Phosphoryl Group."**

---

**VII.A-4** P. Kripylo et al., *J. prakt. Chem.*, **334**, 105.

**Review:    "Structure / Activity Correlations of
Hydrogenation Catalysts."**

---

**VII.A-5** Y.C. Yang et al., *Chem. Rev.*, **92**, 1729.

**Review:    "Decontamination of Chemical Warfare Agents."**

---

**VII.A-6** *Rec. Trav. Chim.*, **111** (6).

**"Computer Assisted Organic Synthesis."**

**VII.A-7**  M.A. Ott and J.H. Noordik, *Rec. Trav. Chim.*, **111**, 239.

> **Review:** "Computer Tools for Reaction Retrieval and Synthesis Planning in Organic Chemistry. A Brief Review of their History, Methods, and Programs."

**VII.A-8**  Y.C. Martin, *J. Med. Chem.*, **35**, 2145.

> 3D Datbase Searching in Drug Design

**VII.A-9**  T.N. Gerasimova and V.V. Shekovnikov, *Russ. Chem. Rev.*, **61**, 55.

> **Review:** "Organic Dyes for Constant Memory Optical Discs."

**VII.A-10**  J. Fabian et al., *Chem. Rev.*, **92**, 1197.

> **Review:** "Near-Infrared Absorbing Dyes."

**VII.A-11**  V.A. Bren' et al., *Russ. Chem. Rev.*, **60**, 451 (1991).

> **Review:** "Norbornadiene-Quadricyclane - an Effective Molecular System for the Storage of Solar Energy."

**VII.A-12**  K. Tokumaru, J.D. Coyle et al., *Pure Appl. Chem.*, **64**, 1343.

> **Technical Report:** "A Collection of Experiments for Teaching Photochemistry."

**VII.A-13**  M. Colonna and M. Poloni, *Gazz. Chim. Ital.*, **121**, 461 (1991).

**Review:**  **"The Role of the Alpha-effect in Thermolysis, Photolysis and Electron-transfer Processes."**

---

**VII.A-14**  S.-T. Chen et al., *J. Org. Chem.*, **57**, 4781.

**Enhanced Coupling Efficiency in Solid-Phase Peptide Synthesis by Microwave Irradiation**

---

**VII.A-15**  J.-C. Bradley and T. Durst, *Tetrahedron Lett.*, **33**, 7733.

**Visualization of Column Chromatography**

---

**VII.A-16**  R.P.W. Scott, *Chem. Soc. Rev.*, **21**, 137.

**Review:**  **"Modern Liquid Chromatography."**

---

**VII.A-17**  V.V. Takhistov et al., *Russ. Chem. Rev.*, **60**, 1101 (1991).

**Review:**  **"Mass Spectrometry of Halogen-Containing Organic Compounds."**

---

**VII.A-18**  A. Nangia, *Ind. J. Chem.*, **30B**, 811 (1991).

**Review:**  **"Advances in Catalytic Antibodies."**

---

**VII.A-19**  K. Faber and S. Riva, *Synthesis*, 895.

**Review:**  **"Enzyme-catalyzed Irreversible Acyl Transfer."**

**VII.A-20**  M.A. Verkhovskaya and I.A. Yamskov, *Russ. Chem. Rev.*, **60**, 1163 (1991).

>   Review:   "Enzymic Methods of Resolving Racemates of Amino Acids and their Derivatives."

## VII.B   Asymmetric Synthesis and Molecular Recognition

**VII.B-1**  A.J. Walker, *Tetrahedron Asymmetry*, **3**, 961.

>   Review:   "Asymmetric Carbon-Carbon Bond Formation Using Sulfoxide-Stabilized Carbanions."

**VII.B-2**  A.I. Meyers, *Tetrahedron*, **48**, 2589.

>   Review:   "Recent Progress Using Chiral Formamidines ." in Asymmetric Syntheses."

**VII.B-3**  H.R. Sonawane et al., *Tetrahedron Asymmetry*, **3**, 163.

>   Review:   "Recent Developments in the Synthesis of Optically Active α-Arylpropanoic Acids: An Important Class of Non-steroidal Anti inflammatory Agents."

**VII.B-4**  R.L. Halterman, *Chem. Rev.*, **92**, 965.

>   Review:   "Synthesis and Applications of Chiral Cyclopentadienylmetal Complexes."

**VII.B-5**  H.B. Kagan and O. Riant, *Chem. Rev.*, **92**, 1007.

>   Review:   "Catalytic Asymmetric Diels-Alder Reactions."

**VII.B-6**  K. Mikami and M. Shimizu, *Chem. Rev.*, **92**, 1021.

Review:  "Asymmetric Ene Reactions in Organic Synthesis."

---

**VII.B-7**  T.H. Chan and D. Wang, *Chem. Rev.*, **92**, 995.

Review:  "Chiral Organosilicon Compounds in Asymmetric Synthesis."

---

**VII.B-8**  B.L. Feringa and J.C. de Jong, *Bull. Soc. Chim. Belg.*, **101**, 665.

Lecture:  "New Strategies in Asymmetric Synthesis Based on γ-Alkoxybutenolides."

---

**VII.B-9**  F.A. Davis and B.-C. Chen, *Chem. Rev.*, **92**, 919.

Review:  "Asymmetric Hydroxylation of Enolates with N-Sulfonyloxaziridines."

---

**VII.B-10**  Y. Inoue, *Chem. Rev.*, **92**, 741.

Review:  "Asymmetric Photochemical Reactions in Solution."

---

**VII.B-11**  C. Palomo et al., *Bull. Soc. Chim. Belg.*, **101**, 541.

Lecture:  "Synthetic Aspects of Homochiral ß-Lactams Derived from N-Protected α-Aminoimines via.Asymmetric [2+2] Cycloaddition Reactions.

**VII.B-12**  V. Schurig and F. Betschinger, *Chem. Rev.*, **92**, 873.

Review:  "Metal-mediated Enantioselective Access to Unfunctionalized Aliphatic Oxiranes: Prochiral and Chiral Recognition."

---

**VII.B-13**  R.M. Williams and J.A. Hendrix, *Chem. Rev.*, **92**, 889.

Review:  "Asymmetric Synthesis of Arylglycines."

---

**VII.B-14**  R.M. Williams, *Aldrichimica Acta*, **25**, 11.

Review:  "Asymmetric Synthesis of α-Amino Acids."

---

**VII.B-15**  Y. Ohfune, *Acc. Chem. Res.*, **25**, 360.

Review:  "Stereoselective Routes Toward the Synthesis of Unusual Amino Acids."

---

**VII.B-16**  E. Santaniello et al., *Chem. Rev.*, **92**, 1071.

Review:  "The Biocatalytic Approach to the Preparation of Enantiomerically Pure Chiral Building Blocks."

---

**VII.B-17**  M. Sawamura and Y. Ito, *Chem. Rev.*, **92**, 857.

Review:  "Catalytic Asymmetric Synthesis by Means of Secondary Interaction Between Chiral Ligands and Substrates."

---

**VII.B-18**  H.-U. Blaser, *Chem. Rev.*, **92**, 935.

Review:  "The Chiral Pool as a Source of Enantioselective Catalysts and Auxiliaries."

**VII.B-19**  J.K. Whitesell, *Chem. Rev.*, **92**, 953.

Review:  "Cyclohexyl-Based Chiral Auxiliaries."

---

**VII.B-20**  G. Chelucci, *Gazz. Chim. Ital.*, **122**, 89.

Review:  "Chiral Ligands Based on the Pyridine Framework: Synthesis and Application in Asymmetric Catalysis."

---

**VII.B-21**  P. Salvadori et al., *Synthesis*, 503.

Review:  "Synthesis and Applications of Binaphthylic C2 Symmetry Derivatives as Chiral Auxiliaries in Enantioselective Reactions."

---

**VII.B-22**  A. Alexakis, *Pure Appl. Chem.*, **64**, 387.

Report:  "Stereochemical Aspects in the Formation of Chiral Allenes from Propargylic Ethers and Epoxides."

---

**VII.B-23**  H. Kotsuki, *Synlett*, 97.

Review:  "Bicyclic Ketals: Versatile Intermediates for the Stereocontrolled Construction of Cyclic Ether Derivatives."

---

**VII.B-24**  E.I. Klabunovskii, *Russ. Chem. Rev.*, **60**, 980 (1991).

Review:  "Asymmetric Hydrogenation on Metals."

**VII.B-25**  K. Achiwa et al., *Synlett*, 169.

  Review:  "Design Concepts for Developing Highly
           Efficient Chiral Biphosphine Ligands in
           Rhodium-Catalyzed Asymmetric
           Hydrogenations."

**VII.B-26**  G. Zassinovich et al., *Chem. Rev.*, **92**, 1051.

  Review:  "Asymmetric Hydrogen Transfer.Reactions
           Promoted by Homogeneous Transition Metal
           Catalysts."

**VII.B-27**  S.G. Mairanovskii, *Russ. Chem. Rev.*, **60**, 1085 (1991).

  Review:  "The Electroreduction of Organic Compounds in
           the Presence of Catalysts Causing Catalytic
           Evolution of Hydrogen and the Electrosynthesis
           of Chiral Compounds."

**VII.B-28**  H.C. Brown and P.V. Ramachandran, *Acc. Chem. Res.*, **25**, 16.

  Review:  "Asymmetric Reduction with Chiral
           Organoboranes Based on $\alpha$-Pinene."

**VII.B-29**  S. Wallbaum and J. Martens, *Tetrahedron Asymmetry*, **3**, 1475.

  Review:  "Asymmetric Synthesis with Chiral
           Organoborolidines."

**VII.B-30**  T. Hayashi et al., *Pure Appl. Chem.*, **64**, 421.

  Report:  "Catalytic Asymmetric Arylation of Olefins."

**VII.B-31** B.E. Rossiter and N.M. Swingle, *Chem. Rev.*, **92**, 771.

**Review:** **"Asymmetric Conjugate Addition."**

**VII.B-32** J. d'Angelo et al., *Tetrahedron Asymm.*, **3**, 459.

**Review:** **"The Asymmetric Michael Addition Reactions Using Chiral Imines."**

**VII.B-33** H.A.J. Carless, *Tetrahedron Asymm.*, **3**, 795.

**Review:** **"The Use of Cyclohexa-3,5-diene-1,2-diols in Enantiospecific Synthesis."**

**VII.B-34** B.B. Lohray, *Tetrahedron Asymm.*, **3**, 1317.

**Review:** **"Recent Advances in the Asymmetric Dihydroxylation of Alkenes."**

**VII.B-35** K. Mikami et al., *Synlett*, 255.

**Review:** **"Asymmetric Catalysis for the Carbonyl-Ene Reaction."**

**VII.B-36** R. Crossley, *Tetrahedron* , **48**, 8155.

**Review:** **"The Relevance of Chirality to the Study of Biological Activity."**

**VII.B-37** S. Terashima, *Synlett*, 691.

**Review:** **"Synthesis of Optically Active and Biologically Active Compounds."**

**VII.B-38**  M. Protiva, *Coll. Czech. Chem. Commun.*, **56**, 2501 (1991).

Review:  "Fifty Years in Chemical Drug Research."

**VII.B-39**  S.L. Schreiber et al., *Tetrahedron*, **48**, 2545.

Review:  "Molecular Recognition of Immunophilins and Immunophilin-Ligand Complexes."

**VII.B-40**  J. Rebek, Jr. et al., *Acta Chem. Scand.*, **46**, 315.

Review:  "Self-Replicating Systems."

## VII.C  Reactions

**VII.C-1**  D.H.R. Barton, *Tetrahedron*, **48**, 2529.

Review:  "The Invention of Chemical Reactions:  The Last Five Years."

**VII.C-2**  W.B. Motherwell, *Aldrichimica Acta*, **25**, 71.

Report:  "A Curiosity Driven Search for New Chemical Reactions."

**VII.C-3**  L.E. Overman, *Acc. Chem. Res.*, **25**, 352.

Review:  "Charge as a Key Component in Reaction Design. The Invention of Cationic Cyclization Reactions of Importance in Synthesis."

**VII.C-4**  D.H.R. Barton and D.Doller, *Acc. Chem. Res.*, **25**, 504.

Review:  "The Selective Functionalization of Saturated Hydrocarbons.  Gif Chemistry."

**VII.C-5**  V.K. Singh, *Synthesis*, 605.

  **Review:**  "Practical and Useful Methods for the
          Enantioselective Reduction of Unsymmetrical
          Ketones."

---

**VII.C-6**  J.-C. Moutat, *Org. Prep. Proced. Int.*, **24**, 309.

  **Review:**  "Electrocatalytic Hydrogenation on Hydrogen
          Active Electrodes.  A Review."

---

**VII.C-7**  C. Chatgilialoglu, *Acc. Chem. Res.*, **25**, 188.

  **Review:**  "Organosilanes as Radical-Based Reducing
          Agents in Synthesis."

---

**VII.C-8**  J.F. Bunnett, *Acc. Chem. Res.*, **25**, 2.

  **Review:**  "Radical-Chain, Electron-Transfer
          Dehalogenation Reactions."

---

**VII.C-9**  D. Ostovic and T.C. Bruice, *Acc. Chem. Res.*, **25**, 314.

  **Review:**  "Mechanism of Alkene Epoxidation by Iron,
          Chromium and Manganese Higher Valent Oxo-
          Metalloporphyrins."

---

**VII.C-10**  S.K. Taylor, *Org. Prep. Proced. Int.*, **24**, 245.

  **Review:**  "Biosynthetic, Biomimetic and Related Epoxide
          Cyclizations.  A Review."

---

**VII.C-11**  J. Skarzewski and R. Siedlecka, *Org. Prep. Proced. Int.*, **24**,
623.

  **Review:**  "Synthetic Oxidations with Hypochlorites."

**VII.C-12**  A.S. Demir and A. Jeganathan, *Synthesis*, 235.

> Review:  "Selective Oxidation of α,β-Unsaturated Ketones at the α'-Position."

---

**VII.C-13**  A.K. Banerjee and N.C. Gonzalez, *Rec. Trav. Chim.*, **110**, 353 (1991).

> Review:  "Methods for Angular Alkoxycarbonylation in Fused Rings and its Application to the Synthesis of Terpenoid Compounds."

---

**VII.C-14**  Z. Rappoport, *Acc. Chem. Res.*, **25**, 474.

> Review:  "The Rapid Steps in Nucleophilic Vinylic 'Addition-Elimination' Substitution.  Recent Developments."

---

**VII.C-15**  L.A. Paquette and C.J.M. Stirling, *Tetrahedron*, **48**, 7383.

> Review:  "The Intramolecular $S_N'$ Reaction."

---

**VII.C-16**  P. Cintas, *Synthesis*, 249.

> Review:  "Addition of Organochromium Compounds to Aldehydes:  The Nojaki-Hiyama Reaction."

---

**VII.C-17**  C.S. Marson, *Tetrahedron*, **48**, 3659.

> Review:  "Reactions of Carbonyl Compounds with (Monohalo) Methyleninium Salts (Vilsmeier Reagents)."

**VII.C-18**  M.A. McClinton and D.A. McClinton, *Tetrahedron*, **48**, 6555.

Review:  "Trifluoromethylations and Related Reactions in Organic Chemistry."

**VII.C-19**  C. Galli, *Org. Prep. Proced. Int.*, **24**, 285.

Review:  "Cesium Ion Effect and Macrocyclization.  A Critical Review."

**VII.C-20**  R.P. Thummel, *Synlett*, 1.

Review:  "The Application of Friedlander and Fischer Methodologies to the Synthesis of Organized Polyaza Cavities."

**VII.C-21**  X. Fu and J.M. Cook, *Aldrichimica Acta*, **25**, 43.

Review:  "The Synthesis of Polyquinanes and Polyquinenes via the Weiss Reaction."

**VII.C-22**  R. Martin, *Org. Prep. Proced. Int.*, **24**, 369.

Review:  "Uses of the Fries Rearrangement for the Preparation of Hydroxyarylketones.  A Review."

**VII.C-23**  Y.G. Gololobov and L.F. Kasukhin, *Tetrahedron*, **48**, 1353.

Review:  "Recent Advances in the Staudinger Reaction."

**VII.C-24**  B.E. Abalonin, *Russ. Chem. Rev.*, **60**, 1346 (1991).

Review:  "Investigations on the Arbuzov and Retro-Arbuzov Reactions Among the Organic Derivatives of Arsenic."

**VII.C-25**  I.A. Rybakova et al., *Russ. Chem. Rev.*, **60**, 1331 (1991).

Review:  "Methods of Replacing Halogen in Aromatic Compounds by RS-Functions."

---

**VII.C-26**  M.K. Dzhafarov, *Russ. Chem. Rev.*, **61**, 363.

Review:  "Retro-Aldol Processes in Steroid Chemistry."

---

**VII.C-27**  A.M. Moiseenkov et al., *Russ. Chem. Rev.*, **60**, 643 (1991).

Review:  "Synthetic Applications of the Pummerer Reaction."

## VII.D   Reactive Intermediates

**VII.D-1**  A. Padwa and K.E. Krumpe, *Tetrahedron*, **48**, 5385.

Review:  "Application of Intramolecular Carbenoid Reactions in Organic Synthesis."

---

**VII.D-2**  B.A. Khaskin et al., *Russ. Chem. Rev.*, **61**, 306.

Review:  "Reactions of Phosphorus Containing Compounds with Diazo Compounds and Carbenes."

---

**VII.D-3**  M.F. Budyka et al., *Russ. Chem. Rev.*, **61**, 25.

Review:  "The Photochemistry of Phenyl Azide."

---

**VII.D-4**  N.P. Gritsan and E.A. Pritchina, *Russ. Chem. Rev.*, **61**, 500.

Review:  "The Mechanism of Photolysis of Aromatic Azides."

**VII.D-5**  V.D. Nefedov et al., *Russ. Chem. Rev.*, **61**, 283.

**Review:**  "Vinyl Cations."

---

**VII.D-6**  V.V. Grushin, *Acc. Chem. Res.*, **25**, 529.

**Review:**  "Carboranylhalonium Ions: From Striking Reactivity to a Unified Mechanistic Analysis of Polar Reactions of Diarylhalonium Compounds."

---

**VII.D-7**  B. Caro et al., *Bull. Soc. Chim. Fr.*, **129**, 121.

**Review:**  "α-Benchrotrenic Carbanions.  Formation, Stabilization and Synthetic Applications."

---

**VII.D-8**  M. Chanon, *Acta. Chem. Scand.*, **46**, 695.

**Review:**  "The Variety of Molecular Schemes at the Border Between Polar and ET Mechanisms."

---

**VII.D-9**  G. Pandey, *Synlett*, 546.

**Review:**  "Synthetic Perspectives of Photoinduced Electron Transfer Generated Amine Radical Cations."

---

**VII.D-10**  R.D. Little, I. Danuecker-Doerig et al., *Synlett*, 107.

**Review:**  "Factors Affecting Regioselectivity in the Intramolecular Diyl Trapping Reaction."

---

**VII.D-11**  J.C. Walton, *Chem. Soc. Rev.*, **21**, 105.

**Review:**  "Bridgehead Radicals."

**VII.D-12**  D.P. Curran et al., *Synlett*, 943.

   Review:   "New Mechanistic Insights into Reductions of
             Halides and Radicals with Samarium (II) Iodide."

---

**VII.D-13**  A. Padwa et al., *Tetrahedron*, **48**, 7565.

   Review:   "Azomethine Ylide Generation via the Dipole
             Cascade."

---

**VII.D-14**  O.I. Kolodyazhnyi, *Russ. Chem. Rev.*, **60**, 391 (1991).

   Review:   "P-Halogen-substituted Phosphorus Ylids."

---

**VII.D-15**  M. Regitz, *Bull. Soc. Chim. Belg.*, **101**, 359.

   Lecture:  "From Phosphaalkynes to Phosphacubanes -
             Adventures in Organophosphorus Chemistry."

## VII.E.  Organo- metallics and metalloids

**VII.E-1**  E. Negishi, *Pure Appl. Chem.*, **64**, 323.

   Review:   "Zipper-mode Cascade Carbometallation for
             Construction of Polycyclic Structures."

---

**VII.E-2**  A Ricci et al., *Pure Appl. Chem.*, **64**, 439.

   Report:   "Organometallic Polysynthons: a Novel
             Strategy for Selectively Polyfunctionalized
             Molecules."

**VII.E-3**  T.J. Marks, *Acc. Chem. Res.*, **25**, 57.

Review:  "Surface-bound Metal Hydrocarbyls.
Organometallic Connections between
Heterogeneous and Homogeneous Catalysis."


**VII.E-4**  F. Zaera, *Acc. Chem. Res.*, **25**, 260.

Review:  "Preparation and Reactivity of Alkyl Groups
Adsorbed on Metal Surfaces."


**VII.E-5**  A.N. Startsev, *Russ. Chem. Rev.*, **61**, 175.

Review:  "Mechanism of the Hydrogenolysis of Thiophene
on Bimetallic Sulphide Catalysts."


**VII.E-6**  D.J. Burton and Z.-Y. Yang, *Tetrahedron*, **48**, 189.

Review:  "Fluorinated Organometallics: Perfluoroalkyl
and Functionalized Perfluoroalkyl Organometallic
Reagents in Organic Synthesis."


**VII.E-7**  D.B. Collum, *Acc. Chem. Res.*, **25**, 448.

Review:  "Is N,N,N',N'-Tetramethylethylenediamine a
Good Ligand for Lithium?"


**VII.E-8**  F. Bickelhaupt, *Acta Chem. Scand.*, **46**, 409.

Review:  "The Importance of Intramolecular Solvation in
Organomagnesium Chemistry."


**VII.E-9**  G.L. Larson, *J. Organomet. Chem.*, **422**, 1.

Review:  "Silicon - the Silicon-Carbon Bond: Annual
Survey for the Year 1989."

**VII.E-10**  R.O. Duthaler and A. Hafner, *Chem. Rev.*, **92**, 807.

Review:  "Chiral Titanium Complexes for Enantioselective Addition of Nucleophiles to Carbonyl Groups."

---

**VII.E-11**  J. Muzart, *Chem. Rev.*, **92**, 113.

Review:  "Cr-Catalyzed Oxidations in Organic Synthesis."

---

**VII.E-12**  S.G. Davies et al., *Pure Appl. Chem.*, **464**, 379.

Report:  "Stereoselective Manipulation of Acetals Derived from *o*-Substituted Benzaldehyde Chromium Tricarbonyl Complexes"

---

**VII.E-13**  A.N. Kasatkin et al., *Russ. Chem. Rev.*, **61**, 537.

Review:  "Organic Derivatives of Manganese (II) in Organic Synthesis."

---

**VII.E-14**  R.C. Kerber, *J. Organomet. Chem.*, **422**, 209.

Review:  "Organoiron Chemistry:  Annual Survey for the Year 1990."

---

**VII.E-15**  S.E. Thomas et al., *Pure Appl. Chem.*, **64**, 671.

Report:  "Synthesis and Reactivity of Iron Tricarbonyl Complexes of Vinyl Ketenes, Vinylketimines and Vinyl Allenes."

---

**VII.E-16**  Y. Kishi, *Pure Appl. Chem.*, **64**, 343.

Review:  "Applications of Ni(II) / Cr(II)-mediated Coupling Reactions to Natural Products Syntheses."

**VII.E-17**   K. Tamao, K. Kobayashi and Y. Ito, *Synlett*, 539.

Review:   "Nickel (0)-mediated Intramolecular Cyclizations of Enynes, Dienynes, Bis-dienes and Diynes."

**VII.E-18**   P.A. Chaloner, *J. Organomet. Chem.*, **432**, 387 and **442**, 271.

Review:   "Nickel, Palladium and Platinum, Survey Covering the Years 1984, 1985 and 1986."

**VII.E-19**   B.H. Lipshutz et al., *Tetrahedron*, **48**, 2578.

Review:   "Controlled "Decomposition" of "Kinetic" Higher Order Cyanocuprates:   a New Route to Unsymmetrical Biaryls."

**VII.E-20**   T. Ibuka and Y. Yamamoto, *Synlett*, 769.

Lectures: "New Aspects of Organocopper Reagents: 1,3- and 1,2- Chiral Induction and Reaction Mechanism."

**VII.E-21**   P. Knochel et al., *Pure Appl. Chem.*, **64**, 361.

Report:   "The Chemistry of Polyfunctionalized Organozinc and Copper Reagents."

**VII.E-22**   E. Erdik, *Tetrahedron*, **48**, 9577.

Review:   "Transition Metal Catalyzed Reactions of Organozinc Reagents."

**VII.E-23** K. Soai and S. Niwa, *Chem. Rev.*, **92**, 833.

Review:     "Enantioselective Addition of Organozinc
            Reagents to Aldehydes."

**VII.E-24** F. Mathey et al., *Synlett*, 363.

Review:     "Synthesis of Organophosphorus Compounds
            via the Coordination Spheres of Transition
            Metals: Some Non-classical Examples."

**VII.E-25** L.S. Hegedus, *J. Organomet. Chem.*, **422**, 301.

Review:     "Transition Metals in Organic Synthesis:
            Annual Survey Covering the Year 1990."

**VII.E-26** M.E. Welker, *Chem. Rev.*, **92**, 97.

Review:     "3 + 2 Cycloaddition Reactions of Transition-
            Metal 2-Alkynyl and η'-Allyl Complexes and
            their Utilization in Five-membered Ring
            Compound Syntheses."

**VII.E-27** S.-I. Murahashi, *Pure Appl. Chem.*, **64**, 403.

Report:     "Biomimetic Oxidation in Organic Synthesis
            using Transition Metal Catalysts."

**VII.E-28** L. Marko and F. Ungvary, *J. Organomet. Chem.*, **432**, 1.

Review:     "Transition Metals in Organic Synthesis:
            Hydroformylation, Reduction and Oxidation."

**VII.E-29**  O.V. Gerasimov and V.N. Parmon, *Russ. Chem. Rev.*, **61**, 154.

Review:  "Photocatalysis by Transition Metal Complexes."

**VII.E-30**  D. Max Roundhill, *Chem. Rev.*, **92**, 1.

Review:  "Transition Metal and Enzyme Catalyzed Reactions Involving Reactions with Ammonia and Amines."

**VII.E-31**  M. Pfeffer, *Pure Appl. Chem.*, **64**, 335.

Report:  "Selected Applications to Organic Synthesis of Intramolecular C-H Activation Reactions by Transition Metals."

**VII.E-32**  W. Wolfsberger, *J. prakt. Chem.*, **334**, 453.

Review:  "Hydrogermylations."

**VII.E-33**  G. Erker, *Pure Appl. Chem.*, **64**, 393.

Report:  "Stereoselective Synthesis with Zirconium Complexes."

**VII.E-34**  W.P. Griffith, *Chem. Soc. Rev.*, **21**, 179.

Review:  "Ruthenium Oxo Complexes as Organic Oxidants."

**VII.E-35**  M.G. Richmond, *J. Organomet. Chem.*, **432**, 215.

Review:  "Annual Survey of Ruthenium and Osmium for the Year 1989."

**VII.E-36**  A.J. Canty, *Acc. Chem. Res.*, **25**, 83.

Review:   "Development of Organopalladium(IV)
          Chemistry: Fundamental Aspects and Systems for
          Studies of Mechanism in Organometallic
          Chemistry and Catalysis."

**VII.E-37**  J.-E. Backvall, *Pure Appl. Chem.*, **64**, 429.

Report:   "Palladium-catalyzed Intramolecular 1,4
          Additions to Conjugated Dienes."

**VII.E-38**  C.G. Frost et al., *Tetrahedron Asymmetry*, **3**, 1089.

Review:   "Selectivity in Pd Catalyzed Allylic
          Substitution."

**VII.E-39**  T.N. Mitchell, *Synthesis*, 803.

Review:   "Palladium Catalyzed Reactions of Organotin
          Compounds."

**VII.E-40**  G. Tagliavini, *J. Organomet. Chem.*, **437**, 15 and W.P.
Neumann, *J. Organomet. Chem.*, **437**, 23.

Review:   "Use of Organotin Halides as Catalytic
          Precursors in Dehydration Processes."

Review:   "Tin for Organic Synthesis. VI. The New
          Role of Organotin Reagents in Organic
          Synthesis."

**VII.E-41**  L.D. Freedman and G.O. Doak, *J. Organomet. Chem.*, **442**, 1.

Review:   "Antimony:  Annual Survey Covering the Year
          1990."

**VII.E-42**  Y.-Z. Huang, *Acc. Chem. Res.*, **25**, 182.

>   **Review:**  "Synthetic Applications of Organoantimony
>   Compounds."

---

**VII.E-43**  G.A. Molander, *Chem. Rev.*, **92**, 29.

>   **Review:**  "Application of Lanthanide Reagents in Organic
>   Synthesis."

---

**VII.E-44**  R.D. Rogers and L.M. Rogers, *J. Organomet. Chem.*, **442**, 83
and 225.

>   **Review:**  "Lanthanides and Actinides, Annual Survey
>   Covering the Years 1987-1989 and 1990."

---

**VII.E-45**  T.R. Lee and G.M. Whitesides, *Acc. Chem. Res.*, **25**, 266.

>   **Review:**  "Heterogeneous, Platinum-catalyzed
>   Hydrogenations of (Diolefin)dialkylplatinum(II)
>   Complexes."

---

**VII.E-46**  G.O. Doak and L.D. Freedman, *J. Organomet. Chem.*, **442**, 61.

>   **Review:**  "Bismuth, Annual Survey Covering the Year
>   1990."


## VII.F.  Halogen Compounds and Halogenation

>   (see also: VI.A.11.)


**VII.F-1**  A. Laurent, *Bull. Soc. Chim. Belg.*, **101**, 45.

>   **Lecture:**  "Some Recent Results in Organofluorine
>   Chemistry."

**VII.F-2**  J.A. Wilkinson, *Chem. Rev.*, **92**, 505.

Review:  "Recent Advances in the Selective Formation of the C-F Bond."

---

**VII.F-3**  V.P. Kukhar et al., *Russ. Chem. Rev.*, **60**, 1050 (1991).

Review:  "Fluorine-containing Aromatic Aminoacids."

---

**VII.F-4**  V.P. Kukhar and V.A. Soloshonok, *Russ. Chem. Rev.*, **60**, 850 (1991).

Review:  "Aliphatic Fluorine-containing Amino Acids."

---

**VII.F-5**  L.E. Deev, *Russ. Chem. Rev.*, **61**, 40.

Review:  "Iodoperfluoroalkanes."

---

**VII.F-6**  E.S. Turbanova and A.A. Petrov, *Russ. Chem. Rev.*, **60**, 501 (1991).

Review:  "Perfluoroalkyl(aryl)acetylenes."

---

**VII.F-7**  I.V. Koval', *Russ. Chem. Rev.*, **60**, 830 (1991).

Review:  "Progress in the Chemistry of Perhalomethane Sulphenyl Halides."

---

**VII.F-8**  A.Yu. Sizov et al., *Russ. Chem. Rev.*, **61**, 517.

Review:  "Polyfluoroalkane Sulphenyl Chlorides."

**VII.F-9**  S.E. Evsyukov et al., *Russ. Chem. Rev.*, **60**, 373 (1991).

Review:   "Chemical Dehydrohalogenation of Halogen containing Polymers."


## VII.G  Natural Products

**VII.G-1**  A. Eschenmoser and E. Loewenthal, *Chem. Soc. Rev.*, **21**, 1.

Review:   "Chemistry of Potentially Prebiological Natural Products."

---

**VII.G-2**  P. Potier, *Chem. Soc. Rev.*, **21**, 113.

Review:   "Rhone-Poulenc Lecture - Search and Discovery of New Antitumour Compounds."

---

**VII.G-3**  K. Mori, *Bull. Soc. Chim. Belg.*, **101**, 393.

Lecture:   "Application of Biochemical Methods in Enantioselective Synthesis of Bioactive Natural Products."

---

**VII.G-4**  A.I. Scott, *Tetrahedron*, **48**, 2559.

Review:   "Genetically Engineered Synthesis of Complex Natural Products."

---

**VII.G-5**  S. Shibuya et al., *Heterocycles*, **33**, 1051.

Review:   "Synthesis of β-Oxygenated γ-Amino Acids and γ-Oxygenated δ-Amino Acids from α-Amino Acids."

**VII.G-6** V.V. Ryakhovskii et al., *Russ. Chem. Rev.*, **60**, 924 (1991).

Review: "Special Features of the Synthesis of Peptides Containing Secondary Amino Acids."

---

**VII.G-7** D. Seebach, *Aldrichimica Acta*, **25**, 59.

Report: "How We Stumbled into Peptide Chemistry."

---

**VII.G-8** P.S. Rutledge et al., *Aust. J. Chem.*, **45**, 483.

Review: "Synthetic Anthracyclines from Anthraquinones."

---

**VII.G-9** A.R. Butler and Y.-L. Wu, *Chem. Soc. Rev.*, **21**, 85.

Review: "Artemisinin (Qinghaosu): a New Type of Antimalarial Drug."

---

**VII.G-10** K. Burgess and I. Henderson, *Tetrahedron*, **48**, 4045.

Review: "Synthetic Approaches to Stereoisomers and Analogues of Castanospermine."

---

**VII.G-11** B. Fraser-Reid et al., *Bull. Soc. Chim. Belg.*, **101**, 617.

Report: "Novel Reactions of Carbohydrates Discovered en Route to Natural Products."

---

**VII.G-12** B. Fraser-Reid et al., *Synlett*, 927.

Review: "n-Pentenyl Glycosides in Organic Chemistry: a Contemporary Example of Serendipity."

**VII.G-13**  J. Banoub et al., *Chem. Rev.*, **92**, 1167.

Review:   "Synthesis of Oligosaccharides of 2-Amino-2-Deoxy Sugars."

---

**VII.G-14**  M.H.D. Posterna, *Tetrahedron*, **48**, 8545.

Review:   "Recent Developments in the Synthesis of C-Glycosides."

---

**VII.G-15**  P.J. Garegg, *Acc. Chem. Res.*, **25**, 575.

Review:   "Saccharides of Biological Importance: Challenges and Opportunities for Organic Synthesis."

---

**VII.G-16**  D.J. Ager and M.B. East, *Tetrahedron*, **48**, 2803.

Review:   "Methodology to Establish 1,2- and 1,3-Difunctionality for the Synthesis of Carbohydrate Derivatives."

---

**VII.G-17**  C. Tamm, *Helv. Chim. Acta*, **75**, 2109.

Review:   "Chemie der Kohlenhydrate der Pflanzeninhaltsstoffe und der Mikrobiellen Stoffwechselprodukte im Spiegel der Helvetica Chimica Acta 1918-1992."

---

**VII.G-18**  K.L. Dueholm and E.B. Pederson, *Synthesis*, 1.

Review:   "2,3-Dideoxyfuranoses in Convergent Synthesis of 2',3'-Dideoxy Nucleosides."

**VII.G-19** D. Bergstrom et al., *Synlett*, 179.

Review: "C-5-Substituted Nucleoside Analogs."

---

**VII.G-20** D.M. Huryn and M. Okabe, *Chem. Rev.*, **92**, 1745.

Review: "AIDS-Driven Nucleoside Chemistry."

---

**VII.G-21** A.D. Borthwick and K. Biggadike, *Tetrahedron*, **48**, 571.

Review: "Synthesis of Chiral Carbocyclic Nucleosides."

---

**VII.G-22** G.M. Whitesides et al., *Acc. Chem. Res.*, **25**, 307.

Review: "Nucleoside Phosphate Sugars: Syntheses on Practical Scales for Use as Reagents in the Enzymatic Preparation of Oligosaccharides and Glycoconjugates."

---

**VII.G-23** S.L. Beaucage and R.P. Iyer, *Tetrahedron*, **48**, 2223.

Review: "Advances in the Synthesis of Oligonucleotides by the Phosphoramidate Approach."

---

**VII.G-24** L. Mayol et al., *Gazz. Chim. Ital.*, **121**, 505 (1991).

Review: "Solid Phase Synthesis of Cyclic Oligodeoxyribonucleotides: Intermolecular versus Intramolecular Coupling."

---

**VII.G-25** M. Julia et al., *Bull. Soc. Chim. Belg.*, **101**, 65.

Lecture: "Total Synthesis of 22, 23-Dihydroavermectin $B_{1b}$ Aglycone."

**VII.G-26**   K.C. Nicolaou and A.L. Smith, *Acc. Chem. Res.*, **25**, 497.

   **Review:**   "Molecular Design, Chemical Synthesis, and
   Biological Action of Enediynes."

**VII.G-27**   D. Ferreira, *Tetrahedron*, **48**, 1743.

   **Review:**   "Diversity of Structure and Function in
   Oligomeric Flavanoids."

**VII.G-28**   E.A. Ruveda et al., *Tetrahedron*, **48**, 963.

   **Review:**   "Synthetic Routes to Forskolin."

**VII.G-29**   L.N. Mander, *Chem. Rev.*, **92**, 573.

   **Review:**   "The Chemistry of Gibberellins: An Overview."

**VII.G-30**   P.J. Scheuer, *Acc. Chem. Res.*, **25**, 433.

   **Review:**   "Isocyanides and Cyanides as Natural Products."

**VII.G-31**   F. Sato and Y. Kobayashi, *Synlett*, 849.

   **Review:**   "Synthesis of Enantiomerically Pure Secondary
   γ-Halo Allylic Alcohols and their Use in the
   Synthesis of Leukotrienes."

**VII.G-32**   M. Kasai and M. Kono, *Synlett*, 778.

   **Review:**   "Studies on the Chemistry of Mitomycins."

**VII.G-33** J.C. Coll, *Chem. Rev.*, **92**, 613.

Review: "The Chemistry and Chemical Ecology of Octocorals (Coelenterata, Anthozoa, Octocorallia)."

**VII.G-34** R.S. Ward, *Synthesis*, 719.

Review: "Synthesis of Podophyllotoxin and Related Compounds."

**VII.G-35** B. Meunier, *Chem. Rev.*, **92**, 1411.

Review: "Metalloporphyrins as Versatile Catalysts for Oxidation Reactions and Oxidative DNA Cleavage."

**VII.G-36** S.M. Shevchenko and A.G. Apushkinskii, *Russ. Chem. Rev.*, **61**, 105.

Review: "Quinone Methides in the Chemistry of Wood."

**VII.G-37** J.D. Winkler (ed.), *Tetrahedron*, **48**, 6953-7056.

"Total and Semi-synthetic Approaches to Taxol."

Tetrahedron Symposia-in-Print Number 48

**VII.G-38** A.J. Birch, *Steroids*, **57**, 363.

Review: "Steroid Hormones and the Luftwaffe. A Venture into Fundamental Strategic Research and Some of its Consequences: The Birch Reduction Becomes a Birth Reduction."

**VII.G-39**  J. Fried, *Steroids*, 57, 384.

> **Review:**  "Hunt for an Economical Synthesis of Cortisol:
> Discovery of the Fluorosteroids at Squibb (a
> Personal Account)."

---

**VII.G-40**  S. Bernstein, *Steroids*, 57, 392.

> **Review:**  "Historic Reflection on Steroids: Lederle and
> Personal Aspects."

---

**VII.G-41**  G. Rosenkranz, *Steroids*, 57, 409.

> **Review:**  "From Ruzicka's Terpenes in Zurich to Mexican
> Steroids via Cuba."

---

**VII.G-42**  R. Hirschmann, *Steroids*, 57, 579.

> **Report:**  "The Cortisone Era: Aspects of its Impact. Some
> Contributions of the Merck Laboratories."

---

**VII.G-43**  J.A. Hogg, *Steroids*, 57, 593.

> **Report:**  "Steroids, the Steroid Community, and Upjohn
> in Perspective: A Profile of Innovation."

---

**VII.G-44**  K. Nakanishi, *Steroids*, 57, 649.

> **Report:**  "Past and Present Studies with Ponasterones, the
> First Insect Molting Hormones from Plants."

---

**VII.G-45**  J. Kalvoda, *Helv. Chim. Acta*, 75, 2341.

> **Review:**  "60 Years of Steroid Chemistry in Helvetica
> Chimica Acta."

**VII.G-46**   F.A. Lakhvich et al., *Russ. Chem. Rev.*, **60**, 658 (1991).

Review:   "The Synthesis of Brassinosteroids, a New Class of Plant Hormones."


## VII.H.  Others


**VII.H-1**   B.L. Erusalimskii, *Russ. Chem. Rev.*, **61**, 75.

Review:   "Recent Trends in Research into Ionic Polymerisation."

---

**VII.H-2**   W.J. Mijs and R. Addink, *Rec. Trav. Chim.*, **110**, 526 (1991).

Review:   "Recent Developments in Polymer Synthesis."

---

**VII.H-3**   M. Ueda, *Synlett*, 605.

Review:   "Direct Polycondensation."

---

**VII.H-4**   P.M. Hergenrother, *Rec. Trav. Chim.*, **110**, 481 (1991).

Review:   "New Developments in Thermally Stable Polymers."

---

**VII.H-5**   *Russ. Chem. Rev.*, **60**, 689-814 (1991).

Reviews: "Nontraditional Methods of Polymer Synthesis."

---

**VII.H-6**   A.L. Rusanov, *Russ. Chem. Rev.*, **61**, 449.

Review:   "New Bis(naphthalic anhydrides) and Polyheteroarylenes Based on them with Improved Processability."

**VII.H-7**  E.G.J. Staring, *Rec. Trav. Chim.*, **110**, 492 (1991).

   Review:   "Non-linear Optical and Electro-optical
             Polymers."

---

**VII.H-8**  U. Scherf and K. Mullen, *Synthesis*, 23.

   Review:   "Design and Synthesis of Extended π-Systems:
             Monomers, Oligomers, Polymers."

---

**VII.H-9**  J.S. Bradshaw et al., *Chem. Rev.*, **92**, 543.

   Review:   "Macropolycyclic Polyethers (Cages) and Related
             Compounds."

---

**VII.H-10**  Yu.M. Shapiro, *Russ. Chem. Rev.*, **60**, 1035 (1991).

   Review:   "Gem-polyols - A Unique Class of Compound."

---

**VII.H-11**  G.R. Newkome et al., *Aldrichimica Acta*, **25**, 31.

   Review:   "Building Blocks for Dendritic Macromolecules."

---

**VII.H-12**  F. Vogtle et al., *Angew. Chem., Int. Ed. Engl.*, **31**, 1571.

   Review:   "Dendrimers, Arborols, and Cascade Molecules:
             Breakthrough into Generations of New Materials"

---

**VII.H-13**  P.A. Vigato et al., *Acta Chem. Scand.*, **46**, 1025.

   Review:   "Synthesis and Application of Macrocyclic and
             Macroacyclic Schiff Bases."

**VII.H-14**  S.M. Weinreb, *Bull. Soc. Chim. Belg.*, **101**, 381.

Lecture:  "Synthetic Applications of N-Sulfonylimines."

---

**VII.H-15**  G. Zecchi, *Synlett*, 858.

Review:  "The Chemistry of C-Azidohydrazones:
Developments and Perspectives."

---

**VII.H-16**  H. Zollinger, *Helv. Chim. Acta*, 75, 1727.

Review:  "Color Chemistry as Reflected in Helvetica
Chimica Acta."

---

**VII.H-17**  G. Ohloff, *Helv. Chim. Acta*, 75, 2041.

Review:  "Jahre Riechstoff- und Aroma- Chemie in Spiegel
der Helvetica Chimica Acta."

---

**VII.H-18**  V. Balzani, *Tetrahedron*, **48**, 10443.

Review:  "Supramolecular Photochemistry."

---

**VII.H-19**  J. Chen, J.R. Scheffer and J. Trotter, *Tetrahedron*, **48**, 3251.

Review:  "Differences in Photochemical Reactivity of
9,10-Ethanoanthracene Derivatives in Liquid and
Crystalline Media."

---

**VII.H-20**  L. Horner, *J. Prakt. Chem.*, **334**, 645.

Review:  "Selectivity: OH-, NH- and SH- Group Specific
Reagents.  The Application in Organic Analysis
and as Protective Groups."

**VII.H-21**  G.A. Tolstikov et al., *Russ. Chem. Rev.*, **60**, 420 (1991).

   **Review:**  "Synthesis and Reactivity of N-Substituted
              Aminoamides, Antiarrhythmic and Local
              Anaesthetic Activity."

**VII.H-22**  S.A. Abdulganeeva and K.B. Erzhanov, *Russ. Chem. Rev.*, **60**,
676 (1991).

   **Review:**  "Acetylenic Amino Acids."

**VII.H-23**  D.L. Comins, *Synlett*, 615.

   **Review:**  "The Synthetic Utility of α-Amino Alkoxides."

**VII.H-24**  J.S. Bradshaw et al., *Tetrahedron*, **48**, 4475.

   **Review:**  "Preparation of Diamino Ethers and Polyamines."

**VII.H-25**  M.V. Vovk and L.I. Samarai, *Russ. Chem. Rev.*, **61**, 297.

   **Review:**  "N-Functionalised Carbodiimides."

**VII.H-26**  G.N. Nikonov and A.S. Bayeva, *Russ. Chem. Rev.*, **61**, 335.

   **Review:**  "Compounds Containing Phosphorus and
              Boron."

**VII.H-27**  B.A. Trofimov et al., *Russ. Chem. Rev.*, **60**, 1360 (1991).

   **Review:**  "Elemental Phosphorus - Strong Base as a
              System for the Synthesis of Organophosphorus
              Compounds."

**VII.H-28** T. Minami and J. Motoyoshiya, *Synthesis*, 333.

Review: "Vinylphosphonates in Organic Synthesis."

---

**VII.H-29** A.P. Brestkin et al., *Russ. Chem. Rev.*, **60**, 885 (1991).

Review: "S-Alkynyl Esters of Phosphorus Thioacids as Inhibitors of Cholinesterases and as Promising Physiologically Active Compounds."

---

**VII.H-30** N.G. Zabirov et al., *Russ. Chem. Rev.*, **60**, 1128 (1991).

Review: "N-Phosphorylated Amides and Thioamides."

---

**VII.H-31** L. Weber, *Chem. Rev.*, **92**, 1839.

Review: "The Chemistry of Diphosphenes and their Heavy Congeners: Synthesis, Structure and Reactivity."

---

**VII.H-32** F. Mathey, *Acc. Chem. Res.*, **25**, 90.

Review: "Expanding the Analogy Between Phosphorus Carbon and Carbon-Carbon Double Bonds."

---

**VII.H-33** A.O. Gukasyan et al., *Russ. Chem. Rev.*, **60**, 1318 (1991).

Review: "The Synthesis, Structure and Reactivity of ($\alpha$-Trihalomethyl)carbinols."

---

**VII.H-34** P.J. Stang, *Angew. Chem., Int. Ed. Engl.*, **31**, 274.

Review: "Alkynyl and Alkenyl(phenyl)iodonium Compounds."

**VII.H-35**  T. Satoh and K. Yamakawa, *Synlett*, **59**, 455.

   **Review:**  "Recent Developments in Organic Synthesis with 1-Haloalkyl Aryl Sulfoxides."

---

**VII.H-36**  B.B. Lohray, *Synthesis*, 1035.

   **Review:**  "Cyclic Sulfites and Cyclic Sulfates: Epoxide like Synthons."

---

**VII.H-37**  J.A. Berson, *Tetrahedron*, **48**, 3.

   **Perspective:** "Discoveries Missed, Discoveries Made: Creativity, Influence, and Fame in Chemistry."

---

**VII.H-38**  B.K. Carpenter, *Acc. Chem. Res.*, **25**, 520.

   **Review:**  "Intramolecular Dynamics for the Organic Chemist."

---

**VII.H-39**  M.J.S. Dewar and C. Jie, *Acc. Chem. Res.*, **25**, 537.

   **Review:**  "Mechanisms of Pericyclic Reactions: The Role of Quantitative Theory in the Study of Reaction Mechanisms."

---

**VII.H-40**  H. Li and W.J. LeNoble, *Rec. Trav. Chim.*, **111**, 199.

   **Review:**  "Competing Concepts in Electronic Control of Face Selection."

---

**VII.H-41**  J.E. McMurry and T. Lectka, *Acc. Chem. Res.*, **25**, 47.

   **Review:**  "Three-Center, Two-Electron C-H-C Bonds in Organic Chemistry."

**VII.H-42**  B.M. Trost, *Pure Appl. Chem.*, **64**, 315-322.

   **Report:**  "Abiological Catalysis for Synthetic Efficiency."

---

**VII.H-43**  J.K. Kochi, *Acc. Chem. Res.*, **25**, 39.

   **Review:**  "Inner-Sphere Electron Transfer in Organic Chemistry. Relevance to Electrophilic Aromatic Nitration."

---

**VII.H-44**  M.T. Reetz, *Pure Appl. Chem.*, **64**, 351-9.

   **Review:**  "Metal, Ligand and Protective Group Tuning as a Means to Control Selectivity."

---

**VII.H-45**  S.M. Hecht, *Acc. Chem. Res.*, **25**, 545.

   **Review:**  "Probing the Synthetic Capabilities of a Center of Biochemical Catalysis."

---

**VII.H-46**  E. Juaristi and G. Cuevas, *Tetrahedron*, **48**, 5019.

   **Review:**  "Recent Studies of the Anomeric Effect."

---

**VII.H-47**  H. Maehr, *Tetrahedron Asymm.*, **3**, 735.

   **Review:**  "A Tribute to the Centennial of the Fischer Projection. Display of Alkene Geometry and Axial Stereogenicity."

---

**VII.H-48**  D.N. Reinhoudt et al., *Org. Prep. Proced. Int.*, **24**, 437.

   **Review:**  "Selective Functionalization and Conformational Properties of Calix[4]Arenes. A Review."

**VII.H-49**  C. Dietrich-Buchecker and J.P. Sauvage, *Bull. Soc. Chim. Fr.*, **129**, 113.

  Review:    "Template Effect Induced by Copper(I):
             Application to the Synthesis of Catenanes and
             Molecular Knots."

**VII.H-50**  F. Toda and P. Garratt, *Chem. Rev.*, **92**, 1685.

  Review:    "Four-Membered Ring Compounds Containing
             Bis(methylene)cyclobutene or Tetrakis-
             (methylene)cyclobutane Moieties.  Benzocyclo-
             butadiene, Benzodicyclobutadiene, Biphenylene
             and Related Compounds."

**VII.H-51**  N.A. Petasis and M.A. Patane, *Tetrahedron*, **48**, 5757.

  Review:    "The Synthesis of Carbocyclic Eight-Membered
             Rings."

**VII.H-52**  P.E. Eaton, *Angew. Chem., Int. Ed. Engl.*, **31**, 1421.

  Review:    "Cubanes: Starting Materials for the Chemistry of
             the 1990's and the New Century."

**VII.H-53**  *Acc. Chem. Res.*, **25**, issue number 3.

  Reviews: "Special Issue on Buckminsterfullerenes."

**VII.H-54**  R. Gleiter, *Angew. Chem., Int. Ed. Engl.*, **31**, 27.

  Review:    "Cycloalkadiynes - From Bent Triple Bonds to
             Strained Cage Compounds."

**VII.H-55**  B.M. Lerman, *Russ. Chem. Rev.*, **60**, 358 (1991).

Review:  "Skeletal Rearrangements of Cage Compounds with Medium Rings."

---

**VII.H-56**  V.N. Kislenko and A.A. Berlin, *Russ. Chem. Rev.*, **60**, 470 (1991).

Review:  "Kinetics and Mechanism of the Oxidation of Organic Compounds with Hydrogen Peroxide."

---

**VII.H-57**  G. Seitz and P. Imming, *Chem. Rev.*, **92**, 1227.

Review:  "Oxocarbons and Pseudooxocarbons."

---

**VII.H-58**  R.M. Ortumo et al., *Tetrahedron*, **48**, 9001.

Review:  "Unsaturated Acid Derivatives in Diels-Alder Cycloadditions: Effect of the Extended or Cross Conjugation."

---

**VII.H-59**  F.I. Bel'skii et al., *Russ. Chem. Rev.*, **61**, 221.

Review:  "Cyclopendant Ligands."

---

**VII.H-60**  P.F. Cirillo and J.S. Panek, *Org. Prep. Proced. Int.*, **24**, 553.

Review:  "Recent Progress in the Chemistry of Acylsilanes.  A Review."

# AUTHOR INDEX

Bordeau, M. -469
Borschberg, H.-J. -80
Borthwick, A.D. -505
Bose, A.K. -310
Bose, R. -98
Bosnich, B. -30, 112, 471
Bourguihnon, J. -254
Bouyssi, D. -169
Bouzard, D. -406
Bovicelli, P. -220
Bradshaw, J.S. -510, 512
Braish, T.F. -434, 437
Brandange, S. -25
Brandi, A. -352
Brandsma, L. -452, 463, 468
Brandt, A. -386
Bravo, P. -65, 370, 387
Bremner, D.H. -334
Bren', V.A. -479
Brestkin, A.P. -513
Bridges, A.J. -334
Brillon, D. -474
Brimble, M.A. -423
Brinkman, H.R. -434
Brion, J.D. -360
Broncato, E. -151
Brossman, R.B. -156
Brown, E. -260
Brown, H.C. -149, 258, 425, 485
Brown, J.M. -276
Broxterman, Q.B. -463
Bruckner, R. -98, 193, 199
Bruice, T.C. -488
Brukner, R. -178
Brunet, J.-J. -450
Brunner, H. -255
Brussee, J. -412, 440
Bruzik, K.S. -418
Bryce, M.R. -400
Bryson, T.A. -101
Buback, M. -122, 404
Buchwald, S.L. -156, 256, 270, 273, 341
Budyka, M.F. -491
Bull, J.R. -200

Bulman, P.C. -209
Bulpin, A. -473
Bumgardner, C.L. -63
Bunnett, J.F. -488
Bunz, U. -97
Buono, G. -262
Burgen, K. -464
Burger, K. -332
Burgess, K. -503
Burk, M.J. -265, 271
Burnell, D.J. -20, 110, 455, 463, 494
Butler, A.R. -503
Butsugan, Y. -44, 168
Byers, J.H. -128
Caballero, M. -267
Cabiddu, S. -16
Cabrera, A. -154, 277
Cabri, W. -139, 164
Cacchi, S. -150, 168, 322, 344
Cadenas, R.A. -466
Cahiez, G. -43
Cai, M. -410
Caine, D. -1, 124
Cainelli, G. -38
Calo, V. -22
Cambie, R.C. -85
Camp, D. -447
Campi, E.M. -342
Canty, A.J. -499
Caple, R. -151
Capozzi, G. -472
Carless, H.A.J. -306, 486
Carlson, R. -297, 459
Caro, B. -492
Caronna, T. -128
Carpenter, B.K. -514
Carrea, G. -429, 449
Carreno, M.C. -109
Carretero, J.C. -145, 449
Carrie, R. -394
Casey, C.P. -101
Casey, M. -35, 321, 382
Casiraghi, G. -347
Castedo, L. -316, 323
Castillon, J. -232

Kume, Y. -418
Kundig, E.P. -159
Kundu, N.G. -98
Kung, H. -423
Kunieda, T. -39, 113, 434
Kuno, H. -281
Kunz, H. -58, 413
Kurihara, T. -94
Kuroboshi, M. -455
Kurosawa, K. -381
Kurosawa, T. -225
Kuroyan, R.A. -408
Kurth, M.J. -387
Kusama, T. -411
Kusumoto, S. -424
Kuwajima, I. -4, 46, 293
Kuwajima, I. -
Kuyl-Yeheskiely, E. -303
Kuznetsova, T.S. -101
Kwon, T.W. -187
L' abbe, G. -398
Laabassi, M. -95
Laborde, E. -337
Ladouceur, G. -190
Lagar, S. -422
Lagow, R.J. -455
Lahiri, S. -124
Lakhvich, F.A. -408
Lakhvich, F.A. -509
Lam, J.N. -405
Lande, B. -370
Langlois, B.R. -3
Langlois, N. -54
Langlois, Y. -14, 455
Larcheveque, M. -433
LaRosa, C. -398
Larson, G.L. -1717
Lassaletta, J.-M. -62
Lau, C.K. -358
Laurent, A. -500
Laurent, A.J. -4
Lauret, A.J. -278
Lautens, M. -21, 118, 280
Le Floc'h, Y. -72
Le Gottic, F. -425
Leanna, M.R. -209
Leardini, R. -391
Leblanc, Y. -314

Lee, A.W. -311
Lee, A.W.M. -332
Lee, E. -326
Lee, J.B. -227
Lee, J.G. -227, 253, 446
Lee, J.H. -387
Lee, S.-J. -108
LeGrand, D.M. -429
Lejeune, L. -464
Lempert, K. -310
Lennox, R.B. -410
LeNoble, W.J. -514
Lerman, B.M. -517
Leroy, J. -108, 455
Lesniak, S. -278
Lesniak, S. -4
Letcher, R.M. -111
Lett, R. -92
Leung, T.W. -433
Leurs, S. -398
Levin, J.I. -321
Lewis, C.N. -118
Lewis, F.D. -363
Lewis, L.N. -470
Ley, S.V. -404, 410
Lhommet, G. -338, 365
Liao, C. -124
Lidgren, G. -404
Liebeskind, L.S. -147, 148
Liedholm, B. -144
Lin, Y. -319
Linderman, R.J. -4, 426
Link, J.O. -254
Linstrumelle, G. -98
Liotta, D.C. -57, 452, 472
Lipshutz, B.H. -329, 496
Lipshutz, B.H. -58, 476
Lipton, M.A. -326
Liskamp, R.M.J. -190
Lissavetzky, J. -333
Lissel, M. -249
Litchtenthaler, F.W. -453
Little, R.D. -492
Liu, C.-Y. -119
Liu, H.-J. -42, 109, 198
Liu, Y. -435
Livinghouse, T. -339
Llama, E.F. -426

Mori, K. -7, 429, 502
Mori, M. -340, 341, 347
Moriarty, R.M. -92, 238,
 390
Morimoto, T. -231
Morin-Fox, M.L. -326
Moriwake, T. -269, 328
Morrow, G.W. -117, 248
Moskowitz, H. -187
Moss, G.P. -357
Mosset, P. -44
Motherwell, W.B. -66, 83,
 84, 101, 394, 487
Motoki, S. -361
Moulines, J. -304
Mouloungui, Z. -74, 453
Mourino, A. -10, 92
Moustafa, A.H. -362
Moutat, J.-C. -488
Moyano, A. -283
Muathen, H.A. -455
Muchowski, J.M. -367,
 375
Mukaiyama, T. -70, 132,
 189, 235, 271, 423, 429,
 449, 452, 471
Mukherjee, D. -2
Mulchandani, N.B. -360
Müllen, K. -476
Muller, P. -99
Mulzer, J. -51
Murahashi, S. -216, 217
Murahashi, S.-I. -146,
 239, 406, 415, 433, 435,
 449, 497
Murai, A. -325, 357
Murai, S. -155
Murakami, Y. -455
Muraoka, O. -131
Murata, J. -364
Murphy, J.A. -336
Murphy, W.S. -134, 336
Murthy, S. -189
Murumoto, S. -293
Musorin, G.K. -83
Muzart, J. -210, 452, 495
Myers, A.G. -239, 169
Myers, A.I. -164, 261,

Nadir, U.K. -332, 372
Nagamatsu, T. -374
Nagao, Y. -9
Nagarajan, K. -399
Nagarajan, S. -437
Nagaraju, S. -175
Nagashima, H. -470
Naito, T. -315, 321
Najera, C. -35, 40
Nakagawa, M. -271
Nakai, T. -31, 48, 179, 470
Nakamura, A. -216
Nakamura, E. -22, 119
Nakamura, H. -110
Nakanishi, K. -423, 508
Nakatani, S. -152
Nakayama, G.R. -269
Nakayama, J. -362
Nakazawa, M. -136
Nangia, A. -480
Nanni, D. -353
Nanninga, T.N. -25
Nantz, M.H. -438
Naoshima, Y. -264
Napolitan, E. -444
Narasaka, K. -5, 46, 86,
 118
Narayana, C. -193
Naso, F. -472
Natsume, M. -112, 136
Nefedov, O.M. -99
Nefedov, V.D. -492
Negishi, E. -172, 493
Neidlein, R. -376, 462
Nelson, W.L. -411
Nesi, R. -352
Netz, D.F. -74
Neumann, W.P. -63, 499
Neunhoeffer, H. -396
Newkome, G.R. -510
Newnann, W.P. -288
Nguyen, N.V. -374
Nicholas, K.M. -438
Niclas, H.J. -134, 251
Nicolaou, K.C. -49, 108,
 450, 506
Nielsen, A.T. -374
Nigishi, E. -104, 148

www.ingramcontent.com/pod-product-compliance
Lightning Source LLC
Chambersburg PA
CBHW060420220326
41598CB00021BA/2234